S0-BLI-603

QUANTUM TUNNELLING IN CONDENSED MEDIA

MODERN PROBLEMS IN CONDENSED MATTER SCIENCES

Volume 34

Series editors

NORTH-HOLLAND
AMSTERDAM · LONDON · NEW YORK · TOKYO

QUANTUM TUNNELLING IN CONDENSED MEDIA

Yu. KAGAN

Kurchatov Institute of Atomic Energy
Kurchatov Square 46
Moscow, Russia

A.J. LEGGETT

Department of Physics
University of Illinois, Urbana-Champaign
1110 West Green Street
Urbana, IL 61801, USA

1992

NORTH-HOLLAND

AMSTERDAM · LONDON · NEW YORK · TOKYO

PHYS
seplac

ELSEVIER SCIENCE PUBLISHERS B.V.
P.O. Box 211, 1000 AE Amsterdam
The Netherlands

Library of Congress Cataloging-in-Publication Data

Quantum tunnelling in condensed media/[edited by] Yu.M. Kagan,
A.J. Leggett. (Modern problems in condensed matter sciences; v. 34)
p. cm.
Includes bibliographical references and indexes.
ISBN 0-444-88041-0 (acid-free paper)
1. Condensed matter. 2. Tunneling (Physics) I. Kagan, Yu.M.
II. Leggett, A.J. III. Series.
QC173.4.C65Q36 1992
530.4′16–dc20 92-18968
CIP

ISBN: 0 444 88041 0

Printed on acid-free paper

Printed in The Netherlands

MODERN PROBLEMS IN CONDENSED MATTER SCIENCES

Vol. 1. SURFACE POLARITONS
V.M. Agranovich and D.L. Mills, *editors*

Vol. 2. EXCITONS
E.I. Rashba and M.D. Sturge, *editors*

Vol. 3. ELECTRONIC EXCITATION ENERGY TRANSFERS IN CONDENSED MATTER
V.M. Agranovich and M.D. Galanin

Vol. 4. SPECTROSCOPY AND EXCITATION DYNAMICS OF CONDENSED MOLECULAR SYSTEMS
V.M. Agranovich and R.M. Hochstrasser, *editors*

Vol. 5. LIGHT SCATTERING NEAR PHASE TRANSITIONS
H.Z. Cummins and A.P. Levanyuk, *editors*

Vol. 6. ELECTRON–HOLE DROPLETS IN SEMICONDUCTORS
C.D. Jeffries and L.V. Keldysh, *editors*

Vol. 7. THE DYNAMICAL JAHN–TELLER EFFECT IN LOCALIZED SYSTEMS
Yu.E. Perlin and M. Wagner, *editors*

Vol. 8. OPTICAL ORIENTATION
F. Meier and B.P. Zakharchenya, *editors*

Vol. 9. SURFACE EXCITATIONS
V.M. Agranovich and R. Loudon, *editors*

Vol. 10. ELECTRON–ELECTRON INTERACTIONS IN DISORDERED SYSTEMS
A.L. Efros and M. Pollak, *editors*

Vol. 11. MEDIUM-ENERGY ION REFLECTION FROM SOLIDS
E.S. Mashkova and V.A. Molchanov

Vol. 12. NONEQUILIBRIUM SUPERCONDUCTIVITY
D.N. Langenberg and A.I. Larkin, *editors*

MODERN PROBLEMS IN CONDENSED MATTER SCIENCES

Vol. 13. PHYSICS OF RADIATION EFFECTS IN CRYSTALS
R.A. Johnson and A.N. Orlov, *editors*

Vol. 14. INCOMMENSURATE PHASES IN DIELECTRICS
(Two volumes)
R. Blinc and A.P. Levanyuk, *editors*

Vol. 15. UNITARY TRANSFORMATIONS IN SOLID STATE PHYSICS
M. Wagner

Vol. 16. NONEQUILIBRIUM PHONONS IN NONMETALLIC CRYSTALS
W. Eisenmenger and A.A. Kaplyanskii, *editors*

Vol. 17. SOLITONS
S.E. Trullinger, V.L. Pokrovskii and V.E. Zakharov, *editors*

Vol. 18. TRANSPORT IN PHONON SYSTEMS
V.L. Gurevich

Vol. 19. CARRIER SCATTERING IN METALS AND SEMICONDUCTORS
V.F. Gantmakher and I.B. Levinson

Vol. 20. SEMIMETALS – 1. GRAPHITE AND ITS COMPOUNDS
N.B. Brandt, S.M. Chudinov and Ya.G. Ponomarev

Vol. 21. SPECTROSCOPY OF SOLIDS CONTAINING RARE EARTH IONS
A.A. Kaplyanskii and R.M. Macfarlane, *editors*

Vol. 22. SPIN WAVES AND MAGNETIC EXCITATIONS
(Two volumes)
A.S. Borovik-Romanov and S.K. Sinha, *editors*

Vol. 23. OPTICAL PROPERTIES OF MIXED CRYSTALS
R.J. Elliott and I.P. Ipatova, *editors*

MODERN PROBLEMS IN CONDENSED MATTER SCIENCES

Vol. 24. THE DIELECTRIC FUNCTION OF CONDENSED SYSTEMS
L.V. Keldysh, D.A. Kirzhnitz and A.A. Maradudin, *editors*

Vol. 25. CHARGE DENSITY WAVES IN SOLIDS
L.P. Gor'kov and G. Grüner, *editors*

Vol. 26. HELIUM THREE
W.P. Halperin and L.P. Pitaevskii, *editors*

Vol. 27. LANDAU LEVEL SPECTROSCOPY
(Two volumes)
G. Landwehr and E.I. Rashba, *editors*

Vol. 28. HOPPING TRANSPORT IN SOLIDS
M. Pollak and B. Shklovskii, *editors*

Vol. 29. NONLINEAR SURFACE ELECTROMAGNETIC PHENOMENA
H.E. Ponath and G. Stegeman, *editors*

Vol. 30. MESOSCOPIC PHENOMENA IN SOLIDS
B.L. Altshuler, P.A. Lee and R.A. Webb, *editors*

Vol. 31. ELECTRIC STRAIN FIELDS AND DISLOCATION MOBILITY
V.L. Indenbom and J. Lothe, *editors*

Vol. 32. ELECTRONIC PHASE TRANSITIONS
W. Hanke and Yu. Kopaev, *editors*

Vol. 33. NONRADIATIVE RECOMBINATION IN SEMICONDUCTORS
V.N. Abakumov, V.I. Perel and I.N. Yassievich

Vol. 34. QUANTUM TUNNELLING IN CONDENSED MEDIA
Yu. Kagan and A.J. Leggett, *editors*

Vol. 35. SPECTROSCOPY OF NONEQUILIBRIUM ELECTRONS
AND PHONONS
C.V. Shank and B.P. Zakharchenya, *editors*

Oh, how many of them there
are in the fields!
But each flowers in its
own way –
In this is the highest achievement
of a flower!

Matsuo Bashó
1644–1694

PREFACE TO THE SERIES

Our understanding of condensed matter is developing rapidly at the present time, and the numerous new insights gained in this field define to a significant degree the face of contemporary science. Furthermore, discoveries made in this area are shaping present and future technology. This being so, it is clear that the most important results and directions for future developments can only be covered by an international group of authors working in cooperation.

'Modern Problems in Condensed Matter Sciences' is a series of contributed volumes and monographs on condensed matter science that is published by Elsevier Science Publishers under the imprint of North-Holland. With the support of a distinguished Advisory Editorial Board, areas of current interest that have reached a maturity to be reviewed, are selected for the series. Both Soviet and Western scholars are contributing to the series, and each contributed volume has, accordingly, two editors. Monographs, written by either Western or Soviet authors, are also included. The complete series will provide the most comprehensive coverage available of condensed matter science.

Another important outcome of the foundation of this series is the emergence of a rather interesting and fruitful form of collaboration among scholars from different countries. We are deeply convinced that such international collaboration in the spheres of science and art, as well as other socially useful spheres of human activity, will assist in the establishment of a climate of confidence and peace.

The publishing house 'Nauka' publishes the volumes in the Russian language. This way the broadest possible readership is ensured.

The General Editors of the Series,

V.M. Agranovich A.A. Maradudin

PREFACE

The problem of a system which has to be described by quantum mechanics and which interacts with an "open" environment is ubiquitous in physics and chemistry. To be sure, in the early days of quantum mechanics, the experiments which established the theory (such as the Stern–Gerlach experiment or studies of atomic spectroscopy) were done on systems which were sufficiently decoupled from their surroundings so that it was plausible to describe them as isolated; indeed, this was probably essential to the quantitative development of the subject. And even today, there are still a few problems in physics, such as the collisions of particles in high-energy accelerators, where such an approximation is sensible. But over the vast bulk of the subject matter of physics and chemistry, it is impossible to neglect the interactions of the quantum system with its surroundings or, as we shall generally refer to it, the "environment." In a few cases, such as much of quantum optics and gas-phase chemical physics, the effects of this interaction are relatively weak and can be handled by one or the other version of perturbation theory; however, when the environment in question is some form of condensed matter, its effect on the behavior of the system can be very strong and even modify it qualitatively.

Generally speaking, given an *isolated* quantum-mechanical system, we may distinguish two different regimes of behavior: a "semiclassical" regime, in which the behavior is at least qualitatively similar to that predicted by classical mechanics, and an "essentially quantum" regime, where the behavior has no classical analog. The borderline between these two regimes is not sharp but, roughly speaking, one can say that most of the "essentially quantum" regime corresponds to behavior involving at least a moderate degree of tunnelling into, or through, classically inaccessible regions of configuration space. It is, of course, of interest to study the way in which the environment influences the behavior of the system in both regimes. However, it usually transpires that in the semiclassical regime its effects are not dramatically different from those expected in classical mechanics; by contrast, in the "essentially quantum" regime the environment can alter the behavior of the system qualitatively and sometimes dramatically. This is the subject of this book.

The essays collected here, therefore, deal with various aspects of the problem of quantum tunnelling, and related behavior, of a system, either microscopic or macroscopic, which interacts strongly with an "environment", which is some form of condensed matter. The "system" in question need not be physically distinct from its environment, but could, for example, be one particular degree of freedom on which we choose to concentrate our attention (as in the case of the Josephson junction studied in several of the chapters). This general problem has been studied in many hundreds, if not thousands, of papers in the literature, in contexts as diverse as biophysics and quantum cosmology; we certainly cannot claim to have done justice here to the whole of this broad swathe of work, rather we present here a group of articles which we hope are reasonably representative of some of the main trends of the last 15 years or so in this general area, and sufficiently close in general spirit and terminology so that common themes will not be too difficult to find. With the exception of the chapter of Devoret et al., all the chapters in this volume are primarily theoretical, although the comparison with experiment is discussed wherever possible.

Since a superficial inspection might suggest that the physical systems discussed in the various chapters in this volume (Josephson junctions, muons in metals, self-trapped electrons, polarons, etc.) do not have a great deal in common with one another, it may be helpful to give a little historical background. Roughly speaking, interest in the kind of work reported here may be traced back to three different, although related, sources. First, in the context of traditional solid-state physics, there has long been an interest in the problem of the motion of an "impurity" of one kind or another (e.g. electron, muon, isotopic defect) in a perfect or imperfect crystalline solid. Typically, in such cases, tunnelling occurs (mainly) between sites which are degenerate for the isolated system; the "environment" is typically the lattice vibrations (phonons) and/or the conduction electrons. Problems of this type are discussed in the chapters by Kagan and Prokof'ev, Ioselevich and Rashba and Zawadowski and Vladár.

Secondly, in the context of quantum field theory, one sometimes encounters Lagrangians which can sustain more than one local minimum; in this case, the system may be trapped in a metastable configuration and subsequently attain the true global minimum by a *collective* quantum tunnelling process. Such a problem was originally studied by Lifshitz and Kagan (1972) in the context of solid-state phase transformations and, subsequently, by particle physicists and cosmologists under the name of the "fate of the false vacuum" (see e.g. Coleman 1979). In such cases, it is possible to regard the "tunnelling system" as some collective coordinate such as the radius of the droplet of stable phase, with the "environment" being constituted by the other degrees of freedom of the field. Although such field-theoretic problems are as such not explicitly represented in this book, the essentially equivalent problem of tunnelling in a (possibly infinitely) many-dimensional space is discussed in detail in the chapter by Eckern and Schmid.

Finally, a third source of interest in these problems in recent years has been the hope of displaying explicitly the quantum-mechanical behavior of a macroscopic variable such as the flux trapped in a SQUID ring, and thereby shedding light on the quantum measurement paradox. The conceptual aspects of this problem are discussed in the chapter by Leggett, while that by Devoret et al. describes some of the experimental tests conducted in this connection over the last ten years; the chapters by Larkin and Ovchinnikov and Ivlev and Mel'nikov go into more detail on the technical aspects of the theory. It should be pointed out that, apart from their interest in the context of the foundations of quantum mechanics, the Josephson systems discussed in these chapters are particularly useful for the comparison of theory and experiment since it is possible both to adjust their parameters by macroscopic-level engineering and to measure them in independent experiments conducted in the semiclassical regime; this is part of the reason for the attention they have received in recent years.

In the light of the above remarks, it is natural to expect that the way in which the questions are posed would depend upon whether we are discussing a "microscopic" tunnelling problem (e.g. the diffusion of a muon in a crystalline solid) or a "macroscopic" one such as the motion of the Cooper-pair phase in a Josephson junction. In the former case, we usually know enough about the microscopic physics to write down a concrete Hamiltonian and discuss its consequences, whereas in the latter, all we have at our disposal is the experimentally observed classical equations of motion, and the problem becomes one of *relating* the quantum to the classical behavior. It turns out that much of the effect of the interaction of the system with its environment consists in an "adiabatic" renormalization of the parameters of the original "isolated" problem, whose original values in the macroscopic case were not usually observable anyway; the directly observable effects in such a case are usually primarily associated with the dissipative part of the interaction; and for this reason, the problem is often posed in the form: What is the effect of *dissipation* on (quantum tunnelling, etc.)? (For a further discussion of this point, see the chapter by Leggett.)

In reading the papers in this volume, a few general remarks may be helpful. First, an important distinction should be made between the problem of tunnelling out of a metastable potential well into an effectively infinite continuum and tunnelling between two or more potential wells which would be degenerate, or nearly so, in the absence of interaction with the environment. In the first case, the effect of quantum mechanics is primarily seen in the occurrence of tunnelling itself, i.e. the possibility of transition through a classically impenetrable barrier. In the latter, by contrast, there exists also the possibility of effects of the quantum-mechanical coherence between the amplitudes in the different wells; well-known examples of such effects are, of course, the "clock-like" behavior associated with the inversion resonance of the NH_3 molecule, or, in the case of many wells, the phenomenon of Bloch waves in crystalline solids. As discussed in the chapter by Leggett, observation of this type of behavior in the case of a macroscopic variable ("macroscopic quantum coherence" or MQC) would be a further significant input to the quantum measurement paradox. When one comes to consider the effects of the environment, one asks rather different questions in the two cases: in the case of tunnelling out of a metastable well into a continuum, the natural question is how the environment affects the rate of escape, whereas in the case of tunnelling between degenerate wells, both the rate of tunnelling and the very existence of coherent behavior may be affected.

A second important distinction, in the case of tunnelling between degenerate wells, is between "extended" and "truncated" descriptions. Consider for definiteness the case of two wells, a situation which occurs e.g. for the NH_3 molecule or for the planned MQC experiment. Then, in reality, the system in question is characterized by some continuous variable q with which is associated a (c-number) potential energy $V(q)$ possessing two degenerate minima. A complete solution of the quantum-dynamical problem, even in the absence of coupling c in environment, must take into account the complete form of $V(q)$. However, many of the qualitative features of the behavior may be seen from a truncated model, in which one confines oneself to the two-dimensional Hilbert space spanned by the two lowest (approximately harmonic-oscillator-like) states centered around the minimum of the two wells. In such a model (known for the two-well case as the "spin–boson" or "molecular polaron" problem), the matrix element for tunnelling between the two degenerate states is taken as a phenomenological input; of course, in any real physical problem it should be calculated as an output of the "extended" problem. Similar remarks apply to the terms coupling the system to its environment. An analogous truncation can be made in the case of many wells: see the chapter by Kagan and Prokof'ev. Note that in the problem of escape from a metastable well there is no (useful) "truncated" description.

A theme which emerges repeatedly, in explicit or implicit form, in the chapters of this volume is the distinct role of the modes of the environment corresponding to different frequency regimes, or, to put it in a different language, the "adiabatic/irreversible" distinction. We restrict ourselves for the moment to zero temperature. For orientation, let us first consider the effect of the environment on the *classical* (or quasiclassical) motion of the system. Suppose the characteristic frequencies of the motion of the isolated system (let us say, the frequencies for which the Fourier transform of the trajectory $q(t)$ has appreciable weight) are $\lesssim \omega_0$. Then, roughly speaking, those environment modes which have frequencies $\gg \omega_0$ will follow the motion adiabatically and can at best renormalize the parameters of the system, while those of frequency $\lesssim \omega_0$ can exchange energy irreversibly with it and, thereby, give rise to dissipation. As an extreme case, if the environment modes have a lower threshold $\omega_\ell \gg \omega_0$, then no dissipation is possible and the only role of the environment is to renormalize the parameters (i.e. the effective mass and/or the effective potential) of the system. These statements can be confirmed by detailed study of exactly solvable models such as the damped quantum harmonic oscillator.

Now let us pass to the "essentially quantum" regime. Consider first the problem of the decay of a metastable state. Although the process of tunnelling of the isolated system through the potential barrier cannot, of course, be described in classical terms, we can nevertheless associate with it a "characteristic frequency" $\tilde{\omega}_0$, which may, to an order of magnitude, be taken to be the inverse of the duration of the (bulk of the) so-called "instanton" trajectory in imaginary time, cf. the chapter by Eckern and Schmid. Roughly speaking, for non-pathological potentials, $\tilde{\omega}_0$ is of the order of the frequency of small oscillations in the metastable potential well. Then we can say that the effect, if any, of the environment modes with frequencies $\gg \tilde{\omega}_0$ is to renormalize the effective mass and/or effective potential. The effect of modes with frequencies of the order of $\tilde{\omega}_0$ is a bit more subtle and changes the actual process of penetration through the barrier, in general reducing the escape rate. Since it is precisely these modes which give rise to dissipation in the classical motion and (more importantly) which can extract energy from the system during the tunnelling process, this effect is sometimes known as the "effect of dissipation on quantum tunnelling."

In the case of tunnelling between equivalent sites, as in a typical solid-state problem, the situation is a bit more complicated. We will assume for the moment that the interaction with the environment is sufficiently weak so that it does not change the qualitative features of the motion. Consider first for definiteness the problem of tunnelling between two equivalent wells ("molecular polaron" problem). We can define at least three characteristic energies (or frequencies: we set $\hbar = 1$) for this problem, apart from the thermal energy kT. First there is the upper cutoff on the environment spectrum (call it ω_c), which is of order the Fermi energy ε_F (or the bandwidth) for interaction with electron–hole pairs and of order of the Debye frequency ω_D for interaction with phonons. Secondly, we have a quantity $\tilde{\omega}_0$ which reflects the characteristic "frequencies" associated with the process of transmission through the barrier between the two wells; as in the metastable-well problem, this is, in general, of the same order as the small-oscillation frequency in either well separately. And thirdly, we have a characteristic frequency which has no analog in the metastable-well problem, namely, the frequency Δ_0 of coherent ("NH_3-type") oscillations of the isolated system between the two wells. In almost all physically realistic problems of this type, the ratio $\Delta_0/\tilde{\omega}_0$ is exponentially small, being given by a WKB-type factor; however, the ratio $\tilde{\omega}_0/\omega_c$ may be either large or small compared to 1 (typically, $\tilde{\omega}_0 \ll \varepsilon_F$ but $\tilde{\omega}_0 \gtrsim \omega_D$). For the problem of tunnelling between many identical wells, the situation is essentially identical except that, strictly speaking, it is necessary to distinguish between the matrix element Δ_0 for tunnelling between a particular pair of wells and the inverse τ^{-1} of the characteristic time spent at a given site (i.e. the "impurity bandwidth"); generally speaking, we have $\tau^{-1} \sim z^{1/2}\Delta_0$, where z is the number of nearest neighbors, which may be as large as 12 for certain crystal structures.

We now examine the effect of the environmental modes in different frequency regimes. Again, the modes (if any) with frequencies $\tilde{\omega}_0 \ll \omega \lesssim \omega_c$ will simply follow adiabatically and renormalize the parameters of the problem. The modes with frequency in the range $\Delta_0 \ll \omega \lesssim \tilde{\omega}_0$ can, as in the metastable-well problem, affect the actual process of transmission through the potential barriers (the effect called by Kagan and Prokof'ev "fluctuational preparation of the barrier"); however, because the tunnelling is now not into a continuum but into an *equivalent* (hence, prima facie energetically degenerate) site, they can no longer give rise to dissipation. Modes which have $\omega \ll \tilde{\omega}_0$ but still $\omega \gg \Delta_0$ can be handled phenomenologically in the truncated (spin–boson) model and then give rise to the well-known Franck–Condon (or polaron) factor – a typical adiabatic renormalization of the parameters. Indeed, the only modes which can give rise to dissipation in the quantum motion are those of frequency $\omega \lesssim \Delta_0$ (or, more accurately, in the many-well case, $\omega \lesssim \tau^{-1}$). Such modes may actually change even the qualitative nature of the motion from coherent to incoherent. The coherent tunnelling is suppressed already at temperature $kT \ll \omega_0$ (in a metal, actually, at $kT \sim \Delta_0$), and the coherent amplitude of the tunnelling transition decreases exponentially with the increase of T ("dynamic destruction of the band"; for details, see the chapter by Kagan and Prokof'ev). In this connection we would like to note that with $kT \gg \Delta_0$ two-mode (multi-mode) terms in a system–environment interaction can become important for the transition to incoherent dynamics. In this case, it is usually possible to absorb an environment mode of frequency $kT \gg \Delta_0$ and emit simultaneously another similar one with a net energy gain or loss of order Δ_0. Thus, such modes can, in this sense, contribute to the dissipation (cf. the chapter by Kagan and Prokof'ev). In principle, in such cases, it is possible to redefine the "modes" of the environment so that by definition the interaction involves excitation of only single modes (cf. Appendix C of Caldeira and Leggett 1983), and then to use the model single-mode representation for the interaction (eq. (4.3) of chapter 1), except that the environment "spectrum" may be explicitly temperature-dependent; but, depending upon the context, this may or may not be the most "natural" way to look at the problem.

The qualitative distinction, implicit in the above discussion, between the effects of environment modes with frequency large compared to the characteristic frequencies of the system, which give rise to adiabatic renormalization of the system parameters, and those with comparable frequencies, which give rise to irreversibility and can change the behavior qualitatively, is an extremely important one, both for the theoretical calculations and (in the context of a macroscopic variable) for the quantum measurement problem; for the latter, see the discussion in the chapter by Leggett.

Strictly speaking, of course, the above discussion is still a bit too simple. In the first place, if ω_0 is the original "characteristic frequency" of the system, then one of the effects of the modes with $\omega \gg \omega_0$ may precisely be to renormalize ω_0 (in practice, almost invariable downwards). For example, in the metastable-well case, the effective "instanton frequency" may itself be suppressed by the interaction with the environment. In the case of originally coherent tunnelling between equivalent wells, the effect may be even more drastic: the coherent motion (frequency $\sim \Delta_0$) may be reduced up to complete system localization in a given well at $T = 0$ (frequency = 0). In either case, it is clear that the boundary between the modes giving rise to "adiabatic" and "irreversible" effects will itself be shifted downwards.

Let us briefly comment on the general question of the modelling of the "bath" (environment). In the "microscopic" case this is not usually a particular problem, since for any specific physical situation (interaction with phonons, electrons, etc.) we usually know the microscopic details well enough to write down a concrete Hamiltonian. In the case of a macroscopic variable, this is usually not feasible and one must employ a more schematic model of the "bath." It is conventional in the theory of MQT, etc., in Josephson junctions to use an "oscillator–bath" model, in which the environment modes are represented as independent harmonic oscillators and the system–environment coupling by a term linear in both the system and the environment coordinates; this is implicitly done, for example, in the chapter of Larkin and Ovchinnikov and in section 8.2 of that of Ivlev and Mel'nikov in this volume. It is, of course, not at all obvious at first sight that this model is even qualitatively reasonable, and it certainly needs detailed argument to justify it. This question is discussed, and references to the literature on it given, in the chapter by Leggett.

Finally, a remark about the formal methods of calculation used in this book. Generally speaking, the problems discussed in this book are formally problems of barrier penetration (or something related) in a many-dimensional space. The most obvious technique for such problems is, of course, the many-dimensional generalization of the well-known WKB technique; this is explored in considerable detail in the chapter by Eckern and Schmid. However, depending upon the situation, other techniques may, in practice, be more convenient. For example, if the double-well or many-well (crystal) problem is considered, or, in general, for tunnelling transitions with a small ($\ll \omega_0$) change of the system energy, one may use a technique based on direct evaluation of overlapping integrals for the system many-particle wave functions, as done in the chapter by Kagan and Prokof'ev. On the other hand, if the environment can be modelled by a set of harmonic oscillators, functional-integral techniques are very useful (since the environment coordinates can be exactly integrated out of the problem). The latter can be used in "real-time" form (Ivlev and Mel'nikov) or in the "imaginary-time" representation (Ioselevich and Rashba, Larkin and Ovchinnikov); in the latter case, a detailed discussion of the justification for the technique is given by Eckern and Schmid.

We conclude this introduction by reviewing briefly the contents of the various chapters of the book. The chapter by Leggett reviews the motivation for the study of the quantum mechanics of a macroscopic variable, and the connection between the problem of dissipation in quantum mechanics and the quantum theory of measurement. The chapter by Kagan and Prokof'ev is devoted to the problem of quantum diffusion of particles in insulators, metals and superconductors in the presence of strong interactions with the phonons and electrons. Special attention is paid to the crossover between coherent and incoherent tunnelling. The quantum diffusion process is studied for both perfect and imperfect crystals; a detailed picture of the localization process is given. A wide spectrum of experimental results displaying tunnelling diffusion in different systems is analyzed. The chapter of Eckern and Schmid is devoted to a careful study of the problem of tunnelling through a classically forbidden region in a many-dimensional space, with particular attention paid to the question of the justification for the "instanton" method, which is used in several of the other chapters as well as in much of the literature. Larkin and Ovchinnikov study the problem of macroscopic quantum tunnelling (MQT) in the case of a Josephson junction. The effective action for the tunnel junction is derived, which allows one to calculate the lifetime of the metastable zero-voltage state. A detailed analysis is given for the cases of strong and weak dissipation. Ivlev and Mel'nikov consider the role of a high-frequency external field in the tunnelling problem. The change of the decay rate accompanied by the absorption of one or more field quanta is studied as a function of the external field intensity. The results are used for the analysis of MQT in a Josephson junction under the influence of a high-frequency electromagnetic field. Next, the chapter by Devoret et al. reviews the experiments of the authors and their collaborators on Josephson systems, which have provided a test for some of the ideas developed in some of the previous chapters. In the chapter by Ioselevich and Rashba the problem of nonradiative capture of free quasiparticles in crystals (electrons, excitons, etc.) into the self-trapped state is studied. The case of a short-range electron–phonon interaction, which is responsible for both the formation of the self-trapping barrier and its penetration, is considered. The relation between adiabatic and nonadiabatic trapping and the crossover from the tunnelling to the Arrhenius activation regime is discussed in detail. Finally, Zawadowski and Vladár examine the problem of a two-state system in a metal interacting with the conduction electrons, with particular reference to metallic glasses; note that in this case it is essential to take into account the interaction mixing different orbital channels for electron scattering in the tunnelling transition.

Yu. Kagan
A.J. Leggett

References

Caldeira, A.O., and A.J. Leggett, 1983, Ann. Phys. (New York) **149**, 374; Erratum, 1984, ibid. **153**, 445.

Coleman, S., 1979, The Uses of Instantons, in: The Whys of Subnuclear Physics, ed. A. Zichichi (Plenum, New York).

Lifshitz, I.M., and Yu. Kagan, 1972, Zh. Eksp. & Teor Fiz. **62**, 403 [Sov. Phys.-JETP **35**, 206].

CONTENTS

CHAPTER 1

Quantum Tunnelling of a Macroscopic Variable

Anthony J. LEGGETT

Department of Physics
Loomis Laboratory of Physics
1110 West Green Street
Urbana, IL 61801, USA

Quantum Tunnelling in Condensed Media
Edited by
Yu. Kagan and A.J. Leggett

Contents

1. Introduction

Since about 1980 there has been a considerable surge of interest, both experimental and theoretical, in the problem of the quantum tunnelling behavior of variables which may by a reasonable criterion be described as "macroscopic", such as the phase variable in a current-biased Josephson junction, the trapped flux in a so-called rf SQUID ring or the magnetization of ferromagnetic grains. The experimental aspects of this problem are reviewed by Devoret et al. in chapter 6 of this volume. In this chapter, I shall try to provide an overview of the theoretical background to this surge of interest and the principal conceptual questions on which it touches. The technical details of the calculational methods which have been developed in this area are reviewed in this volume by Eckern and Schmid (chapter 3), by Larkin and Ovchinnikov (chapter 4), and by Ivlev and Mel'nikov (chapter 5); so I shall mention them only briefly.

Interest in the quantum tunnelling behavior of macroscopic variables has, I believe, at least two distinct roots. The first is a technical one internal to quantum mechanics. The original development of the quantum theory was motivated by experiments which were mainly if not entirely at the atomic level, the famous Stern–Gerlach experiment on the magnetic deflection of atomic beams being fairly typical. In such experiments one is usually dealing with an ensemble of microscopic systems (atoms, electrons, nowadays neutrons . . .) which are to a very satisfactory degree of approximation isolated from other systems which might have to be described by quantum mechanics (needless to say, they cannot be isolated in all respects, otherwise we could not gain any information on them, but the point is that the external probes we use to influence them, such as the magnetic field gradient in a Stern–Gerlach experiment, can be satisfactorily described in classical terms). Thus, the idea of assigning a complete quantum-mechanical description to the system in isolation (or more accurately to the ensemble of such systems) is a natural and attractive one. In many cases of practical interest we believe we know enough about the ensemble that the appropriate quantum-mechanical description is in terms of a pure state.

The question of how we actually extract the information contained in the quantum-mechanical description is, of course, a somewhat different one, which has been discussed since the earliest days of the quantum theory. We will have to return to this question later, but for the moment let us just note that in these "classical" experiments in atomic physics there is usually a natural separation between two stages of the experiment: a stage at which the particle in question is more or less isolated and is apparently well described by single one-particle quantum mechanics, and a "detection" stage at which one infers the results of stage 1 by causing the particle to interact strongly with, and thereby trigger, some kind of detector (e.g., a photographic plate). The salient point is that in a well-designed experiment the interpretation of the "raw" output in terms of a particular model of the stage-1 behavior does not appear to be sensitive to the details of the counter, etc. (perhaps this is an implicit definition of "well-designed"!), and, therefore, it is natural to regard stage 2 as simply "reading off" results which had already been established at stage 1. Thus, while we certainly expect that in principle it ought to be possible to give a description of the working of the counter in entirely quantum-mechanical terms, it seems at first sight that the details of such a description would be merely of technical interest and that nothing is lost by describing the counter and its output in ordinary classical language.

In the sixty-odd years since the inception of the quantum theory it has, of course, been applied to a larger and larger area of physics – not only to those ones, such as particle physics, which we naturally regard as owing their very existence to quantum mechanics, but also to subjects like biophysics, the theory of condensed matter and general relativity where there was already a well-developed theoretical framework formulated in classical language. Indeed, in the last few years it has become quite fashionable in some circles to apply the formalism of quantum mechanics to the universe as a whole, or more accurately to a few selected degrees of freedom of the general-relativistic metric tensor which describes it ("quantum cosmology"). Such extensions of the formalism from its original experimental base in atomic physics clearly raise a number of questions of principle, of which I will here confine myself to one particular one which is of direct relevance to the subject of the present chapter.

One feature which many, indeed most, of these extensions of the domain of quantum theory have in common is that it is no longer possible to regard the system of primary interest as "quantum-mechanically isolated". That is, it will very likely interact with other systems, or other degrees of freedom, which are not of primary interest to us and whose behavior we do not directly sample; as has become customary in this area of research, we shall refer to these other degrees of freedom as the "environment". Moreover, unlike the typical case in atomic physics, these other degrees of freedom cannot be represented by some external c-number force controlled by the experimenter; they are intrinsic to the nature of the system studied and represent a sort of irreducible "noise" (although see below). As an example, when we measure the electrical conductivity of a metal we may regard the "system" of primary interest as the conduction electrons. However, these electrons are continually interacting with the phonons of the metal, which in this context constitute (part of) the "environment".

The existence of an "environment" which exerts an influence on the system under study but is not itself subject to direct measurement is, of course, in no way peculiar to quantum mechanics. However, there is one very fundamental respect in which the situation differs in classical and in quantum physics. In classical physics, however much the system may be influenced by its environment, there is no question that it can possess properties in its own right. As an example, let us consider a heavy particle executing Brownian motion in a liquid suspension, or more precisely an ensemble of such particles; in this case the particle is the "system" and the atoms of the liquid, the "environment". The conventional classical description of this situation is by a Langevin equation: i.e., the coordinate $\boldsymbol{x}(t)$ of the particle is assumed to satisfy an equation of motion of the form

$$M\ddot{\boldsymbol{x}}(t) + \eta\dot{\boldsymbol{x}}(t) + \nabla V(\boldsymbol{x}) = \boldsymbol{F}_{\mathrm{r}}(t), \tag{1.1}$$

where η is the phenomenological viscosity, $V(\boldsymbol{x})$ is any external c-number potential acting on the particle (e.g., electric or gravitational fields) and $\boldsymbol{F}_{\mathrm{r}}(t)$ is a Gaussian random force whose properties are specified only statistically:

$$\langle F_{\mathrm{r}}(t)\rangle = 0, \qquad \langle F_{\mathrm{r}}(t)F_{\mathrm{r}}(t')\rangle = 2\eta kT\delta(t-t'). \tag{1.2}$$

The salient point about the above description, which in the context of classical physics is so self-evident that it does not normally even occur to us to comment on it, is that it is perfectly consistent with the assignment to each particle of the ensemble at any given time a definite, although unknown, value of the variable $\boldsymbol{x}(t)$. We can, in fact, regard each individual ensemble member as having been acted on by a particular sequence of forces $\boldsymbol{F}_{\mathrm{r}}(t)$; since we have no information about these forces other than the statistical information contained in eq. (1.2), our final description of the ensemble has to be in statistical terms, but that in no way prevents us from ascribing to the members of the ensemble properties such as $\boldsymbol{x}(t)$ in their own right. The role of the environment is simply to act as a random-noise generator, and nothing more.

In quantum mechanics the situation is fundamentally different, at least if we accept the interpretation of the formalism which is favored by most physicists. For in quantum mechanics, when two subsystems interact (or even when they have interacted in the past, even though they may now be spatially separated) there is in general no question of either subsystem possessing properties in its own right: the description is *nonseparable*. Technically, if the set of variables describing subsystem 1 is collectively denoted q_1 and that describing subsystem 2 q_2, then under certain circumstances the correct description of the ensemble of combined systems (1 + 2) is by a pure state

$$\Psi = \Psi(q_1, q_2), \tag{1.3}$$

which *cannot* in general be factored into a product description $[\Psi(q_1, q_2) \neq \chi_1(q_1)\chi_2(q_2)]$. This means that in general there is simply *no description* of subsystem 1, or of subsystem 2 in its own right. More accurately, there exists no description of 1 and of 2 such that combining them would give complete information on the properties of the combined system (1 + 2): needless to say, we can form the reduced density matrix $\rho(q_1, q'_1)$ of subsystem 1 by the prescription

$$\rho_1(q_1, q'_1) \equiv \int \mathrm{d}q_2\, \Psi^*(q_1, q_2)\Psi(q'_1, q_2),$$

and this will be adequate for the calculation of the expectation value of any operator $\hat{A}$ which depends only on the variable q_1; similarly, we can form $\rho_2(q_2, q'_2)$ and this will permit the calculation of the expectation value of any operator $\hat{B}$ which depends only on q_2. However, a knowledge of ρ_1 and ρ_2 will *not*, in general, permit the calculation of the correlation $\langle \hat{A}\hat{B} \rangle$. Moreover – and this is the subtle, but crucial, point – the missing information is not simply of a classical statistical nature, in the form of classical correlations between the behavior of the subsystems. Technically, we *cannot* in general write the density matrix $\rho(q_1, q_2, q'_1, q'_2)$ as a sum of products, i.e., in the form

$$\sum_n p_n \rho_{n1}(q_1, q'_2)\rho_{n2}(q_2, q'_2), \quad 0 \leqslant p_n \leqslant 1,$$

whatever the choice of the functions ρ_{n1}, ρ_{n2}. This is essentially the content (as applied to the formalism of quantum mechanics itself) of the famous theorem of Bell (for a lucid nontechnical discussion, see d'Espagnat 1979, Mermin 1985). It should be noted that recent experimental work in this connection (Aspect et al. 1982) has shown that, if the results are interpreted within the framework of the quantum formalism, the fundamental property of nonseparability persists even under very extreme conditions, when the subsystems are not only no longer interacting but are separated by a space-like interval. How much more so should we expect nonseparability to be important when the "subsystems" are actually in close and continual interaction – which is exactly the case when one "subsystem" is what we have called above our "system" and the other the "environment"! It follows that, in general, in quantum mechanics a system interacting with an environment *has no (complete) description in its own right.*

At this stage, it is tempting to make the following objection: Included in our definition of what we meant by "environment" was the condition that no direct measurements are to be made on the environment; the only actual measurements are made, by definition, on the "system". On the other hand, we saw that if we are interested only in the properties of the system, a knowledge of the reduced-density matrix $\rho_1(q_1, q'_1)$ is perfectly adequate. Consequently, the values of all variables which are of experimental interest are determined by ρ_1; therefore, *for our purposes* it forms a complete description.

This argument is incorrect. What has been left out is the temporal aspect. It is perfectly true that a knowledge of the reduced-density matrix $\rho_1(q_1, q'_1 : t)$ at time t is sufficient to determine uniquely the expectation value of any operator $\hat{A}(q_1)$ *at time t*; however, it will *not*, in general, uniquely determine the value of $\hat{A}(q_1)$ at times $t' > t$. The point at issue is a subtle one: the relation between the expectation value at $\hat{A}$ at time t' and the value of $\hat{\rho}$ at time t involves the unitary evolution operator $\hat{U}(t) = \exp(-\mathrm{i}\hat{H}t)$ and, since the Hamiltonian $\hat{H}$, in general, couples the system and environment it automatically introduces the question of correlation between them. Consider, e.g., a particle moving between two identical wells and interacting with a bath of oscillators ("spin-boson" problem; see, e.g., Leggett et al. 1987). If the particle starts in one well at $t = 0$, then after an appropriate time its density matrix will, in general, be, to within exponentially small terms, an incoherent mixture of states in the two wells with equal weights. If this were the case for an *isolated* particle, we could conclude that thereafter no interesting oscillatory behavior is possible – the probability of finding the particle in either of the wells is $\frac{1}{2}$ forever after. For the interacting system, however, no such conclusion is possible – under appropriate conditions it will eventually return to the original well with 100% probability. (For further discussion related to this question, see Leggett 1980.) We conclude that, in general, a knowledge of the reduced-density matrix of the system is *not* sufficient to predict its future behavior; a full quantum-mechanical description of the "universe", i.e., the system *plus* its environment, is, in general, essential. Thus, the application of the quantum formalism to the kind of systems typically met with in, e.g., condensed-state physics raises conceptual and technical issues far beyond those encountered in the simple original applications to atomic systems.

Or does it? At this stage it might well occur to us to retrace our steps and ask: Are the atomic or elementary-particle systems which are the subject of the "classical" experiments actually as isolated as we thought? To avoid inessential technical complications in the ensuing argument, let me consider as a specific example a rather more recent experiment, namely one on neutron interferometry. In such an experiment, which is fairly closely analogous to a "Young's slits" experiment, one detects an interference pattern due to neutron beams which have at an intermediate stage been separated by a difference of the order of several centimeters. The standard theory of this experiment (see, e.g., Greenberger 1983) treats the ensemble of neutrons as propagating in free space, uncoupled to anything else in the world (except at the brief stages where Bragg diffraction by a deflecting crystal or detection occurs).

Is this in fact a correct description? Actually, no, because quite apart from anything else the neutron interacts, by virtue of its magnetic moment, with the modes of the vacuum electromagnetic field. Thus, a neutron traversing the upper of the two paths polarizes the vacuum in a way different from one traversing the lower path. Moreover – and this is perhaps the surprising aspect – the interaction, though weak by atomic-physics standards, is still strong enough that with a reasonable upper cutoff on the energies of the electromagnetic modes involved the states of the vacuum induced on the two paths are very close to orthogonal. Now, at first sight, this is a worrying state of affairs: for it is a well-known result of the quantum theory of measurement that, whereas for a simple system (S) with a wave function of the form

$$\psi = a\psi_1 + b\psi_2,$$

the interference can be detected by measuring the expectation value of any operator $\hat{A}$ such that $\mathrm{Re}\{a^*b\langle\psi_1|\hat{A}|\psi_2\rangle\} \neq 0$, once the two branches have been correlated with orthogonal states χ_1, χ_2 of something else (call it the "environment" E) such that

$$\psi = a\psi_1\chi_1 + b\psi_2\chi_2, \quad (\chi_1, \chi_2) = 0, \tag{1.4}$$

then only a measurement of the *correlations* between S and E can distinguish the linear superposition from a mixture. Thus, at first sight, the effects of interference should be, if not identically zero, at least very small, and the neutron interferometer should not work!

Again, the solution to this apparent paradox lies in consideration of the time variable. It is true that a measurement *at time* t_p (when the two neutron beams are well separated) could not reveal the effects of interference (but then, it is difficult to think of an experiment which would do this even for the completely isolated ensemble of neutrons!). However, the time evolution operator $\exp(-\mathrm{i}\hat{H}t)$ ensures that the "environment" (in this case the vacuum electromagnetic field) is as it were dragged along with the neutron, so that by the time the latter arrives at the final (detection) screen, where the two beams are recombined, the associated states of the vacuum are also very nearly coincident, i.e., the wave function is crudely of the form

$$\psi = (a\psi_1 + b\psi_2)\chi_0, \tag{1.5}$$

and the interference term is once more detectable by measurement of an operator on S alone. The crucial point is that given typical neutron velocities, etc., the interaction between the neutron magnetic moment and (most of) the radiation field is effectively *adiabatic* in nature; there is no possibility of real excitation of a field mode, because of considerations of conservation of energy and momentum. (A very few of the lowest-frequency modes may in some circumstances evade those considerations, but the probability of their excitation is so small as to have negligible effects.) If one thinks about it carefully, this is the case in a great many other examples of atomic-level systems which we have always thought of as "isolated"; it turns out that our instinct in this respect was essentially correct, although the justification for it is more subtle then one might have guessed. If the interaction with the environment is genuinely "adiabatic", then its effects are limited to renormalizing the parameters of the system (e.g., in the above example, the neutron mass can be renormalized as a result of the interaction with the electromagnetic field – a phenomenon well-known in quantum electrodynamics). As we shall see, it is this crucial point which saves the whole subject of the quantum mechanics of a macroscopic variable from being an experimental nonsense.

Thus, the only case in which the "environment" can *qualitatively* affect the behavior of the "system" is when the adiabatic condition fails, i.e., when there is an appreciable probability amplitude for the motion of the system to excite the degrees of freedom of the environment. This clearly requires that the environment should have appreciable spectral weight in the region of frequencies corresponding to a typical frequency of motion of the system, and, moreover, that the Hamiltonian should have nonzero matrix elements coupling the relevant modes to the system.

There are, of course, various microscopic systems which meet these conditions – e.g., an electron undergoing a transfer reaction in a molecule and interacting with the vibronic modes. It is, indeed, well-known that for such systems the effects of the interaction with the environment can sometimes have a drastic qualitative effect, e.g., changing the oscillatory behavior characteristic of a degenerate two-level system into an incoherent relaxation process. Needless to say, it is possible to do specific quantum-mechanical calculations for each such microscopic system piecemeal. However, if one wishes to study the effects of the system–environment interaction systematically, there are great advantages in using systems where the variable in question is *macroscopic* – e.g., superconducting devices based on the Josephson effect, where the variable in question is either the trapped flux (in an rf SQUID ring) or the difference in Cooper-pair phase (in a current-biased junction) (cf. below, and the chapter 6 by Devoret et al.). The main advantages of using a macroscopic variable are, first, that one can actually monitor individual events rather than having to observe only the averaged properties of a large ensemble, and secondly that one can not only often adjust the parameters by merely "turning a knob", but, perhaps more importantly, can establish the values of those parameters in experiments which can be interpreted in classical terms (i.e., which are conducted in the limit in which the predictions of quantum and classical mechanics are indistinguishable). As an example, in a current-biased Josephson junction it is possible, by appropriate tuning of the (directly measured) bias current and/or temperature, to first explore a situation where quantum effects are negligible and thereby determine the (usually not directly measurable) capacitance and shunting impedance of the junction, then go to a situation where quantum effects are dominant and compare the experimental results with the predictions of a calculation which uses as inputs the said capacitance and impedance and, thus, has no adjustable parameters; see chapter 6, by Devoret et al.

I have tried to make, above, the case that the quantum mechanics of a macroscopic variable is a particularly informative way of exploring the more general problem of the quantum behavior of a system strongly coupled to its environment. But this volume is specifically devoted to the problem of quantum *tunnelling* and associated phenomena. What is special about tunnelling in this context? As remarked in the Preface, the point is that we are primarily interested in those aspects of quantum mechanics which have no analog in classical physics. Indeed, as we shall see in more detail below, in so far as the behavior of the system is semiclassical the effects of the environment are, unsurprisingly, little different from what they would be in classical physics; to see spectacular effects we need the "least classical" behavior possible, and the phenomenon of tunnelling fills the bill.

In this section I have implicitly assumed that we believe that quantum mechanics is the ultimate truth about the physical world, and merely wish to explore its more surprising consequences. In the next section I turn to a quite different motivation for exploring the quantum behavior (or not!) of a macroscopic variable.

2. *The quantum measurement paradox*

It is arguable that the most fundamental conceptual problem in the whole of the physical sciences is the question of the apparent transition from "potentiality" to "actuality" which characterizes most interpretations of the formalism of quantum mechanics. To explain this, let us consider an ensemble of microscopic systems (e.g., neutrons) described by quantum mechanics, which make transitions between various possible states as indicated in fig. 1. (The different states may, but need not, correspond to different spatial positions; they could also, e.g., correspond to different values of some internal quantum number such as strangeness.) We assume that the intermediate states B and/or C can be "shut off" at will and that in shutting off (say) B we do not physically affect the probability of arriving at C or the transmission through C, at least in any sense that would be recognized by the average experimental physicist. Let $P_{\mathrm{A}\to\mathrm{B}\to\mathrm{E}}$ be the probability that a system starting at A reaches E when only intermediate state B is open (i.e., the number of systems reaching E, divided by the number which left A); define $P_{\mathrm{A}\to\mathrm{C}\to\mathrm{E}}$ similarly, and let $P_{\mathrm{A}\to(\mathrm{B\ or\ C})\to\mathrm{E}}$ be, by definition, the probability of reaching E from A when *both* of the intermediate states B and

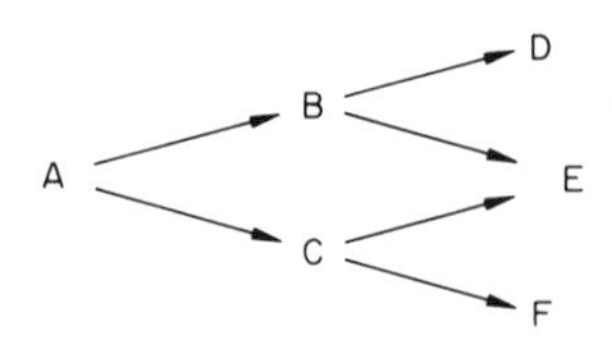

Fig. 1. A quantum system making transitions between various possible states.

C are open. All these quantities are directly measurable in experiment. Now, if our ensemble is indeed correctly described by quantum mechanical (QM) propagation amplitudes in the standard way, then we should expect the possibility, under appropriate conditions, of *interference* between the two paths $A \to B \to E$ and $A \to C \to E$, i.e., that in general we have

$$P_{A \to (B \text{ or } C) \to E} \neq P_{A \to B \to E} + P_{A \to C \to E}. \tag{2.1}$$

Let us suppose that we can find conditions under which this inequality is experimentally verified (as is well known, there are certainly realistic cases, such as neutron interferometry or the neutral K-meson system, where this is true). What are the implications of this *purely experimental* result?

Suppose that we believe that for each individual member of the ensemble one or other of the intermediate states B or C was actually realized. That is, the total ensemble can be broken up into two subensembles consisting of those members which went through intermediate state B or C. Now, we have postulated that there is no physical mechanism (of any type known to experimentalists) by which the opening or shutting of possibility B could affect the propagation via state C or vice versa. Consequently, the number of systems reaching E through B should be exactly the same whether or not possibility C is left open, and vice versa. Such an argument would predict that the inequality (2.1) is replaced by an equality, in contradiction to experiment, and as a consequence we are apparently forced to reject the premises that for each member of the ensemble one or other of the possibilities was realized. (Needless to say, this is just a slightly generalized version of the familiar Young's-slits argument.)

Can we go beyond this negative statement? At first sight, it is extremely tempting to do so; namely, to argue that for each system there must, positively, be some "element of reality" associated with *each* of the intermediate states B and C. Indeed, the formalism of quantum mechanics strongly biases us in this direction, since it associates a finite transition amplitude (or wave function) with each of the intermediate states, and explains the inequality (2.1) in terms of the interference of these amplitudes. It is, of course, extremely tempting to regard this probability amplitude as itself constituting an "element of reality" or, more precisely, as the representation within the formalism of such an element (whose physical nature we do not, of course, at present know). However, adherents of the Copenhagen interpretation and certain other interpretations of the QM formalism warn us forcefully against succumbing to this temptation: according to them, the quantum-mechanical probability amplitude itself corresponds to nothing at all in the physical world, its *only* significance being as an auxiliary mathematical quantity in the calculation of physically observable probabilities. Whether or not this is a metaphysically satisfactory point of view is, of course, a hotly debated question, to which I return below in the context of the so-called statistical interpretation; for present purposes I merely want to emphasize that the further development of my argument does *not* rest on the positive assertion that there is, at the microscopic level, some element of reality associated with each of the possibilities* B and C, but merely on the negative statement that neither experiment, nor the formalism of QM, is apparently consistent with the claim that one or other of the possibilities is realized for each member of the ensemble.

Were this the whole story, there would be no obvious conceptual difficulty. What throws the cat among the pigeons is, of course, the observation that we can perfectly well arrange to observe (by techniques familiar to all working experimentalists) whether any particular member of the ensemble actually occupied state B and/or state C; and that if we do this, we invariably find that any particular system did indeed pass through *one and only one* of the two states. (Needless to say, under these conditions there no longer appears any interference between the two paths in $P_{A\to E}$.) Thus, if *not* observed the system appears to choose not to "realize" one or other of the two possibilities; if observed, it immediately does so!

The standard response to this apparently paradoxical state of affairs is of course to point out that in quantum mechanics "observation" can never be simply a passive process; it necessarily involves causing the system in question to interact with a macroscopic apparatus, and, plausibly, to induce a macroscopically irreversible process in this apparatus (see, e.g., Daneri et al. 1962). Thus, we have radically changed the experimental environment of the system, and it is, therefore, not surprising (or so it is said) that its behavior is radically different from what it would be were it left unobserved. One could try to sum up this point of view (in a way which might perhaps not be acceptable to all of its adherents, but I believe would be assented to by most) by saying that it is the coupling of the microscopic system to a macroscopic apparatus, and the consequent occurrence of a macroscopic event, which induces the transition from "potentiality" to 'actuality".

Now I want to emphasize strongly that so long as this claim is made in isolation from any particular theoretical prescriptions, it involves no obvious conceptual problems; indeed, it is the basis of the point of view which I call below "macrorealism". It causes difficulty – and that in extreme form – only to the extent that it is combined with the belief that quantum mechanics is a complete theory of the physical world. Since, however, this belief appears to be held explicitly or implicitly by most physicists, this is not a major constraint on the present line of argument.

The source of the problem was exhibited long ago by Schrödinger (1935) and may be put succinctly, and in extremely oversimplified form, as follows. Consider for simplicity an ensemble of microscopic systems which have available two (microscopically) different states ψ_1 and ψ_2 (corresponding, e.g., to spin up and spin down, respectively). To measure which of these states is realized for

*Although such an assertion would further strengthen the argument, cf. Leggett (1987).

a particular system of the ensemble, we couple it to a macroscopic measuring apparatus, which we assume also to be described by quantum mechanics, in such a way that state ψ_1 of the system will induce final state Ψ_1 of the apparatus, while ψ_2 will induce the *macroscopically distinguishable* final state of the apparatus Ψ_2 (these assertions can be checked in a preliminary experiment). That is, under the influence of the coupling we have (Ψ_0 = initial state of apparatus)

$$\psi_1 \Psi_0 \to \psi_1 \Psi_1, \qquad \psi_2 \Psi_0 \to \psi_2 \Psi_2. \tag{2.2}$$

Now suppose we prepare an ensemble of the microsystems which is described by the *linear superposition* state $a\psi_1 + b\psi_2$, $a, b \neq 0$ (e.g., for the spin-$\frac{1}{2}$ case we could prepare it in an eigenstate of $\hat{\sigma}_x$). Following the passage of the first microsystem of this ensemble through the measuring apparatus, what is the correct description of the state of the latter? Actually this question as stated has no answer, since the apparatus by itself may not have a definite description (see section 1).. However, we can legitimately ask for the correct description of the "universe" consisting of the apparatus plus the microsystem in question, and the answer is then immediate: by the linearity of quantum mechanics (which is a fundamental and nonnegotiable property of the theory) we see at once that the correct prescription is

$$(a\psi_1 + b\psi_2)\Psi_0 \to a\psi_1 \Psi_1 + b\psi_2 \Psi_2. \tag{2.3}$$

That is, it is (at least prima facie!) a linear superposition of states of the universe corresponding to *macroscopically distinct* behavior of the measuring apparatus.

Let us just pause for a moment at this stage to make the standard disclaimer that the above presentation of the argument is, of course, naive in the extreme; apart from anything else, the microsystem-plus-apparatus is itself not a closed system and, in any case, it is most unlikely that we would ever in practice be able to determine even the initial properties of a macroscopic body to the extent necessary to assign to it a pure quantum state.* While these considerations are of crucial importance for attempts at practical realization of experiments which could throw light on the paradox (see below) they in no way affect the *conceptual* difficulty, namely that the characteristic "lack of definiteness" of the quantum description at the microlevel has now propagated up to the macrolevel. (For a technical discussion of these complications, see, e.g., Leggett 1980.)

Let us try to make more precise the reason why there is a prima facie difficulty with eq. (2.3). The point is that we decided earlier on that when a microscopic system (or more correctly an ensemble of such systems, is described by a linear superposition of the type $a\psi_1 + b\psi_2$, we cannot, in general, interpret this as meaning that each member of the ensemble has selected either state 1 or state 2. If, therefore, we consistently carry through the same interpretation to the

*A consideration which appears not always to be remembered in papers on "quantum cosmology".

macroscopic level, there we are forced to say that we cannot interpret eq. (2.3) as saying that a definite *macroscopic* outcome, corresponding* to Ψ_1 or Ψ_2, has been selected; even at the macroscopic level, no definite one of the possibilities has been realized! Yet, of course, we know that if we actually inspect the state of the counter in question with the naked eye, we will certainly find a result corresponding to either Ψ_1 or Ψ_2, and it seems difficult (at least at first sight) to believe that at the *macroscopic* level the distinction between a system which is or is not observed by a human observer is significant.

Of the many "solutions" of the paradox which have appeared in the literature over the last fifty years I shall mention only two which are particularly relevant to the subject of this chapter (for a discussion of others, see, e.g., Leggett 1987a). The first is the so-called "statistical" interpretation which one might, I believe, legitimately regard as the carrying of the somewhat ill-defined collection of remarks and precepts known as the "Copenhagen interpretation" to its logical conclusion. According to this interpretation, one should take utterly seriously the idea that the quantum-mechanical wave function corresponds to no reality in the physical world whatever, *even at the macroscopic level.* Thus, the *only* meaning of a final-state wave function of the form eq. (2.3) is that it permits the calculation of the probabilities of various *directly observable* macroscopic events, and it is pointless (and dangerous) to attribute any other significance to it. (For an extended defence of this point of view, see, e.g., Ballentine 1970.)

Now it seems to me that adherents of this interpretation are not always clear about exactly which kinds of event the final state eq. (2.3) is supposed to allow one to calculate. To clarify this consideration, let us call $\hat{A}$ the operator whose eigenvectors are Ψ_1 and Ψ_2, and $\hat{B}$ some operator which has *nonzero* matrix elements $\langle \Psi_1 | \hat{B} | \Psi_2 \rangle$ (though see below). Is the wave function (2.3) supposed to permit only the calculation of the probability of obtaining one or other eigenvalue of $\hat{A}$, or does it also allow a similar calculation for $\hat{B}$? If the former, then at first sight it would seem that quantum mechanics cannot be a universal description; while if the latter, the interpretation, which no doubt internally consistent, has the very surprising feature that *even at the macroscopic level* one is obtaining physical interference effects from quantities (the probability amplitudes Ψ_1 and Ψ_2) which correspond to no "element of reality" in the physical world; call this for future reference "the unreality of macroscopically interfering amplitudes." A possible reply to this objection is that the premise of the question is false, in that when Ψ_1 and Ψ_2 correspond to macroscopically distinct states there simply is no realistic operator $\hat{B}$ with the property that $\langle \Psi_1 | \hat{B} | \Psi_2 \rangle$ is nonzero (or at any rate has any observable magnitude). If this claim is made then the statistical interpretation comes very close to the interpretation which I shall call the "orthodox" one, to which I now turn.

*In this and the next two paragraphs only, I use Ψ_i as a shorthand for $\psi_i \Psi_i$ $(i = 1, 2)$.

The "orthodox" solution of the quantum measurement problem (by which I mean the solution, in so far as any exists, which is probably embraced* by the majority of practising physicists who have devoted serious thought to the problem) goes roughly as follows: Let us admit (it is said) that the linear superposition of macroscopically distinct states on the right-hand side of eq. (2.3) is a technically correct description of the final state of the universe. Now let us assert that no remotely conceivable experiment could ever distinguish the predictions of the right-hand side of eq. (2.3) from that of a classical statistical mixture of the states $\psi_1 \Psi_1$ and $\psi_2 \Psi_2$ with probability $|a|^2$ and $|b|^2$, respectively. This assumption is crucial to the argument: let us call it "premise P." Once we grant premise P, it is said, we may as well say that for each member of the ensemble one or other of these states has actually been realized by this stage; the transition from microscopic to macroscopic has indeed induced a transition from "potentiality" to "actuality".

Now even if for the moment we indeed grant "premise P" for the sake of argument, it is not at all clear that this line of argument is logically satisfactory. Essentially, the reasoning is that while we cannot interpret the microscopic superposition $(a\psi_1 + b\psi_2)$ as implying that each member of the ensemble realizes one and only one of the alternatives ψ_1 or ψ_2 (because of the possibility of observing the effects of interference between the two), we are entitled to interpret the *macroscopic* superposition $a\psi_1 \Psi_1 + b\psi_2 \Psi_2$ (or the analogous density-matrix description) as implying that each ensemble member realizes one and only one of the macroscopically distinct states $\psi_1 \Psi_1$ and $\psi_2 \Psi_2$, because at this level there is now no possibility (according to premise P) of observing any interference between them. But the formalism of QM is a seamless whole, and changes not at all on the passage from microscopic to macroscopic; what entitles us to change our interpretation of it in midstream? The above argument seems to me to embody a fundamental confusion between the interpretation itself, and the *evidence* that that interpretation is correct. After all, if during a murder trial the evidence of the accused's guilt suddenly disappears, the jury may have to declare him legally innocent, but that does not mean that he has thereby *become* innocent!

At first sight the arguments of this section are entirely on the plane of metaphysics rather than physics. Indeed, probably the most striking feature of the quantum measurement paradox over the last fifty years has been that it has seemed totally impervious to any kind of *experimental* resolution. And it is indeed clear that, as long as we grant premise P, this will continue forever to be so. If, therefore, we have any hope at all that experiment may be able to provide some useful input to this problem (and, thereby, in the language of the average

* As evidenced, inter alia, by the regularity with which it is rediscovered and republished as new in the literature.

practising physicist, raise it from the lowly plane of "philosophy" to the exalted status of "physics") it is essential to challenge premise P.

We, therefore, ask the crucial question: Is it in fact true that it is always and inevitably impossible to see the effects of quantum interference between macroscopically distinct states (hereafter QIMDS)? If the answer is yes, the most fundamental problem of physics will remain forever inaccessible to any techniques which physicists can devise. If it is no, then we have at least some hope that with sufficient ingenuity we may be able to devise experiments which shed some light on the paradox. Ideally, what we would like to do is to prepare an experimental situation where quantum mechanics unambiguously predicts (a) that a superposition of macroscopically distinct states should occur, and (b) that the interference between these states should (in violation of premise P) give rise to experimentally observable effects. If these effects are indeed seen, then we should presumably have to adopt something like the statistical interpretation of QM, and swallow the very surprising (and disturbing) "unreality of macroscopically interfering amplitudes" mentioned above. If the predicted effects do *not* occur, then we should have to conclude that the description of the world embodied in standard quantum mechanics breaks down at some stage on the road from the microscopic level of electrons and atoms, which provided its original base, to the macroscopic level at which it was until recently completely untested.

All of this strongly suggests that it is of interest to investigate the predictions of quantum mechanics as they apply to the (possible) interference of macroscopically distinct states, i.e., in effect, to the dynamics of *macroscopic variables.* This will be done in general terms in the next section; then, after discussing the implications of existing experiments on this subject (some of which are reviewed by Devoret et al. in chapter 6), I will discuss a specific experiment, at present in the construction stage, which would not only test the principle of QIMDS, but, if the latter is seen, definitively refute an alternative "common-sense" view of the macroscopic world.

The reader will notice that up to this point I have deliberately avoided any discussion of the plausibility or otherwise of premise P. The reason is that it is more natural, in the present context, to place this discussion in the general framework of the quantum mechanics of a macroscopic variable in the next section. However, it is necessary to make one remark at this stage: If we interpret the words "macroscopically distinct states" in what I believe would be the most natural common-sense way, so that, e.g., the experiments discussed below and in chapter 6 by Devoret et al. indeed relate to such states, then it will turn out that premise P is, or may be, false. However, it is certainly possible to reinterpret the words in question, or reformulate premise P, in such a way that the latter remains true, and it could perhaps be argued (and no doubt will be by some!) that such a reinterpretation may be appropriate in the context of the measurement paradox. For example, if we regard two states as "macroscopically distinct" only if there has been a macroscopic amount of irreversible dissipation in one or both of them, then it will be implicit in the results of section 3 that under such circumstances interference is indeed unobservable. Since there are strong arguments that anything we are prepared to regard as a true "measurement" must involve a macroscopic degree of irreversibility (even if only at the level of the optic nerve!), it might be argued that in this context this is really the crucial characteristic of a "macroscopic" process. All I can say in response to this argument is that I believe it is a meaningful contribution to the *conceptual* debate about the quantum measurement paradox if and only if it is believed that irreversibility is a basic and irreducible new component of physical processes, rather than (as most physicists believe it to be) simply the macroscopic manifestation of a set of processes which at the fundamental level are no different from any other exemplification of the workings of quantum mechanics and, therefore, have no specific conceptual status. In the absence of such a belief, recourse to reliance on irreversibility seems to me just a not particularly interesting case of "moving the goalposts": it does nothing to blunt the assertion that the search for evidence of QIMDS (in the sense understood in the rest of this chapter) is of fundamental interest, and that reliable observation of its absence would be revolutionary.

To summarize the argument of this section: The alleged solution to the quantum measurement paradox which I have called the "orthodox" one derives much, if not all, of its plausibility from the fact that it embodies "premise P", i.e., that quantum interference of macroscopically distinct states (QIMDS) never occurs in real life and, therefore, unlike (some versions of) the "statistical" interpretation, does not have to face up to the possibility that even at the macroscopic level physical effects may be obtained from the 'interference" of quantities occurring in the formalism (namely, the macroscopic amplitudes Ψ_1, Ψ_2) which, allegedly, correspond to nothing at all in the physical world ("unreality of macroscopically interfering amplitudes"). Reliable observation of the phenomenon of QIMDS would, therefore, definitively refute this "solution" and force us back, at best, to a version of the statistical interpretation which recognizes the need to face up explicitly to this very disturbing state of affairs.

3. The QM of a macroscopic variable

We now turn to the details of the description in quantum-mechanical terms of the dynamics of a variable some of whose eigenvalues correspond, by some reasonable criterion, to "macroscopically distinct" states. This discussion will make more quantitative many of the points already brought up in a qualitative way in section 1; of course, not all of them are peculiar to the case of macroscopic variable. At the same time, we shall see along the way both the reasons why "premise P" of section 2 is prima facie plausible, and the circumstances under which it may be false. In what follows, as above, we shall use the word "system" to refer to the physical degree of freedom characterized by the macroscopic variable, which we shall denote by q, and the word "environment" to refer to what is described by all the other degrees of freedom, here labelled* $\{x_i\}$.

Consider a system described by N microscopic coordinates ξ_i and conjugate momenta p_i. These coordinates might include positions of individual atoms, Fourier components of the electromagnetic field, spin degrees of freedom and much else. In many cases of interest we can introduce a coordinate transformation

$$\{\xi_i\} \to q, \{x_i\}: \qquad \{p_i\} \to p, \{\pi_i\}, \tag{3.1}$$

with $N-1$ new microscopic variables $\{x_i\}$ and their conjugate momenta π_i and a single variable q (conjugate momentum p) some of whose eigenvalues are, in some "everyday" sense, macroscopically distinct. A very familiar example of such a transformation is of course that from atomic to center-of-mass and relative coordinates for a solid body; however, there are many other cases where such a transformation is possible, e.g., in the case of an rf SQUID ring we can transform to a system of coordinates which includes the total magnetic flux Φ trapped through the ring. The Hamiltonian $\hat{H}$ can be reexpressed as a function of the new coordinates $(q, \{x_i\})$ and the corresponding conjugate momenta $(p, \{\pi_i\})$, and we then have in the Schrödinger representation the usual time-dependent Schrödinger equation (TDSE)

$$i\hbar \frac{\partial \Psi}{\partial t}(q, \{x_i\}:t) = \hat{H}\Psi(q, \{x_i\}:t), \quad \hat{H} \equiv \hat{H}(p, q:\{\pi_i, x_i\}). \tag{3.2}$$

So far, the description is completely general. Let us now suppose for a moment that the Hamiltonian actually separates into a sum of a term $\hat{H}_S(p, q)$ depending only on the macroscopic ("system") variable and a term $\hat{H}_E\{\pi_i, x_i\}$ depending only on the environmental coordinates and momenta:

$$\hat{H}(p, q:\{\pi_i, x_i\}) = \hat{H}_S(p, q) + \hat{H}_E\{\pi_i, x_i\}. \tag{3.3}$$

This is actually a very rare case, probably the only reasonably realistic example being the motion of a solid body in a *uniform* gravitational field; however, we will use it for the moment to illustrate an important point. With the separation eq. (3.3) a possible class of solutions to the TDSE (3.2) is of the form

$$\Psi(q, \{x_i\}:t) = \psi(q, t)\chi(\{x_i\}, t), \tag{3.4}$$

and inserting this in eq. (3.2) we see that $\Psi(q, t)$ obeys its own Schrödinger equation with no reference to the state of the environment

$$i\hbar \frac{\partial}{\partial t}\psi(q, t) = \hat{H}_S(p, q)\psi(q, t). \tag{3.5}$$

* Much of the discussion of this section follows fairly closely that of Leggett (1987b), with some changes in notation.

Thus, in this special case we can prima facie treat the motion of the macroscopic variable in a way exactly analogous to that which we would use for a microscopic one such as the coordinate of an electron.

It is convenient at this stage to examine one argument which is sometimes given for "premise P" of section 2 (which, we recall, is the assertion that quantum interference between macroscopically distinct states is experimentally unobservable). The argument is simply that if two states Ψ_1 and Ψ_2 of a macroscopic body are macroscopically distinct, then by definition a macroscopic number of particles are behaving differently in the two: crudely and schematically we might think of writing, in our original coordinates

$$\Psi_1 \sim \prod_{i=1}^{N} \phi_1(\xi_i), \qquad \Psi_2 \sim \prod_{i=1}^{N} \chi_i(\xi_i), \tag{3.6}$$

where the overlap (ϕ_i, χ_i) is typically appreciably less than unity ($\sim c$, say) (and could even be zero). It is now clear that if in a state $a\Psi_1 + b\Psi_2$ we try to measure any operator which involves only a few particles, or any sum of such operators, we will get a large number of factors of (ϕ_i, χ_i) and, hence, the expectation value will be of order $|c|^{-N}$, which for large N is unobservably small. Since it is experimentally impossible to measure any operator which involves a product of more than a few (and, in particular, $\sim N$) of the ξ_i, premise P follows.

How does this argument look in the context of the above example? Let us consider for simplicity the case where the ξ_i represent the atomic coordinates and the macroscopic variable q, therefore, the COM coordinate. A typical case in which the wave function might be of the form $a\Psi_1 + a\Psi_2$ would be a "Young's slits" diffraction experiment on a polyatomic molecule.* In that case, since the relative wave function may be taken to be the ground state $\chi_i(\{x_i\})$ of the molecule** and, hence, insensitive to which slit it passes, we can write the complete wave function at the time t_p the molecule passes the intermediate screen in the form (in the (q, x_i) representation)

$$\Psi(q, \{x_i\}:t_p) = [a\psi_1(q) + b\psi_2(q)]\chi_G\{x_i\}, \tag{3.7a}$$

where ψ_1 and ψ_2 are localized close to the upper and lower slits, respectively. Alternatively, in the original coordinates $\{\xi_i\}$ we can write *schematically*

$$\Psi(\{x_i\}:t_p) = a\prod_i \phi_i(\xi_i) + b\prod_i \chi_i(\xi_i). \tag{3.7b}$$

We assume that the spread of the single-particle wave functions ϕ_i, χ_i corresponding to passage through the upper (lower) slits is of order $\varepsilon \lesssim d$ (for the

* In the ensuing argument it is implicit that the slits, and the associated screen, are treated as purely classical and fixed objects – they in no sense themselves constitute a "measuring apparatus." (Recall that in the standard account of the measurement process (section 2) the "measuring apparatus" has to change its state drastically in response to the state of the microscopic system.)

** This assumption may be justified in terms of the adiabatic approximation discussed below.

purposes of the argument we need not have $\varepsilon \ll d$, although in practice, of course, this is likely to be so). The order of magnitude of the spread of $\psi_1(q)$ and $\psi_2(q)$ is now $\sim N^{-1/2}\varepsilon \ll d$, so that ψ_1 and ψ_2 are to all intents and purposes orthogonal (even if, in a similar geometrical situation, the corresponding wave functions of a *single* diffracted particle ($\sim \phi_i, \chi_i$) would *not* be orthogonal). It is then clear that, e.g., a measurement of the expectation value of (a sum of single-particle operators) q at the time of passing the slits will not distinguish the linear superposition of Ψ_1 and Ψ_2 from a classical mixture, since any terms of the form

$$\int \psi_1(q,t) q \psi_2(q,t)$$

will be unobservably small. So far, the argument is in agreement with the one given above. However, the conclusion that no effects of interference between wave packets propagating through the two slits are observable is clearly wrong; they are no more and no less present then in the diffraction of a single particle.* We can now see, more explicitly than in section 1, that the reason that such effects can occur is that even in the simplest (not totally trivial) case the Hamiltonian contains (at least) a term $p^2/2M + V(q) \equiv (2M)^{-1} (\sum_i p_i)^2 + V(\sum_i \xi_i/N)$ and the unitary evolution operator $\hat{U}(t) \equiv \exp(-\mathrm{i}\hat{H}t)$, therefore, contains correlations between arbitrarily many particles (for long enough t). Thus, while $\langle \Psi_1(t_p)|q|\Psi_2(t_p)\rangle$ is unobservably small, the quantity corresponding to observation of interference at the *final* (detecting) screen, $\langle \Psi_1(t_f)|\hat{q}|\Psi_2(t_f)\rangle$ is not:

$$\begin{aligned}\langle \Psi_1(t_f)|\hat{q}|\Psi_2(t_f)\rangle &\equiv \langle \hat{U}(t_f - t_p)\Psi_1|\hat{q}|\hat{U}(t_f - t_p)\Psi_2\rangle \\ &= \langle \Psi_1|\hat{U}(t_f - t_p)\hat{q}\hat{U}(t_f - t_p)|\Psi_2\rangle \neq 0.\end{aligned} \tag{3.8}$$

Thus, this particular argument in favor of premise P is invalid, provided that we are prepared to detect the effects of interference at a time *later* than the interference itself occurs. In effect, we do not attempt to measure the interference directly but let Nature do the work for us.**

Although, as mentioned, cases where the separation eq. (3.3) is valid are extremely rare, a much more common situation is that it is possible to write the Hamiltonian in the form

$$\hat{H}(p, q, \{x_i, p_i\}) = \hat{H}_S(p, q) + \hat{H}_E(\{p_i, x_i\}:q). \tag{3.9}$$

Such a representation is possible, e.g., for a solid body moving (nonrelativistically) in an *inhomogeneous* gravitational or electric field. This is a case familiar

* These may, of course, be practical difficulties in detecting them, connected with the short period of the diffraction pattern for a macroscopic object.

** If the reader objects that the considerations given here are so trivially obvious as to be superfluous, I would personally agree, but a survey of the literature on quantum measurement theory will reveal that they are clearly not "obvious" to many practitioners in this area!

from the theory of molecular vibrations, and just as in that case we can try to write a class of approximate solutions to the time-dependent Schrödinger equation in the form

$$\Psi(q, \{x_i\}:t) = \psi(q, t)\phi(\{x_i\}: q), \tag{3.10}$$

where $\phi(\{x_i\}, q)$ is the solution of the *time-independent* Schrödinger equation, in which q enters as a parameter

$$\hat{H}_{\mathrm{E}}(\{p_i, x_i\}:q)\phi(\{x_i\}; q) = U(q)\phi(\{x_i: q\}), \tag{3.11}$$

where $U(q)$ is the energy eigenvalue, which depends parametrically on q. This equation does not fix the normalization of ϕ, and we choose the latter so that it is normalized for each value of q, i.e., $\int|\phi(\{x_i\}, q)|^2\,\mathrm{d}\{x_i\} = 1, \forall q$. To the extent that eq. (3.10) is valid, the system wave function $\psi(q, t)$ satisfies the simple equation

$$\mathrm{i}\hbar\frac{\partial\psi}{\partial t}(q, t) \equiv \hat{H}_{\mathrm{S}}(p, q)\psi(q, t), \quad \hat{H}_{\mathrm{S}}(p, q) \equiv \hat{H}_{\mathrm{S}}(p, q) + U(q). \tag{3.12}$$

In principle, the quantity $U(q)$ can depend on the environment state ϕ; however, as we shall see below, for a *macroscopic* variable q the effect of this is negligible.

The important point to note about the form of solution (3.10) is that, *provided we are interested only in the expectation values of system operators which are diagonal in* q, the environment state $\phi(\{x_i\}, q)$ falls right out of the problem. The reason is clear: the expectation value of any such operator $\hat{\Omega}(q)$ can be expressed in the form ($\rho(q, q':t) \equiv$ reduced-system density matrix)

$$\begin{aligned}\langle\hat{\Omega}\rangle(t) &= \int\Omega(q)\rho(q, q: t)\,\mathrm{d}q \\ &\equiv \int\!\!\int \mathrm{d}q\,\mathrm{d}\{x_i\}|\psi(q, t)|^2|\phi(\{x_i\}:q)|^2\Omega(q),\end{aligned} \tag{3.13}$$

and in view of the normalization condition on ϕ (above) this reduces to the simple expression

$$\langle\hat{\Omega}\rangle(t) = \int \mathrm{d}q\,\Omega(q)|\psi(q, t)|^2, \tag{3.14}$$

just as in a simple one-particle problem. Thus, to the extent to which eq. (3.10) is valid, and provided we are interested only in operators of this class, the problem of the motion of a macroscopic variable reduces to a simple one-particle quantum-mechanical problem, the only effect of the environment being to provide a contribution $U(q)$ to the potential energy seen by the system variable.

There is still the difference, however, that in general the quantity $U(q)$ will depend on the particular state $\phi(\{x_i\}, q)$ which describes the environment. If we were dealing with a microscopic system such as a diatomic molecule, such an effect could be very important; it is well known, e.g., that the vibrational frequency of such a molecule will depend on the degree of electronic excitation. Fortunately, at this point the macroscopic nature of our system comes to the rescue. In such a system, the coupling of each individual environmental mode to the system variable q is of order $N^{-1/2}$ (where N is the total number of degrees of freedom of the environment) (cf. below); hence, provided the fluctuations in the state of the environment have any normal thermodynamic origin, the fluctuations in $U(q)$ relative to its average value will be at most of order $N^{-1/2}$. Consequently, it is a good approximation to regard $U(q)$ as a fixed function of q (which may, however, depend on the temperature of the environment and other external parameters characterizing it).

At first sight, therefore, the totality of the quantum mechanics of a macroscopic variable consists in finding an effective Hamiltonian $\hat{H}(p, q)$ which is a function only of the variable q and its conjugate momentum, writing down and solving the Schrödinger equation (3.5), and using the solution $\psi(q, t)$ to evaluate the expectation value of any desired operator $\hat{\Omega}(q)$ according to eq. (3.14). Plenty of papers can be found in the literature, particularly on superconducting devices, which do just that; indeed, in certain very special cases it turns out to be a good approximation. However, in general the situation is both more complicated and more interesting than this. In fact, it is intuitively clear that the zeroth-order adiabatic approximation we have made implies, among other things, the complete neglect of any irreversible transfer of energy from the system to its environment; it corresponds, therefore, to the limit in which (inter alia) the motion of the macroscopic variable is completely *nondissipative*. Such a limit is never strictly speaking encountered in real life, although with sufficient experimental care it may be possible to reduce the dissipation to a low enough value that its effects are negligible (cf. below).

Before proceeding to the more realistic case, let us make a few remarks. First, the argument given above concerning "premise P" goes through unchanged for the adiabatic case. Secondly, we should emphasize that even in this simple case, the business of finding the appropriate momentum conjugate to q and, hence, the effective system Hamiltonian $\hat{H}_S(p, q)$ may not be entirely trivial. Consider, e.g., the case of an rf SQUID ring (superconducting ring interrupted by a single Josephson junction). In this case the obvious choice of "macroscopic" variable q is the total flux Φ trapped through the ring (to be distinguished from the externally imposed flux Φ_x). The potential energy associated with this flux is believed, not only on the basis of microscopic considerations but, more relevantly for our purposes, directly from experiment, to have the form

$$V(\Phi) = -\frac{I_c \phi_0}{2\pi} \cos(2\pi \Phi/\phi_0) + \frac{(\Phi - \Phi_x)^2}{2L}, \tag{3.15}$$

where I_c is the critical current of the Josephson junction, L is the self-inductance of the ring and $\phi_0 \equiv h/2e$ is the superconducting flux quantum. The question now arises: What is the correct physical definition of the momentum p_Φ conjugate to Φ, and what is the Hamiltonian expressed in terms of Φ and p_Φ?

It is possible to try to answer this question on at least three levels. The most phenomenological (and, hence, the simplest) is as follows: It is known both from classical phenomenological considerations and from experiment that in the absence of dissipation, and in the regime where one expects classical mechanics to be a good approximation, the dynamics of the flux Φ is well described by the equation

$$C\ddot{\Phi} = -\frac{\partial V}{\partial \Phi}, \tag{3.16}$$

where C is an appropriate capacitance (usually taken to be that of the Josephson junction: but see below). The classical Lagrangian which would give rise to this equation would contain a "kinetic-energy" term of the form $\frac{1}{2}C\dot{\Phi}^2$; therefore, following the usual prescriptions of classical mechanics, the appropriate definition of the conjugate momentum p_Φ is $\partial L/\partial\dot{\Phi} = C\dot{\Phi}$. If one neglects the capacitance of the bulk ring, then $\dot{\Phi}$ is the voltage developed across the junction and $C\dot{\Phi} = CV = Q_J$ is the charge imbalance across it. Hence, we make the identification (up to a possible sign)

$$p_\Phi = Q_J. \tag{3.17}$$

We then write the Hamiltonian in the form

$$H(\Phi, p_\Phi) = \frac{p_\Phi^2}{2C} + V(\Phi). \tag{3.18}$$

Finally, we make the transition to quantum mechanics by interpreting the quantity p_Φ in the usual way as the operator $-i\hbar\partial/\partial\Phi$ and H as $i\hbar\partial/\partial t$. Thus, we obtain the Schrödinger equation

$$i\hbar\frac{\partial\Psi(\Phi{:}t)}{\partial t} = \left[-\frac{\hbar^2}{2C}\frac{\partial^2}{\partial\Phi^2} + V(\Phi)\right]\Psi(\Phi{:}t), \tag{3.19}$$

with $V(\Phi)$ given by eq. (3.15).

A similar argument may be carried out for the case of an isolated Josephson junction biased by a fixed external current I_x (the so-called "current-biased junction" discussed by Devoret et al. (chapter 6), Larkin and Ovchinnikov (chapter 4), and Ivlev and Mel'nikov (chapter 5). In this case the relevant "classical" variable is the relative phase $\delta \equiv 2\phi$ of the quantum-mechanical Cooper-pair wave function on the two sides of the junction; the classical equation of motion is (cf. eq. (1) of Larkin and Ovchinnikov in chapter 4)

$$(2\pi/\phi_0)C\ddot{\delta} = I_x - I_c \sin\delta \tag{3.16a}$$

[cf. eq. (3.16)] and this may be derived from a Hamiltonian

$$H(\delta, p_\delta) = \frac{p_\delta^2}{2C} - V(\delta), \quad V(\delta) = -I_c/\phi_0/2\pi \cos\delta - I_x\delta. \tag{3.18a}$$

Formally, the current-biased junction may be regarded as the limit of the rf SQUID ring as $L \to \infty$, with the replacements*

$$\Phi \to (\phi_0/2\pi)\delta, \qquad \Phi_x \to LI_x.$$

The appropriate Schrödinger equation is then the corresponding limit of eq. (3.19).

The above argument, while direct and simple, clearly begs a number of questions, and, in addition, gives no insight into the nature of the full many-body wave function of the system. An alternative argument goes schematically as follows (for details see Leggett 1987b): Let us imagine that the trapped flux Φ is a c-number and solve for the many-electron wave function $\Psi(r_1 r_2 \ldots r_N; \sigma_1 \ldots \sigma_N)$ in the presence of this flux. The important dependence is through the angular variable θ_i corresponding to the angular position of the ith electron in the ring and is given by a factor which in the bulk ring is of the form

$$\Psi_\Phi\{\theta_i\} = \prod_i \exp\left[\tfrac{1}{2}\mathrm{i}\left(\frac{\Phi}{\phi_0}\right)\theta_i\right]. \tag{3.20}$$

(The factor of $\frac{1}{2}$ enters because θ_i is the angular position of a single electron whereas the flux quantum ϕ_0 as defined contains the charge $2e$ of a Cooper pair.) In the junction itself the form of Ψ_Φ may be more complicated and cannot, in general, be determined without a detailed knowledge of the geometry etc. We now regard Φ as no longer a c-number but a quantum-mechanical variable, and form a general linear superposition of eigenfunctions of $\hat{\Phi}$ with different eigenvalues Φ':

$$\begin{aligned}\Psi(\Phi:\{\theta_i\}) &\equiv \int a(\Phi')\Psi_{\Phi'}\{\theta_i\}\,\mathrm{d}\Phi' \\ &\equiv \int \mathrm{d}\Phi'\, a(\Phi') \prod_i \exp\left(\tfrac{1}{2}\mathrm{i}\frac{\Phi'}{\phi_0}\theta_i\right),\end{aligned} \tag{3.21}$$

where $a(\Phi')$ is a complex amplitude (which would, of course, normally be written as $\psi(\Phi')$: for notational clarity we do not do this at this stage). It may now be seen that the operator $\partial/\partial\Phi$, acting on the "universe" wave function

*These replacements raise a delicate conceptual point since, while values of the flux Φ in a SQUID ring which differ by $n\phi_0$ clearly correspond to physically distinct states, values of the phase δ differing by $2\pi n$ need not obviously do so. This point has been extensively discussed in the literature, see, e.g., Likharev and Zorin (1984); I shall not discuss it here as it is not directly relevant to the topics discussed in this book.

$\Psi(\Phi:\theta_i)$, gives in the bulk region*, where Ψ_Φ is given by eq. (3.20), the result

$$-i\hbar\frac{\partial\Psi}{\partial\Phi}=\left(-\frac{e}{2\pi}\sum_{i=1}^{N}\theta_i\right)\Psi, \tag{3.22}$$

and hence the momentum conjugate to Φ is to be interpreted, apart from a factor, as the *angular displacement of the electronic charge around the ring.* Note that this is *not* in the general case simply equal to the charge displaced across the junction itself; however, in the common case where the capacitance of the junction dominates that of the bulk ring, the two are approximately equal. To this extent, this intermediate-level argument may be regarded as a justification for the more phenomenological approach used above.

Finally, as a third possibility, if we know the detailed microscopic Hamiltonian for the many-electron system interacting with the electromagnetic field, it may be possible to derive a Hamiltonian, or what for our purposes is actually more useful, an effective action, as a function of the macroscopic variable (flux) by explicit elimination of the microscopic variables. For the SQUID case this has been done by Eckern et al. (1984); we postpone a discussion of the results until we have discussed the question of dissipation, since this is automatically included in their treatment.

4. Dissipation

An absolutely essential feature of any viable theory of the quantum mechanics of a macroscopic variable must be a correct account of the effects of dissipation, i.e., of the phenomenon of irreversible transfer of energy between the system (i.e., the motion of the macroscopic variable q) and the environment. Until we have the effects of dissipation under control, calculations based on the simple one-particle-like picture given in section 3 are essentially meaningless. Indeed, as we shall see, it has been a repeated theme of the quantum theory of measurement that for a macroscopic variable these effects will inevitably totally invalidate even the qualitative results of such a calculation. While it will turn out that this need not always be true, it is essential that we understand why.

There is an obvious way for dissipation to arise as a correction to the simple adiabatic picture of section 3. Let us suppose for simplicity that for the system in question the Hamiltonian can indeed be written in the form eq. (3.9).** Now, in fact, a state of the form eq. (3.10) is *not* an exact solution of eq. (3.9); it is only a solution to the extent that terms in the Hamiltonian proportional to

* In the junction itself there are some difficulties; see Leggett (1987b).

** Although the possible presence of terms of the form $H_{\text{int}}(p,q:\{x_i\},\{p_i\})$, which cannot be written simply as $H_E(p_i,x_i:q)$, leads to complications, these are not related to the effects of dissipation as such and will be ignored here.

$(\partial\psi/\partial q)(\partial\phi/\partial q)$ and $\psi\partial^2\phi/\partial q^2$ are ignored. In the standard application of the adiabatic approximation to (light) diatomic molecules one normally treats such terms as a small perturbation, and this is justified by the consideration that the spacing between the "environment" (here electronic) energy levels is very large compared to the characteristic energies of the "system" (here nuclear) motion; under these circumstances, no irreversible transfer of energy from the nuclear to the electronic degrees of freedom is possible, and the only effect of the perturbation is a renormalization of energy levels and matrix elements (which, moreover, turns out to be small, although this is not obvious without some calculation). When the variable q is macroscopic, however, the situation is very different: in general the environment will likewise be "macroscopic" and will, therefore, have energy levels which are extremely closely spaced, certainly on a scale much less than the typical system energies. Thus, the conditions for applicability of the adiabatic approximation are, in general, not met.

Despite the fact that wave functions of the type given by eq. (3.10) are not, in general, even approximate eigenfunctions of the system, they can still form a perfectly good *basis of description*. Using this basis, we can obviously write the total Hamiltonian without approximation in the general form

$$\hat{H} = \tilde{H}_S(p, q) + \hat{H}_E(\{x_i, p_i\}) + \hat{H}_{int}(p, q{:}\{x_i, p_i\}),$$

where we have incorporated the zeroth-order adiabatic energy in $\tilde{H}_S$. This form is, of course, so generic as to be in itself of little use.

We now claim that insofar as we are interested only in the *dynamics of the macroscopic variable* q, the quantities $\hat{H}_E$ and $\hat{H}_{int}$ can usually be taken to be of a remarkably simple form which at once permits useful calculations to be done. Namely, the "environmental" Hamiltonian $\hat{H}_E$ can be taken to be that of an assembly of simple harmonic oscillators, so that with x_i and p_i (formerly π_i) now taken to be the coordinates and momenta, respectively, of those oscillators we have

$$\hat{H}_E = \sum_i (p_i^2/2m_i + \tfrac{1}{2}m_i\omega_i^2 x_i^2). \tag{4.1}$$

Moreover, the interaction term $\hat{H}_{int}$ can be taken in the form

$$\hat{H}_{int} = -\sum_i f_i(q)x_i + \tfrac{1}{2}\sum_i f_i^2(q)/m_i\omega_i^2, \tag{4.2}$$

i.e.,

$$\hat{H} = \tilde{H}_S(p, q) + \sum_i \left[p_i^2/2m_i + \tfrac{1}{2}m_i\omega_i^2\left(x_i - \frac{f_i(q)}{m_i\omega_i^2}\right)^2 \right]. \tag{4.3}$$

(The function of the second term in eq. (4.2), the so-called "counterterm", is to ensure that the renormalization of the potential seen by the system in the static limit is not counted twice. See, e.g., Leggett 1987b.) We will call the model Hamiltonian (4.3) the "oscillator-bath model". In general, to get a sensible result in the thermodynamic limit, the spacing of the oscillator frequencies ω_i must be proportional to N^{-1} and the coupling constants f_i to $N^{-1/2}$, where N is the total number of environmental degrees of freedom.

The justification of the simple form eq. (4.1) for H_E is straightforward, at least at zero temperature, and is similar to the argument which shows in retrospect why the "oscillator models" used to describe the spectroscopic behavior of atoms in the late nineteenth century worked so surprisingly well, despite the fact that we now know that the states of electrons in atoms are very far from harmonic-oscillator states. The argument is simply that in a macroscopic system, barring pathologies, the coupling to any individual environmental mode will be $\propto N^{-1/2}$ and, hence, we can ignore double excitation.* But if we deal only with the ground state and the first excited state of each mode, these can clearly be put into one–one correspondence with the two lowest states of a harmonic oscillator, and the fact that higher states do not preserve the correspondence is irrelevant. For formal details, see Caldeira and Leggett (1983), Appendix C.

It should be strongly emphasized that the assumption that any one degree of freedom is only weakly excited does *not* imply that the total effect of the dissipation on the system is weak; in fact, we will see below that in certain cases it can change even the qualitative aspects of the behavior. Again, the analogy with the theory of light interacting with the atoms of a gas is helpful; the fact that any one atomic level is sufficiently weakly excited that it can be treated as an oscillator does not prevent a light wave from being totally absorbed in the gas!

The justification of eq. (4.2) for H_{int} is a bit more tricky. It can be carried out explicitly for the case where the Hamiltonian is of the form of eq. (3.9), so that the "zeroth-order approximation" (which, we repeat, need not be at all a good approximation!) is of the familiar adiabatic form. For the argument in this case, see Appendix C of Caldeira and Leggett (1983), or Leggett (1990); note that the "counterterm" [second term in eq. (4.2)] arises automatically. In the more general case, where the original system–environment coupling was of the most general form $H(p, q\colon \{x_i\}, \{p_i\})$ and cannot be reduced to $H(x_i, p_i\colon q)$, eq. (4.2) can be justified only if, in addition to the assumption of "weak excitation", it is assumed that the system–environment coupling is linear in p and/or q and, moreover, is invariant under time reversal; for details, see Leggett (1984, section IIB). Beyond this no general justification is attainable (as far as I know); in Appendix C of Caldeira and Leggett (1983) it is argued that provided the classical equation of motion corresponds to linear dissipation (cf. below), it is highly plausible that the coupling can be put in the form of eq. (4.2), but one

* For the relation between this statement and the calculation of two-phonon processes by Kagan and Prokof'ev (chapter 2), cf. the remarks in the Preface.

cannot claim this as rigorous result. While the absence of a general proof of the adequacy of eq. (4.3) as a general description of dissipation is a lacuna in the theory of the quantum dynamics of a macroscopic variable, it is reassuring that (a) no one has, to the present author's knowledge, produced an experimentally realistic case where eq. (4.3) is in appreciable error (for an example where there may be a small correction to it, which probably has negligible effects in practice, see Leggett 1988), and (b) in a number of specific cases it has been shown to be valid even though the original "environment" bears no recognizable resemblance to an oscillator bath. In particular, this has been done by Chen (1987) for a Fermi bath and by Eckern et al. (1984) for a tunnel junction described by the standard Bardeen–Josephson model.

Even granted that eq. (4.3) is a correct general description of the effects of the interaction of the system with its environment, the following question clearly arises: How do we know the parameters $(m_i, \omega_i, f_i(q))$ which go into this description? In some cases, no doubt, we could calculate them with a fair degree of confidence from a microscopic model; e.g., in the case of a current-biased tunnel oxide junction this has been done by Eckern et al. (1984). To the extent that we are interested in the agenda of the first section, i.e., in exploring the more exotic predictions of quantum mechanics, such an approach is quite reasonable. However, if we are interested in fundamental *tests* of quantum mechanics, as discussed in section 2, then it is clear that it is undesirable to rely on any specific microscopic model (even if one should be readily available, which is not always the case) since, should the experiments turn out to contradict the quantum-mechanical predictions based on such a model, the natural assumption would be that it is the model, and not the quantum formalism itself, which is at fault. Indeed, if we really had to rely exclusively on a priori microscopic knowledge of the parameters, I suspect that it would be in practice impossible to convince the majority of the physics community that any experimental result at all obtained on a macroscopic system could be taken as an evidence for a breakdown of the quantum theory.

An alternative, and in this context more promising, line of approach is to try to relate the microscopic parameters entering eq. (4.3) to the *classical* behavior of the system in question, which can be directly observed in experiment. It is clear that a complete knowledge of the microscopic parameters will uniquely determine the behavior in the classical limit as well as in the quantum regime. For our purposes, however, more important is the converse question: Does the classical behavior uniquely determine the parameters of eq. (4.3)? Or, if not, does it at least determine them to the extent that is needed to predict uniquely the behavior in the quantum regime? If the answer is yes, then we expect to be able, in principle at least, to use the behavior observed in classical experiments to make firm predictions about the behavior with respect to essentially quantum-mechanical phenomena such as tunnelling, *without* making any particular assumptions about the correct microscopic description of the system in question. Then, if the predicted behavior is not seen experimentally, we are on much firmer ground in using this result to question quantum mechanics.

Here, again, unfortunately, the situation is not entirely clear-cut. Let us first consider the case where it is known on some a priori ground that the coupling coefficients $f_i(q)$ in eq. (4.2) have the simple form qC_i. This case is actually not so artificial as it seems. For example, if the system is a SQUID ring, then in the context of realistic experiments (see section 5) one is often interested in a situation where all the action takes place over a range of the macroscopic variable, the trapped flux Φ, which is small compared to the flux quantum ϕ_0. On the other hand, irrespective of the microscopic details of the description of the ring, there are excellent general arguments that the scale of dependence of the coupling constants f_i on Φ must be $\sim\phi_0$. Consequently, for the purpose of describing the realistic experiments in question, it should be a very good approximation to take $f_i(q)$ be equal to $f_i(q_0) + (q - q_0)f'_i(q_0)$ and ignore the higher terms in the Taylor expansion; the constant term is then clearly irrelevant and a redefinition of q produces the required form $f_i(q) = qC_i$. The same type of general argument applies to a current-biased Josephson junction in which the macroscopic variable is the phase difference of the Cooper pairs on the two sides of the junction (see section 3 and chapter 6, by Devoret et al.).

If the coupling coefficients $f_i(q)$ are indeed linear in q, then it can be shown by a rather straightforward direct calculation that experiments conducted in the classical regime can uniquely determine, not indeed the quantity C_i itself, but rather the so-called "environmental spectral density" $J(\omega)$ defined by

$$J(\omega) \equiv \tfrac{1}{2}\pi \sum_i (C_i^2/m_i\omega_i)\delta(\omega - \omega_i). \tag{4.4}$$

In fact, if the classical equation of motion is written in the general form [with $q(\omega)$ as the Fourier transform (FT) of $q(t)$, and $(\partial V/\partial q)(\omega)$ the FT of $(\partial V/\partial q)(t)$]

$$K(\omega)q(\omega) = -(\partial V/\partial q)(\omega), \tag{4.5}$$

then we have simply $J(\omega) = \operatorname{Im} K(\omega)$. For example, in the case of simple ohmic friction, where the equation of motion is

$$M\ddot{q} + \eta\dot{q} + \partial V/\partial q = 0, \tag{4.6}$$

the quantity $J(\omega)$ is simply $\eta\omega$. For the details of the argument, see Leggett (1984).* Now it turns out that $J(\omega)$ is precisely the combination of parameters which we need to calculate the behavior also in the quantum regime; in fact, given normal boundary conditions, e.g., thermal equilibrium of the environment at time $-\infty$, it is clear that the only way in which the environment parameters can affect the dynamics of the system at all is in the combination $J(\omega)$. Thus, in

* In the case of a current-biased junction the role of the "friction coefficient" η is played by the conductance R^{-1} shunting the junction: cf. eq. (4) of Larkin and Ovchinnikov in chapter 4.

this case a knowledge of the classical behavior indeed permits a unique prediction of the quantum behavior, independently of any microscopic model. This is one of the most important results in the theory of the quantum mechanics of a macroscopic variable.

If the coupling coefficients $f_i(q)$ do not have the simple form qC_i, then the situation is a bit more complicated. It, of course, remains true that the behavior of the system under arbitrary conditions (classical, quantum, or whatever) is determined by the generalized spectral density

$$J(\omega:q) \equiv \tfrac{1}{2}\pi \sum_i \{[f_i(q)/q]^2/m_i\omega_i\}\delta(\omega-\omega_i). \tag{4.7}$$

However, the business of inferring $J(\omega:q)$ from the classical motion is then much more complicated and may not even have a unique solution. Nevertheless, one can make some nontrivial statements: e.g., it is shown in Appendix C of Caldeira and Leggett (1983) that if the phenomenological classical equation of motion is of the form (4.6) (which is, in principle, possible even though $f_i(q) \neq qC_i$) then we can use this information, not indeed to calculate an exact value of the quantum tunnelling rate, but at least to set a lower limit on it.

The "oscillator-bath" Hamiltonian (4.3) has been widely used in the literature (usually with the special form $f_i(q) = qC_i$, in which case it is often called the "Caldeira–Leggett model") to treat the quantum dynamics, and in particular the quantum tunnelling behavior, of a macroscopic variable: see, e.g., section 13 of chapter 3, by Eckern and Schmid. Before discussing the results of these calculations, let us conclude this section by returning briefly to the question of premise P of section 2 to see how it looks in this framework.

The claim made by premise P was that it is always impossible to see the effect of quantum interference between macroscopically distinct states. In the framework of the Hamiltonian (4.3), what this means is that sufficiently different values of q will, because of the coupling term $\hat{H}_{\text{int}}$, become *irreversibly* correlated to different states of the oscillator bath in such a way that not just at the time in question but ever afterwards the interference terms in the density matrix vanish. Now, the great advantage of the effective Hamiltonian (4.3) is that it is simple enough to enable one, under many circumstances, to eliminate the oscillators entirely and actually *calculate* the form of the reduced-density matrix $\rho(q, q':t)$ or some related quantity, thereby testing premise P. The result is unambiguous: despite the prejudice to the contrary embodied in five decades of the literature on quantum measurement theory, for sufficiently (and not obviously unattainably) weak values of the classical dissipation *premise P is demonstrably false*. It is this conclusion, and this alone, which has made the whole subject of the quantum mechanics of a macroscopic variable of more than merely technical interest.

To summarize the essential results of this and the last section: To the extent that the adiabatic approximation is valid, there is no difficulty, in principle, in seeing the phenomenon of QIMDS (given, of course, a suitable physical system). Now, in fact, the adiabatic *approximation* is practically never satisfied for any degree of freedom which can be reasonably called "macroscopic." However, this does not prevent us from using the adiabatic *basis* as a basis for description, and thereby obtaining an effective Hamiltonian for the system–environment interaction which has the form of eq. (4.3). When we calculate with this Hamiltonian, we sometimes find in practice that even if the "adiabatic approximation" in the usual sense has broken down (i.e., the effects of the coupling terms $(\partial\psi/\partial q)(\partial\phi/\partial q)$, etc. are not small), these effects are overwhelmingly still "adiabatic" in the sense that they do not lead to *irreversible* correlations of the system with its environment, i.e., to dissipation,* and, thus, do not necessarily prevent us from observing the phenomenon of QIMDS. Only dissipation suppresses QIMDS in an essential way! Thus, we see the essential connection between the technical problem of dissipation in quantum mechanics and the "philosophical" problem posed by the traditional quantum measurement paradox.

Let me finally note one slightly paradoxical point which may otherwise lead to confusion: In many of the traditional discussions of the quantum measurement paradox, the "system" in question is microscopic, and it is the interaction with a *macroscopic* system, the measurement apparatus, which destroys the coherence of its wave function. In the discussion of this and the last section, by contrast, we have considered a "system" (degree of freedom) which is itself *macroscopic*, and the coherence of whose wave function may be destroyed by dissipation, i.e., by interaction with a host of *microscopic* degrees of freedom ("environment"). Needless to say, we could, if we wished, discuss the problem of a microscopic degree of freedom, e.g., that of a tunnelling muon, interacting with a microscopic environment, e.g., phonons – cf. chapter 2, by Kagan and Prokof'ev – or that of a macroscopic degree of freedom interacting with a macroscopic measuring apparatus; but from the point of view of the quantum measurement paradox neither of these situations would have the same degree of relevance.

5. *Experiments and their significance*

In this section I will review the principal types of experiment which have been conducted or are being planned on the quantum-mechanical behavior, in particular tunnelling and related phenomena, of a macroscopic variable, in particular in Josephson devices. Since the existing experiments themselves are

* An extended discussion of the analogous questions in the (microscopic) context of diffusion in metals, including the important question of the choice of effective cutoff, is given in the chapter by Kagan and Prokof'ev, section 2.2.

described by Devoret et al. (chapter 6), and the details of the theoretical calculations are given by Eckern and Schmid (chapter 3), Larkin and Ovchinnikov (chapter 4), and Ivlev and Mel'nikov (chapter 5), I shall concentrate mainly on the significance of these experiments in the context of the goals outlined in sections 1 and 2.

Let us first briefly consider the question of which systems are suitable for such experiments, recalling that we are specifically interested in those phenomena, such as tunnelling, which have no classical analog. The first point is that there is a very old argument, which goes back to Bohr, to the effect that it will never be possible to see any kind of characteristically quantum behavior in macroscopic systems, simply because this requires the action S to be not too large compared to $\hbar$, and this condition can never be met when the variable in question is macroscopic. Let us examine this argument a bit more closely. To an order of magnitude, the relevant action is of order a typical system energy E times a typical period $\tau \sim \omega_0^{-1}$ of classical motion in the relevant potential. Thus, the claim is that $E/\hbar\omega_0$ is always $\gg 1$ for any macroscopic variable. Although this is generally true, there are exceptions, and it is precisely these exceptional systems, and only these, which are useful for our purposes. A number of systems are candidates: e.g., the transport properties of quasi-one-dimensional charge density wave systems have been interpreted (Bardeen 1979) by a model involving quantum tunnelling as an essential ingredient, and specific proposals have been made (Chudnovsky and Gunther 1988) to look for tunnelling of the macroscopic magnetization in small ferromagnetic grains. However, the only class of systems in which both (a) the description of the behavior in the classical limit is generally regarded as uncontroversial and (b) a substantial program of experimental work has already been carried out is the class of systems known as Josephson devices – typically, single Josephson junctions biased by a fixed external current or so-called rf SQUID rings (isolated superconducting rings containing a single junction), but including also "dc SQUID" (rings with *two* junctions and biased by an external current). I shall, therefore, confine the discussion here to these systems.

With the exception of a group of experiments in which a SQUID ring is coupled to a tank circuit as in standard magnetometer operation and the characteristics of the latter monitored, which are (in my opinion) difficult to interpret unambiguously (Prance et al. 1983, Dmitrenko et al. 1984), existing and planned experiments on Josephson junctions and SQUID rings fall into two main classes. The first are experiments in which the macroscopic variable (the phase difference for a current-biased junction, the trapped flux for an rf SQUID) are prepared in a metastable state and the tunnelling rate of escape is observed (see fig. 2a). In the literature this kind of experiment is known as "macroscopic quantum tunnelling" or MQT. In a very elegant variant of this experiment conducted by the Berkeley group, rf radiation is shone on the system in the metastable state and the dependence of the escape rate on the frequency and

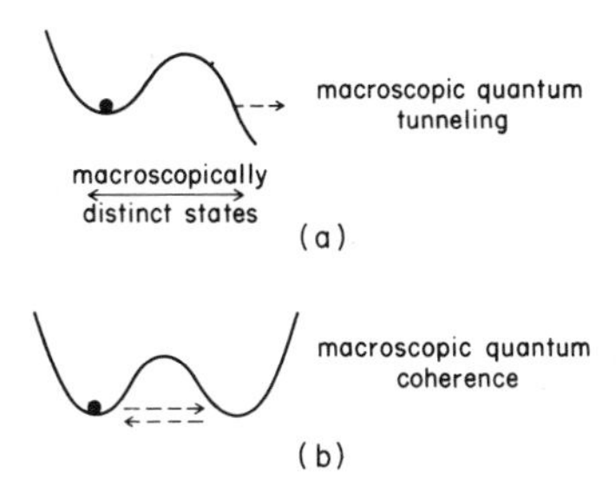

Fig. 2. Illustration of (a) macroscopic quantum tunnelling (MQT) and (b) macroscopic quantum coherence (MQC).

power of this radiation measured ("resonant activation", see chapter 6, by Devoret et al.). A second group of experiments detects the motion of the system between two nearly degenerate potential wells, a situation possible only in a SQUID ring (fig. 2b); in the existing experiments of this type (de Bruyn Ouboter 1984 and references therein, Han et al. 1991) the motion has apparently been of the incoherent hopping type described by the classical rate theories of chemical physics, but the hope is that one can do this type of experiment in a region of the parameter space when the theory predicts coherent "clock-like" oscillation similar to that observed in the NH_3 molecule. Such (currently unobserved) behavior is known as macroscopic quantum coherence or MQC.

From the point of view of the considerations of sections 1 and 2 of this chapter, the MQT (and resonant activation) experiments on the one hand, and the MQC experiment on the other, have a rather different significance. A major point which should be stressed about the MQT experiments is that, since the relevant parameters [in effect, the spectral density $J(\omega)$ defined by eq. (4.4)] can be, and in some cases have been, measured in independent experiments which operate in the classical limit, the application of quantum mechanics according to the scheme of section 4 leads to a prediction of the quantum tunnelling behavior *which contains no adjustable parameters.* Thus, any substantial failure of the experiments to agree with the theoretical predictions would cast doubt on whether quantum mechanics still works at this level; conversely, agreement with the theory is strong circumstantial evidence in favor, not just of the continued validity of quantum mechanics, but also of the general correctness of the scheme laid out in section 4 for handling the effects of dissipation. As discussed in the chapter 6, by Devoret et al., the agreement is in fact very satisfactory, even in cases where the effect of dissipation is substantial, and this gives one a good deal of confidence that the scheme of section 4 is not seriously in error.

Thus, the MQT experiments clearly make it very plausible that QM is still working at levels as "macroscopic" as that of the phase variable in a Josephson junction or the trapped flux in an rf SQUID. However, they do not *directly* display the effect of interference between macroscopically distinct states. Moreover, it is not clear that the experimental results could not be mimicked by a theory which rejected the existence of such interference and claimed, rather, that a macroscopic system must always "be" in a definite macroscopic state (cf. the considerations advanced in section 2). The point about the MQC experiment is that, if feasible, it could block this loophole and indeed force Nature to choose between the predictions of QM and those of a whole class of theories embodying the principle of realism at a macroscopic level.

The argument goes schematically as follows (for details see Leggett and Garg 1985, Leggett 1987b): Assume, as is, in fact, the case, that in the situation shown in fig. 2b one can measure, at an arbitrary time, which of the two wells the system is in.* It is known from existing experiments (de Bruyn Ouboter 1984, Han et al. 1991) that such an experiment always gives a unique result (just as did the measurement of which alternative B or C was realized for a particular system in section 2). Define *operationally* a variable $Q(t)$ which takes the value $+1$ (-1) if the system is measured at time t and found to be in the right (left) well; note that if no measurement is made at time t, then $Q(t)$ is so far undefined. Experiment can measure (inter alia) the correlations $\langle Q(t_1)Q(t_2)\rangle$ of Q at arbitrary pairs of times t_1 and t_2, on an ensemble prepared in an identical way at $t = 0$.

We now make the following two assumptions, which define a class of theories we shall call "macrorealistic":

(1) The quantity $Q(t)$ exists at all times t, irrespective of whether or not it is measured at that time, and takes the values ± 1 (with the caveat given in the second footnote).

(2) It is in principle possible to measure $Q(t)$ in such a way that the subsequent behavior of the system is unaffected by the fact that a measurement has been made ("noninvasive measurability").

It should be strongly emphasized that postulate (2) makes sense *only* in conjunction with postulate (1); in quantum mechanics (2) would, of course, be false, but then so would be (1).** In theories satisfying (1), (2) can be made very plausible – in fact, almost a corollary of (1) – by envisaging an "ideal-negative-result" experiment, in which we arrange to interact physically with the system only if it has (say) $Q(t) = +1$, and then keep only the results of those runs in which $Q(t)$ is observed to be -1.

It is now very easy to demonstrate that in any theory satisfying conditions (1) and (2), the quantity K defined by the combination of *experimentally measured correlations*

$$K \equiv \langle Q(t_1)Q(t_2)\rangle + \langle Q(t_2)Q(t_3)\rangle + \langle Q(t_3)Q(t_4)\rangle - \langle Q(t_1)Q(t_4)\rangle, \quad (5.1)$$

* Of course, there is a finite probability of finding it under the barrier, but this is exponentially small and can be taken into account in the analysis (Leggett and Garg 1985).

** For a rather different point of view on this experiment, see Foster and Elby (1991).

satisfies the Bell-type inequality

$$K \leqslant 2 \quad \forall t_1, t_2, t_3, t_4. \tag{5.2}$$

Now the punch-line is that for an ideal isolated two-state system such as an NH_3 molecule, it is easy to choose values of t_1, etc., so that the inequality (5.2) is violated by the quantum-mechanical predictions. If, therefore, the *macroscopic* variable q (e.g., the flux in a SQUID ring) behaves exactly or approximately like a simple isolated system, then we can set up an experiment whose results, if they agree with the quantum-mechanical predictions, must inexorably refute the hypothesis of macrorealism, i.e. (given the corollary of noninvasive measurability) prove that the quantity $Q(t)$ could not have "had" a definite value when it was not measured.

The crucial question, therefore, is: To what extent does the behavior of the macroscopic variable in a realistic experiment resemble that of a simple two-state system, and in particular to what extent does it exhibit the characteristic NH_3-type oscillations of the latter? The traditional answer implicit in most of the literature on the quantum theory of measurement, which embodies premise P, is "not in the least": the coupling to the environment, it is argued, will automatically destroy the "two-state" oscillations and reduce the behavior to the classical rate-theory type, i.e., incoherent relaxation, which follows when one assumes that there is no definite phase relation between the parts of the system wave function corresponding to the two wells (or, more correctly, that the off-diagonal terms in the reduced density matrix are zero). The crucial point, however, is that with the help of the techniques developed in sections 3 and 4, *this is no longer a matter of speculation*: for any given experimental setup we can obtain the necessary parameters of eq. (4.4), i.e., the spectral density $J(\omega)$, from experiment and then as the basis of eq. (4.3) *do a quantitative calculation* of the quantity $\rho(q, q' : t)$ (or, at least, of the elements of it which we need).

Such a calculation has been carried out by the author and his co-workers (Leggett et al. 1987). The general results are complicated, but in the present context the salient point is that for not obviously unreasonable values of the experimental parameters the quantum-mechanical prediction is indeed that $K > 2$, in contrast to the predictions of any macrorealistic theory. Thus, the MQC experiment could indeed force Nature to choose between QM and the macrorealism which seems so natural to our common sense.

This experiment is currently at the design and construction stage (Tesche 1990). I believe that a clear-cut result in either direction would have profound significance. If the quantum-mechanically predicted results are not found, the most immediate and natural reaction of most physicists will, of course, be that there must be something wrong with the experiment or (more likely) with the theory. However, the latter is now sufficiently well developed that most of the obvious loopholes can be identified. If they can be plugged, and the discrepancy remains, then clearly it would be of revolutionary significance. If, on the other hand, the experiment comes out in favor of quantum mechanics, I believe that, quite apart from being a really spectacular example of the counterintuitive aspects of the theory, it will be a significant input into the arguments concerning the measurement paradox. In particular, I believe that it would strongly suggest (although it would not, of course, prove) that there is indeed an "element of reality" corresponding to *each* of the interfering macroscopically distinct stages, and would motivate us to think much harder about the possible nature of this "reality".

References

Aspect, A., P. Grangier and G. Roger, 1982, Phys. Rev. Lett. **49**, 91, 1804.

Ballentine, L.E., 1970, Rev. Mod. Phys. **42**, 358.

Bardeen, J., 1979, Phys. Rev. Lett. **42**, 1498.

Caldeira, A.O., and A.J. Leggett, 1983, Ann. Phys. (New York) **149**, 374; Erratum, 1984, ibid. **153**, 445.

Chen, Y.C., 1987, J. Stat. Phys. **47**, 17.

Chudnovsky, E.M., and L. Gunther, 1988, Phys. Rev. Lett. **60**, 661.

Daneri, A., A. Loinger and G.M. Prosperi, 1962, Nucl. Phys. **33**, 297.

De Bruyn Ouboter, R., 1984, Proc. Int. Symp. on the Foundations of Quantum Mechanics in the Light of New Technology, eds S. Kamefuchi, H. Ezawa, Y. Murayama, M. Namiki, S. Nomura, Y. Ohnuki and T. Yajima (Japanese Physical Society, Tokyo) p. 83.

Dmitrenko, I.M., V.A. Khlus, G.M. Tsoi and V.I. Shnyrkov, 1985, Fiz. Nizkh. Temp. **11**, 146 [Sov. J. Low Temp. Phys. **11**, 77].

Eckern, U., G. Schön and V. Ambegaokar, 1984, Phys. Rev. B **30**, 6419.

Foster, S., and A. Elby, 1991, Foundation Phys. **21**, 773.

Greenberger, D.M., 1983, Rev. Mod. Phys. **55**, 875.

Han, S., J. Lapointe and J.E. Lukens, 1991, Phys. Rev. Lett. **66**, 810.

Leggett, A.J., 1980, Prog. Theor. Phys. Suppl. **69**, 80.

Leggett, A.J., 1984, Phys. Rev. B **30**, 1208.

Leggett, A.J., 1987a, in: Quantum Implications: Essays in Honor of David Bohm, eds B.J. Hiley and F.D. Peat (Routledge and Kegan Paul, London).

Leggett, A.J., 1987b, in: Chance and Matter, eds J. Souletie, J. Vannimenus and R. Stora (North-Holland, Amsterdam).

Leggett, A.J., 1988, in: Frontiers and Borderlines in Many-Particle Physics, eds R.A. Broglia and J.R. Schrieffer (Italian Physical Society, Bologna).

Leggett, A.J., 1990, in: Proc. Nato ASI, Vol. 214, Applications of Statistical and Field Theory Methods to Condensed Matter, Evora, Portugal, May 22–June 2, 1989 (Plenum Press, New York).

Leggett, A.J., and A. Garg, 1985, Phys. Rev. Lett. **54**, 857.

Leggett, A.J., S. Chakravarty, A.T. Dorsey, M.P.A. Fisher, Anupam Garg and W. Zwerger, 1987, Rev. Mod. Phys. **59**, 1.

Likharev, K.K., and A.B. Zorin, 1985, J. Low Temp. Phys. **59**, 347.

Mermin, N.D., 1985, Phys. Today (April 1985) p. 38.

Prance, R.J., J.E. Mutton, H. Prance, T.D. Clark, A. Widom and G. Megaloudis, 1983, Helv. Phys. Acta **56**, 789.

Schrödinger, E., 1935, Naturwiss. **23**, 824.

Tesche, C.D., 1990, Phys. Rev. Lett. **64**, 2358.

CHAPTER 2

Quantum Tunneling Diffusion in Solids*

Yu. KAGAN and N.V. PROKOF'EV

Kurchatov Institute of Atomic Energy
Moscow, Russia

* Translated from the Russian by Artavaz Beknazarov

Quantum tunnelling in condensed media
Edited by
Yu. Kagan and A.J. Leggett

Contents

List of symbols

a	lattice constant
$a_{k\sigma}$	electron annihilation operators
$A(\omega)$	one-phonon spectral function
$b_\beta, b_\beta^{(0)}, \bar{b}_\beta$	phonon annihilation operators
c_i	particle annihilation operator
$D(T), D_{\alpha\beta}$	diffusion coefficients
$E, E^{(1),(2)}$	energy, energy corrections; activation energy
$\mathscr{E}_{\mathrm{p}}$	particle energy
E_k	electron energy in a superconductor
$E_n(\boldsymbol{R})$	adiabatic energy
$\boldsymbol{E}$	full elliptic integral of the second kind
$\boldsymbol{f}$	particle density matrix
$f(\omega)$	spectral function of excitations
g	nearest-neighbor index
$g(\varepsilon)$	phonon density-of-states
$G(T)$	fluctuational preparation of the barrier
$\mathscr{H}$	Hamiltonian
$\mathscr{H}'$	interaction Hamiltonian
$\mathscr{H}_{\mathrm{int}}$	interaction Hamiltonian with an external field
$\mathscr{H}_{\mathrm{p}}$	particle Hamiltonian
$\mathscr{H}_{\mathrm{ex}}$	excitation Hamiltonian
I, I_{cd}	collision integral
J_0	tunneling amplitude
K	trapping rate
m	electron mass
M, M_*	particle mass
M_{nm}^{12}	transition matrix element
n_k	Fermi occupation numbers
N	number of unit cells
N_α	phonon occupation numbers
$\boldsymbol{r}$	set of electronic coordinates
$\boldsymbol{R}, R$	particle coordinate
$\boldsymbol{R}_{1,2}$	center of well coordinate
R_0	radius of effective interaction $\mathscr{U}(R_0) = \Delta$
R_{t}	trapping radius
S	tunneling action
T_{c}	superconducting transition temperature
T_*	crossover temperature between tunneling and classical diffusion motion
T'	crossover temperature between coherent and incoherent motion
T_{min}	crossover temperature between the overdamped band and phonon-induced diffusion
$U(\boldsymbol{R})$	potential
$\mathscr{U}(\boldsymbol{R})$	impurity potential
$\mathscr{V}, \mathscr{V}_{0,x,y,z}$	interaction Hamiltonians
$V_{kk'}, \Delta V_{kk'}$	electron–particle interaction matrix elements
V_0	unit cell volume
$W, W^{(0),(1),(2)}$	transition probability
x	impurity concentration
x_{p}	particle concentration
x_{c}	critical concentration for particle self-localization
$\boldsymbol{x}$	set of phonon coordinates
z	number of nearest neighbors
γ	Euler number
γ_1	relaxation rate
γ_2	phase relaxation rate
Γ	gamma function
δ, δ_j	phase shifts
Δ_0	renormalized tunneling amplitude

Δ_c	coherent tunneling amplitude	σ	spin variable; critical exponent
Δ_c^s	coherent tunneling amplitude in a superconductor	$\sigma(T)$	effective depolarization rate
		$\sigma_{x,y,z}$	Pauli matrices
Δ_s	superconducting gap	τ	particle lifetime in the well
Δ	bandwidth	$\varphi_s^{(i)}(\boldsymbol{R})$	particle site wave function
$\varepsilon_{k,p}$	electron energy	$\phi(T)$	phonon polaron exponent
$\tilde{\varepsilon}$	renormalized energy splitting	$\Phi_{sn}^{(0)}$	adiabatic wave function
Θ_D	Debye temperature	$\Phi_{sn}^{(i)}$	site wave function of the system
$\Lambda, \Lambda_{1,2}$	polaron operators, depolarization rate	ψ	digamma function
		$\Psi_n^{(i)}(\boldsymbol{r}, \boldsymbol{R})$	electron site wave function
ξ	energy difference between ground states in adjacent wells	ω	level separation in the well; external field frequency
$\xi_*(T)$	crossover energy difference between one-phonon and two-phonon scattering regimes of diffusion	ω_0	inverse subbarrier time
		ω_α	phonon frequencies
		Ω, Ω^{2ph}	phase damping frequency
		Ω_p	interparticle scattering frequency
$\rho(\varepsilon_F)$	density of electron states at ε_F		
$\rho_n^{(0)}$	equilibrium density matrix of excitations	Ω_{im}	particle–impurity scattering frequency

1. Introduction

The key point in many phenomena occurring in a solid is the tunneling motion of heavy particles or collective excitations with a heavy effective mass. The word "heavy" is used here only in comparison with the electron mass; in all other aspects light particles are implied. As striking examples we may mention the tunneling transport in quantum crystals, in particular, the diffusion of He^3 in solid He^4, hydrogen subbarrier motion in insulators and metals, quantum diffusion of muonium in insulators and of μ^+-muons in metals, the tunneling motion in insulating, metallic and superconducting glasses, the thermodynamics and kinetics of heavy electrons.

The starting qualitative picture can be traced on the simple example of a perfect crystal at $T = 0$ with a single impurity atom in the lattice site. In the process of impurity-neighboring matrix atom exchange the energies of the initial and final states will be equal. This is a typical resonance process, but any path for such an exchange is associated with the subbarrier motion. Although the transition amplitude may be very small, it will inevitably take place, resulting in delocalization of an impurity particle. Such a picture may also be realized in the simpler case of an interstitial atom or vacancy and in more complex cases when a single resonance transition requires a series of correlated tunneling transitions. Here, in all cases we shall come to the concept of the particle band motion with a bandwidth Δ governed by the scale of the resonance transition amplitude. This concept has been consistently developed by Andreev and Lifshitz (1969) and by Guyer and Zane (1969).

However, the picture outlined above is oversimplified. In fact, the tunneling motion occurs under the conditions of the coupling with virtual and real (with $T \neq 0$) excitations of the medium. Because of the smallness of Δ this coupling is practically always strong. As a result, the tunneling kinetics in a solid, in fact, is the problem of motion in a system with weak quantum correlations and a strong dynamic interaction with excitations. This is the most characteristic feature of the phenomena under consideration.

While analyzing the tunneling in a solid we often deal with two channels of the interaction with the medium. The first is due to the intrawell interaction, which gives rise to a polaron effect. In insulators it is an ordinary phonon polaron effect characteristic of a small polaron. In metals there arises a particular electron polaron effect (Kondo 1976a, 1984) associated with the narrow band of nonadiabatic electron–hole excitations in the vicinity of the Fermi surface with $\varepsilon < \omega_0$, where ω_0 is the inverse time of the passage of a particle under the barrier (Kagan and Prokof'ev 1986b). The smallness of the phase volume, however, is compensated by a sharp enhancement of the polaron effect due to the familiar infrared divergence near ε_F (Mahan 1981, Nozieres and de Dominicis 1969) or due to the so-called orthogonality catastrophe (Anderson 1967a, b). The nonlinear character of Δ renormalization in this case (e.g., Leggett et al. 1987) may, in principle, lead to localization in a crystal as $T \to 0$, despite the presence of the translational symmetry (Schmid 1983).

The second channel is the interaction of a particle with barrier fluctuations during the passage through the subbarrier region. This interaction is responsible for the appearance of the effect of fluctuational preparation of the barrier (FPB) (Kagan and Klinger 1976, Kagan and Prokof'ev 1989a), which manifests itself in an effective decrease of the barrier for the optimal path of the subbarrier motion. The effect leads to an increase in Δ, which is enhanced with increasing T. In the case of phonon coupling it may become dominant, and the interaction itself will cause an increase in the bandwidth, instead of an ordinary polaron narrowing. In the case of electron coupling the FPB effect is limited in magnitude and insensitive to an infrared singularity (Kagan and Prokof'ev 1989a). Therefore, the determining factor in this case is the electronic polaron effect.

One of the most significant features of the coupling with medium excitations is a very early destruction of coherent tunneling. The basic kinetic characteristic here is the frequency Ω of phase correlations damping at neighboring equivalent positions of the particle. Even at low T, the frequency Ω reaches the scale of Δ. At higher temperature, one gets an exponentially rapid disappearance of the coherent tunneling transition (Kagan and Prokof'ev 1987a, b, 1990a). In a metal this occurs practically at $T \sim \Delta$ and in an insulator, a priori, when $T \ll \Theta_D$ (Θ_D is the Debye temperature). The quantity Ω depends only on relative fluctuations of the interaction at neighboring sites and is independent of the overlap integral or the value of Δ. Therefore, as the particle mass increases the exponential drop of Δ results in an exponential decrease of the temperature of the coherent-to-incoherent tunneling transition. This circumstance may prove essential in revealing the coherent tunneling of a heavy macroscopic object, this being the key topic within the sphere of ideas advanced by A. Leggett (see chapter 1 of this book and also Caldeira and Leggett 1981, Leggett et al. 1987).

Although the physical picture of coherent band motion and incoherent tunneling, when the band is dynamically destroyed, is quite different, the smallness of Δ allows us to find a common solution for the quantum diffusion coefficient (Kagan and Maksimov 1973, Kagan and Klinger 1974). In the case of phonon coupling at low temperature the kinetics is governed by a two-phonon interaction. This interaction is responsible for the destruction of the coherent band motion (Kagan and Prokof'ev 1989b). The one-phonon coupling is essential at high temperature.

For the case of particle–electron coupling the quantum diffusion coefficient in the incoherent region assumes, as a result of a strong electronic polaron effect, an unusual power dependence on T with an interaction-dependent exponent (Kondo 1984, Yamada 1984).

The smallness of Δ with respect to all the energy parameters in a solid renders quantum diffusion very sensitive to crystal imperfections. The static destruction of the band and the particle localization occur at a relatively low defect concentration. Now the interaction with excitations removes the localization. A rather nontrivial picture occurs in a metal with the transition to the superconducting state. The freezing of normal excitations with $T \ll T_c$ leads to a complete localization even if such localization is absent in the normal state.

Perhaps, the most striking result in this region is the localization in a perfect crystalline matrix due to the interaction between diffusing particles (Kagan 1981, Kagan and Maksimov 1983b, 1984).

The present work is concerned with the entire range of problems mentioned above within the framework of a unified physical and formal approach. We shall use here a method based on a direct determination of the overlap integral for the many-particle wave functions of the medium formed during the time of the particle stay in a single potential well. This method is adequate for describing the tunneling motion of a heavy particle strongly coupled with medium excitations both for the case of the two-well problem and the coherent and incoherent quantum diffusion in a crystal. We will also analyze the experimental findings that revealed the basic features of tunneling motion in insulators, metallic systems and superconductors.

2. *Quantum tunneling with medium excitations coupling*

2.1. *Basic considerations*

The central point in the problem of the tunneling motion of heavy particles in a solid is the subbarrier motion between two equivalent positions in the nearest unit cells. In this case, a problem equivalent to that of the elementary act of particle tunneling is the problem of the motion in a two-well potential with arbitrary coupling. Here, the detailed picture of a tunneling transition, however complicated, manifests only in quantitative characteristics. The solution of the two-well problem is sufficient for describing both the coherent (band) motion of particles in a crystal and the incoherent motion when the phase memory is lost at each translational step.

There are three energy parameters characteristic for the two-well problem (see fig. 1): J_0 is the transition amplitude from one well into the other; ω is the

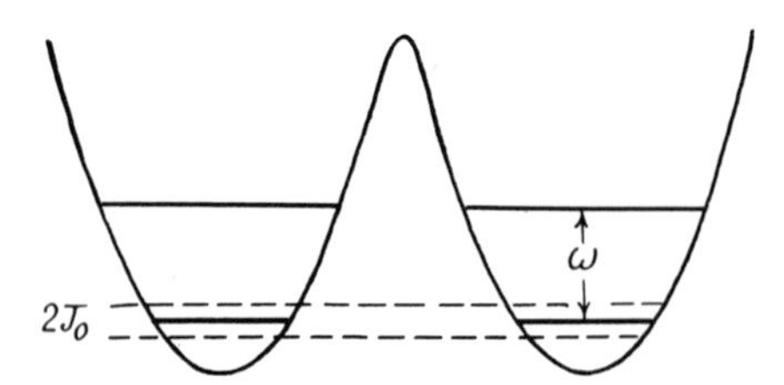

Fig. 1. Double well potential.

characteristic energy level separation in the well; ω_0/π is the inverse time of passage of the particle under the barrier (this time is determined by the quasiclassical motion in an inverted potential; ω_0 is the level separation in this potential). In the general case there is a fourth parameter ξ – the energy difference between the lowest levels in separate wells. Note that the frequencies ω and ω_0 are close to each other for the ordinary nonexotic form of potential in the crystal. In our estimates no distinction is made between these parameters.

Let us consider the region of low temperatures:

$$T \ll \omega_0/2\pi \quad (\hbar = 1). \tag{2.1}$$

In this case we can study tunneling motion only along the lowest level.

The problem under consideration is characterized by two strongly different time scales: the lifetime in a single well, τ, and the time of the under barrier motion π/ω_0. Since

$$\tau^{-1} \lesssim J_0 \approx \frac{\omega}{2\pi} \mathrm{e}^{-S},$$

where S is the quasiclassical action, the ratio of the times $\tau\omega \gtrsim \mathrm{e}^S$. The tunneling of heavy particles is characterized by $S \gg 1$, so that

$$J_0 \ll \omega, \quad \tau\omega \gg 1. \tag{2.2}$$

Thus, the particle resides in the well for a long time and passes under the barrier relatively rapidly. This circumstance as well as the large parameter [eq. (2.2)] is essential for the consideration of the effect of the medium on the particle tunneling dynamics. As for the ratio J_0/T, it is assumed to be arbitrary. This makes it possible to cover both low-temperature and high-temperature limits without violating the inequality (2.1).

The Hamiltonian of the two-well problem may be written in the general case as follows:

$$\mathscr{H} = \mathscr{H}_{\mathrm{p}} + \mathscr{H}_{\mathrm{ex}} + \mathscr{V}, \quad \mathscr{H}_{\mathrm{p}} = \frac{M\dot{\boldsymbol{R}}^2}{2} + U(\boldsymbol{R}). \tag{2.3}$$

Here $\mathscr{H}_{\mathrm{p}}$ is the Hamiltonian of a particle of mass M in a fixed potential $U(\boldsymbol{R})$, $\mathscr{H}_{\mathrm{ex}}$ is the Hamiltonian of the medium excitations, $\mathscr{V}$ is the interaction of the particle with the excitations. To the condition $S \gg 1$ or $J_0 \ll \omega$ there corresponds an exponentially weak overlap of the intrasite wave functions of the particle $\varphi_s^{(i)}(\boldsymbol{R})$ (the indices i and s refer to the energy level ε_s in the ith well; $i = 1, 2$). This allows us to use the representation of the eigenfunctions $\varphi_s^{(i)}(\boldsymbol{R})$ and to rewrite the Hamiltonian in the form

$$\mathscr{H} = \mathscr{H}_{\mathrm{ex}} + \sum_{iss'} (\varepsilon_{is}\delta_{ss'} + \mathscr{V}^{is,\,is'})c_{is}^{+} c_{is'}$$
$$+ \sum_{i \neq i',\, ss'} (\mathscr{V}^{is,\,i's'} + J_0^{is,\,i's'})c_{is}^{+} c_{i's'}, \tag{2.4}$$

where $\mathscr{V}$ is still an operator over the medium variables.

Bearing in mind the inequality (2.1) and the condition $\xi \ll \omega$, which is always assumed, it seems to be natural to retain in eq. (2.4) only the matrix elements with $s, s' = 0$, corresponding to the lowest levels. Using an equivalent form of the spin Hamiltonian, we get

$$\mathscr{H} = \tfrac{1}{2}\xi\sigma_z + J_0\sigma_x + \mathscr{V}_z\sigma_z + \mathscr{V}_x\sigma_x + \mathscr{V}_0 + \mathscr{H}_{\mathrm{ex}}, \tag{2.5}$$

$$\mathscr{V}_z = \tfrac{1}{2}(\mathscr{V}^{10,\,10} - \mathscr{V}^{20,\,20}), \qquad \mathscr{V}_x = \mathscr{V}^{10,\,20},$$
$$\mathscr{V}_0 = \tfrac{1}{2}(\mathscr{V}^{10,\,10} + \mathscr{V}^{20,\,20}), \tag{2.6}$$

where σ_i are Pauli matrices for the resultant two-level problem.

In fact, in the general case the transformation of eq. (2.4) to eqs. (2.5) and (2.6) is invalid. If for the high-energy cutoff of the excitation spectrum, ε_0, we have the inequality $\varepsilon_0 \gg \omega$, in describing the interaction of a tunneling particle with excitations the virtual intrawell transitions become essential. As a result, it is of fundamental importance to incorporate intrawell degrees of freedom, although the tunneling of a heavy particle still occurs along the lowest levels. The Hamiltonian [eq. (2.5)] implies that the next level in the well lies above ε_0 and thereby introduces the concept of the nonadiabatic nature of the particle–excitations coupling.

For a variety of problems, primarily for the tunneling of heavy particles in a metal it is exactly the relation $\omega \ll \varepsilon_0$ that is valid, and adiabatic readjustment of excitations to the moving particle is very important. Therefore, first we are to solve the problem of adiabaticity and try to take into account, in an explicit form, the intrawell degrees of freedom and only then to introduce the effective Hamiltonian for the two-well problem.

2.2. *The adiabatic problem*

For quantum diffusion of a heavy particle in a metal the question of whether or not there is an adiabatic readjustment of the electrons to the moving particle is of fundamental significance. Kondo was the first to consider this problem (Kondo 1976a, 1984). In the absence of adiabaticity the tunneling matrix element of the transition from well 1 to well 2 is given by the following expression:

$$\tilde{\Delta}_0 = J_0 \Lambda, \tag{2.7}$$

where

$$\Lambda = \langle \Psi_n(\boldsymbol{r}, \boldsymbol{R}_1) | \Psi_n(\boldsymbol{r}, \boldsymbol{R}_2) \rangle \tag{2.8}$$

is the overlap integral for the perturbed electronic states Ψ_n, corresponding to the position of the particle in wells 1 and 2 (the set of electronic coordinates is denoted by convention as $\boldsymbol{r}$). By analogy with a small polaron in the phonon system eq. (2.8) specifies the electron polaron effect (EPE). However, in the case of the EPE we encounter an unusual picture. The overlap integral [eq. (2.8)] formally vanishes with macroscopic accuracy. This result is a consequence of the so-called orthogonality catastrophe introduced by Anderson (1967a, b) and is clearly seen from the explicit form of the overlap integral [eq. (2.8)] found by Yamada and Yosida (1982) at $T = 0$:

$$\Lambda = \exp\{-\alpha(\boldsymbol{R}_1 - \boldsymbol{R}_2)^2 \ln N\}, \quad |\boldsymbol{R}_1 - \boldsymbol{R}_2| k_{\mathrm{F}} \ll 1, \tag{2.9}$$

where $\alpha > 0$ and finite, N is the number of electrons in the system, and $\boldsymbol{R}_i$ are the center-of-well coordinates.

Equations (2.7) and (2.8) imply that there is no readjustment of the electron cloud to the particle and that the remaining electronic wave function in well 1 must be projected to its counterpart in well 2. If, on the contrary, we adopt the adiabatic approximation, then the perturbed electronic wave function will move continuously with the particle and Λ in eq. (2.7) will turn to unity. Here, of course, it is the screened particle that is moving under the barrier in the renormalized potential with a slightly different mass.

In a real case, although $\omega \ll \varepsilon_0$, inevitably there is an energy band near the Fermi energy ε_{F} of width ω, in which the excitations must not follow adiabatically the tunneling particle, despite the fact that the electron mass is small compared to the particle mass. The role of these excitations is singled out because they give rise to an electron polaron effect (Kagan and Prokof'ev 1986b) and govern the scattering in tunneling kinetics at low temperatures.

In order to elucidate the role of adiabaticity, let us return to the initial Hamiltonian [eq. (2.3)] and make use of the scheme of the adiabatic theory of metals developed by Brovman and Kagan (1967). In doing so, we shall use, whenever possible, the general notations to extend directly the results obtained to the cases of phonon and other excitations coupling. Within the scope of this theory the adiabatic wave function (the zeroth approximation) has the form

$$\Phi_{sm}^{(0)} = \varphi_{sm}(\boldsymbol{R}) \Psi_m(\boldsymbol{r}, \boldsymbol{R}), \tag{2.10}$$

where Ψ_m is the solution of the Schrödinger equation for the electrons at a fixed $\boldsymbol{R}$

$$[\mathscr{H}_{ex} + \mathscr{V}(\boldsymbol{r}, \boldsymbol{R})]\Psi_m(\boldsymbol{r}, \boldsymbol{R}) = E_m(\boldsymbol{R})\Psi_m(\boldsymbol{r}, \boldsymbol{R}) \tag{2.11}$$

and $\varphi_{sm}(\boldsymbol{R})$ is the eigenfunction of the Hamiltonian $\mathscr{H}_p + E_m(\boldsymbol{R})$. At first, we shall neglect the weak tunneling between the potential wells in the potential $U(\boldsymbol{R}) + E_m(\boldsymbol{R})$. Then, the wave functions $\varphi_{sm}(\boldsymbol{R})$ will refer to an individual well. The system of functions $\Phi^{(0)}_{sm}$ is complete in $(\boldsymbol{r}, \boldsymbol{R})$ space. Therefore, an arbitrary single-well solution will be as follows:

$$\Phi(\boldsymbol{r}, \boldsymbol{R}) = \sum_{sn} \xi_{sn}\varphi_{sn}(\boldsymbol{R})\Psi_n(\boldsymbol{r}, \boldsymbol{R}). \tag{2.12}$$

By operating on eq. (2.12) with the initial Hamiltonian [eq. (2.3)], we arrive at a system of equations for the coefficients ξ_{sn}:

$$(E - E_{sn})\xi_{sn} = \sum_{ms'} C_{sn,s'm}\xi_{s'm}, \tag{2.13}$$

where for the nonadiabaticity operator $\mathscr{C} = \mathscr{A} + \mathscr{B}$ we have

$$A_{sn,s'm} = \left\langle \varphi_{sn}\right| - \left\langle n\left|\frac{\partial_R}{M}\right|m\right\rangle \nabla_R\left|\varphi_{s'm}\right\rangle,$$

$$B_{sn,s'm} = \left\langle \varphi_{sn}\right| - \left\langle n\left|\frac{\partial_R^2}{2M}\right|m\right\rangle \left|\varphi_{s'm}\right\rangle. \tag{2.14}$$

The operator ∂_R differs from ∇_R in that it acts only on the variable $\boldsymbol{R}$ incorporated into the adiabatic electronic wave function.

Equation (2.11) allows us to transform the matrix element $\langle n|\partial_R|m\rangle$ to a form convenient for further analysis. To do this, we let the operator ∇_R act on both sides of eq. (2.11) and after multiplying by Ψ_n^* we integrate over the electron coordinates. Thus, we find

$$\langle n|\partial_R|m\rangle = -\frac{\langle n|\partial\mathscr{V}(\boldsymbol{r}, \boldsymbol{R})/\partial\boldsymbol{R}|m\rangle}{E_n - E_m} \quad (n \neq m) \qquad \langle n|\partial_R|n\rangle = 0. \tag{2.15}$$

The solution of the system of equations (2.13) makes it possible to reconstruct the wave function nonadiabatically and to find the corresponding energy corrections.

In the formalism of second quantization the electron–particle coupling operator may be written in the form

$$\mathscr{V}(\boldsymbol{r} - \boldsymbol{R}) = \sum_{kk'\sigma} V_{kk'}(\boldsymbol{R})a^+_{k\sigma}a_{k'\sigma}. \tag{2.16}$$

If we assume, for the sake of simplicity, that the electron fluid is homogeneous, then

$$V_{kk'}(\boldsymbol{R}) = e^{-i(\boldsymbol{k}-\boldsymbol{k}')\cdot\boldsymbol{R}}V_{kk'}. \tag{2.17}$$

From eq. (2.16) it follows that the matrix elements [eq. (2.15)] are nonzero only for transitions involving the creation of one electron–hole pair:

$$\langle n|\partial_R|m\rangle \longrightarrow -\frac{d_{kk'}}{\varepsilon_k - \varepsilon_{k'}},$$

$$d_{kk'} = \left\langle k \left| \frac{\partial}{\partial R} \mathscr{V}(r, R) \right| k' \right\rangle. \tag{2.18}$$

Taking into account a small relative displacement of the particle from the equilibrium position in the potential well, we may assume $d_{kk'}$ to have been determined at the point of equilibrium R_1. Here the operator $\mathscr{B}$ is diagonal with respect to the particle states in the well and the matrix elements of $\mathscr{A}$ factorize into

$$A_{sn,s'm} = -\langle n|\partial_R|m\rangle \cdot \left\langle s \left| \frac{\nabla_R}{M} \right| s' \right\rangle. \tag{2.19}$$

In this expression we have omitted the index n for the particle states since a nonadiabatic perturbation leads to a slight energy reconstruction of the electronic state (see below), which practically has no effect on the particle wave function in the well.

We shall now calculate the correction to the energy of the adiabatic state. A first-order correction $E^{(1)}$ is determined by a diagonal matrix element of the operator $\mathscr{B}$. It can be found directly by using the completeness of the electronic wave functions and also the equality of the diagonal matrix element of ∂_R to zero:

$$\langle n|\partial_R^2|n\rangle = \sum_m \langle n|\partial_R|m\rangle \cdot \langle m|\partial_R|n\rangle.$$

Taking advantage of eq. (2.18), we find that

$$E^{(1)} = \frac{1}{2M} \sum_{kk'\sigma} \frac{|d_{kk'}|^2}{(\varepsilon_k - \varepsilon_{k'})^2} n_{k'}(1 - n_k). \tag{2.20}$$

This expression contains a second-order pole with $\varepsilon_k - \varepsilon_{k'} \to 0$, which gives rise to a logarithmic divergence ($T = 0$):

$$E^{(1)} = \alpha' \ln N, \quad \alpha' = \frac{1}{M} \rho^2(\varepsilon_F) \overline{|d_{kk'}|^2}, \tag{2.21}$$

where $\rho(\varepsilon_F)$ is the density of electronic states on the Fermi surface; the bar signifies an averaging over the Fermi surface. The divergence $E^{(1)}$ has been discussed in an early work by Brovman and Kagan (1967). An important point is that for a one-level well, in the neglect of the intrawell degrees of freedom, this divergence is irremovable. It is interesting that the results [eqs. (2.20) and (2.21)] are associated with the specific behavior of the overlap integral [eq. (2.9)] noted above. Indeed, the expression for $E^{(1)}$ may be rewritten, using eq. (2.9), as follows (Kondo 1984):

$$E^{(1)} = \frac{1}{2M} \partial^2_{R'} \langle \Psi_n(\boldsymbol{r}, \boldsymbol{R}) | \Psi_n(\boldsymbol{r}, \boldsymbol{R}') \rangle |_{R'=R} = \frac{3\alpha}{M} \ln N. \tag{2.22}$$

A comparison with eq. (2.21) immediately shows that in the general case we can find a relationship between the coefficient α in the exponent of the overlap integral [eq. (2.9)] and the parameters of the initial Hamiltonian. If the electron–particle coupling is determined only by a phase of s-scattering, δ_0, then in the single-particle representation of the eigenfunctions of the Hamiltonian, $\mathscr{H}_{\text{ex}} + \mathscr{V}(\boldsymbol{r}, \boldsymbol{R})$, the vertex $V_{kk'}$ in eq. (2.16) may be written in the form

$$V_{kk'} \to (\sin \delta_0)/\pi\rho(\varepsilon_{\text{F}}). \tag{2.23}$$

As a result, we have

$$\alpha = \frac{2}{3\pi^2} k_{\text{F}}^2 \sin^2 \delta_0. \tag{2.24}$$

This value coincides with that found by Yamada and Yosida (1982). The extension to the case of an arbitrary interaction can be made in a straightforward manner.

While considering the intrawell degrees of freedom, we have to take into account the energy correction due to the nonadiabatic operator $\mathscr{A}$, which starts contributing only at second order of perturbation theory. To calculate $E^{(2)}$, we employ the following general relation:

$$\left\langle s \left| \frac{\nabla_R}{M} \right| s' \right\rangle \cdot \left\langle s' \left| \frac{\nabla_R}{M} \right| s \right\rangle = \frac{\mathrm{i}\varepsilon_{ss'}}{2M} \{ \langle s | \boldsymbol{p} | s' \rangle \cdot \langle s' | \boldsymbol{R} | s \rangle - \langle s | \boldsymbol{R} | s' \rangle \cdot \langle s' | \boldsymbol{p} | s \rangle \}. \tag{2.25}$$

Then, using eq. (2.19), we find that

$$E^{(2)} = -\frac{1}{2M} \sum_{kk'\sigma} \frac{|\boldsymbol{d}_{kk'}|^2}{(\varepsilon_k - \varepsilon_{k'})^2} n_{k'}(1 - n_k) \sum_{s'} \frac{\mathrm{i}\varepsilon_{ss'}}{\varepsilon_{k'} - \varepsilon_k + \varepsilon_{ss'}} \times \{ \langle s | \boldsymbol{p} | s' \rangle \cdot \langle s' | \boldsymbol{R} | s \rangle - \langle s | \boldsymbol{R} | s' \rangle \cdot \langle s' | \boldsymbol{p} | s \rangle \}. \tag{2.26}$$

With the energy of the electron–hole pair tending to zero the sum over s' retains only the expression in the curly brackets. It reduces to the commutator of the momentum and coordinate and the sum over s' in eq. (2.26) is equal to unity. Comparing this result with eq. (2.22), we see that in the limit $|\varepsilon_k - \varepsilon_{k'}| \ll \omega$ the first-order correction is compensated. In this way the divergence is removed. Such a compensation, which occurs in all orders of perturbation theory, is far from accidental. It reflects the fact that both nonadiabatic operators $\mathscr{A}$ and $\mathscr{B}$ are of the same origin and must always be taken into account jointly.

Adding up $E^{(1)}$ and $E^{(2)}$, we get the following estimate for the leading correction to the adiabatic energy (Kagan and Prokof'ev 1986b):

$$\Delta E \sim \varepsilon_{\mathrm{F}}(m/M)\ln(\varepsilon_{\mathrm{F}}/\omega) \sim \omega(\omega/\varepsilon_{\mathrm{F}})\ln(\varepsilon_{\mathrm{F}}/\omega). \tag{2.27}$$

Thus, the nonadiabatic energy corrections are not only finite but also small with respect to the adiabatic parameter $\omega/\varepsilon_{\mathrm{F}}$. The concept of the universal violation of adiabaticity in a metal, which evolves with the neglect of the intrawell degrees of freedom on the basis of eqs. (2.20) and (2.21), is fictitious.

2.3. *Nonadiabatic reconstruction of the wave function*

We shall now consider the nature of the reconstruction of the adiabatic wave function brought about by a nonadiabatic perturbation. To do this, we go back to the representation of the wave function in the form of an expansion in adiabatic basis [eq. (2.12)] and will look for the coefficients ξ_{sn} through the solution of eq. (2.13). For simplicity, we adopt an oscillator form for the renormalized potential of an individual well. Let the initial unperturbed state have the indices $s = 0$ and m_0. Since the matrix element of $\boldsymbol{V}_R$ in eq. (2.19) is nonzero only for transitions between the nearest levels, we have on the basis of first-order perturbation theory with respect to $\mathscr{A}$

$$\xi_{1,m_1} = -(\omega/2M)^{1/2} \frac{\boldsymbol{e}_\lambda \cdot \boldsymbol{d}_{kk'}}{(\varepsilon_k - \varepsilon_{k'} + \omega)(\varepsilon_k - \varepsilon_{k'})}. \tag{2.28}$$

Here the state m_1 differs from m_0 by one electron–hole pair $(\boldsymbol{k}, \boldsymbol{k}')$; $\boldsymbol{e}_\lambda$ is the polarization vector.

In this approximation the admixture of the states with the first-level excitation in the well is determined by the quantity

$$\zeta_1 = \sum_{\lambda m_1} |\xi_{1m_1}|^2 \approx \frac{\omega}{\varepsilon_{\mathrm{F}}}(V\rho(\varepsilon_{\mathrm{F}}))^2 \iint \mathrm{d}\varepsilon\, \mathrm{d}\varepsilon' \frac{\omega^2 n_{\varepsilon'}(1-n_\varepsilon)}{(\varepsilon-\varepsilon')^2(\varepsilon-\varepsilon'+\omega)^2}. \tag{2.29}$$

Here V is the characteristic scale of the particle–electron coupling.

From the above expression we see that the role of virtual excitations with an energy $\delta E > \omega$ is insignificant, in spite of a large energy interval of order ε_{F}. This is because for high frequencies the adiabatic approximation describes the wave function adequately.

The role of low-frequency excitations, with $\delta E < \omega$, in the integral (2.29) is enhanced by a logarithmic divergence when $\delta E \to 0$. The question of how this divergence is removed will be discussed in detail below. Assuming the parameter $\omega/\varepsilon_{\mathrm{F}}$ to be sufficiently small, we may assert that the inequality $\zeta_1 \ll 1$ is satisfied. This enables us to develop a perturbation theory with respect to this parameter. According to Kagan and Prokof'ev (1986b), the resulting solution has the form

$$\xi_{sm_s} = \langle sm_s| \exp[-(\boldsymbol{R} - \boldsymbol{R}_1)\tilde{\partial}_{\mathrm{R}}]|0m_0\rangle. \tag{2.30}$$

Here

$$\tilde{\partial}_x = \mathscr{P}\partial_x\mathscr{P}, \tag{2.31}$$

where $\mathscr{P}$ is the projection operator, which separates from a complete set of electronic states $\{m_s\}$ (at a fixed $\boldsymbol{R}$) only those states which differ from the initial one, $|m_0\rangle$, by electron–hole pairs with an energy less than ω. As follows from eq. (2.29) and will be seen below, the cutoff at frequency ω has logarithmic accuracy.

In fact, the result [eq. (2.30)] is more general and has nothing to do with the smallness of the parameter ζ_1. Indeed, knowing from the outset that electronic excitations with an energy small compared to ω are important in the structure of perturbation theory, we get

$$(E - E_{sn})\xi_{sn} = \sum_{s'm} (A + B)_{sn,\,s'm}\xi_{s'm}. \tag{2.32}$$

It is not difficult to show (e.g., Kagan and Prokof'ev 1986b) that eq. (2.30) is the solution of eq. (2.32) with $E = E_{s=0n} = E_{0n}$. To do this, it will suffice to substitute eq. (2.30) and to transform the left- and right-hand sides of eq. (2.32) separately. Here it becomes clear that the solution in the form of eq. (2.30) is valid for both an arbitrary ζ_1 and an arbitrary form of the potential well and also that it is not associated with the approximate representation of the matrix elements of the operator $\mathscr{A}$ [eq. (2.19)].

We substitute eq. (2.30) into the definition of the perturbed wave function [eq. (2.12)]. From the resultant expression it follows that the solution may be given in the form

$$\Phi(\boldsymbol{r}, \boldsymbol{R}) = \mathrm{e}^{-(\boldsymbol{R}-\boldsymbol{R}_1)\cdot\tilde{\partial}_R}\Phi^{(0)}_{0m_0}(\boldsymbol{r}, \boldsymbol{R}). \tag{2.33}$$

Here, the function $\Phi(\boldsymbol{r}, \boldsymbol{R})$ is automatically normalized.

If the operator $\tilde{\partial}_R$ were acting throughout the entire space of the adiabatic functions, i.e., if $\mathscr{P}$ were equal to unity in eq. (2.31), the operator $\exp[-(\boldsymbol{R} - \boldsymbol{R}_1)\cdot\partial_R]$ would be an ordinary shift operator. In such a case,

$$\Phi(\boldsymbol{r}, \boldsymbol{R}) = \varphi_0(\boldsymbol{R})\Psi_{m_0}(\boldsymbol{r}, \boldsymbol{R}_1), \tag{2.34}$$

i.e., the initial adiabatic function would become a typical antiadiabatic function with the electrons being oriented to the center of the well rather than adjusted to the position of the particle.

In the situation considered, however, the excitation spectrum, which extends to $\delta E \sim \varepsilon_F$, is split into two regions (see fig. 2). The virtual excitations in the

Fig. 2. Energy scale for the adiabatic and nonadiabatic excitations of electron–hole pairs in a Fermi liquid.

interval $\omega < \delta E < \varepsilon_F$ form an adiabatic electronic wave function, which is moving together with the particle. This is the so-called fast excitations. To within the minor factor ω/ε_F, this part of the perturbed wave function creates a screening for the moving particle. "Slow" excitations with $\delta E < \omega$ give rise to an antiadiabatic wave function, which is oriented to the center of the well and which cannot follow the particle. With the particle tunneling from one well to another this part of the wave function governs the appearance of an electronic polaron effect (EPE). As a matter of fact, this statement implies that $\omega \approx \omega_0$, which we assumed from the very outset. In the general case, we must consider the adiabatic problem for a purely subbarrier motion. In this case, the upper cutoff ω_0 of the nonadiabatic excitation band is determined by the level separation in an inverted quasiclassical barrier. In analyzing tunneling dynamics, we shall use the symbol ω_0 for the cutoff.

Despite the relatively small phase volume, the nonadiabatic excitations play an important role. This is primarily associated with an infrared divergence due to electron–hole pairs with an energy close to zero (e.g., Nozieres and de Dominicis 1969, Mahan 1981, Kondo 1988). This infrared divergence manifests itself in divergences in eqs. (2.20) and (2.29). Since T, ξ, $\tilde{\Delta}_0 \ll \omega_0$, all scattering processes in tunneling are associated only with excitations within the same frequency range.

Note that the finite lifetime τ of the particle in the well leads to the cutoff, from below, of the nonadiabatic part of the spectrum. Indeed, excitations with $\delta E < \tau^{-1}$ cannot form before the particle leaves the potential well. In this case, the polaron effect will be governed by a frequency range (τ^{-1}, ω_0), whose limits are described by the strong inequality (2.2).

We have so far analyzed the adiabatic problem for the interaction of a heavy particle with electronic excitations in a metal. It is not difficult, however, to see that the result [eqs. (2.30) and (2.33)] is rather general for interaction with any excitations. Naturally, the separation of the nonadiabatic interval with $\delta E < \omega_0$, which is responsible for the polaron effect and inelastic scattering, is also general. In a limiting case when the particle mass is small compared to the mass of atoms in a crystal and when $\omega_0 > \Theta_D$ (Θ_D is the characteristic frequency of the phonon spectrum), the entire phonon spectrum proves non-adiabatic with respect to the tunneling particle. In the opposite limiting case of a heavy particle in a light matrix, it is the interval (τ^{-1}, ω_0) that will be nonadiabatic, just as with the electrons. An essential difference in the case of phonons is the fact that the density of low-frequency one-phonon excitations $g(\varepsilon)$ is proportional to ε^{d-1}, where d is the dimensionality of the crystal. In two- and three-dimensional cases this density turns to zero when $\varepsilon \to 0$, which removes the infrared divergence in the formation of a polaron effect, and the smallness of the phase volume is not compensated. Only in the one-dimensional case in the absence of the transport effect (see below and also Kagan and Prokof'ev 1986a) does the infrared divergence occur. In this sense, this case is an analog of tunneling in an electron liquid. Note that practically all the results depend but very slightly on the upper cutoff of the nonadiabatic band ω_0. Indeed, the diffusion kinetics is governed by excitations $\delta E \leqslant T, \tilde{\Delta}_0, \xi$. On the other hand, when $\omega_0 \ll \Theta_D$, the polaron effect at $d = 3, 2$ is negligibly small, and in a special one-dimensional case the polaron effect depends logarithmically slightly on the frequency ω_0.

2.4. The transition matrix element

From the picture outlined above it is clear that the tunneling "particle" is the original "particle" dressed in a shielding "coat" of fast excitations. Such a particle is moving in a renormalized potential $\tilde{U}(\boldsymbol{R})$, with the mass undergoing a slight change. During the movement in this potential the "particle" is interacting only with excitations lying in the nonadiabatic frequency range. We now introduce a nondecay well $\tilde{U}^{(i)}(\boldsymbol{R})$ in the potential $\tilde{U}(\boldsymbol{R})$, using the notation $\mathscr{H}'(\boldsymbol{R})$ to designate the difference $\tilde{U}(\boldsymbol{R}) - \tilde{U}^{(i)}(\boldsymbol{R})$. So, for the matrix element of the transition from one well to another we have

$$M_{nm}^{12} = \langle \Phi_n^{(1)}(\boldsymbol{r}, \boldsymbol{R}) | \mathscr{H}'(\boldsymbol{R}) | \Phi_m^{(2)}(\boldsymbol{r}, \boldsymbol{R}) \rangle. \tag{2.35}$$

Here $\Phi_n^{(i)}$ is the solution of the one-well Schrödinger equation with a potential $\tilde{U}^{(i)}$. The accuracy of eq. (2.35) is specified by the inequality (2.2).

The fact that eq. (2.35) incorporates only the coupling with slow excitations signifies the possibility of seeking the eigenstates of the one-well problem within the framework of the inverse adiabatic approximation:

$$\Phi_n^{(i)}(\boldsymbol{r}, \boldsymbol{R}) = \varphi_0^{(i)}(\boldsymbol{R}, \boldsymbol{r}) \Psi_n^{(i)}(\boldsymbol{r}), \tag{2.36}$$

where the wave functions $\varphi_0^{(i)}$ and $\Psi_n^{(i)}$ are solutions of the following equations:

$$\left[\frac{M\dot{\boldsymbol{R}}^2}{2} + \tilde{U}^{(i)}(\boldsymbol{R}) + \tilde{\mathscr{V}}(\boldsymbol{r}, \boldsymbol{R}) \right] \varphi_0^{(i)}(\boldsymbol{R}, \boldsymbol{r}) = \varepsilon_0^{(i)}(\boldsymbol{r}) \varphi_0^{(i)}(\boldsymbol{R}, \boldsymbol{r}), \tag{2.37}$$

$$[\mathscr{H}_{\text{ex}}(\boldsymbol{r}) + \varepsilon_0^{(i)}(\boldsymbol{r})] \Psi_n^{(i)}(\boldsymbol{r}) = E_n^{(i)} \Psi_n^{(i)}(\boldsymbol{r}). \tag{2.38}$$

The particle now is moving in a perturbed potential created by quasistatic fluctuations due to slow excitations. This is described by eq. (2.37).

The characteristic scale of the displacement of a particle residing at the lowest level in the well is small compared to the interatomic distance a. Therefore,

$$\begin{aligned} \varepsilon_0^{(i)}(\boldsymbol{r}) &= \varepsilon_0^{(i)} + \langle \varphi_0^{(i)}(\boldsymbol{R}, \boldsymbol{r}) | \tilde{\mathscr{V}}(\boldsymbol{r}, \boldsymbol{R}) | \varphi_0^{(i)}(\boldsymbol{R}, \boldsymbol{r}) \rangle \\ &\approx \varepsilon_0^{(i)} + \tilde{\mathscr{V}}(\boldsymbol{r}, \boldsymbol{R}_i). \end{aligned} \tag{2.39}$$

From eqs. (2.39) and (2.38) it can be seen that upon formation of the wave function Ψ_n the excitations see effectively the particle residing in the center of the potential well $\Psi_n^{(i)}(\boldsymbol{r}) = \Psi_n(\boldsymbol{r}, \boldsymbol{R}_i)$. As has been pointed out in the previous section, this is a typical result of the adiabatic picture.

As regards the function $\varphi_0^{(i)}$, it is necessary to make an important remark. If we were interested in the behavior of this function at small displacements of the particle in the well from the equilibrium position, then $\varphi_0^{(i)}$ would be dependent on $\boldsymbol{R}$ alone and practically independent of $\boldsymbol{r}$. This is exactly the result that we obtained in the preceding section when in relations (2.18) and (2.19) we ignored the corrections due to particle displacements. However, the tunneling matrix element [eq. (2.35)] is naturally determined by large values of $\boldsymbol{R} - \boldsymbol{R}_i$, of the order of the half-distance between the wells. Therefore, the dependence on $\boldsymbol{r}$ should be preserved in the wave function $\varphi_0^{(i)}$ in the sub-barrier region. At the same time it is easy to see that taking account of the behavior of $\varphi_0^{(i)\infty}$ in this region has only a slight effect on $\varepsilon_0^{(i)}(\boldsymbol{r})$.

Substituting eq. (2.36) into the definition (2.35) for the transition matrix element, we get

$$M_{nm}^{12} = \langle \Psi_n^{(1)}(\boldsymbol{r}) | \mathscr{J}(\boldsymbol{r}) | \Psi_n^{(2)}(\boldsymbol{r}) \rangle,$$
$$\mathscr{J}(\boldsymbol{r}) = \langle \varphi_0^{(1)}(\boldsymbol{R}, \boldsymbol{r}) | \mathscr{H}'(\boldsymbol{R}) | \varphi_0^{(2)}(\boldsymbol{R}, \boldsymbol{r}) \rangle. \tag{2.40}$$

The expression for $\mathscr{J}(\boldsymbol{r})$ reflects the dependence of the tunneling amplitude on distortions of the barrier due to the interaction with fluctuations in the excitation system. Equation (2.40) may be given in the form

$$\mathscr{J}(\boldsymbol{r}) = J_0 \mathrm{e}^{-\mathscr{B}(\boldsymbol{r})}, \tag{2.41}$$

where J_0 is the tunneling amplitude in the potential $\tilde{U}(\boldsymbol{R})$ in the absence of fluctuations. Taking into account that barrier fluctuations are small, the following relation is valid for $\mathscr{B}(\boldsymbol{r})$,

$$\mathscr{B} = \int_{R_1}^{R_b} \sqrt{2\tilde{M}(\tilde{U}(\boldsymbol{R}) + \tilde{\mathscr{V}}(\boldsymbol{r}, \boldsymbol{R}) - \tilde{\mathscr{V}}(\boldsymbol{r}, \boldsymbol{R}_1))}\, \mathrm{d}\boldsymbol{R}$$
$$+ \int_{R_b}^{R_2} \sqrt{2\tilde{M}(\tilde{U}(\boldsymbol{R}) + \tilde{\mathscr{V}}(\boldsymbol{r}, \boldsymbol{R}) - \tilde{\mathscr{V}}(\boldsymbol{r}, \boldsymbol{R}_2))}\, \mathrm{d}\boldsymbol{R} - \int_{R_1}^{R_2} \sqrt{2\tilde{M}\tilde{U}}\, \mathrm{d}\boldsymbol{R}$$

or

$$\mathscr{B} \approx \int_{R_1}^{R_2} \frac{\tilde{\mathscr{V}}(\boldsymbol{r}, \boldsymbol{R})}{v(\boldsymbol{R})} \mathrm{d}\boldsymbol{R} - \frac{\pi}{2} \frac{\tilde{\mathscr{V}}(\boldsymbol{r}, \boldsymbol{R}_1) + \tilde{\mathscr{V}}(\boldsymbol{r}, \boldsymbol{R}_2)}{\omega_0}$$
$$\approx \frac{\pi}{2} \frac{2\tilde{\mathscr{V}}(\boldsymbol{r}, \boldsymbol{R}_b) - \tilde{\mathscr{V}}(\boldsymbol{r}, \boldsymbol{R}_1) - \tilde{\mathscr{V}}(\boldsymbol{r}, \boldsymbol{R}_2)}{\omega_0}. \tag{2.42}$$

Here $v(\boldsymbol{R})$ is the velocity of the particle in the inverted barrier (here we do not distinguish the entry and exit points from $\boldsymbol{R}_1$ and $\boldsymbol{R}_2$).

The structure of eqs. (2.40) and (2.41) reflects the fact that to the optimum tunneling path there corresponds an effective lowering of the barrier due to fluctuations. This effect, which has come to be known as the "fluctuational preparation of the barrier", was first considered by Kagan and Klinger (1976) in dealing with the tunneling problem for the coupling with phonons of the medium. Later this effect was studied for the coupling with conduction electrons of a metal by Kondo (1976b), Vladar and Zawadowski (1983a, b, c), Vladar et al. (1988a, b), Kagan and Prokof'ev (1989a).

In the case of electron coupling, the small phase volume of the band of nonadiabatic excitations, which define $\mathscr{B}$, leads to an estimate $\mathscr{B} \lesssim 1$. For phonon coupling, if $\omega_0 \gtrsim \Theta_D$, there are no fundamental restrictions on the value of the quantity $\mathscr{B}$ and the exponent in eq. (2.41) may be larger than unity.

The transition matrix element fully governs the particle tunneling dynamics both for a purely coherent motion, when the state of the medium is not changed ($n = m$) and which is responsible for the formation of a band motion in a crystal, and for incoherent tunneling when the phase memory is lost at each step ($n \neq m$).

In describing the particle tunneling dynamics, the general expressions (2.40) and (2.41) may be compared with an equivalent Hamiltonian. If the particle–excitations coupling is taken into account in accordance with perturbation theory, this Hamiltonian may be written in the form of eq. (2.5) with the replacement

$$\mathscr{V}_z = \tfrac{1}{2}(\tilde{\mathscr{V}}(\boldsymbol{r}, \boldsymbol{R}_1) - \tilde{\mathscr{V}}(\boldsymbol{r}, \boldsymbol{R}_2)), \qquad \mathscr{V}_x = J_0(\mathrm{e}^{-\mathscr{B}(\boldsymbol{r})} - 1),$$
$$\mathscr{V}_0 = \tfrac{1}{2}(\tilde{\mathscr{V}}(\boldsymbol{r}, \boldsymbol{R}_1) + \tilde{\mathscr{V}}(\boldsymbol{r}, \boldsymbol{R}_2)). \tag{2.43}$$

In contrast to eq. (2.6), in eq. (2.43) the coupling with excitations ($\mathscr{V}_z$, $\mathscr{V}_x$, $\mathscr{V}_0$) is cut off at the energies of phonons and electron–hole pairs $\delta E < \omega_0$. The operator $\mathscr{V}_x$ now contains a diagonal (with respect to the state of the medium) component, which leads to renormalization of J_0.

If we abandon the perturbation theory, the problem of obtaining an equivalent Hamiltonian will require the renormalization of the interaction vertices. To render the quantum diffusion problem as transparent as possible, we at first confine ourselves to perturbation theory and then show how the results are changed with an arbitrary scale of the interaction $\mathscr{V}(\boldsymbol{r}, \boldsymbol{R}_i)$. From these results it will be seen that all the answers are changed quantitatively rather than qualitatively.

2.5. *The polaron operator*

The transition matrix element [eq. (2.40)] has been defined on eigenfunctions corresponding to different Hamiltonians. In such cases it is always convenient to introduce a unitary operator Λ, which relates the representations of the wave functions $\Psi_n^{(1)}$ and $\Psi_n^{(2)}$. We take advantage of the fact that for the transport of the particle the symmetry of the problem allows the nomenclature of the states to be preserved. Then

$$\Psi_n^{(2)} = \Lambda \Psi_n^{(1)} \equiv \Lambda |n\rangle. \tag{2.44}$$

Accordingly,

$$M^{12}_{nm} = J_0 \langle n|\mathrm{e}^{-\mathscr{B}(r)}\Lambda|m\rangle. \tag{2.45}$$

Here the indices n and m refer to the eigenstates of the same Hamiltonian $\mathscr{H}^{(1)} = \mathscr{H}_{\mathrm{ex}} + \tilde{\mathscr{V}}(\boldsymbol{r}, \boldsymbol{R}_1)$ [cf. eqs. (2.38) and (2.39)]. The general expression for the polaron operator Λ takes the form

$$\Lambda = \mathscr{T}_\tau \exp\left[-\mathrm{i}\int_{-\infty}^{0} \mathscr{V}(\tau)\,\mathrm{d}\tau\right], \tag{2.46}$$

where

$$\mathscr{V} = \tilde{\mathscr{V}}(\boldsymbol{r}, \boldsymbol{R}_2) - \tilde{\mathscr{V}}(\boldsymbol{r}, \boldsymbol{R}_1) \tag{2.47}$$

and $\mathscr{T}_\tau$ is the standard chronological operator.

We begin with the case of the interaction with the electronic subsystem. Using eq. (2.16), we rewrite the interaction [eq. (2.47)] as follows:

$$\begin{aligned} \mathscr{V} &= \sum_{p>p'} \mathscr{V}_{pp'}, \\ \mathscr{V}_{pp'} &= \Delta V_{pp'} a_p^+ a_{p'} + \Delta V_{p'p} a_{p'}^+ a_p, \\ \Delta V_{pp'} &= V_{pp'}(1 - \mathrm{e}^{-\mathrm{i}(\boldsymbol{p}-\boldsymbol{p}')\cdot(\boldsymbol{R}_1 - \boldsymbol{R}_2)}). \end{aligned} \tag{2.48}$$

Here $p(p') \equiv \boldsymbol{p}(\boldsymbol{p}'), \sigma$. While considering perturbation theory with respect to the electron–particle coupling, we may neglect electron rescattering to higher orders, which, in fact, leads to the replacement of the Born amplitude by a true vertex. In this approximation we may regard the individual terms $\mathscr{V}_{pp'}$ in eq. (2.48) as mutually commuting, which in turn corresponds in the polaron operator [eq. (2.46)] to the independent creation or absorption of electron–hole pairs. Here Λ may be represented in the form of the following product (Kagan and Prokof'ev 1986b):

$$\begin{aligned} \Lambda = \prod_{p>p'} \Lambda_{pp'}, \quad \Lambda_{pp'} &= \mathscr{T}_\tau \exp\left\{-\mathrm{i}\int_{-\infty}^{0} \mathscr{V}_{pp'}(\tau)\,\mathrm{d}\tau\right\} \\ &= \exp\left\{-\frac{\Delta V_{pp'}}{\varepsilon_p - \varepsilon_{p'}} a_p^+ a_{p'} + \frac{\Delta V_{p'p}}{\varepsilon_p - \varepsilon_{p'}} a_{p'}^+ a_p\right\}. \end{aligned} \tag{2.49}$$

To the same accuracy

$$\Lambda = \exp\left\{\sum_{pp'} -\frac{\Delta V_{pp'}}{\varepsilon_p - \varepsilon_{p'}} a_p^+ a_{p'}\right\}. \tag{2.50}$$

From this expression it can be seen that the approximate value of Λ differs from the exact value simply by the removal of the time-ordering operator.

In order to be sure about the nature of the approximation made it will suffice to consider the difference between the exact value of the polaron operator [eq.

(2.46)] and its approximate value [eq. (2.50)]. To second order in the interaction $\mathscr{V}$ this difference may be written as

$$-\int_{-\infty}^{0} d\tau \int_{-\infty}^{\tau} d\tau' \mathscr{V}(\tau)\mathscr{V}(\tau') + \tfrac{1}{2}\int_{-\infty}^{0} d\tau \int_{-\infty}^{0} d\tau' \mathscr{V}(\tau)\mathscr{V}(\tau')$$

$$= -\tfrac{1}{2}\int_{-\infty}^{0} d\tau \int_{-\infty}^{\tau} d\tau' [\mathscr{V}(\tau), \mathscr{V}(\tau')].$$

A direct calculation gives the following result:

$$-\tfrac{1}{2}\sum_{pp's} \frac{\Delta V_{ps}\Delta V_{sp'}}{\varepsilon_p - \varepsilon_{p'}} \left\{ \frac{1}{\varepsilon_p - \varepsilon_s} - \frac{1}{\varepsilon_s - \varepsilon_{p'}} \right\} a_p^+ a_{p'}.$$

Comparing the value obtained with the term Λ linear in $\mathscr{V}$, we see that the inclusion of this correction to Λ is equivalent to the following simple replacement:

$$\Delta V_{pp'} \to \Delta V_{pp'} + \tfrac{1}{2}\sum_{s} \Delta V_{ps}\Delta V_{sp'} \left\{ \frac{1}{\varepsilon_p - \varepsilon_s} - \frac{1}{\varepsilon_s - \varepsilon_{p'}} \right\}. \tag{2.51}$$

Note that $\varepsilon_{p'} = \varepsilon_p$ at the mass surface and such a replacement really corresponds to the switch from the Born scattering amplitude to the amplitude calculated to second order.

As pointed out above, the particle resides in a single well for a finite time τ. In constructing the intrawell wave functions the finite τ may be taken into account by introducing the time dependence of the interaction $\mathscr{V} \to \mathscr{V}e^{t/\tau}$. With this circumstance taken into consideration, eq. (2.49) must be replaced by

$$\Lambda_{pp'} = \exp\left[-\left(\frac{\Delta V_{pp'}}{\varepsilon_p - \varepsilon_{p'} - i/\tau} a_p^+ a_{p'} - \text{h.c.} \right) \right]. \tag{2.52}$$

After going over to eqs. (2.49) and (2.52) the calculation of the matrix elements of the polaron operator presents no difficulty. Indeed, in expanding $\Lambda_{pp'}$ into a series in powers of the exponent it is sufficient to retain only the first three expansion terms, which is valid to within macroscopic accuracy.

We shall now calculate the diagonal matrix element $\Lambda_{nn'}$ which defines the overlap integral for coherent tunneling of a heavy particle. For a macroscopic system it is equivalent to the thermodynamic average $\langle \Lambda \rangle$:

$$\langle \Lambda \rangle = \prod_{p>p'} \left(1 - \frac{1}{2} \frac{[n_p(1-n_{p'}) + n_{p'}(1-n_p)]|\Delta V_{pp'}|^2}{(\varepsilon_p - \varepsilon_{p'})^2 + 1/\tau^2} \right) = e^{-Z}, \tag{2.53}$$

$$Z = \sum_{pp'} \frac{|\Delta V_{pp'}|^2}{(\varepsilon_p - \varepsilon_{p'})^2 + 1/\tau^2} n_p(1 - n_{p'}). \tag{2.54}$$

We now rewrite the expression for Z in the form

$$Z = \frac{1}{\pi}\int_0^{\omega_0} \frac{\mathrm{d}u\,\Omega(u)}{u^2 + 1/\tau^2}, \tag{2.55}$$

$$\Omega(u) = \pi \sum_{pp'} |\Delta V_{pp'}|^2 n_p(1 - n_{p'})(\delta(\varepsilon_p - \varepsilon_{p'} + u) + \delta(\varepsilon_p - \varepsilon_{p'} - u)). \tag{2.56}$$

The calculation of the quantity $\Omega(u)$ is not difficult:

$$\Omega(u) = \pi\, bu \coth\left(\frac{u}{2T}\right), \quad b = \rho^2(\varepsilon_{\mathrm{F}})\overline{|\Delta V_{pp'}|^2}. \tag{2.57}$$

Inserting the above expression into eq. (2.55), we find (Kagan and Prokof'ev 1990a)

$$Z = b\left[\ln\frac{\omega_0}{2\pi T} - \Psi\left(1 + \frac{1}{2\pi T\tau}\right) + \pi T\tau\right], \tag{2.58}$$

where Ψ is a logarithmic derivative of the gamma function.

Equation (2.58) characterizes the overlap integral over the entire temperature range. With $T \to 0$, bearing in mind that $\Psi(x \to \infty) \approx \ln x$, we have

$$Z = b\ln(\omega_0\tau). \tag{2.59}$$

For the set of problems under discussion, the parameter $\omega_0\tau$ is the major parameter of the problem [see eq. (2.2)]. Since perturbation theory imposes restrictions only on the value of b, the quantity Z may assume an arbitrary value in a general case. This predetermines the appearance of a strong polaron effect.

At high temperatures $T\tau > 1$, taking into account that $\Psi(1) = -\ln\gamma$ (γ is Euler's constant), the quantity Z takes on the value given by (Kagan and Prokof'ev 1987a, b)

$$Z = b\ln\left(\frac{\gamma\omega_0}{2\pi T}\right) + \pi bT\tau. \tag{2.60}$$

From this expression it may be concluded that with rise of T there occurs a drastic exponential decrease of the overlap integral, which starts at relatively low temperatures $\tau^{-1} < T \ll \omega_0$.

The appearance of large logarithms in eqs (2.59) and (2.60) is associated with the infrared divergence characteristic of the Fermi liquid and caused by the presence of a sharp Fermi surface and constant density of states of single-particle excitations. The appearance of the second term in eq. (2.60) is associated with a different infrared divergence, which occurs in eq. (2.54) at a finite temperature. In this case there takes place a strong reconstruction of the wave function due to a quasielastic scattering of the electrons ($\varepsilon_{p'} \approx \varepsilon_p$) present in an energy band of order T [for them $n_p(1 - n_p) \neq 0$]. From eqs. (2.54) and (2.55) it follows that for the contribution of an electron we have

$$\rho(\varepsilon_F)\Delta V^2 \int \frac{du}{u^2 + 1/\tau^2} \sim \rho(\varepsilon_F)\Delta V^2 \tau.$$

If we now multiply this expression by the number of electrons in a temperature range equal to $T\rho(\varepsilon_F)$, we will obtain, in order of magnitude, the second term in eq. (2.60). It should be stressed that this outcome is not connected qualitatively with the sharp boundary of the Fermi distribution and in this sense it is universal.

Let us now consider the interaction of a tunneling particle with the phonon subsystem of a crystal. We return to the system of equations (2.37) and (2.38), which describe the particle–slow-excitations coupling. Here we are speaking of phonons with frequencies $\omega_\beta < \omega_0$, whose set of coordinates is denoted by $\boldsymbol{x}$. The potential $\mathscr{V}(\boldsymbol{x}, \boldsymbol{R}_i)$ can be found in this case by expanding the energy $\varepsilon_0(\boldsymbol{x})$ [eq. (2.39)] in a series in powers of the displacement of atoms from the equilibrium positions:

$$\mathscr{V}(\boldsymbol{x}, \boldsymbol{R}_i) = \varepsilon_0^{(i)}(x) - \varepsilon_0^{(i)}(0) = \sum_\beta \gamma_\beta^{(i)} x_\beta + \tfrac{1}{2}\sum_{\beta\beta'} \gamma_{\beta\beta'}^{(i)} x_\beta x_{\beta'} + \cdots, \tag{2.61}$$

where x_β are normal phonon coordinates. Usually, in determining the phonon wave function Ψ_n account was taken only of one-phonon processes [the first term in eq. (2.61)]. Although the next terms of the expansion (2.61) are small in powers of the ratio of the displacements to the interatomic distance a, going beyond the framework of the one-phonon approximation is very important (as can be seen below) for describing the quantum diffusion in a crystal. Here it is sufficient to retain only two-phonon processes in eq. (2.61), since the inclusion of higher-order terms introduces no qualitative changes. Below we will confine ourselves to this approximation.

In the case under consideration the matrix element of a tunneling transition is defined, as before, by eq. (2.40) or (2.45) with a respective definition of the polaron operator Λ [eq. (2.44) or (2.46)]. Taking eq. (2.61) into account and passing over to the second-quantization formalism, the expression for the difference in the interactions $\mathscr{V}$ [eq. (2.47)] may be written as follows:

$$\mathscr{V} = \sum_\beta C_\beta(b_\beta + b^+_{-\beta}) + \tfrac{1}{2}\sum_{\alpha\beta} C_{\alpha\beta}(b_\alpha + b^+_{-\alpha})(b_\beta + b^+_{-\beta}) + \cdots. \tag{2.62}$$

Here $C_\beta = C_\beta^{(2)} - C_\beta^{(1)}$, $C_{\alpha\beta} = C_{\alpha\beta}^{(2)} - C_{\alpha\beta}^{(1)}$, the index $\pm\beta \equiv \pm\boldsymbol{q}, \lambda$, where $\boldsymbol{q}$ is the wave vector and λ is the number of the branch. As is known, by virtue of the translational symmetry for $q \to 0$ the vertices $C_\beta^{(i)} \sim \sqrt{\omega_\beta}$ and $C_{\alpha\beta}^{(i)} \sim \sqrt{\omega_\alpha\omega_\beta}$. If the two wells are equivalent, then in the expression for C_β there appears an additional transport factor of the form

$$(e^{i\boldsymbol{q}_\beta\cdot\boldsymbol{R}_1} - e^{i\boldsymbol{q}_\beta\cdot\boldsymbol{R}_2}),$$

which with $q \to 0$ leads to an additional power of ω_β. Analogously, the vertex $C_{\alpha\beta}$ contains the transport factor

$$(e^{i(q_\alpha+q_\beta)\cdot R_1} - e^{i(q_\alpha+q_\beta)\cdot R_2}).$$

But if the wells are energetically equivalent but nonidentical with respect to the elastic-deformation tensor, then no transport effect is present. Recall that in the determination of the polaron operator [eq. (2.44)] we must employ a representation corresponding to the phonon functions $\Psi_n^{(1)}$. Below we will restrict ourselves to perturbation theory with respect to two-phonon interaction, assuming that the following inequality is satisfied:

$$NC_{\alpha\beta}^{(i)}/(\omega_\alpha\omega_\beta)^{1/2} \ll 1. \tag{2.63}$$

For the initial phonon system of a crystal, with the leading one-phonon interaction taken into account, the presence of a particle in the well leads only to the shift of normal oscillators without a change of the phonon spectrum. Therefore, the second-quantization operators in eq. (2.62) are related to the corresponding operators of an unperturbed crystal, $b_\beta^{(0)}$, by a linear relation:

$$b_\beta = b_\beta^{(0)} + C_{-\beta}^{(1)}/\omega_\beta.$$

We add to eq. (2.62) the Hamiltonian

$$\mathscr{H}_{\text{ph}} = \sum_\beta \omega_\beta(b_\beta^+ b_\beta + \tfrac{1}{2})$$

and employ the shift operator of normal oscillators (e.g., Appel 1968, Klinger 1968), which eliminates the one-phonon processes from eq. (2.62) through a unitary transformation:

$$\bar{\Psi}_n^{(2)}(x) = \Lambda_1 \Psi_n^{(1)}(x), \quad \Lambda_1 = \exp\left\{ -\sum_\alpha \left(\frac{A_{-\alpha}}{\omega_\alpha} b_\alpha^+ - \frac{A_\alpha}{\omega_\alpha} b_\alpha \right) \right\} \equiv \prod_\alpha \Lambda_{1\alpha}. \tag{2.64}$$

The total Hamiltonian is thus transformed to the form

$$\mathscr{H}_{\text{ph}} + \mathscr{V} = \sum_\beta \omega_\beta(\bar{b}_\beta^+ \bar{b}_\beta + \tfrac{1}{2}) + \tfrac{1}{2}\sum_{\alpha\beta} C_{\alpha\beta}(\bar{b}_\alpha + \bar{b}_{-\alpha}^+)(\bar{b}_\beta + \bar{b}_{-\beta}^+). \tag{2.65}$$

Here $\bar{b}_\beta = b_\beta + A_{-\beta}/\omega_\beta$ are the annihilation operators of new normal oscillators. By direct substitution it is easy to see that the coefficients A_β must satisfy the equation

$$A_\beta = C_\beta - 2\sum_\alpha \frac{A_\alpha C_{-\alpha\beta}}{\omega_\alpha}. \tag{2.66}$$

Within perturbation theory in a two-phonon interaction we shall ignore the difference between A_β and C_β.

Let us now consider the reconstruction of the wave function $\Psi_n^{(2)}$ due to two-phonon processes. In a first-order perturbation theory the overall admixture of states to the initial state is characterized by the quantity ζ defined as

$$\zeta = \sum_m \frac{|\mathscr{V}_{nm}^{(2\text{ph})}|^2}{(E_n - E_m)^2}.$$

The creation and absorption of two phonons make a contribution equal to

$$\zeta' = \tfrac{1}{2} \sum_{\alpha\beta} \frac{|C_{\alpha\beta}|^2}{(\omega_\alpha + \omega_\beta)^2} [(N_\alpha + 1)(N_\beta + 1) + N_\alpha N_\beta], \tag{2.67}$$

where N_α are occupation numbers in state n. The expression under the summation sign has a singularity at $\omega_\alpha = \omega_\beta = 0$ ($N_\alpha \sim T/\omega_\alpha$), which, however, is removed when account is taken of the frequency dependence of the vertex $C_{\alpha\beta}^{(i)} \sim (\omega_\alpha \omega_\beta)^{1/2}$ and density of states $g(\omega_\alpha) \sim \omega_\alpha^{d-1}$ in a crystal.

The situation is quite different with two-phonon processes of the scattering type, to which there correspond terms of the type $\bar{b}_\alpha^+ \bar{b}_\beta$ in eq. (2.65). In this case

$$\zeta'' = \sum_{\alpha\beta} \frac{|C_{\alpha\beta}|^2}{(\omega_\alpha - \omega_\beta)^2} (N_\alpha + 1) N_\beta. \tag{2.68}$$

This expression has an irremovable singularity when $\omega_\alpha = \omega_\beta \neq 0$ at a finite temperature. At low frequencies, the contribution to eq. (2.68) from the region of $|\omega_\alpha - \omega_\beta| \gtrsim T$ is proportional to $(T/\Theta_\mathrm{D})^4 \ll 1$ and may be neglected. But the frequency region $|\omega_\alpha - \omega_\beta| \ll T$ leads to the divergence of eq. (2.68). This singularity is of the same nature as the one arising in an electronic system due to quasielastic scattering of excitations. From eq. (2.68) it can be seen that we are again dealing with an excitation band in the interval T.

Had we introduced the finite particle lifetime τ in a single well, the divergence in eq. (2.68) would have been removed, just as in the case of electron coupling. Here

$$\begin{aligned} \zeta'' &= \sum_{\alpha\beta} \frac{|C_{\alpha\beta}|^2}{(\omega_\alpha - \omega_\beta)^2 + 1/\tau^2} (N_\alpha + 1) N_\beta \\ &= \frac{2}{\pi} \int_0^{\omega_0} \frac{\mathrm{d}u}{u^2 + 1/\tau^2} \Omega^{(\mathrm{ph})}(u), \end{aligned} \tag{2.69}$$

$$\Omega^{(\mathrm{ph})}(u) = \frac{\pi}{2} \sum_{\alpha\beta} |C_{\alpha\beta}|^2 (N_\alpha + 1) N_\beta [\delta(u - \omega_\alpha + \omega_\beta) + \delta(u + \omega_\alpha - \omega_\beta)]. \tag{2.70}$$

When $\tau T \gg 1$, the integral in eq. (2.69) has its principal contribution from frequencies $u \ll T$ and we may put $u = 0$ in the argument of the δ-function.

Then,

$$\zeta'' = \Omega^{(\mathrm{ph})}\tau, \quad \Omega^{(\mathrm{ph})} \equiv \Omega^{(\mathrm{ph})}(0). \tag{2.71}$$

Although, at low temperatures, the frequency $\Omega^{(\mathrm{ph})}$ is low, the parameter ζ'' may be arbitrarily large.

Thus, the reshaping of the wave function due to scattering processes is always strong in the general case, irrespective of the two-phonon interaction scale. This makes it necessary to take into account two-phonon processes in dealing with the particle tunneling dynamics. If higher-order perturbation theories are taken into account, the picture of the reshaping of the wave function is completely retained qualitatively. This can also be seen from the general solution obtained when $T \ll \Theta_{\mathrm{D}}$ and arbitrary $C_{\alpha\beta}$ value (see Kagan and Prokof'ev 1989b).

At low temperatures, the condition (2.63) allows us to ignore the reshaping of the wave function associated with the creation or absorption of phonon pairs [see eq. (2.67)]. In such a case, in the Hamiltonian (2.65) we may retain only the terms that describe the scattering of phonons, whose role is anomalous.

Returning to the general definition of the polaron operator [eq. (2.46)], we see that in dealing with two-phonon scattering processes we have formally a structure quite analogous to the electron coupling. Using within perturbation theory the same idea of the independence now of pair phonon modes of excitation, it is possible to remove the chronological operator and to perform an integration over time in the exponential. As a result, we shall arrive at an expression analogous to eq. (2.52) for the two-phonon polaron operator Λ_2:

$$\Lambda_2 = \prod_{\alpha>\beta} \Lambda_{2\alpha\beta}, \quad \Lambda_{2\alpha\beta} = \exp\left\{ -\left(\frac{C_{\alpha\beta}}{\omega_\alpha - \omega_\beta - \mathrm{i}/\tau} \bar{b}^{+}_{-\alpha}\bar{b}_{\beta} - \mathrm{h.c.} \right) \right\}. \tag{2.72}$$

The difference between the exact value of Λ_2 and eq. (2.72) reduces to the renormalization of the vertex $C_{\alpha\beta}$, which to second order coincides with eq. (2.51).

Thus, in the case under discussion the full phonon polaron operator in eq. (2.44) may be written as

$$\Lambda = \Lambda_2 \Lambda_1. \tag{2.73}$$

In this expression there are no restrictions on the one-phonon coupling scale. The representation of both operators in the form in eqs. (2.64) and (2.72) is very convenient for direct calculations.

3. *Quantum diffusion with phonon coupling*

3.1. *Coherent tunneling*

The coherent tunneling process corresponds to the subbarrier motion of a particle without a change in the phonon state. To this process there corresponds the diagonal (with respect to excitation occupation numbers) matrix element [eq. (2.45)]:

$$\Delta_c = J_0 \langle e^{-\mathscr{B}} \Lambda_2 \Lambda_1 \rangle. \tag{3.1}$$

Here the symbol $\langle \ldots \rangle$ signifies an averaging of the diagonal matrix element over the thermodynamic equilibrium. In a regular crystal Δ_c determines the bandwidth in full analogy with the formation of an electronic band in the tight-binding approximation. Since within the framework of perturbation theory we do not consider rescattering due to two-phonon processes, only the diagonal matrix element of the operator Λ_2 in eq. (3.1) is important:

$$\Delta_c = J_0 \langle e^{-\mathscr{B}} \Lambda_1 \rangle \langle \Lambda_2 \rangle. \tag{3.2}$$

If relatively small displacements of atoms from the equilibrium position appear to be the decisive factor in the formation of an optimum fluctuation of the barrier, then in defining the operator $\mathscr{B}$ [eq. (2.42)] we may confine ourselves to a linear expansion in displacements:

$$\mathscr{B} = \frac{1}{\omega_0} \sum_\beta (B_\beta b_\beta^{(0)} + B_{-\beta} b_\beta^{(0)+}). \tag{3.3}$$

Here the operators $b_\beta^{(0)}$ and $b_\beta^{(0)+}$ refer to an unperturbed crystal. Passing over to the shifted operators b_β and b_β^+ in the polaron transformation [eq. (2.64)], we get

$$e^{-\mathscr{B}} \Lambda_1 = e^{B_0} \prod_\alpha \exp\left\{ \left(\frac{C_\alpha}{\omega_\alpha} - \frac{B_\alpha}{\omega_0} \right) b_\alpha - \left(\frac{C_{-\alpha}}{\omega_\alpha} + \frac{B_{-\alpha}}{\omega_0} \right) b_\alpha^+ \right\}. \tag{3.4}$$

Here

$$B_0 = \sum_\beta \frac{B_{-\beta}(C_\beta^{(1)} + C_\beta^{(2)})}{\omega_0 \omega_\beta} \tag{3.5}$$

is a quantity due to a change in the barrier upon displacement of the matrix atoms to positions equal to the half-sum of displacements that occur when the particle resides in wells 1 and 2.

Let us now calculate the average of the operator (3.4). We expand each exponential into a series, retaining the first three terms. After performing standard transformations we have

$$\langle e^{-\mathscr{B}} \Lambda_1 \rangle = e^{B_0} e^{-\phi(T) + G(T)}, \tag{3.6}$$

$$\phi(T) = \frac{1}{2} \sum_\beta \frac{|C_\beta|^2}{\omega_\beta^2} \coth\left(\frac{\omega_\beta}{2T} \right), \tag{3.7}$$

$$G(T) = \frac{1}{2} \sum_\beta \frac{|B_\beta|^2}{\omega_0^2} \coth\left(\frac{\omega_\beta}{2T} \right). \tag{3.8}$$

When $T \to 0$, the average $\langle \Lambda_2 \rangle \to 1$ and, hence, eq. (3.6), in fact, defines the renormalized coherent transition amplitude. As can be seen from eq. (3.6), a competition takes place between the ordinary polaron effect [the factor $\phi(T)$] and the fluctuational preparation of the barrier $G(T)$ (Kagan and Klinger 1976). In the general case the relationship between ϕ and G is quite arbitrary, despite the difference between ω_0 and Θ_D. This is because the major contribution to eqs. (3.7) and (3.8) is associated with the displacements of various atomic groups in the crystal. Indeed, the vertex B_β is determined by the particle–phonon coupling when the former is residing in the subbarrier region. Therefore, it is not directly associated with the difference $C_\beta^{(1)} - C_\beta^{(2)}$, which governs the polaron effect.

Since the quantities ϕ and G have different signs, then, contrary to the generally accepted viewpoint, one-phonon coupling may bring about not only a decrease but also an increase in the tunneling amplitude. If the fluctuational preparation of the barrier prevails, we will have, instead of the polaron narrowing, an increase in the bandwidth with rise of temperature.

Even from purely geometric considerations it is clear that the fluctuational preparation of the barrier will play a decisive role if the particle–lattice-atom interaction strongly depends on the distance between them and the tunneling process leads to their approach. Presumably, the interaction with barrier fluctuations predominates in a case where the diffusing particle is residing in the lattice sites and the elementary tunneling act is of the exchange type. It is especially obvious in the case of isotopic diffusion. Here, the static displacements of the matrix atoms are small when the particle is present in the well, whereas the factor G contains no special smallness.

Note that in many cases the optimum fluctuation may be associated with a noticeable displacement of the atoms of the nearest environment, which brings about a drastic lowering of the barrier. In such cases, we cannot limit ourselves to the linear expansion [eq. (3.8)]. A discussion of this more general case can be found in Kagan and Klinger (1976). The ideas developed have been used in an analysis of tunneling motion in insulating glasses (e.g., Fleurov and Trakhtenberg 1983, Goldanski et al. 1986).

For the tunneling motion of a heavy particle, when $\omega_0 \ll \Theta_D$, all the three quantities, ϕ, B and G, are small. This is because now the nonadiabatic spectral band is bounded from above by the value ω_0, which in turn causes a decrease in the effective density-of-states $g(\omega_0)$.

We have, so far, considered only the contribution of one-phonon coupling to the formation of Δ_c. Let us now analyze the role of two-phonon coupling in eq. (3.2). We expand each factor $\Lambda_{2\alpha\beta}$ [eq. (2.72)] in a series in powers of the exponential and retain, to macroscopic accuracy, the first three terms of the expansion. So, we obtain

$$\langle \Lambda_2 \rangle = \mathrm{e}^{-\zeta''/2}, \tag{3.9}$$

where ζ'' coincides with eq. (2.69).

At $T = 0$, we have $\zeta'' = 0$ and the two-phonon processes play no role. But even at sufficiently low temperatures dictated by the long lifetime of the particle in an individual well, ζ'' assumes the value in eq. (2.71) and, accordingly,

$$\Delta_c = \tilde{\Delta}_0 \exp(-\tfrac{1}{2}\Omega\tau), \tag{3.10}$$

$$\tilde{\Delta}_0 = J_0 e^{B_0} e^{-\phi(T)+G(T)}. \tag{3.11}$$

For simplicity, we omit the index at the quantity Ω^{2ph} in this section.

From the above expressions it can be seen that the two-phonon coupling leads to an exponential decrease of the coherent transition amplitude with an exponent proportional to the particle lifetime in the well (Kagan and Prokof'ev 1989b). For the band motion of the particle $\tau = (\mu\Delta_c)^{-1}$, where the numerical coefficient μ depends on the lattice type. For crystals of cubic symmetry $\mu \sim \sqrt{z}$ (e.g., Petzinger 1982, Kagan and Prokof'ev 1987b). Substituting this expression into eq. (3.10), we obtain a self-consistent equation for Δ_c. It is easy to show that, with $\Omega > \Omega_c = 2\mu\tilde{\Delta}_0/e$, the only solution provided by this equation is $\Delta_c = 0$, indicating that the band (coherent) motion is suppressed. With $\tilde{\Delta}_0 \ll \Theta_D$ this occurs when $\tilde{\Delta}_0 \ll T \ll \Theta_D$. It follows readily from the low-temperature limit of eq. (2.70) that

$$\Omega(T) \sim 10^{4+(2)}\Theta_D \left(\frac{T}{\Theta_D}\right)^{7+(2)}, \tag{3.12}$$

where the 2 in parentheses corresponds to a case where the two wells are identical; so, $C_{\alpha\beta}^{(1)}$ and $C_{\alpha\beta}^{(2)}$ in the definition of $C_{\alpha\beta}$ in eq. (2.62) differ only in the phase factor.

Thus, the two-phonon polaron effect causes narrowing, this being followed by the disappearance of the coherent band in the temperature region where the one-phonon polaron effect practically does not change [ϕ and G in eq. (3.11) undergo substantial change only when $T \sim \Theta_D$]. At a sufficiently small $\tilde{\Delta}_0$, the result obtained is universal.

The nature of the destruction of the coherent band becomes clear if we assume that phonon fluctuations entail relative phase fluctuations in the neighboring unit cells. This results in a damping of the off-diagonal elements f_{12} of the particle density matrix defined in the site representation. As has been shown by Kagan and Maksimov (1973), $f_{12} \sim e^{-\Omega t}$, exactly with a frequency Ω coinciding with eq. (2.70). It is this loss of phase memory which causes the dynamic destruction of the coherence.

A strict vanishing of Δ_c occurs in the case of the purely coherent nature of the particle's motion. However, with $T \neq 0$ there appears an incoherent tunneling channel where during the subbarrier motion the particle–excitations coupling leads to a change in the state of the medium. The lifetime in a single well remains finite when $\Omega > \Omega_c$, though still very long (see below). In this case, according to the law (3.10), the coherent amplitude Δ_c is finite, but it becomes negligibly small when $\Omega \gg \Omega_c$.

Thus, when $\Omega \gg \mu\tilde{\Delta}_0$ the band motion is practically absent and the tunneling transport becomes purely incoherent.

3.2. Incoherent tunneling

In describing the incoherent motion for $\Omega \gg \mu\tilde{\Delta}_0$, when there are no coherent correlations, the site representation is evidently adequate to the problem. Such a statement may also be made if the asymmetry of the neighboring wells $\xi \gg \mu\tilde{\Delta}_0$, irrespective of the value of Ω. In both cases the probability of the transition from well 1 to well 2 is given by

$$W = 2\pi \sum_{nm} \rho_n^{(0)} |M_{nm}^{12}|^2 \delta(\xi + E_n - E_m). \tag{3.13}$$

If we employ eq. (2.45) and use the time representation of the δ-function, the above expression may then be cast in the form

$$W = J_0^2 \int_{-\infty}^{+\infty} \mathrm{d}t\, \mathrm{e}^{\mathrm{i}\xi t} \operatorname{Tr}\{\rho^{(0)} \mathrm{e}^{-\mathscr{B}(t)} \Lambda(t) \Lambda^+(0) \mathrm{e}^{-\mathscr{B}(0)}\}. \tag{3.14}$$

Just as in the calculation of the coherent tunneling amplitude, with the rescattering due to two-phonon coupling being neglected, we can write the correlator in eq. (3.14) in the following form:

$$\begin{aligned} &\langle \mathrm{e}^{-\mathscr{B}(t)} \Lambda_1(t) \Lambda_2(t) \Lambda_2^+(0) \Lambda_1^+(0) \mathrm{e}^{-\mathscr{B}(0)} \rangle \\ &\quad = \langle \mathrm{e}^{-\mathscr{B}(t)} \Lambda_1(t) \Lambda_1^+(0) \mathrm{e}^{-\mathscr{B}(0)} \rangle \langle \Lambda_2(t) \Lambda_2^+(0) \rangle. \end{aligned} \tag{3.15}$$

Substituting eq. (3.4) into the first of the correlators in eq. (3.15) and performing ordinary transformations, we find

$$\langle \mathrm{e}^{-\mathscr{B}(t)} \Lambda_1(t) \Lambda_1^+(0) \mathrm{e}^{-\mathscr{B}(0)} \rangle = \mathrm{e}^{\chi_1(t)}, \tag{3.16}$$

$$\begin{aligned} \chi_1(t) = {} & -2B_0 + 4G(T) \\ & - \int_0^{\omega_0} \frac{\mathrm{d}\omega}{\omega} A(\omega) \left[1 - \cos(\omega t) \coth\left(\frac{\omega}{2T}\right) + \mathrm{i} \sin(\omega t) \right] \\ & + \int_0^{\omega_0} \frac{\mathrm{d}\omega}{\omega} h(\omega) \left[\cos(\omega t) - \mathrm{i} \sin(\omega t) \coth\left(\frac{\omega}{2T}\right) \right], \end{aligned} \tag{3.17}$$

where

$$A(\omega) = \sum_\alpha \left(\frac{|C_\alpha|^2}{\omega_\alpha} + \frac{\omega_\alpha}{\omega_0^2} |B_\alpha|^2 \right) \delta(\omega - \omega_\alpha), \tag{3.18}$$

$$h(\omega) = \sum_\alpha \left(\frac{C_\alpha B_{-\alpha} + C_{-\alpha} B_\alpha}{\omega_0} \right) \delta(\omega - \omega_\alpha). \tag{3.19}$$

In order to calculate the two-phonon correlator in eq. (3.15), we use the representation (2.72). Expanding again each of the factors $(\Lambda_2^+(t))_{\alpha\beta}$ and $(\Lambda_2)_{\alpha\beta}$ in a series in powers of the exponent and retaining only the first three terms, we find

$$\langle \Lambda_2(t)\Lambda_2^+(0)\rangle = \mathrm{e}^{\chi_2(t)},$$

$$\chi_2(t) = -\sum_{\alpha\beta} \frac{|C_{\alpha\beta}|^2}{(\omega_\alpha - \omega_\beta)^2 + 1/\tau^2}(N_\alpha + 1)N_\beta(1 - \mathrm{e}^{-\mathrm{i}(\omega_\alpha - \omega_\beta)t}). \tag{3.20}$$

The expression for χ_2 has no singularities when $|\omega_\alpha - \omega_\beta| \to 0$. The characteristic times in the integral (3.14) are defined now by the quantity Ω^{-1} or ξ^{-1}. Therefore, the characteristic minimum values of the difference $|\omega_\alpha - \omega_\beta|$, which are important in eq. (3.20), are of order $(\Omega, \xi)_{\max}$. Physically, it is clear (and it will be demonstrated below) that the particle lifetime in the well in the incoherent region $\tau \gg 1/(\Omega, \xi)_{\max}$. This allows us to drop the term $1/\tau$ from the denominator of eq. (3.20).

Making in this expression the substitution of the indices $\alpha, \beta \to \beta, \alpha$, we finally find for the half-sum

$$\chi_2(t) = -\int_0^{\omega_0} \frac{\mathrm{d}\omega}{\omega}\frac{\Omega(\omega)}{\pi T}\left[(1 - \cos(\omega t))\coth\left(\frac{\omega}{2T}\right) + \mathrm{i}\sin(\omega t)\right], \tag{3.21}$$

where

$$\Omega(\omega) = \pi T \sum_{\alpha\beta} |C_{\alpha\beta}|^2 \frac{N_\beta - N_\alpha}{(\omega_\alpha - \omega_\beta)}\delta(\omega_\alpha - \omega_\beta - \omega). \tag{3.22}$$

The expression enclosed in square brackets in the first integral in eq. (3.17) [and also in the integral in eq. (3.21)] may be rewritten as

$$\frac{\cosh(\omega/2T) - \cosh(\omega(\mathrm{i}t - 1/2T))}{\sinh(\omega/2T)}.$$

Analogously, for the expression within the square brackets in the second integral in eq. (3.17) we have

$$-\frac{\sinh(\omega(\mathrm{i}t - 1/2T))}{\sinh(\omega/2T)}.$$

Let us now go back to the initial expression [eq. (3.14)] and substitute the values $\chi_1(t)$ and $\chi_2(t)$ into the integrand. We use the standard procedure (e.g., Appel 1968) of replacement of the variable $t + \mathrm{i}/2T \to t$ with subsequent shift of the integration contour again to the real axis. As a result, we find

$$W = J_0^2 e^{\xi/2T} \int_{-\infty}^{+\infty} dt\, e^{i\xi t} \exp\left\{2B_0 + 4G(T) - \int_0^{\omega_0} \frac{d\omega}{\omega}\left[f(\omega)\frac{(\cosh(\omega/2T) - \cos(\omega t))}{\sinh(\omega/2T)} + ih(\omega)\frac{\sin(\omega t)}{\sinh(\omega/2T)}\right]\right\}, \tag{3.23}$$

$$f(\omega) = A(\omega) + \frac{\Omega(\omega)}{\pi T}. \tag{3.24}$$

We consider first the one-phonon contribution to the exponent in eq. (3.23). Separating the t-independent term with account taken of the definition (3.19), we obtain (bearing in mind the shift of the integration contour)

$$\chi_1 = 2\{B_0 - \phi(T) + G(T)\} + \bar{\Psi}(t) - i\Psi(t), \tag{3.25}$$

where

$$\bar{\Psi}(t) = \int_0^{\omega_0} \frac{d\omega}{\omega} A(\omega)\frac{\cos(\omega t)}{\sinh(\omega/2T)}, \qquad \Psi(t) = \int_0^{\omega_0} \frac{d\omega}{\omega} h(\omega)\frac{\sin(\omega t)}{\sinh(\omega/2T)}. \tag{3.26}$$

The first term in eq. (3.25) presumably defines the renormalization $J_0 \to \tilde{\Delta}_0$ [see eq. (3.11)].

In separating out the t-independent term in the same integral in the case of two-phonon coupling, it is necessary to be cautious against divergence when $\omega \to 0$. In order to avoid this, we rewrite the two-phonon contribution in the form

$$\chi_2 = -\int_0^{\omega_0} \frac{\Omega(\omega)}{\pi T\omega}\frac{(\cosh(\omega/2T) - 1)}{\sinh(\omega/2T)} d\omega - \int_0^{\omega_0} \frac{\Omega(\omega)}{\pi T\omega}\frac{(1 - \cos(\omega t))}{\sinh(\omega/2T)} d\omega. \tag{3.27}$$

The first term in this expression describes the correction to the polaron effect. It is small at low temperatures, but it may be neglected when $T \sim \Theta_D$ in view of eq. (2.63). The second integral is sensitive to the value of t. It is small when $tT < 1$ and its contribution may also be ignored. In the opposite limit $tT \gg 1$ the frequencies $\omega \ll T$ become essential in the integral; at these frequencies $\Omega(\omega)$ is no longer dependent on the frequency:

$$\Omega(0) \equiv \Omega = \pi \sum_{\alpha\beta} |C_{\alpha\beta}|^2 (N_\alpha + 1) N_\beta \delta(\omega_\alpha - \omega_\beta) \tag{3.28}$$

[cf. eqs. (2.71) and (2.70)]. It is easy to see that a divergence here arises in the second integral at low frequencies, and it is cut off only when $\omega \sim 1/t$. Replacing the upper limit of integration by ∞, we arrive at a tabulated integral

(Gradshteyn and Ryzhik 1965):

$$\chi_2 \approx -\frac{\Omega}{\pi T}\ln(\cosh(\pi T t)). \tag{3.29}$$

Taking into account eqs. (3.25), (3.29) and (3.11), the final expression for the probability assumes the form

$$W = \tilde{\Delta}_0^2 e^{\xi/2T} \int_{-\infty}^{+\infty} dt\, e^{i\xi t} e^{\chi_2(t) + \bar{\Psi}(t) - i\Psi(t)}. \tag{3.30}$$

The resultant expression contains all the phonon coupling channels involved in incoherent tunneling of the particles.

The special role of two-phonon coupling shows up in scattering processes, leading to an infrared divergence when $|\omega_\alpha - \omega_\beta| \to 0$. Within the perturbation theory, with account taken of eq. (2.63), in $C_{\alpha\beta}$ the creation and absorption of phonon pairs at low temperatures could be neglected, although in the region of $T \sim \Theta_D$ the role of such processes is already comparable with phonon scattering. In a quantitative analysis of experimental findings over the entire temperature range it is expedient to keep in mind that the inclusion of the terms omitted does not alter the general form of eq. (3.23) and reduces only to the substitution (Kagan and Prokof'ev 1989b)

$$\Omega(\omega) \to \Omega(\omega) + \Omega'(\omega),$$

$$\Omega'(\omega) = \pi T \sum_{\alpha\beta} |C_{\alpha\beta}|^2 \frac{1 + N_\alpha + N_\beta}{2(\omega_\alpha + \omega_\beta)} \delta(\omega_\alpha + \omega_\beta - \omega). \tag{3.31}$$

When $\omega \to 0$, the frequency $\Omega'(\omega) \to 0$ and the analysis given above remains unchanged.

3.3. Quantum diffusion in a perfect crystal

We begin our consideration of quantum diffusion with the case of incoherent tunneling when $(\Omega, \xi)_{\max} \gg \mu\Delta_c$ and, as has been found out in section 3.1, there is practically no purely coherent tunneling.

Let us start with the symmetrical case $\xi = 0$. It is easy to see that with $T \ll \Theta_D$, the functions $\bar{\Psi}(t)$, $\Psi(t) \ll 1$. In this case the probability [eq. (3.30)] does not diverge and remains finite only by virtue of two-phonon coupling. Since $\Omega \ll T$ according to eq. (3.12), the integral in eq. (3.30) has its principal contribution from $t \sim 1/\Omega$, at which $\pi T t \gg 1$ and the function χ_2 [eq. (3.29)] assumes its asymptotic value:

$$\chi_2 \approx -\Omega|t|. \tag{3.32}$$

Such asymptotics reflect the process of tunneling under the regime of ohmic dissipation (Caldeire and Leggett 1981, Leggett et al. 1987). It is largely associated with the fact that the two-phonon part of the function $f(\omega)$ [eq. (3.24)] behaves as follows:

$$f(\omega) \xrightarrow[\omega \to 0]{} \text{const.} \tag{3.33}$$

As for one-phonon processes, it is not difficult to show that the corresponding contribution to $f(\omega)$, defined by the function $A(\omega)$ [eq. (3.18)] at low frequencies in a crystal of dimensionality d behaves as (Kagan and Prokof'ev 1986a, 1989b)

$$A(\omega) \approx \lambda_1 \left(\frac{\omega}{\Theta_D} \right)^{d-1+(2)}. \tag{3.34}$$

The additional (2) in the exponent appears whenever $C_\beta^{(1)}$ and $C_\beta^{(2)}$ differ only in the phase

$$e^{i\boldsymbol{q}\cdot(\boldsymbol{R}_1 - \boldsymbol{R}_2)}$$

in the phonon field with a wave vector $\boldsymbol{q}$, which predetermines the appearance of a "transport effect" for the vertex in the interaction [eq. (2.62)]. In this case, the second term in eq. (3.18), which is associated with barrier fluctuations, has the same frequency dependence. From eq. (3.34) it follows that the limit (3.33) proves valid only in a one-dimensional crystal in the absence of the transport effect. In all the remaining cases, consideration of one-phonon coupling to any order with respect to the vertex C_β does not lead to tunneling motion in the ohmic (viscous) regime, at least under the approximation of $\sim \tilde{\Delta}_0^2$. In this sense, two-phonon (multiple-phonon) coupling plays a special role, restoring the "tunneling-with-dissipation" regime (Kagan and Prokof'ev 1989b).

In the presence of the transport effect all results for crystals of lesser dimensionality remain valid provided that we properly recalculate $\Omega(T)$, which at low temperatures has the dependence $\Omega(T) \propto T^{2d+3}$. Below we shall confine ourselves to the case $d = 3$. It should only be noted that in the absence of the transport effect all results obtained for the cases $d = 2$ and $d = 1$ must be modified (for more detail, see Kagan and Prokof'ev 1989b). Here the case $d = 1$ formally does not differ from the case of the electron coupling, which will be discussed in detail in section 5.

It should be emphasized that two-phonon processes eliminate the divergence of the integral in eq. (3.30), which inevitably appears if one-phonon coupling alone is taken into consideration. This enables us to avoid the substitution procedure $e^{\bar{\Psi}} \to e^{\bar{\Psi}} - 1$ known from the theory of polarons (Holstein 1959, Flynn and Stoneham 1970) (in the absence of the barrier preparation effect) with simultaneous inclusion, together with eq. (3.30), of the coherent band motion. As a matter of fact, from the results obtained above it follows that eq. (3.30) is valid only under the conditions in which the coherent transition amplitude practically vanishes. The coherent and incoherent tunneling regimes are not realized simultaneously; actually, they only replace each other.

The leading term in the transition probability [eq. (3.30)] arises when $T \ll \Theta_D$ even to zero order in $\bar{\Psi}$, Ψ, i.e., when the one-phonon excitations are neglected:

$$W^{(0)} = \frac{2\tilde{\Delta}_0^2}{\Omega}. \tag{3.35}$$

The transition from this expression to the diffusion coefficient in a regular crystal is trivial. In the case of cubic symmetry

$$D = \frac{za^2}{3} \frac{\tilde{\Delta}_0^2}{\Omega}. \tag{3.36}$$

Equation (3.36) was originally derived by Kagan and Maksimov (1973) (see also Kagan and Klinger 1974).

A relative increase of phase fluctuations in neighboring cells, which continues with rise of T, apart from the dynamic destruction of the band, leads also to a decrease in the probability of incoherent transitions. This involves the appearance of an extra factor $\tilde{\Delta}_0/\Omega$ in eqs. (3.35) and (3.36). As the particle lifetime in the unit cell increases, the effective mean free path $l_{\text{eff}} \sim a\tilde{\Delta}_0/\Omega$ becomes much smaller than a.

The temperature dependence of the diffusion coefficient with $T \ll \Theta_D$ is dictated by the behavior of the quantity $\Omega(T)$ [eq. (3.12)]:

$$D \propto \frac{1}{T^{7+(2)}}. \tag{3.37}$$

It should be emphasized that this dependence holds only at sufficiently low $T < \frac{1}{10}\Theta_D - \frac{1}{20}\Theta_D$. In a general case, to find $\Omega(T)$, we must use the basic expression (3.28), which can be conveniently reduced to the form

$$\Omega(T) = \pi \int_0^{\omega_0} \lambda(\omega) g^2(\omega) \frac{e^{\omega/T}}{(e^{\omega/T} - 1)^2}\, d\omega, \tag{3.38}$$

$$\lambda(\omega) = \overline{|C_{\alpha\beta}^{(2)} - C_{\alpha\beta}^{(1)}|^2}. \tag{3.39}$$

The bar in the last expression signifies averaging over the phase space and the branches of the phonon spectrum provided that $\omega_\alpha = \omega_\beta = \omega$. For $\omega \ll \Theta_D$

$$\lambda(\omega) = \lambda_2 \omega^2 (\omega/\Theta_D)^{(2)}. \tag{3.40}$$

Note that in the general case the temperature dependence $\Omega(T)$ is very sensitive to the real dependence of the function $g(\omega)$. That this is actually the case will be illustrated below with the example of quantum diffusion of muonium in KCl and NaCl crystals (see section 4.1).

In the classical temperature limit $\Omega(T) \propto T^2$.

If we return to the definition of Δ_c in eq. (3.10) and use the value of τ given by eq. (3.35), we find in a straightforward way that

$$\Delta_c = \tilde{\Delta}_0 \exp\left[-\frac{1}{2}\left(\frac{\Omega}{\mu\tilde{\Delta}_0}\right)^2 \right]. \tag{3.41}$$

This outcome demonstrates a rapid fall of Δ_c in the incoherent region and leads parametrically to the same estimate for the replacement of the coherent interwell transition regime by the incoherent one.

The expression for the diffusion coefficient [eq. (3.36)] is, in fact, valid within a much wider temperature range and requires only the condition $T \gg \Delta_c$ be satisfied. This will be shown in a rigorous manner in the next section. For the moment we only note that, generally speaking, this outcome is nontrivial since, with $\Omega < \mu\Delta_c < T$, a coherent band motion takes place. Physically, this is associated with the fact that when $T \gg \mu\Delta_c$ the wave packet for a particle can be formed on the scale of a single unit cell. Therefore, the loss of phase coherence is determined, as before, by the damping of the correlations at the neighboring sites, although now it is relatively small at each of the steps. (With $T \ll \Delta_c$ the wave packet for the particle can be created only in a region with a linear scale much larger than a.) The diffusion coefficient for the band motion with the dependence $D \propto T^{-9}$ was first obtained by Andreev and Lifshitz (1969).

Let us now define the probability of a transition accompanied by the emission or absorption of one phonon. To do this, it is sufficient to retain the term linear in $\bar{\Psi} - \mathrm{i}\Psi$ in the expansion of the exponential $\mathrm{e}^{\bar{\Psi} - \mathrm{i}\Psi}$ in eq. (3.30). It is not difficult to see that the integral of the function Ψ over $\mathrm{d}t$ identically turns to zero. Therefore, the one-phonon contribution to W reduces to the following expression:

$$W^{(1)} = \tilde{\Delta}_0^2 \int_0^{\omega_0} \frac{\mathrm{d}\omega\, A(\omega)}{\omega \sinh(\omega/2T)} \int_{-\infty}^{+\infty} \cos(\omega t) \exp(-\Omega|t|)\, \mathrm{d}t, \tag{3.42}$$

which is readily transformed to the form

$$W^{(1)} = 2\pi \tilde{\Delta}_0^2 T \lim_{\omega \to 0} \frac{A(\omega)}{\omega^2}. \tag{3.43}$$

If use is made of eq. (3.34), it appears that $W^{(1)}$ is nonzero only in a case where the transition occurs between nonequivalent wells and, hence, the transport effect is absent. In this case, $A(\omega) = \lambda_1 (\omega/\Theta_\mathrm{D})^2$ [see eq. (3.34)] and

$$W^{(1)} = 2\pi\lambda_1 \frac{\tilde{\Delta}_0^2 T}{\Theta_\mathrm{D}^2}. \tag{3.44}$$

This result has been obtained by Kagan and Maksimov (1980) and by Teichler and Seeger (1981).

Comparing eq. (3.44) with eq. (3.35), we see that at low temperatures the probability of one-phonon transitions is invariably small compared to $W^{(0)}$. Moreover, the result [eq. (3.44)] itself is valid only for a dynamic destruction of the band coherent motion, which, as we have seen earlier, is realized only due to two-phonon processes. But if two-phonon processes are ignored, then even at $\xi = 0$ we will have a level splitting equal to $2\Delta_c$ in the two-well problem. In this case, $W^{(1)}$ [eq. (3.44)] describes the relaxation between the levels with $T \gg \Delta_c$, a result known in the low-temperature kinetics of amorphous systems (e.g., Jäckle et al. 1976, Black 1981).

Let us calculate a contribution quadratic in $\bar{\Psi} - \mathrm{i}\Psi$:

$$W^{(2)} = \frac{1}{2}\pi\tilde{\Delta}_0^2 \int_0^{\omega_0} \frac{\mathrm{d}\omega}{\omega^2 \sinh^2(\omega/2T)} (A^2(\omega) - h^2(\omega)). \tag{3.45}$$

With account taken of the definitions (3.18) and (3.19), $W^{(2)} > 0$ and with $T \ll \Theta_D$ the above relation leads to the following estimate:

$$W^{(2)} \sim \frac{\tilde{\Delta}_0^2}{T}\left(\frac{T}{\Theta_D}\right)^{4+(4)}. \tag{3.46}$$

If there is no transport effect, the leading term in $W^{(2)}$ is related to a polaron shift of normal oscillators:

$$W^{(2)} \sim 10\lambda_1^2 \frac{\tilde{\Delta}_0^2}{\Theta_D}\left(\frac{T}{\Theta_D}\right)^3 \tag{3.47}$$

(see Fujii 1979). Comparison of $W^{(2)}$ with $W^{(1)}$ shows that at low temperatures the contribution of higher-order terms is small compared to the probability due to one-phonon processes. However, if the transport effect is operative, then $W^{(1)} = 0$ and corrections to $W^{(0)}$ are introduced, beginning with the second-order terms. Here, the temperature dependence undergoes change and the leading term in $W^{(2)}$ is given by

$$W^{(2)} \sim 10^4 \lambda_1^2 \frac{\tilde{\Delta}_0^2}{\Theta_D}\left(\frac{T}{\Theta_D}\right)^7. \tag{3.48}$$

This result coincides with that obtained by Holstein (1959) for small polarons. An analogous relation has been derived for hopping diffusion by Flynn and Stoneham (1970). Note that the result [eq. (3.48)] may, in fact, be due equally well to both intrasite scattering and interaction with barrier fluctuations. This follows from eq. (3.18), where both contributions have the same frequency dependence if the transport effect is operative.

It should be noted that $W^{(0)}$ and $W^{(1)}$, $W^{(2)}, \ldots$ have an opposite temperature dependence. This is because in the second case we are dealing with phonon-induced tunneling transitions, while in the first case the particle is moving in a band destroyed by dynamic fluctuations, the scale of which grows with temperature. It should be stressed that in all cases at $\xi = 0$ and low temperatures the decisive factor for the transition probability is the term $W^{(0)}$, and that the ratio $[W^{(1)}, W^{(2)}]/W^{(0)}$ is proportional to a high power of (T/Θ_D).

3.4. Quantum diffusion in a crystal with defects

Let us consider the case $\xi \neq 0$, more exactly $\xi > \mu\Delta_c$. We return to the original expression (3.30) and calculate $W^{(0)}$. The corresponding integral in eq. (3.30) can be evaluated (see Gradshteyn and Ryzhik 1965):

$$W^{(0)} = \frac{2\tilde{\Delta}_0^2 \Omega}{\xi^2 + \Omega^2} e^{\xi/2T} \frac{|\Gamma(1 + (\Omega + i\xi)/2\pi T|^2}{\Gamma(1 + \Omega/\pi T)}, \tag{3.49}$$

where Γ is the gamma function. Since $\Omega \ll T$, we have

$$W^{(0)} \approx \frac{2\tilde{\Delta}_0^2 \Omega}{\xi^2 + \Omega^2} \frac{\xi/T}{1 - e^{-\xi/T}}. \tag{3.50}$$

This outcome coincides with the result obtained by Kagan and Maksimov (1980, 1983a) (see also Kagan 1981) within the framework of the kinetic equation for the density matrix.

With $\xi < \Omega$ all the preceding results remain unchanged. But if $\xi \gg \Omega$, then

$$W^{(0)} \sim \Omega/\xi^2 \propto T^{7+(2)} \tag{3.51}$$

and we have an opposite temperature dependence with respect to eq. (3.35). Now, as T decreases there is a tendency towards localization. In this case, the transitions induced by one-phonon processes may become important. Indeed, under these conditions

$$W^{(1)} = \frac{2\tilde{\Delta}_0^2}{\xi^2} \pi T(A(|\xi|) + h(|\xi|)) \frac{\xi/T}{1 - e^{-\xi/T}}. \tag{3.52}$$

If we make use of eqs. (3.18) and (3.19), we get

$$A(|\xi|) + h(|\xi|) = \sum_\alpha \frac{1}{2\omega_\alpha} \left| C_\alpha + \frac{|\xi|}{\omega_0} B_\alpha \right|^2 \delta(\omega_\alpha - |\xi|) \equiv A'(\xi). \tag{3.53}$$

From eqs. (3.52) and (3.50) it follows that with large level shifts one-phonon processes begin to play a leading role. This occurs when $\xi > \xi_*(T)$, where $\xi_*(T)$ is defined by the expression

$$A'(\xi_*) = \Omega/\pi T$$

or, taking into account eq. (3.34),

$$\xi_*(T) \approx \Theta_D \left(\frac{\Omega}{\pi \lambda_1 T} \right)^{1/[2+(2)]} \tag{3.54}$$

It is not difficult to see that higher-order terms in $\bar{\Psi} - \mathrm{i}\Psi$ lead again to small corrections, this time with respect to $W^{(1)}$ or, in the general case, with respect to the sum $W = W^{(0)} + W^{(1)}$. It is exactly this sum that determines in all cases the tunneling transition probability for a phonon coupling when $T \ll \Theta_{\mathrm{D}}$. It can be written with high accuracy in the form

$$W \approx \frac{2\tilde{\Delta}_0^2}{\Omega^2 + \xi^2}(\Omega + \pi T A'(\xi))\frac{\xi/T}{1 - \mathrm{e}^{-\xi/T}}. \tag{3.55}$$

When $\xi > \xi_*$ (see Kagan and Maksimov 1980),

$$W \approx 2\pi\lambda_1 \frac{\tilde{\Delta}_0^2 T}{\Theta_{\mathrm{D}}^2}\left(\frac{\xi}{\Theta_{\mathrm{D}}}\right)^{(2)} \quad (T > \xi > \xi_*) \tag{3.56}$$

[when $\xi > T$, this expression contains the same statistical factor as eq. (3.55)]. Comparing with eq. (3.51), we see that the dependence on the level shift and the temperature dependence are drastically changed. The quantity W as a function of ξ passes through a minimum at $\xi = \xi_*$ if the transport effect is operative or evolves to a plateau if this effect is absent $[(\xi/\Theta_{\mathrm{D}})^{(2)} \to 1]$ (see fig. 3). Note that with $T \to 0$ the probability $W \to 0$ if $\xi < 0$ and tends to a finite value when $\xi > 0$.

The results obtained allow us to come to the conclusion as to the role of the fluctuational preparation of the barrier at low temperatures. If the transport effect is operative in phonon-stimulated transitions, the interaction with these fluctuations and the intrawell interaction make a comparable contribution. In the absence of the transport effect the role of barrier fluctuations is insignificant. However, in all cases the effect due to the renormalization of the transition amplitude $\tilde{\Delta}_0$ [eq. (3.11)] remains to be substantial.

If in a crystal the characteristic values for relative level shifts in neighboring wells are larger than the coherent bandwidth, we shall not be able to switch

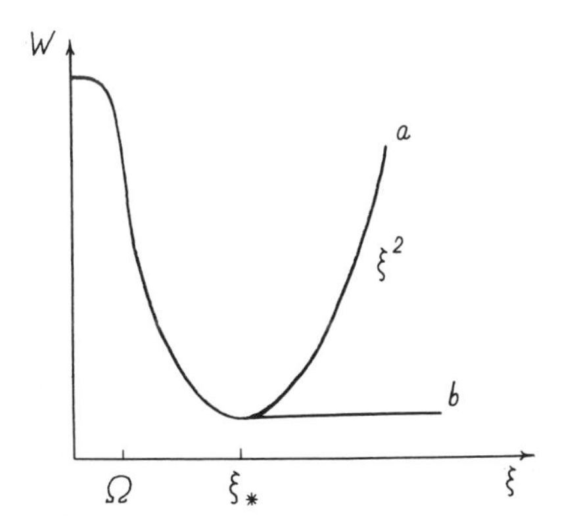

Fig. 3. Transition probability W versus energy splitting ξ.

simply from the solution of the two-well problem to the diffusion coefficient, just as in the case of the transition from eq. (3.35) to eq. (3.36). Taking advantage of the fact that interwell transitions realize a typical Markovian process, we may write a kinetic equation. This equation assumes an obvious form in the site representation:

$$\frac{\partial f_i}{\partial t} + \sum_g (f_i W_{i,i+g} - f_{i+g} W_{i+g,i}) = 0. \tag{3.57}$$

Here f_i is the particle distribution function over the lattice sites.

If the relative level shifts in neighboring wells are random, then eq. (3.57) with $W_{i,i+g}$ given by eq. (3.55) is the equation for the diffusion on a three-dimensional lattice with random bonds between sites. The problem simplifies in limiting cases. Thus, when $\Omega \gg \xi$, the probability $W_{i,i+g}$ takes on a constant value [eq. (3.35)] and eq. (3.57) leads to a macroscopic diffusion coefficient defined by eq. (3.36). In the opposite case, when the characteristic level shifts $T > \xi \gg \Omega$, the dependence of the transition probability on temperature and level shift is factorized. Here, in the case of $\xi < \xi_*(T)$ the averaged diffusion coefficient D is proportional to $\Omega \propto T^{7+(2)}$ and the proportionality factor is determined by the conductivity of the three-dimensional lattice with bonds of $\sim 1/\xi_{i,i+g}^2$.

When $T > \xi > \xi_*(T)$, two different cases occur. If the transport effect is absent, the coefficients $W_{i,i+g}$ in eq. (3.57) cease to be dependent on the level shift (see fig. 3). Paradoxical as it may be, a space-homogeneous diffusion takes place in this case. Here, the diffusion coefficient assumes the following value [see eq. (3.44)]:

$$D = \frac{za^2}{3} \pi \lambda_1 \frac{\tilde{\Delta}_0^2 T}{\Theta_D^2}. \tag{3.58}$$

But if the transitions occur with the transport effect remaining operative, then, as before, $D(T) \propto T$ with a coefficient defined by the conductivity of the three-dimensional lattice with bonds of $\sim \xi_{i,i+g}^2$. In contrast to the case of $\xi < \xi_*(T)$, the optimum paths prefer regions with a large ξ. Note that the finding of the averaged macroscopic diffusion coefficient, in fact, has much in common with percolation problems (e.g., Shklovsky and Efros 1979).

The question of the temperature dependence of the diffusion coefficient for an irregular crystal may, in fact, be found to be more complicated. Let us consider, for definiteness, a crystal with a low concentration of point defects.

If the impurity potential corresponds to attraction, then each of the impurities is a trap with a binding energy ε_d. From statistical considerations it is clear that the particle will reside for the greater part of the time in traps with a concentration x if $x \exp(|\varepsilon|_d/T) \gg 1$. Here it is evident that

$$D(T) \propto \exp(-|\varepsilon_d|/T). \tag{3.59}$$

In the opposite limiting case when $T > |\varepsilon_d|/|\ln x|$, the particle is moving practically bypassing the traps. Regardless of the character of the dependence of the interaction on the distance from the impurity center, the energy value at a mean distance, $\varepsilon(\bar{R})$, will definitely be lower than T.

3.5. Band tunneling diffusion

In order to describe the band motion ($\xi = 0, \Omega < \Delta$) of a tunneling particle in a crystal, where the delocalization of the particle wave function becomes significant, we employ the formalism of the kinetic equation for the density matrix. Extending the Hamiltonian of the two-well problem [eqs. (2.5) and (2.43)] to the case of the tunneling motion in the crystal, we have for the total Hamiltonian

$$\mathscr{H} = \mathscr{H}_{ph} + \sum_i \varepsilon_i c_i^+ c_i + J_0 \sum_{ig} c_i^+ c_{i+g} + \sum_i \tilde{\mathscr{V}}(\boldsymbol{x}, \boldsymbol{R}_i) c_i^+ c_i + J_0 \sum_{ig} (e^{-\mathscr{B}(x)} - 1) c_i^+ c_{i+g}, \tag{3.60}$$

where the phonon coupling $\tilde{\mathscr{V}}$ is defined in accordance with eq. (2.61).

We now transform this Hamiltonian by performing a polaron transformation with respect to a one-phonon interaction. As we have seen (see section 3.3), all inelastic transitions associated with the off-diagonal elements of the operators Λ_1 and $\mathscr{B}$ are negligibly small at low temperatures compared to two-phonon processes. Therefore, one-phonon coupling, in fact, reduces to the renormalization of the coherent transition amplitude:

$$\mathscr{H} = \mathscr{H}_0 + \mathscr{H}', \quad \mathscr{H}_0 = \mathscr{H}_{ph} + \sum_i \varepsilon_i c_i^+ c_i + \tilde{\Delta}_0 \sum_{ig} c_i^+ c_{i+g}, \tag{3.61}$$

$$\mathscr{H}' = \sum_i \tilde{\mathscr{V}}_2(\boldsymbol{x}, \boldsymbol{R}_i) c_i^+ c_i, \tag{3.62}$$

where $\tilde{\mathscr{V}}_2$ is the two-phonon part of the interaction [eq. (2.61)]. The inclusion of $\mathscr{H}'$ will be carried out according to perturbation theory, just as in the previous sections.

The basic equation for the density matrix has the form

$$\frac{\partial \rho}{\partial t} + i[\mathscr{H}_0, \rho] = -i[\mathscr{H}', \rho] \tag{3.63}$$

or, in a Fourier representation,

$$\omega\rho(\omega) - [\mathscr{H}_0, \rho(\omega)] = [\mathscr{H}', \rho(\omega)]. \tag{3.64}$$

We introduce the density matrix for the particle:

$$f(\omega) = \mathrm{Tr}_{ph}\, \rho(\omega). \tag{3.65}$$

Then

$$\omega \boldsymbol{f}(\omega) - [\mathscr{H}_0, \boldsymbol{f}(\omega)] = \mathrm{Tr}_{\mathrm{ph}}[\mathscr{H}', \rho(\omega)]. \tag{3.66}$$

If we substitute into the right-hand side a formal solution of eq. (3.66), the resultant equation may be written in the form

$$\omega \boldsymbol{f}(\omega) - [\mathscr{H}_0, \boldsymbol{f}(\omega)] = I, \tag{3.67}$$

$$I = \mathrm{Tr}_{\mathrm{ph}}\left[\mathscr{H}', \frac{1}{\omega - \mathscr{L}}[\mathscr{H}', \rho(\omega)]\right]. \tag{3.68}$$

Here we have adopted a convention (cf. Kagan and Maksimov 1980), the meaning of which is clear from the definition of the matrix element in the representation of the eigenfunctions of the Hamiltonian $\mathscr{H}_0$:

$$\left(\frac{1}{\omega - \mathscr{L}}\mathscr{A}\right)_{ss'} = \frac{1}{\omega - E_{ss'}} A_{ss'}, \quad (E_{ss'} = E_s - E_{s'}), \tag{3.69}$$

where $E_s = \varepsilon_a + E_n$ are the eigenvalues of $\mathscr{H}_0$ (E_n is the energy of the phonon system and ε_a is the particle's energy).

Within the framework of perturbation theory in $\mathscr{H}'$ we may ignore in eq. (3.68) the correlation between the particle and phonon states, this being equivalent to the following approximation:

$$\rho(\omega) = \rho_{\mathrm{ph}}^{(0)} \boldsymbol{f}(\omega).$$

As a result, eq. (3.67) becomes closed with respect to $\boldsymbol{f}(\omega)$.

The matrix elements of the collision integral could be written in an arbitrary representation as follows:

$$I_{cd} = \sum_{ab} \Omega_{ab}^{cd} f_{ab}. \tag{3.70}$$

A direct transformation of eq. (3.68) yields the following expression for the matrix elements Ω:

$$\Omega_{ab}^{cd} = \mathrm{Tr}_{\mathrm{ph}}\left\{\rho_{\mathrm{ph}}^{(0)} \langle b| \left[\frac{1}{\omega - \mathscr{L}}[|d\rangle\langle c|, \mathscr{H}'], \mathscr{H}'\right] |a\rangle\right\}. \tag{3.71}$$

A direct calculation of eq. (3.71) in the representation of the Hamiltonian $\mathscr{H}_0$ presents no difficulties:

$$\Omega_{ab}^{cd} = \sum_{nm} \rho_n^{(0)} \left\{ \delta_{bd} \sum_h \frac{(\mathscr{H}')_{ch}^{nm}(\mathscr{H}')_{ha}^{mn}}{\omega - E_{hb}^{mn}} + \delta_{ca} \sum_h \frac{(\mathscr{H}')_{bh}^{nm}(\mathscr{H}')_{hd}^{mn}}{\omega - E_{ah}^{nm}} \right.$$
$$\left. - (\mathscr{H}')_{bd}^{nm}(\mathscr{H}')_{ca}^{mn}\left(\frac{1}{\omega - E_{cb}^{mn}} + \frac{1}{\omega - E_{ad}^{nm}}\right)\right\}. \tag{3.72}$$

Note that to the adiabatic turn-on of the interaction there corresponds the replacement $\omega \to \omega + \mathrm{i}0$. The real part of the matrix elements [eq. (3.72)] governs the renormalization of the particle's spectrum due to two-phonon coupling. This renormalization will be neglected in what follows. The imaginary part of eq. (3.72) defines the collision integral in eq. (3.67).

We begin by considering the region of $T \gg \Delta, \xi$. In this case, the particle energy may be eliminated from the denominator of eq. (3.72). This renders the matrix Ω_{ab}^{cd} invariant with respect to the choice of the representation for the particle, which enables us to deal from the outset with the kinetic equations (3.67) and (3.70) in the site representation (Kagan and Maksimov 1973, 1983a), in which case the matrix Ω becomes very simple. Only the elements $\Omega_{i,i'}^{i,i'}$ appear to be different from zero:

$$\Omega_{i,i'}^{i,i'} = -\mathrm{i}\pi \sum_{\alpha\beta} (N_\alpha + 1) N_\beta |C_{\alpha\beta}^{(i)} - C_{\alpha\beta}^{(i')}|^2 \, \delta(\omega_\alpha - \omega_\beta) \equiv -\mathrm{i}\Omega_{i,i'}. \tag{3.73}$$

The characteristic relaxation time for the diffusion process is a priori much less than T. For this reason, the solution of the kinetic equation in the limit $\omega \to 0$ is important. This was used in passing over to eq. (3.73). Substituting eq. (3.73) into eqs. (3.70) and (3.67) and returning to the time representation, we have

$$\frac{\partial f_{i,i'}}{\partial t} + \mathrm{i}\xi_{i,i'} f_{i,i'} + \mathrm{i}\tilde{\Delta}_0 \sum_g (f_{i+g,i'} - f_{i,i'+g}) = -\Omega_{i,i'} f_{i,i'}, \quad \xi_{i,i'} = \varepsilon_i - \varepsilon_{i'}. \tag{3.74}$$

In particular, for the diagonal element of the density matrix we have

$$\frac{\partial f_{i,i}}{\partial t} + \mathrm{i}\tilde{\Delta}_0 \sum_g (f_{i+g,i} - f_{i,i+g}) = 0. \tag{3.75}$$

From eq. (3.74) it follows that the nondiagonal elements are damped according to the law $f_{i,i'} \propto \exp(-\Omega_{i,i'} t)$. Since the logarithmic decrement is defined by the quantity Ω, it becomes clear why it is the ratio $\tilde{\Delta}_0/\Omega$ that determines the dynamic destruction of the band.

Let $|\xi_{i,i'}|$ or $\Omega_{i,i'}$ be large compared to $\tilde{\Delta}_0$. Then, for large times characteristic of the time evolution of the diagonal elements of the density matrix, we may omit the derivative with respect to time in eq. (3.74) and retain only the diagonal elements of the density matrix in the third term on the left-hand side. Then,

$$f_{i,i+g} \approx \frac{-\mathrm{i}\tilde{\Delta}_0}{\mathrm{i}\xi_{i,i+g} + \Omega_{i,i+g}} (f_{i+g,i+g} - f_{i,i}).$$

Inserting this expression in eq. (3.75), we obtain eq. (3.57) with the interwell transition probability given by

$$W_{i,i+g} = \frac{2\tilde{\Delta}_0^2 \Omega}{\xi_{i,i+g}^2 + \Omega^2}, \tag{3.76}$$

where $\Omega = \Omega_{i,i+g}$ identically coincides with eq. (3.28).

Considering the diffusion in a perfect crystal ($\xi_{i,i'} = 0$), we can solve the equation, discarding the condition $\Omega \ll \tilde{\Delta}_0$ (Kagan and Maksimov 1973, Kagan and Klinger 1974). To do this, we treat $f_{i,i'}$ as a function of the variables i and $i' - i = l$ and perform a Fourier transformation of the variable i. In this mixed representation eqs. (3.74) and (3.75) are written for $l = g, 0$ in the form

$$\frac{\partial f_q(\boldsymbol{k})}{\partial t} + \mathrm{i}\tilde{\Delta}_0 f_0(\boldsymbol{k})[\mathrm{e}^{\mathrm{i}\boldsymbol{k}\cdot\boldsymbol{a}_g} - 1] + \mathrm{i}\tilde{\Delta}_0 \sum_{g'(\neq -g)} f_{g+g'}(\boldsymbol{k})[\mathrm{e}^{\mathrm{i}\boldsymbol{k}\cdot\boldsymbol{a}_{g'}} - 1]$$

$$= -\Omega f_g(\boldsymbol{k}), \tag{3.77}$$

$$\frac{\partial f_0(\boldsymbol{k})}{\partial t} + \mathrm{i}\tilde{\Delta}_0 \sum_g f_g(\boldsymbol{k})[\mathrm{e}^{\mathrm{i}\boldsymbol{k}\cdot\boldsymbol{a}_g} - 1] = 0, \tag{3.78}$$

where $f_0(\boldsymbol{k}) = n(\boldsymbol{k})$ is the Fourier component of the particle density.

The diffusion in a macroscopic system is governed by the relaxation of long-scale density fluctuations with $k \ll 1/a$. The relationship between the off-diagonal and diagonal elements of $\boldsymbol{f}$ is now given by

$$f_g(\boldsymbol{k}) \approx \frac{\tilde{\Delta}_0}{\Omega}(\boldsymbol{k}\cdot\boldsymbol{a}_g) f_0(\boldsymbol{k}),$$

in which we have omitted the terms in the parameter $\tilde{\Delta}_0 ka/\Omega \ll 1$. The smallness of this ratio is realized due to the smallness of the parameter ka and, therefore, it is valid with an arbitrary $\tilde{\Delta}_0/\Omega$.

Substituting the resultant relation into eq. (3.78) for $g = 0$, we arrive at a diffusion equation

$$\frac{\partial n(\boldsymbol{k})}{\partial t} + D\boldsymbol{k}^2 n(\boldsymbol{k}) = 0, \tag{3.79}$$

with a diffusion coefficient coinciding with eq. (3.36).

It is interesting that within the solution of the kinetic equation for the density matrix we obtain a unified solution for the diffusion coefficient on the sole condition $T > \Delta$, which holds for both $\Omega < \mu\Delta_c$ and $\Omega > \mu\Delta_c$ (Kagan and Maksimov 1973, Kagan and Klinger 1974). In the former case we have a purely coherent band motion with a large mean free path. A continuous transition through the region of $\Omega \sim \mu\Delta_c$ is an indication that the quasiband motion persists also when $\Omega > \mu\Delta_c$, despite the fact that purely coherent tunneling is destroyed by dynamic fluctuations.

To solve the problem for $T < \Delta$, we can take advantage of the fact that the representation of the band wave functions with a quasimomentum $\boldsymbol{p}$ and an energy ($\varepsilon_i = 0$)

$$\mathscr{E}_p = \Delta_c \sum_g \mathrm{e}^{\mathrm{i}\boldsymbol{p}\cdot\boldsymbol{a}_g} \tag{3.80}$$

characteristic of the tight-binding model is adequate. Here, in the kinetic equation we are interested in the evolution of the elements of the density matrix $f_{p+k,p} \equiv f_p(\boldsymbol{k})$, with k tending to zero. Naturally, in this case we immediately obtain an ordinary Boltzmann equation:

$$\frac{\partial f_p(\boldsymbol{k})}{\partial t} + \mathrm{i}\boldsymbol{k} \cdot \boldsymbol{v}_p f_p(\boldsymbol{k}) = \sum_q \{ f_{p+q}(\boldsymbol{k}) W_{p+q,p} - f_p(\boldsymbol{k}) W_{p,p-q} \}. \tag{3.81}$$

The expression for the scattering probability is found directly from eqs. (3.70) and (3.72):

$$W_{p,p'} = 2\pi \sum_{\alpha\beta} |C^{(1)}_{\alpha\beta}|^2 (N_\alpha + 1) N_\beta \delta(\omega_\alpha - \omega_\beta - \mathscr{E}_p + \mathscr{E}_{p'}) \delta_{p+q_\beta, p'+q_\alpha} \tag{3.82}$$

The momentum conservation law in eq. (3.82) corresponds to the assumption that the vertex of the two-phonon coupling $|C^{(i)}_{\alpha\beta}|$ has an invariable value on a lattice of equivalent sites across which the particle is moving.

As usual, the solution of the equation will be sought in the form

$$f_p(\boldsymbol{k}) = n(\boldsymbol{k}) f^{(0)}_p (1 + \chi_p(\boldsymbol{k})), \tag{3.83}$$

where $f^{(0)}_p$ is the equilibrium distribution function for the particle. Then, in the stationary case we have

$$-\mathrm{i}\boldsymbol{k} \cdot \boldsymbol{v}_p = \sum_q W_{p,p-q} \{ \chi_p(\boldsymbol{k}) - \chi_{p-q}(\boldsymbol{k}) \}. \tag{3.84}$$

In a narrow band, for which the relation $\Delta/\Theta_\mathrm{D} \ll 1$ is a priori valid, the characteristic momentum of phonons $q \sim (1/a)\ T/\Theta_\mathrm{D} \ll p$. This allows us to ignore the difference $\mathscr{E}_p - \mathscr{E}_p$ in the argument of the δ-function in eq. (3.82) and to switch from eq. (3.84) to the Fokker–Planck equation by carrying out an expansion in $\boldsymbol{q}$:

$$\mathrm{i}\boldsymbol{k} \cdot \boldsymbol{v}_p = \frac{\Omega}{a^2} \left(-\frac{\boldsymbol{v}_p}{T} \frac{\partial}{\partial \boldsymbol{p}} \chi_p(\boldsymbol{k}) + \frac{\partial^2}{\partial \boldsymbol{p}^2} \chi_p(\boldsymbol{k}) \right). \tag{3.85}$$

Here

$$\Omega = \tfrac{1}{6} \sum_q (qa)^2 W_{p,p-q} \tag{3.86}$$

and, as follows from eq. (3.82), this p-independent quantity coincides with the two-phonon scattering frequency introduced earlier [see eq. (3.28)]. Knowing $\chi_p(\boldsymbol{k})$, it is easy to find the diffusion coefficient from the relation for the Fourier component of the current:

$$j_k = \sum_p \boldsymbol{v}_p f_p(\boldsymbol{k}) \equiv -\mathrm{i}D\boldsymbol{k}n(\boldsymbol{k}). \tag{3.87}$$

With $T \ll \Delta$ for the spectrum $\mathscr{E}_p$ [eq. (3.80)] near the band bottom, we may confine ourselves to the quadratic momentum dependence of the energy. Then,

$\boldsymbol{v}_p = \boldsymbol{p}/M_*$ and the direct solution of eq. (3.85) will look like

$$\chi_p(\boldsymbol{k}) = -\mathrm{i}\boldsymbol{k}\cdot\boldsymbol{p}\,\frac{a^2 T}{\Omega}. \tag{3.88}$$

Here the diffusion coefficient is given by

$$D = \frac{a^2}{3}\frac{\langle \boldsymbol{p}\cdot\boldsymbol{v}\rangle T}{\Omega(T)} = \frac{a^2 T^2}{\Omega(T)}. \tag{3.89}$$

Thus, in the region of extremely low temperatures [see eq. (3.37)]

$$D(T) \propto 1/T^7. \tag{3.90}$$

This result was first obtained by Gogolin (1980).

It should be emphasized that the temperature law [eq. (3.90)] is universal and holds even if we abandon the condition $|C^{(i)}_{\alpha\beta}| = \text{const}$. Indeed, in the general case, we have a lattice with a smaller volume of the reciprocal unit cell. However, this circumstance does not manifest itself with $T \ll \Delta$, when the scattering problem involves only small values of the momenta $\boldsymbol{p}, \boldsymbol{p}'$ and $\boldsymbol{q}$, and we again have the exact momentum conservation law in eq. (3.82). As a result, the transport effect is restored.

If $T > \Delta$, then in eq. (3.85) we may neglect the first term in parentheses and the solution, with account taken of eq. (3.80), assumes the form

$$\chi_p(\boldsymbol{k}) = -\frac{\mathrm{i}\boldsymbol{k}\cdot\boldsymbol{v}_p}{\Omega}.$$

Substituting this relation into eq. (3.87), we obtain a value of diffusion coefficient coinciding with eq. (3.36). Thus, with $T > \Delta$ the solutions obtained in the site and band representations coincide.

For a large number of problems the lowest level in the well is close to being degenerate and, therefore, in a real case it is split with a level separation smaller than ω_0 (for example, the diffusion of ortho-hydrogen in para-hydrogen). In this case, the off-diagonal elements of the density matrix are damped not only due to the frequency of dynamic fluctuations $\Omega_{i,i+g}$ but also due to intrawell transitions, which render the lifetime of a single level finite. This predetermines the effect of intrawell transitions on the kinetics of interwell transitions. Suppose that because of the transitions between the levels a thermodynamic equilibrium with a distribution function $f^{(0)}_{is}$ is set up in the well. So, the extension of eq. (3.50) to the case under consideration may be written in the form (Kagan 1981):

$$W_{i,i+g} = 2\sum_{ss'} f^{(0)}_{is}|\tilde{\Delta}^{ss'}_0|^2 \frac{\tilde{\Omega}^{ss'}_{i,i+g}}{(\xi^{ss'}_{i,i+g})^2 + (\tilde{\Omega}^{ss'}_{i,i+g})^2}\,\frac{\xi^{ss'}_{i,i+g}/T}{1-\exp(-\xi^{ss'}_{i,i+g}/T)}, \tag{3.91}$$

$$\tilde{\Omega}^{ss'}_{i,i+g} = \Omega^{ss'}_{i,i+g} + \tfrac{1}{2}(\Gamma^s_i + \Gamma^{s'}_{i+g}), \tag{3.92}$$

$$\Gamma_i^s = \pi \sum_{s' \neq s} \sum_{nm} \rho_n^{(0)} |\mathscr{V}_{is,\, is'}^{nm}|^2 \{\delta(E_{nm} + \varepsilon_{is} - \varepsilon_{is'}) + \delta(E_{nm} + \varepsilon_{is'} - \varepsilon_{is})\}. \quad (3.93)$$

In deriving eq. (3.91) the role of transitions between the levels was taken into account in the relaxation time approximation.

Although the structure of the diffusion equation (3.57) remains valid in this case, there is one circumstance on which we wish to focus our attention. If at low temperatures $\Omega \ll \Gamma \ll \xi_{i,\, i+g'}^{ss'}$ then the localization is removed only due to intrawell relaxation.

We shall now briefly consider the role of impurity scattering for a band motion. In the case of point defects, when $T > \Delta$, the diffusion coefficient may be given as follows:

$$D = \frac{za^2}{3} \frac{\Delta_c^2}{\Omega_{im} + \Omega}, \quad (3.94)$$

where

$$\Omega_{im} \approx (a\mu\Delta_c)\left(\frac{x}{V_0}\right)\sigma_{eff}. \quad (3.95)$$

Here the first factor is the band-averaged modulus of group velocity for the particle and x is the relative impurity concentration (V_0 is the volume of the unit cell). For very narrow bands the effective scattering cross-section σ_{eff} is determined by the size of the region of $R_0 \gg a$ around a defect, in which the level shift is larger than the bandwidth Δ:

$$\sigma_{eff} \approx \pi R_0^2, \qquad |\mathscr{U}(R_0)| = \Delta, \quad (3.96)$$

where $\mathscr{U}(\boldsymbol{R})$ is the interaction potential with a defect.

It is natural that the size of the defect region is determined by a long-range interaction. Such an interaction, at least in insulating crystals, is the interaction caused by a deformation field, which at distances of $R \gg a$ behaves as

$$|\mathscr{U}(R_0)| = \mathscr{U}_0\left(\frac{a}{R}\right)^3. \quad (3.97)$$

If we adopt this expression, then

$$\sigma_{eff} \sim \pi a^2 \left(\frac{\mathscr{U}_0}{\Delta}\right)^{2/3}. \quad (3.98)$$

The result for $\Omega \ll \Omega_{im}$ is (Kagan and Klinger 1974)

$$D \propto \frac{\Delta^{5/3}}{x}. \quad (3.99)$$

Because of the inequality $R_0 \gg a$ the spheres of radius R_0 begin to overlap even at low defect concentrations. In the neglect of inelastic processes, with such an overlap a localization phenomenon arises and the coherent band motion is confined to a finite region. The corresponding critical concentration is given by

$$x_0 = \nu\left(\frac{\Delta}{\mathcal{U}_0}\right), \tag{3.100}$$

where ν is a numerical coefficient characteristic of the percolation problem (e.g., Shklovsky and Efros 1979). Near x_0 the diffusion coefficient will behave as

$$D \propto (x - x_0)^{\sigma}, \tag{3.101}$$

in which the critical index σ, by virtue of $R_0 \gg a$, will be close to the classical value.

With account taken of the inelastic processes the result becomes sensitive to the sign of the interaction $\mathcal{U}(R)$. If it corresponds to repulsion, then with $x < x_0$ all the results remain valid. The critical picture actually holds in the attraction potential as well (and, hence, with an alternating interaction with respect to the angles, which is characteristic of a crystal), if the relation $x_0 \exp(|\varepsilon_{\mathrm{d}}|/T) \ll 1$ is valid. In the opposite limiting case the diffusion coefficient assumes an activation temperature dependence with an activation energy close to ε_{d}.

3.6. *Hopping diffusion at high temperatures*

Let us consider the region of high temperatures where it may be assumed that $\bar{\Psi} \gg 1$. With a large polaron effect or a large scale of barrier fluctuations, this equality holds already at relatively moderate temperatures. In the integral (3.30) the region of small t is important in this case. Assuming $T > \Theta_{\mathrm{D}}/2$ and expanding the exponent up to quadratic terms, we find

$$W = J_0^2 \exp\left\{2B_0 - \frac{E}{T} + \frac{T}{E_{\mathrm{b}}}\right\} \int_{-\infty}^{+\infty} \mathrm{d}t \exp\{-4(E+\gamma)Tt^2 - \mathrm{i}\zeta_{\mathrm{b}} Tt\}. \tag{3.102}$$

Here

$$E = \frac{1}{4}\sum_{\alpha} \frac{|C_\alpha|^2}{\omega_\alpha}, \qquad E_{\mathrm{b}}^{-1} = 4\sum_{\alpha} \frac{|B_\alpha|^2}{\omega_\alpha \omega_0^2},$$

$$\zeta_{\mathrm{b}} = 4\sum_{\alpha} \frac{C_\alpha B_{-\alpha}}{\omega_\alpha \omega_0}, \qquad \gamma = \frac{1}{4}\sum_{\alpha} \omega_\alpha \frac{|B_\alpha|^2}{\omega_0^2}. \tag{3.103}$$

Bearing in mind the weakness of two-phonon coupling, we have omitted $\chi_2(0)$ in the exponent of eq. (3.30). After integration we finally have

$$D = \frac{za^2}{6} J_0^2 \mathrm{e}^{2B_0} \sqrt{\frac{\pi}{4(E+\gamma)T}} \exp\left[-\frac{E}{T} + \frac{T}{E_{\mathrm{b}}} - \frac{(\zeta_{\mathrm{b}})^2}{16(E+\gamma)} T\right]. \tag{3.104}$$

Let us show that the overall term linear in T in the exponent has a positive coefficient. To do this, we make use of the well-known Schwartz inequality. Then,

$$\frac{(\zeta_b)^2}{16} \leqslant \sum_\alpha \frac{|C_\alpha|^2}{\omega_\alpha} \sum_\alpha \frac{|B_\alpha|^2}{\omega_\alpha \omega_0^2} = \frac{E}{E_b}. \tag{3.105}$$

The answer is then obvious:

$$\frac{T}{E_b} - \frac{(\zeta_b)^2}{16(E+\gamma)} T \geqslant \frac{T}{E_b} \frac{\gamma}{E+\gamma}. \tag{3.106}$$

Taking this result into account, we see that as the temperature increases, the transition probability increases exponentially due to the polaron effect and barrier fluctuations.

The temperature dependence in these two cases is different. If the determining factor is the polaron effect, we obtain the activation dependence well known in the theory of small polarons (e.g., Appel 1968). If the fluctuational preparation of the barrier prevails, then the exponent increases with temperature by a linear law. From a formal viewpoint, with a sufficiently large value of the ratio T/Θ_D the leading role will always be played by the fluctuation mechanism.

Note that in the general case a drastic increase of the quantum diffusion coefficient with temperature is an alternative of vacancy diffusion.

If the diffusing particle has several levels in a single well, then when we depart from the low-temperature region these levels begin to be populated with a corresponding Gibbs factor. Here, a question immediately arises as to the competition between the levels since an increase in the transition probability can, in principle, compensate for a decrease in the Gibbs factor. The optimal path in this case will be associated with the extremum (minimum value) of the overall exponent:

$$2S(\varepsilon) + \frac{\varepsilon}{T}. \tag{3.107}$$

The value of $S(\varepsilon)$ is sensitive to the shape of the barrier and, therefore, strictly speaking the question has no unique answer. However, a quasiclassical analysis of simple model barriers has led to an interesting conclusion: up to a certain temperature T_* quantum diffusion occurs at the lowest level. At higher temperatures, the quasiclassical overbarrier motion become discontinuously more favorable (Lifshitz and Kagan 1972, Kagan and Klinger 1974). Since within the quasiclassical approximation $-\partial S/\partial\varepsilon \approx \pi/\omega_0$, the extremum [eq. (3.107)] determines the characteristic value

$$T_* = \frac{\omega_0}{2\pi}. \tag{3.108}$$

Compare with the discussion in this volume (chapters 3 and 4) [cf. eq. (2.1)]. When $T \gtrsim T_*$, there arises an activation temperature dependence of the diffusion coefficient:

$$D = D_0 \exp\left[-\frac{U_0}{T}\right]. \tag{3.109}$$

Here D_0 is a quantity which characterizes the diffusion coefficient at an energy close to the top of the barrier (Kramers 1940) and U_0 is the barrier height with respect to the lowest level in the well. We shall not dwell here on the special behavior of classical overbarrier diffusion, referring the reader for details, e.g., to Alefeld and Völkl (1978), Mel'nikov (1984).

If $T_* > \alpha\Theta_D$ [$\alpha < 1$; with $\phi(0)$ or $G(0) \gg 1$ the coefficient α is much less than unity], then with rise of temperature there first sets in an exponential quantum diffusion regime [eq. (3.104)] and only then the classical overbarrier diffusion regime is realized. The two cases can be distinguished by the behavior of the pre-exponential term. Indeed, in the case of classical diffusion the inverse lifetime in the unit cell invariably has a pre-exponential term of the order of the particle vibration frequency $\sim \omega_0$ in the well. But in the case of the regime [eq. (3.104)], the pre-exponential term depends on the tunneling transition amplitude and, therefore, it may be much lower. The first manifestation of this has been observed for muon diffusion [see the pioneer work on the diffusion of μ^+-muon in copper by Gurevich et al. (1972)], which is not accidental since the small mass of muon gives rise to a large value of T_*. Of the recent works reporting an analogous result, mention should be made of the very interesting investigations of muonium diffusion in KCl (Kiefl et al. 1989) and in NaCl (Kadono et al. 1990) (see section 4.1).

3.7. *A general picture of quantum diffusion in an insulator*

The results obtained in the previous sections enable us to construct a general picture of the temperature dependence of the quantum diffusion in a crystal. In a pure crystal the dependence $D(T)$ has a form shown in fig. 4. When $T < T'$, where T' is determined by the condition $\Omega(T') = \mu\Delta_c$, band coherent diffusion occurs with a mean free path $l > a$. Here, with T lower than the bandwidth Δ, the diffusion coefficient behaves as $D(T) \propto T^{-7}$ [see eq. (3.90)]. When $T > \Delta$, the scattering regime is changed and the diffusion coefficient behaves as $D(T) \propto T^{-9}$ or retains the dependence T^{-7} in the absence of the transport effect.

For $T \sim T'$ the tunneling regime is drastically changed – it becomes purely incoherent when $T > T'$. However, the motion is of the quasiband nature and the dependence $D(T)$ continuously passes through this region, retaining the previous temperature dependence. When $T > \Delta$, during the band or quasiband motion there arises a contribution from stimulated phonon diffusion, which

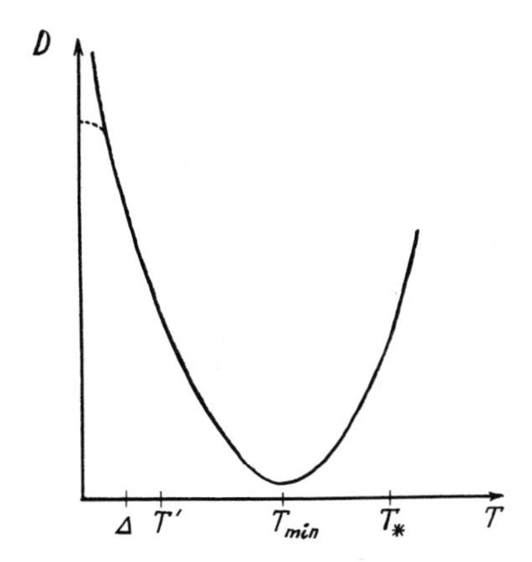

Fig. 4. Temperature dependence of the diffusion coefficient $D(T)$ in a perfect crystal. The dashed line corresponds to some low defect concentration.

increases with increasing temperature. However, over a wide low-temperature region this contribution is insignificant.

The change of sign of the $D(T)$ derivative at point $T_{\min}$ is associated either with the enhanced role of stimulated quantum diffusion [eq. (3.104)] or with a direct crossover to the classical diffusion regime [eq. (3.109)] (fig. 4 shows the former case). If the role of barrier fluctuations is insignificant, the polaron effect may cause a more drastic decrease of the left branch of the curve with temperature and a lower value of temperature $T_{\min}$. On the contrary, when the role of barrier fluctuations is dominant, the decrease of $D(T)$ with temperature is slowed down since $\tilde{\Delta}_0(T)$ increases. At rather high temperatures, $D(T)$ will always be associated with classical overbarrier diffusion (the region of $T > T_*$).

For low defect concentrations the defects influence only the band diffusion regime and impose a limitation on $D(T)$ with T tending to zero.

For high defect concentrations, which lead to localization in the absence of inelastic processes, the temperature dependence of the diffusion coefficient undergoes a substantial change (see fig. 5). With $T \to 0$ there inevitably exists a region of an exponential temperature dependence of D. With relatively weak level shifts, which though exceed but remain within the scale of Δ, this temperature region is relatively narrow. As T increases the exponential dependence is replaced by a linear dependence $D(T) \propto T$ associated with one-phonon processes (region 2). At higher temperatures (region 3), two-phonon processes become dominant and $D(T) \propto T^{7+(2)}$. When the value of Ω becomes larger than the characteristic level shifts, $D(T)$ returns to the ideal curve with an inverse temperature dependence, $D(T) \propto 1/T^{7+(2)}$.

The picture outlined above occurs in the presence of attraction to the defects at a relatively low binding energy, when the condition $x \exp(\varepsilon_d/T) < 1$ is achieved at really low temperatures. Otherwise, this inequality itself determines

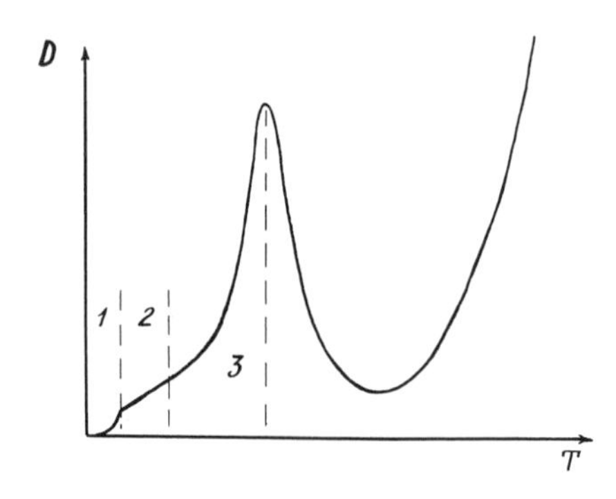

Fig. 5. Temperature dependence of the diffusion coefficient $D(T)$ in a crystal with defects. Regions 1, 2, 3 are explained in the text.

the temperature below which the exponential dependence is valid. At large ε_d, this temperature may "cover" partly or completely region 2 or even region 3.

3.8. Localization in a system of interacting particles diffusing in a perfect crystal

In dealing with quantum diffusion in a perfect crystal we are faced with the fact that, even at a small particle concentration x_p, the average distance between the particles becomes smaller than R_0, where R_0 is specified by the condition (3.96) if $\mathscr{U}(\boldsymbol{R})$ is the pair interaction potential between the particles. Here we find ourselves in a region of strong interaction in a subsystem of diffusing particles. With $x_p > x_0$ [eq. (3.100)] an individual particle appears to be in a potential, which corresponds to localization in terms of single-particle language. The random field, however, is now dynamic rather than static. This is associated with the occurrence of two-particle or many-particle excitations responsible for an energy diffusion of particles and a local change in the potential relief. As a result, with $x_p > x_0$, instead of localization, there occurs a crossover from the gas to the liquid regime of motion.

However, as has been predicted theoretically (Kagan 1981, Kagan and Maksimov 1983b, 1984) and observed experimentally (Mikheev et al. 1982, 1983a, b) (see section 4.2), in the neglect of the interaction with phonons, a self-localization phenomenon must occur in this liquid with further increase of the concentration. The corresponding characteristic value of particle concentration x_{00} is determined from the condition that the average scale of level shifts in neighboring wells exceeds the bandwidth:

$$a|\nabla\mathscr{U}(\bar{R} = R_{00})| = \zeta z \Delta_c, \tag{3.110}$$

where $\bar{R}$ is the average distance between the particles. The value of the coefficient $\zeta > 1$ depends on the lattice type. It is assumed that $T \ll \Theta_D$, but at the same time the temperature is sufficiently high in order to avoid the formation of bound pairs in the case of particle attraction (see section 3.4). In any case, presuming a random particle distribution, we have to assume that $T > \mathscr{U}(\bar{R})$. The condition $T > \Delta$ automatically means that the particles obey classical statistics.

The onset of self-localization is largely associated with the discreteness of space imposed by the periodic structure of the crystalline matrix. The discreteness gives rise to a number of important features that distinguish the system in question from a homogeneous medium:

(1) The occurrence of Umklapp processes and, as a consequence, the impossibility of the drifting state.

(2) Resonance tunneling occurs only to a finite distance a.

(3) There exist a finite number of discrete directions for resonance tunneling associated with the number of equivalent sites in the nearest coordination sphere.

(4) The kinetic energy of particles is limited by the bandwidth scale.

In a homogeneous liquid the diffusing particle can be displaced in a resonant manner in an arbitrary direction to an arbitrarily small distance with the environment being adiabatically readjusted. The fundamental difference between the two cases can be traced for the He^3/He^4 system, where the effective mass of He^3 atoms in liquid and solid solutions with close densities (the effective mass in the solid phase is easily found if the bandwidth is known) differs by four orders of magnitude and the diffusion coefficient, by eight orders. To do this, it is sufficient to compare the results obtained by Richards et al. (1976), Mikheev et al. (1977) and Eselson et al. (1973).

When the concentration $x_p \sim x_{00}$, a strongly pronounced tendency is observed towards the formation of particle clusters, in which the distribution of level shifts leads, due to the interaction of particles, to suppression of the single-particle motion. Statistically, the many-particle motions in the cluster are also suppressed in this case. In order to be sure that this is actually the case, we are to find the amplitude of two-particle motion. Treating the kinetic energy associated with translational transport as a perturbation, we have

$$\Delta_2 = \frac{\Delta_c^2}{\delta\varepsilon_1} + \frac{\Delta_c^2}{\delta\varepsilon_2} = \frac{\Delta_c^2}{\delta\varepsilon_1\,\delta\varepsilon_2}\left[\varepsilon_{fi} - (\boldsymbol{a}_1\cdot\boldsymbol{\nabla}_1)(\boldsymbol{a}_2\cdot\boldsymbol{\nabla}_2)\mathscr{U}(\boldsymbol{R}_1 - \boldsymbol{R}_2)\right]. \tag{3.111}$$

Here $\delta\varepsilon_s$ is the change of the cluster energy with displacement of the sth particle to a_s, ε_{fi} is the difference between the initial and final cluster energies. The second term in parentheses contains, as compared to $\delta\varepsilon_s$, an additional small parameter a/R ($R = |\boldsymbol{R}_1 - \boldsymbol{R}_2|$) and in a general case is smaller than ε_{fi}. Therefore,

$$\left|\frac{\Delta_2}{\varepsilon_{fi}}\right| \sim \left(\frac{\Delta_c}{\delta\varepsilon}\right)^2 \ll 1. \tag{3.112}$$

An accidental realization of the resonance situation $|\varepsilon_{fi}| < |\Delta_2|$ required for a real transport of two particles simultaneously is much less probable than in the case of single-particle transitions.

Analogously, we can find an expression for the k-particle excitation amplitude in the cluster and make certain that the probability of such excitation falls off progressively with increasing k (Kagan and Maksimov 1984).

An isolated cluster consisting of n particles can move as a whole. However, the amplitude of such a transfer with a shift of the center of gravity to the interatomic distance falls off exponentially with increasing n (Kagan and Maksimov 1984):

$$\Delta_n \propto \exp[-n \ln(ez)] \tag{3.113}$$

If the particle, while being within a sphere of radius R_{00} [eq. (3.110)] near another particle, loses the ability to move, then at a concentration x_∞ of the same scale as x_{00} there appears an infinite cluster (e.g., Shklovsky and Efros 1979), which as a whole undergoes no translational motion and in which the internal motions are suppressed. With $x > x_\infty$ the system of particles in an ideal crystalline matrix breaks up into two subsystems: an immobile subsystem incorporating particles that belong to the infinite cluster, and a mobile subsystem consisting of separate particles and small clusters, which retain the ability to move. Because of the limited kinetic energy of mobile particles their interaction with the immobile cluster cannot induce the excitation of the latter. Therefore, the immobile cluster behaves, with respect to mobile particles, as a static defect entity. It immediately becomes clear that at a certain concentration $x_c > x_\infty$ the mobile particles become deprived of the ability to move off to infinity and that, neglecting the phonon coupling, complete localization occurs in the system. The analysis carried out by Kagan and Maksimov (1984) has shown that x_c lies relatively close to x_{00}.

In considering the overall concentration dependence $D(x_p)$ in a perfect crystal we are, in fact, dealing with three characteristic regions. When $x_p < x_0$, the motion of particles is of the band nature. As a result of the occurrence of strongly pronounced Umklapp processes ($T > \Delta$), the collision frequency $\Omega_p(x_p)$ between the particles does not differ from Ω_{im} [eq. (3.95)]. Here, if $\Omega_p \gg \Omega(T)$ in eq. (3.94), then

$$D \propto 1/x_p. \tag{3.114}$$

When $x_p \gtrsim x_0$, the transition to the quasiliquid motion regime alters the concentration dependence as compared to eq. (3.114). The corresponding law can easily be established on the basis of the following simple reasonings. The mean free path for the particle in an instantaneous potential relief is given by the condition

$$l|\nabla \mathscr{U}|_{R=\bar{R}} = \Delta \tag{3.115}$$

(the change in potential energy equals the maximum kinetic energy). At a distance of order l, the particle is displaced in time determined by the quasiband motion:

$$\tau = \Omega_{\mathrm{p}}^{-1} \approx l/v_{\mathrm{g}}, \quad v_{\mathrm{g}} \approx a\mu\Delta_{\mathrm{c}}. \tag{3.116}$$

It is easy to see that the same time is characteristic for the variation of the potential on the scale of l due to the motion of other particles and, hence, due to the shift of the band to a scale of just Δ. As a result, for the diffusion coefficient we have approximately

$$D(x_{\mathrm{p}}) \approx \frac{1}{3} l^2/\tau \sim za^2 \frac{\Delta_{\mathrm{c}}^2}{\Omega_{\mathrm{p}}}. \tag{3.117}$$

Taking into account that in the case at hand, $\Omega_{\mathrm{p}} \sim a|\nabla \mathscr{U}|_{R=\bar{R}}$, and using the law $\mathscr{U}(R) \propto R^{-\alpha}$, we have

$$D \propto x_{\mathrm{p}}^{-(\alpha+1)/3}. \tag{3.118}$$

The possibility of appearance of such a dependence for the interaction through a deformation field ($\alpha = 3$) has been discussed by Landesman (1975) and Andreev (1975).

The third region is the one of critical behavior of $D(x_{\mathrm{p}})$ near x_{c}. Expression (3.117) is now replaced by the relation

$$D(x_{\mathrm{p}}) \approx \frac{za^2}{3} \frac{\Delta_{\mathrm{c}}^2}{\Omega_{\mathrm{p}}} Q(x_{\mathrm{p}}), \quad Q(x_{\mathrm{p}}) = A(x_{\mathrm{p}}) \left(\frac{x_{\mathrm{c}} - x_{\mathrm{p}}}{x_{\mathrm{c}}} \right)^{\sigma}, \tag{3.119}$$

where $A(x_{\mathrm{p}})$ is a smooth function of x_{p} tending to unity with $x \ll x_{\mathrm{c}}$. The quantity $Q(x)$ characterizes the fraction of particles not belonging to the infinite cluster and still capable of moving off to a macroscopic distance. The critical behavior of $D(x)$, just as with the single-particle localization due to defects [eq. (3.101)], must be close to the classical behavior predicted by percolation theory. In this particular case it is tied up with two circumstances. On the one hand, the particle wavelength is small compared to the size of the region of effective interaction R_0 [eq. (3.96)]. On the other hand, the inelastic nature of the interaction of mobile particles is preserved, leading to the destruction of the phase correlations.

One remark should be made at this point. In a recent work Burin et al. (1989) has considered the energy diffusion in an ensemble of two-level systems with account taken of four-particle excitations. If we use the results obtained by these authors, we may come to the conclusion that these excitations play no role with $\alpha + 2 \geqslant 6$, leaving the localization picture unchanged. At $\alpha = 3$, the possibility that in the immediate vicinity of x_{c} the critical behavior will be replaced by a very steep power dependence is not excluded. On the $D(x_{\mathrm{p}})$ curve this would lead to the appearance of a weak tail shown in fig. 6 by a dashed line.

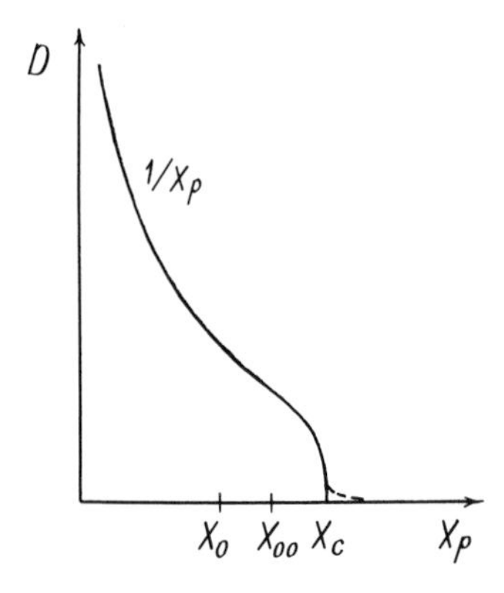

Fig. 6. The dependence of the diffusion coefficient D on the particle concentration x_p; x_0 is the gas–liquid crossover concentration, x_c is the critical concentration for the particle self-localization.

Strictly speaking, the interaction with phonons at a finite temperature removes the localization when $x > x_c$. The local motion of particles belonging to an immobile cluster is governed by the probability [eq. (3.55)]. A specific feature of the problem in question is that we simultaneously follow the motion of all particles, each moving in a potential varying with time. Physically it is clear that in this case the macroscopic diffusion coefficient is determined by a value of W averaged over all possible level-shift configurations. If two-phonon processes are dominant (and $\Omega < \xi$), the problem simply reduces to finding the value of $\langle 1/\xi^2 \rangle$.

The particles that do not belong to the immobile cluster and whose number is small when $x > x_c$ cannot move to a macroscopic distance without joining the immobile cluster on the way due to phonons. Therefore, taking them into account affects little the averaging result.

3.9. *Diffusion kinetics in processes of particle recombination and trapping*

A wide class of phenomena presuppose the approach of particles diffusing in a crystal to the interatomic distance. We may speak of the recombination of atomic particles and, in general, of a chemical reaction in the solid phase or else of the formation of bound pairs, in particular, of the trapping of particles by defect centers. The basic feature of such problems at low temperatures is the inevitable passage of the particle through a region where the levels in neighboring wells are shifted appreciably relative to one another. Such level shifting cannot be avoided since it is associated with the interaction between particles or between a particle and a defect, which may be either direct or indirect (via a deformation field). Beginning with a relatively large distance R_0

[see eq. (3.96)], the particles can approach one another only due to the coupling with excitations of the medium. At a rather low T, the tunneling motion in the region of $R < R_0$ is very slow and when $x(R_0/a)^3 < 1$ it is exactly this process which, as a rule, gives the characteristic reaction time.

If we assume that $R_0 \gg a$ and, hence, that the change of the level energy is small at the displacement to the interatomic distance, then eq. (3.57) may be transformed to a differential form:

$$\frac{\partial f}{\partial t} + \operatorname{div} \boldsymbol{j} = 0, \quad j_\alpha = -D_{\alpha\beta}\frac{\partial f}{\partial R} + v_\alpha f,$$

with a local diffusion tensor

$$D_{\alpha\beta}(\boldsymbol{R}) = \tfrac{1}{2}\sum_g (a_g)^\alpha (a_g)^\beta W_{i,i+g}$$

and a "hydrodynamic" velocity of motion

$$v_\alpha(\boldsymbol{R}) = -\frac{D_{\alpha\beta}(\boldsymbol{R})}{T}\frac{\partial \mathscr{U}(\boldsymbol{R})}{\partial R^\beta}.$$

If we adopt, for simplicity, the spherical symmetry of the problem, then, using a quasistationary solution with the boundary condition $f(a) = 0$, we have for the reaction rate, K,

$$K \approx \frac{4\pi n}{\displaystyle\int_a^\infty \frac{dR}{R^2 D(R)} \exp(\mathscr{U}(R)/T)}. \tag{3.120}$$

If the reaction is associated with the mutual diffusion of identical particles, then here and henceforth it is necessary to make the substitution $D \to 2D$. We will assume that attraction occurs between the particles. If the interaction via the deformation field is dominant, then, as is known, in a crystal there are always angular sectors, to which there corresponds an attraction. In the case under consideration, presuming that $D(R)$ is a monotonic function which does not decrease with increasing R, eq. (3.120) yields the universal relation

$$K \approx 4\pi c_1 R_t n D(R_t), \tag{3.121}$$

where R_t is given by the condition

$$|\mathscr{U}(R_t)| = T \tag{3.122}$$

and plays the role of the effective reaction radius. The numerical coefficient c_1 depends on the interaction law $\mathscr{U}(R)$. If formally R_t is of the order of a, then in formula (3.121) R_t should be replaced by a.

In the absence of the transport effect this expression is valid in the general case (see curve b in fig. 3). With $\xi(R_t) > \xi_*(T)$, which is a priori valid at sufficiently low temperatures [see the definition (3.54)], the diffusion coefficient in relation

(3.121) depends linearly on T:

$$D(R_t) \propto T. \tag{3.123}$$

If the transport effect is operative, the diffusion coefficient varies with ξ sharply and nonmonotonically (see curve a in fig. 3) and if the above condition, $\xi(R_t) > \xi_*(T)$, is fulfilled, the bottleneck effect occurs (Kagan et al. 1982). It is easy to see that this kinetic regime will always set in with decrease of temperature.

Let us denote by R_* the distance at which the two-phonon regime is replaced by the one-phonon regime, whose value is found from the condition $\xi(R_*) \approx \xi_*(T)$. Then, taking into account eqs. (3.54) and (3.56), we obtain

$$K \approx 4\pi c_2 R_* n D(R_*),$$

$$D(R_*) = \frac{za^2}{3} \frac{\tilde{\Delta}_0^2 T}{\Theta_D^2} \pi\lambda_1 \left(\frac{\xi_*(T)}{\Theta_D}\right)^2 \propto T^5. \tag{3.124}$$

The role of the effective reaction radius is now played by the quantity $R_*(T)$.

4. Experimental study of quantum diffusion in an insulator

4.1. Quantum diffusion of muonium in KCl and NaCl crystals

We begin this chapter with an analysis of the recent results obtained in a study of the tunneling motion of muonium in KCl and NaCl crystals (Kiefl et al. 1989, Kadono et al. 1990). In these experiments the authors were concerned with a direct determination of the inverse time of escape from the unit cell. The experimental findings on the dependence $\tau^{-1}(T)$ over a wide temperature range are given in figs. 7 and 8. What immediately attracts attention is that these results are in full qualitative agreement with the quantum diffusion picture that has been revealed in the previous chapter. The most striking result is undoubtedly the increase of τ^{-1} by 2.5–3 orders of magnitude with decreasing temperature, which replaces the usual exponential decay at higher temperatures. This is a striking manifestation of quantum diffusion. The experimental data allow us to state that the high-temperature course of the curves is also associated with the subbarrier motion of muonium. It immediately follows from the fact that the pre-exponential term in the corresponding activation dependence $\tau^{-1}(T)$ is less by three orders of magnitude than the frequency $\omega_0/2\pi \sim 10^{13}\,\mathrm{s}^{-1}$. The wide temperature range and the scale of variation of $D(T)$ make it possible to perform a general quantitative comparison of the theoretical and experimental data.

Let us begin with the low-temperature region $T < T_{\min}$. According to the results of section 3.3, two-phonon coupling prevails in this region, while the

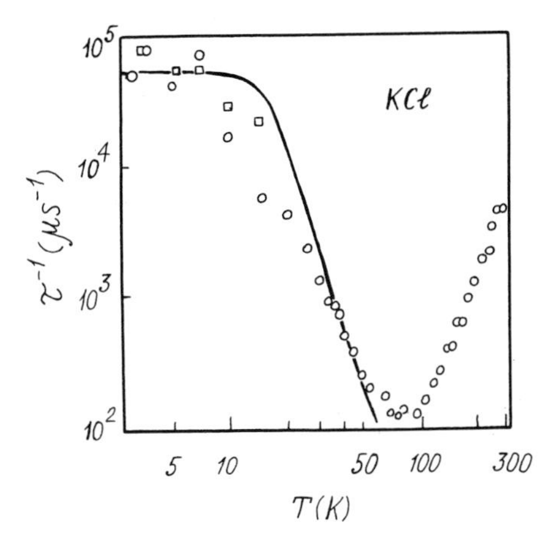

Fig. 7. Temperature dependence of the inverse correlation time for muonium in KCl. The experimental points are: □ Kiefl et al. (1989); ○ Kadono et al. (1990). The fitting parameters for the theoretical curve are $\Delta_c = 0.11$ K, $\lambda_2 = 0.11$.

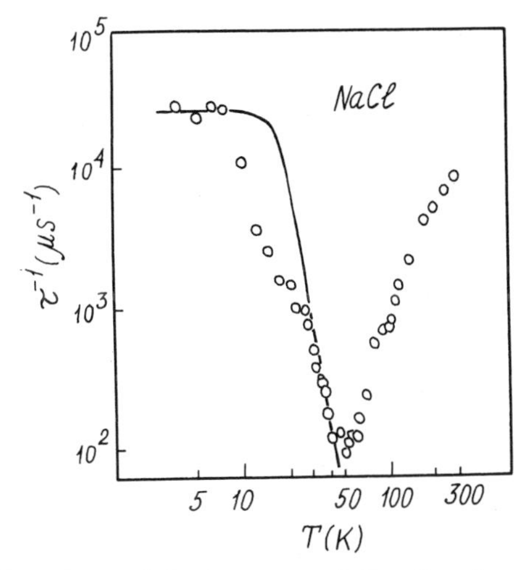

Fig. 8. Temperature dependence of the inverse correlation time for muonium in NaCl. The experimental points are after Kadono et al. (1990). The fitting parameters for the theoretical curve are $\Delta_c = 0.07$ K, $\lambda_2 = 0.25$.

one-phonon interaction enters only via the renormalized value of $\tilde{\Delta}_0(0)$ [eq. (3.11)]. This quantity can be determined directly via the limiting value of τ_0^{-1}, which is assumed by τ^{-1} when $T \to 0$.

Indeed, for the purely band motion of the particle the time of escape from the unit cell of a simple cubic lattice is given by (e.g., Petzinger 1982)

$$\tau_0^{-1} = 2\sqrt{2}\tilde{\Delta}_0(0). \tag{4.1}$$

From the data presented in figs. 7 and 8 it immediately follows that

$$\tilde{\Delta}_0(0) \approx \begin{cases} 0.13\,\mathrm{K} & (\mathrm{KCl}), \\ 0.07\,\mathrm{K} & (\mathrm{NaCl}). \end{cases} \tag{4.2}$$

We emphasize that with $T \gg \tilde{\Delta}_0$ it is exactly the time given by eq. (4.1) that determines the correlation time measured by experiment. When $\Omega > \tau_0^{-1}$, the tunneling regime changes to the incoherent regime; using eq. (3.36), we have

$$\tau^{-1} \approx \frac{4D(T)}{a^2} = \frac{8\tilde{\Delta}_0^2}{\Omega(T)}. \tag{4.3}$$

Here Ω has the value given in eq. (3.38) and the coefficient 4 is the outcome of direct numerical calculations for a simple cubic lattice.

The experimentally found temperature dependence in this region is considerably weaker than that predicted by the limiting law [eq. (3.37)]. This is, however, associated not with the loss of the dominant role by two-phonon processes but with the real structure of the phonon spectrum of these crystals (Kagan and Prokof'ev 1990). In order to make certain that this is so in reality, we may adopt the simplest form for the vertex of the two-phonon coupling in eq. (3.38) (in the KCl lattice the transport effect is absent because of the nonequivalency of the neighboring wells from the standpoint of the symmetry of the phonon coupling):

$$|C_{\alpha\beta}|^2 = \lambda_2 \omega_\alpha \omega_\beta \tag{4.4}$$

and make use of the experimentally determined phonon density function $g(\omega)$ (e.g., Bilz and Kress 1979). The transition from eq. (4.3) to eq. (4.1) is smooth and may be described by an interpolation formula:

$$\tau^{-1} = \frac{\tau_0^{-1}}{1 + \Omega\tau_0}. \tag{4.5}$$

The results of a theoretical calculation of τ^{-1} as a function of T at $\tilde{\Delta}_0 = \mathrm{const}$ are shown in figs. 7 and 8. Use was made of only one fitting parameter, λ_2, which, in fact, controls only the scale rather than the relative temperature course. It is seen that there is a good agreement between the experimental and theoretical results on the low-temperature branch immediately beyond T_{min}. The theoretical results lead to the dependence $D(T) \propto T^{-3.5}$ in KCl [Kiefl et al. (1989) report the experimental dependence $D(T) \propto T^{-3.3}$] and $D(T) \propto T^{-4.3}$ in NaCl. Note that the irregular character of the experimental data for NaCl is most likely associated with a change in the hyperfine interaction (Kadono et al. 1990). The nature of this phenomenon remains obscure, which is why in what follows we will restrict ourselves to a quantitative comparison of the theoretical and experimental data for KCl [see the discussions in Kagan and Prokof'ev (1990b)].

From the analysis given in section 3.6 it follows that one-phonon coupling becomes significant when $T \sim T_{\min}$ and decisive at large T. For this reason, while considering the region of $T \gtrsim T_{\min}$, we return to the general expression (3.30). Separating out from it the two-phonon part, to which there explicitly correspond relations (3.35) and (4.3) and retaining only one-phonon coupling in the region indicated, we get

$$\tau^{-1} \approx 4W = 4J_0^2 \int_{-\infty}^{\infty} \mathrm{d}t\, (\mathrm{e}^{\bar{\Psi}(t) - \mathrm{i}\Psi(t)} - 1). \tag{4.6}$$

The high-temperature asymptotics of this expression coincide with eqs. (3.103) and (3.104). Equations (4.5) and (4.6) describe the true inverse lifetime in the unit cell.

We first assume that barrier fluctuations play no role and put $B_\alpha = 0$ in eqs. (3.103) and (3.104). One-phonon coupling then is of the intrawell nature. The vertex of this interaction is chosen in a simple form:

$$|C_\alpha|^2 = \lambda_1 \omega_\alpha \omega_{\mathrm{m}}, \tag{4.7}$$

where ω_{m} is the cutoff frequency of the phonon spectrum.

If we use the value of activation energy found for KCl by Kiefl et al. (1989), $E = 390$ K, and the experimentally found phonon spectrum (e.g., Bilz and Kress 1979), then from eq. (3.103) we shall directly find the value of λ_1. Substituting eq. (4.7) into eq. (3.7), we determine the polaron exponent of $\varphi(T)$ and also the value of J_0 from eqs. (3.11) and (4.2). As a result, all the parameters in eq. (4.6) appear to have been determined and we can find $\tau^{-1}(T)$ for the entire temperature range. Having done so, we become convinced that there is complete disagreement between the theoretical and experimental results in the region of $T \sim T_{\min}$, where the difference reaches a few orders of magnitude.

Let us consider the opposite case, when the intrawell interaction may be neglected, i.e., we set $C_\alpha \approx 0$. The vertex of the interaction with barrier fluctuations in eq. (3.3) is chosen to be

$$\frac{|B_\alpha|^2}{\omega_0^2} = \lambda_{\mathrm{b}} \frac{\omega_\alpha}{\omega_{\mathrm{m}}} \tag{4.8}$$

The constant λ_{b} may again be defined from the high-temperature limit of eqs. (3.104) and (3.103), which now has the form

$$\tau_{\mathrm{c}}^{-1} = 2(J_0 \mathrm{e}^{B_0})^2 \sqrt{\frac{\pi}{\gamma T}}\, \mathrm{e}^{T/E_{\mathrm{b}}}. \tag{4.9}$$

Now $G(T)$ [eq. (3.8)], which determines the renormalization of J_0 [eq. (3.11)], is found unambiguously. Without introducing any additional parameters, we have found the entire temperature dependence $\tau^{-1}(T)$, which is shown in fig. 9 by the solid line. Here we have an excellent agreement between the experimental and

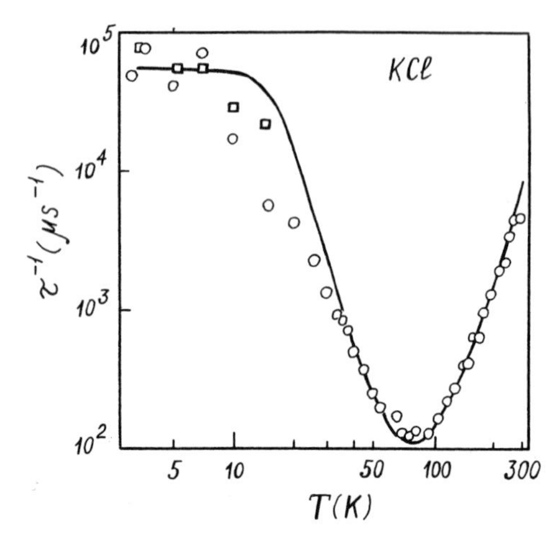

Fig. 9. Temperature dependence of the inverse correlation time for muonium in KCl. The experimental points are: □ Kiefl et al. (1989); ○ Kadono et al. (1990). The fitting parameters for the theoretical curve are $\Delta_c = 0.11$ K, $\lambda_2 = 0.11$, $\lambda_b = 0.25$.

theoretical results, including the position of the minimum and the quantity $\tau^{-1}(T_{\min})$ (Kagan and Prokof'ev 1990b). There is no doubt that in the case under discussion the interaction with barrier fluctuations plays a dominant role.

Note that the result obtained has a physical explanation. In an ionic crystal the interaction between a neutral muonium and ions depends strongly on their separation. This interaction reaches a maximum value when the tunneling particle crosses the atomic plane between the neighboring wells.

Comparison of theory with experiment leads to the value of $J_0 \approx 0.1$ K, to which there corresponds the value of the tunneling exponent $S \sim 5$–6. Under these conditions, going over from muonium to hydrogen leads to such a large value of S that J_0 becomes negligibly small. This does not enable us to observe the growth of the diffusion coefficient for hydrogen in KCl with decrease of temperature. On the other hand, at high T we can observe only the classical overbarrier diffusion. This explains why the activation energy for hydrogen (Ikeya et al. 1978) is much greater than for muonium.

4.2. Quantum diffusion and self-localization of He^3 atoms in a He^4 crystal

The study of quantum diffusion in a solid He^3–He^4 solution has made it possible to reveal experimentally the fundamental features characteristic of the tunneling motion in insulators. As a matter of fact, this was the first system in which long-range quantum diffusion was detected. Investigations in this area were mostly initiated by the theoretical work of Andreev and Lifshitz (1969) and also by the works of Guyer and Zane (1969), Guyer et al. (1971), in which the authors introduced the concept of the band motion of defects in quantum crystals (called defectons).

The He^3–He^4 system has been found to be uniquely adequate with respect to the parameters for the study of both quantum diffusion and self-localization phenomena. Indeed, the interaction between He^3 atoms when they are at the interatomic distance from one another is only $\mathscr{U}_0 \approx 10^{-2}$ K (Slyusarev et al. 1977). This means that, even at very low temperatures, not only the condition $T \gg \mathscr{U}(\bar{R})$ is realized but also $T \gg \mathscr{U}_0$. Since the interaction between He^3 atoms occurs through a deformation field, the latter inequality renders the picture insensitive towards the alternating nature of this interaction as a function of the direction in the crystal. On the other hand, the bandwidth scale $\Delta \sim 10^{-4}$ K (Richards et al. 1976, Mikheev et al. 1977). Here, the concentration x_{00} is on the order of a few percent and $x_c \ll 1$. Thus, the self-localization phenomenon and the assumption of the small value of critical concentration appear to be consistent with each other. With $x_p \sim x_{00}$ for the phase separation to be absent it is sufficient that $T > 0.2$ K. At such temperatures, phonon coupling is sufficiently weak and is determined by two-phonon processes.

The authors of early experiments (e.g., Richards et al. 1972, Grigoriev et al. 1973 and also Mikheev et al. 1977 and Allen et al. 1982) have clearly revealed that, despite the smallness of Δ, the motion of He^3 atoms is a band motion at a low concentration, the mean free path being determined by collisions between the particles. Figure 10 presents the results obtained by two experimental groups for the dependence $D(x_p)$ for $T < 0.8$ K. The curves given demonstrate the dependence $D \propto 1/x_p$, which is characteristic of the band motion [eq. (3.114)]. It

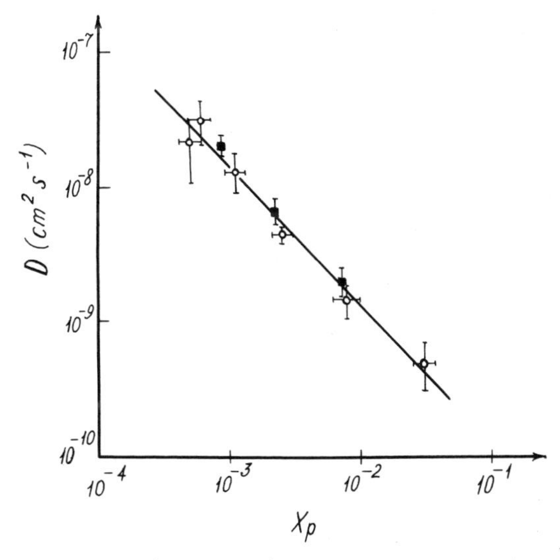

Fig. 10. Diffusion coefficient D of He^3 in solid He^4 versus particle concentration x_p; molar volume 21 cm^3; ○ Richards et al. (1972); ■ Grigoriev et al. (1973).

is interesting that in the same experiments there was detected a nonlinear behavior of the dependence of the diffusion coefficient on Δ, which according to eq. (3.98) is associated with the dependence of the scattering cross-section on R_0 [eq. (3.96)] and, hence, on the bandwidth scale.

The most clear-cut manifestation of the tunneling motion in the system considered has been the increase of the diffusion coefficient with decreasing temperature. According to eq. (3.94), this required lower concentrations. Such a temperature dependence was first found for $x_{He^3} < 10^{-4}$ in the work of the Kharkov group (Mikheev et al. 1977). The same result was obtained later by the Sussex group (Allen and Richards 1978, Allen et al. 1982). Figure 11 gives the results of these experiments. The analysis performed by the authors led to the conclusion that the behavior of the diffusion coefficient obeys relation (3.94), with the dependence $\Omega(T) \propto T^n$ ($n = 9 \mp 2$) being close to that predicted by Andreev and Lifshitz (1969).

The introduction of the concepts of self-localization and phonon-stimulated particle delocalization (Kagan 1981, Kagan and Maksimov 1983b, 1984) predetermined the second stage in the investigation of the quantum diffusion of He^3 in a He^4 crystal. The experimental studies carried out by the Kharkov group (Mikheev et al. 1982, 1983a, b) led to the detection of a very pronounced

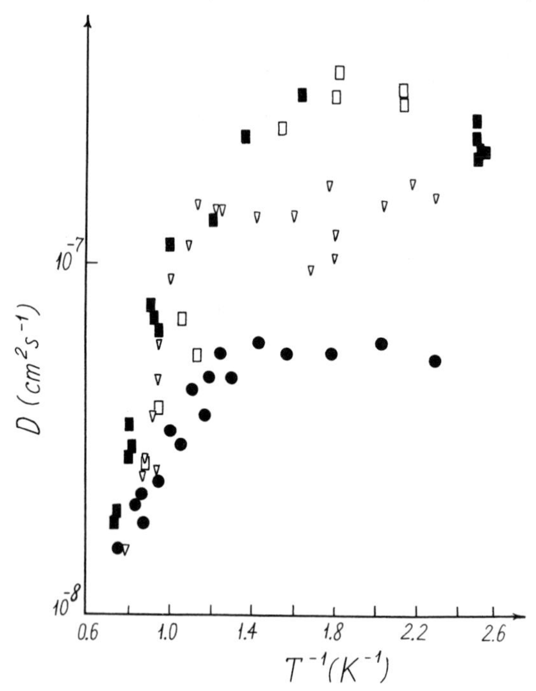

Fig. 11. Temperature dependence of the diffusion coefficient of He^3 in solid He^4: ■ $x = 6 \times 10^{-5}$ (Mikheev et al. 1977); □ $x = 10^{-4}$; ▽ $x = 2 \times 10^{-4}$; ● $x = 5 \times 10^{-4}$ (Allen et al. 1982). Molar volume 21 cm^3.

tendency towards localization even at a concentration of $x_p \sim 2\text{–}5 \times 10^{-2}$, depending upon the pressure. Figure 12 shows the behavior of the $D(x_p)$ curve at several densities. Perhaps, the most significant experiment was the one involving a single sample with a fixed concentration, where the dependence D versus Δ was obtained (Δ was governed by the crystal density) (see fig. 13). An increase in the pressure causes a decrease in Δ_c, which in accordance with eq. (3.110) must lead to a decrease in x_{00} and x_c.

The clear-cut critical picture allowed the authors to evaluate the critical index σ in eq. (3.119), which proved close to the value of $\sigma \approx 1.7$ correlating with the value characteristic of the percolation problem (see the discussion in section 3.8).

A measurement of the dependence $D(T)$ on samples with a minimal value of $D(x_p)$ (Mikheev et al. 1983a, b) revealed a crossover in the temperature

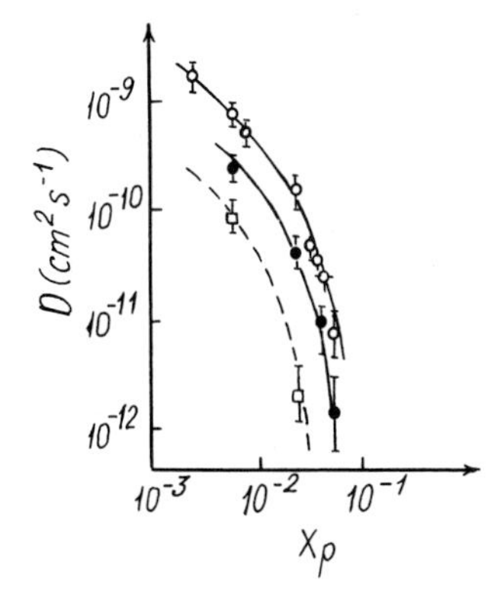

Fig. 12. Concentration dependence $D(x_p)$ of He^3 in solid He^4: ○ $V = 20.7\ cm^3\,mol^{-1}$; ● $V = 20.5\ cm^3\,mol^{-1}$; □ $V = 19.9\ cm^3\,mol^{-1}$ (Mikheev et al. 1983a).

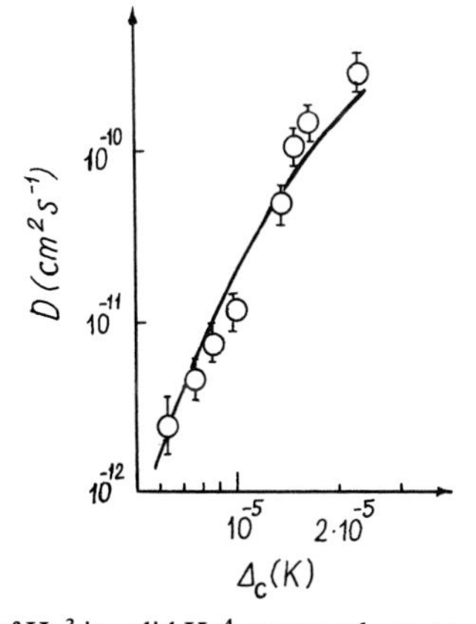

Fig. 13. Diffusion coefficient D of He^3 in solid He^4 versus coherent tunneling amplitude Δ_c (Mikheev et al. 1984).

dependence $D(T)$ (see fig. 14). The authors cited have now found the law $D \propto T^9$ characteristic of the removal of localization due to two-phonon processes [see eq. (3.51)].

Mikheev et al. (1983b) have studied the dependence $D(x_p, T)$ over a wide range of variation of the particle concentration from the purely band motion to the self-localization region. These findings are summarized in fig. 15. For a comparison with theory (the solid lines in fig. 15) use was made of an

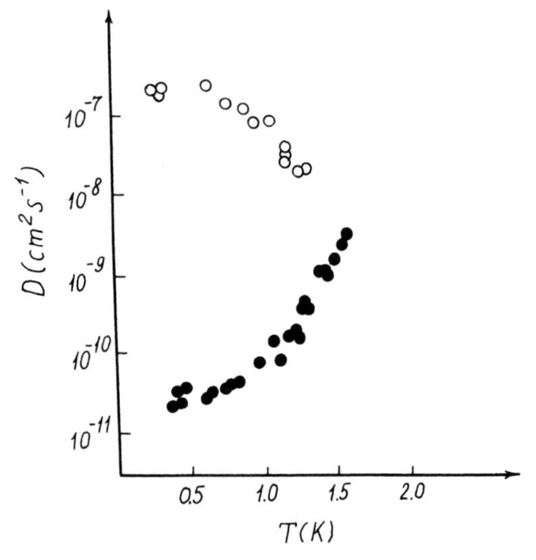

Fig. 14. Temperature dependence of the diffusion coefficient of He^3 in solid He^4: ○ $x_p = 6 \times 10^{-5}$, $V = 21.0\ cm^3\ mol^{-1}$; ● $x_p = 4 \times 10^{-2}$, $V = 20.7\ cm^3\ mol^{-1}$ (Mikheev et al. 1983b).

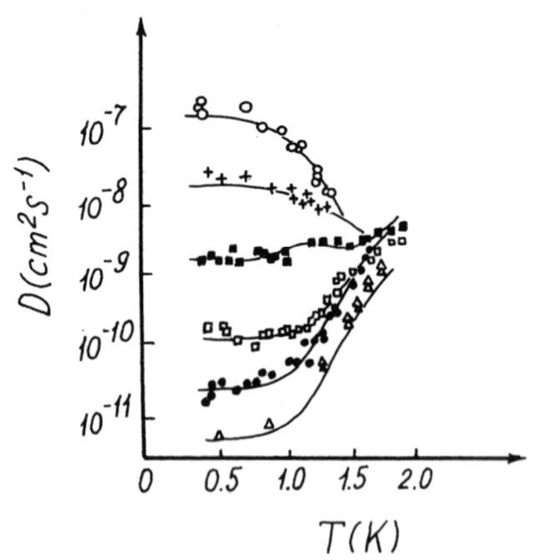

Fig. 15. Temperature dependence of the diffusion coefficient D for different concentrations of He^3: ○ $x_p = 6 \times 10^{-5}$, $V = 21.0\ cm^3\ mol^{-1}$; + $x_p = 5 \times 10^{-4}$, $V = 21.0\ cm^3\ mol^{-1}$; ■ $x_p = 2.5 \times 10^{-3}$, $V = 20.7\ cm^3\ mol^{-1}$; □ $x_p = 2.17 \times 10^{-2}$, $V = 20.7\ cm^3\ mol^{-1}$; ● $x_p = 4 \times 10^{-2}$, $V = 20.7\ cm^3\ mol^{-1}$; △ $x_p = 4.98 \times 10^{-2}$, $V = 20.5\ cm^3\ mol^{-1}$ (Mikheev et al. 1983b).

interpolation representation for $D(x_p, T)$ proposed by Kagan and Maksimov (1983b):

$$D(x_p, T) = \frac{za^2\Delta_c^2}{3}\left[\frac{Q(x_p)}{\Omega_p + \Omega(T)} + \frac{\Omega(T)(1 - Q(x_p))}{\xi^2(x_p) + \Omega^2(T)}\right]. \tag{4.10}$$

The first term in this expression is tied up with the transport of particles along the percolation paths [$Q(x)$ is defined by eq. (3.119)]. The quantity $\xi(x_p)$ was used, with account taken of $\alpha = 3$, in the universal form $\xi(x_p) = \bar{\xi}x_p^{4/3}$.

4.3. *Low-temperature recombination of atomic hydrogen in a H_2 crystal*

A striking example of low-temperature reaction in a crystal limited by the quantum diffusion of particles is the recombination of hydrogen atoms in a H_2 crystal. This phenomenon has been explored in a number of works (Katunin et al. 1981, 1982, Ivliev 1982, 1983; see also Miyazaki et al. 1984). These authors studied the recombination kinetics as a function of temperature at an initial atomic hydrogen concentration of the order of $x_H \sim 10^{-4}$. With such a concentration the recombination process required the diffusion of particles to relatively large distances. The reaction rate was determined from a decrease in the number of free electron spins using the EPR technique. The characteristic times were equal to 10^3–10^4 s.

Figure 16 shows the experimental dependence of the logarithm of the reaction rate constant $\tilde{K}(T)$. This constant was determined from the equation $\dot{n}_H = -2\tilde{K}(T)n_H^2$ and, hence, is connected with the quantity K [eq. (3.120)] by the relation $\tilde{K} = K/n_H$. The behavior of the temperature dependence $K(T)$ clearly reveals a crossover of the activation dependence to a weak power dependence at low T. The latter dependence is close to a linear one. This is evidence for the theory that the quantum diffusion corresponding to the

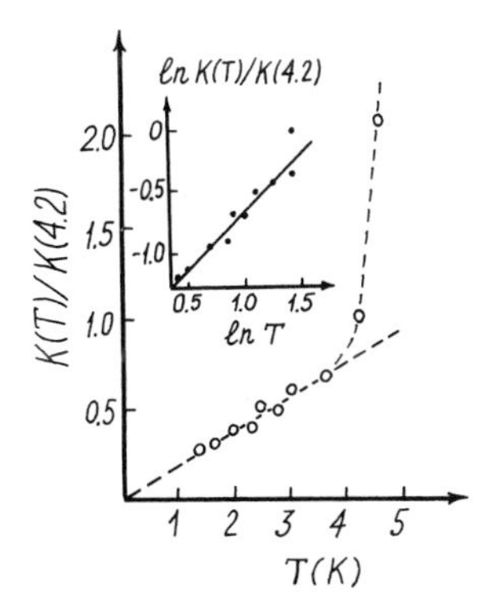

Fig. 16. Recombination rate $K(T)$ of H atoms in solid H_2 versus temperature (Ivliev et al. 1982).

tunneling motion of two atoms towards each other is governed by one-phonon processes.

As follows from the results of section 3.9, with $\xi(a) > \xi_*(T)$, since $R_t < R_*$ at sufficiently low temperatures, this mechanism predominates whenever the transport effect is absent. The latter condition is automatically satisfied if the hydrogen atom tunnels across the interstitials of the hexagonal close-packed lattice of the H_2 crystal. It is easy to assess from relation (3.54) that with $T < 4\,K$, where a power law holds, $\xi_*(T)$ is a priori lower than the quantity $\xi(a)$, which exceeds the value of 1 K even due to the direct interaction of two hydrogen atoms. At the same time, for the experimental interval $1.5\,K < T < 4\,K$ the radius R_t is close to a in scale. This keeps the temperature dependence $K(T)$ close to a linear one. Note that the linear temperature law also holds with the transport effect being operative if the characteristic level shifts ξ_d due to extraneous defects exceed $\xi_*(T)$. In this case, we should make the substitution $\xi_*(T) \to \xi_d$ in eq. (3.124).

The replacement of the activation regime of classical diffusion by a slower power law has also been detected for the recombination of atomic deuterium in a D_2 crystal (Iskovskih et al. 1986), although the crossover is less distinct because of the specific experimental difficulties.

5. *Quantum diffusion with electron coupling*

5.1. *Coherent tunneling*

In this chapter we shall consider the tunneling motion of a particle in a crystal interacting with conduction electrons. The general expression for the matrix element of the transition between two equivalent wells has the form shown in eq. (2.45) with the polaron operator Λ defined according to eqs. (2.49) and (2.52). In order to simplify the picture, we first neglect the barrier fluctuations. In section 5.6 we shall return to the analysis of the role of this effect.

The coherent transition amplitude, when tunneling occurs without excitation of the electron subsystem, is defined by

$$\Delta_c = J_0 \langle \Lambda \rangle. \tag{5.1}$$

The quantity $\langle \Lambda \rangle$ has been calculated in section 2 [see eqs. (2.53) and (2.58)]. With $T \to 0$, when Z assumes the value given in eq. (2.59),

$$\Delta_c = J_0 \mathrm{e}^{-b \ln(\omega_0 \tau)}. \tag{5.2}$$

As noted earlier, the parameter $\omega_0 \tau$ is the major parameter of the problem [see eq. (2.2)]. Since the dimensionless parameter b has no special smallness, this results in a strong narrowing of the coherent band due to the electronic polaron effect. At $T = 0$ in a perfect lattice τ and Δ_c are unambiguously interrelated by

$\tau^{-1} = \mu\Delta_c$ [as regards the quantity μ, see the remark following eq. (3.11)]. Therefore, eq. (5.2) provides a self-consistent solution for Δ_c, which with $b < 1$ has the form

$$\Delta_c(0) = J_0\left(\frac{\mu J_0}{\omega_0}\right)^{b/(1-b)}. \tag{5.3}$$

With $b \geqslant 1$, the solution of eq. (5.2) yields the following result:

$$\Delta_c(0) = 0, \tag{5.4}$$

i.e., it leads to a localization of the particle, despite the existence of translational symmetry. The results analogous to eqs. (5.3) and (5.4) have been obtained by a number of authors, who dealt with different model systems. The relevant references can be found in a review by Leggett et al. (1987). The authors of these works were concerned with a single-particle interaction with bosons, which is equivalent to the first term in eq. (2.62) and the simultaneous requirement in eq. (3.33). As we have seen, in the case of phonons this requirement is valid only in a one-dimensional crystal in the absence of the transport effect. It is essential that such an interaction does not lead to the renormalization of the vertex. For the interaction of a particle with a real electronic system [eq. (2.16)] the condition (3.33) is fulfilled automatically for electronic excitations near the Fermi surface (see below). It is exactly for this reason that our treatment, which is based on perturbation theory with respect to the particle–electron coupling $\mathscr{V}$ and which neglects rescattering, leads to the results (5.3) and (5.4).

In fact, eq. (5.2) retains its form with an arbitrary value of the vertex [eq. (2.16)] (Yamada et al. 1986), this being a consequence of the universal properties of the Fermi liquid. It is only the expression for the dimensionless parameter b that undergoes change, being now dependent on the scattering phases and the geometry of the two-well system. Thus, if the scattering of electrons on a particle is governed by a single orbital s-scattering channel (Yamada 1984, Yamada et al. 1986), then

$$b = \frac{2}{\pi^2}\left\{\tan^{-1}\left[\frac{\sqrt{1-j_0^2}\tan\delta_0}{\sqrt{1+j_0^2\tan^2\delta_0}}\right]\right\}^2, \tag{5.5}$$

where $j_0 = 1 - [\sin(k_F R)/k_F R]^2$. Analysis of this expression shows that $b \leqslant \frac{1}{2}$ and, hence, in this case no localization occurs at $T = 0$. The question of whether it is possible to satisfy the condition $b \geqslant 1$ in a real metal remains open so far. In all experimentally explored and model cases known up to the present time, b has never exceeded $\frac{1}{2}$.

The expression derived for Δ_c presupposes the absence of a gap in the electronic excitation spectrum near the Fermi level. However, if we consider a metal in the superconducting state and if the magnitude of the superconducting gap $\Delta_s > 1/\tau$, then since the minimum energy of electron–hole excitations is equal to $2\Delta_s$, the cutoff at frequency $1/\tau$ in the divergent integral [eq. (2.55)] is not important and

$$\Delta_c^s = J_0 e^{-b\ln(2\omega_0/e\Delta_s)} = J_0 \left(\frac{e\Delta_s}{2\omega_0}\right)^b. \tag{5.6}$$

The coefficient e/2 appears in place of 2 if account is taken of the density of states and coherence factors characteristic of a superconductor (see Kagan and Prokof'ev (1990a) and also section 5.5). The linear dependence of the coherent band on J_0 proves important to a number of phenomena, say the renormalization of sound in an amorphous metal (Kagan and Prokof'ev 1988). It is interesting that the transition to the superconducting state in a perfect lattice, in principle, eliminates the localization when $b \geqslant 1$ and in any case increases the coherent bandwidth if $\Delta_s \tau > 1$.

One significant circumstance deserves attention. If in the crystal the renormalization of the coherent amplitude Δ_c occurs simultaneously due to electron and phonon coupling, the nonlinear character of this renormalization in the electron case enhances the renormalization due to phonons. Indeed, here from the outset we can make the substitution $J_0 \to \tilde{\Delta}_0^{ph}$ [eq. (3.11)], which, via eq. (5.2), leads to the following relation:

$$\Delta_c(0) = \tilde{\Delta}_0^{ph} \left(\frac{\mu \tilde{\Delta}_0^{ph}}{\omega_0}\right)^{b/(1-b)}. \tag{5.7}$$

In contrast to eq. (3.11), we now have

$$\Delta_c(0) \sim [\tilde{\Delta}_0^{ph}]^{1/(1-b)}.$$

With $T \neq 0$ the amplitude $\Delta_c(T)$ is defined by eq. (5.1), the diagonal element of the operator Λ being specified by eqs. (2.53) and (2.58). At temperatures of $\pi T \tau > 1$ the exponent Z in eq. (2.53) has the value given by eq. (2.60), and we get

$$\Delta_c(T) = \tilde{\Delta}_0(T) e^{-(1/2)\Omega\tau}, \quad \tilde{\Delta}_0(T) = \tilde{\Delta}_0^{ph} \left(\frac{2\pi T}{\gamma \omega_0}\right)^b. \tag{5.8}$$

Here, in accordance with eq. (2.57),

$$\Omega \equiv \Omega(0) = 2\pi b T. \tag{5.9}$$

Just as with two-phonon processes, electron coupling leads, at a finite temperature, to an exponential decay of the coherent transition amplitude. This result was first obtained by Kagan and Prokof'ev (1987a, b). In contrast to phonon coupling, destruction of the coherent band occurs already at a temperature of $T > \mu \Delta_c(0)/2\pi b$ comparable with the bandwidth. This outcome is of fundamental importance to the tunneling motion in a metal. Naturally, all the remarks pertaining to the nature of the dynamic destruction of the coherent band given in section 3.1 remain valid to the full extent. This refers, in particular, to the statement that with $T > T' \approx \mu\tilde{\Delta}_0(T')/e\pi b$ the self-consistent solution of eq. (5.8) leads to $\Delta_c = 0$.

5.2. *Incoherent tunneling*

When $T \gg T'$, the tunneling motion between neighboring wells is of a purely incoherent nature. The same is true with the level shift $\xi \gg \mu\tilde{\Delta}_0$ at an arbitrary value of T. The quantum correlations are now destroyed at each step. In order to describe the diffusive motion in a crystal in this case, it will suffice to compute the transition probability from well 1 to well 2. The appropriate general expression has the form shown in eq. (3.14). Using the explicit expression for the electron polaron effect Λ, eqs. (2.49) and (2.52), and expanding the exponential function in eq. (2.52) with retention of the first three terms, we obtain for the average value of the product $\Lambda_{pp'}(t)\Lambda^{+}_{p'p}(0)$:

$$1 - \frac{|\Delta V_{pp'}|^2}{(\varepsilon_p - \varepsilon_{p'})^2 + 1/\tau^2}\,(n_p(1 - n_{p'})(1 - e^{i(\varepsilon_p - \varepsilon_{p'})t})$$
$$+ n_{p'}(1 - n_p)(1 - e^{-i(\varepsilon_p - \varepsilon_{p'})t})).$$

Inserting this result into eq. (3.14), we find

$$W = (\tilde{\Delta}_0^{\mathrm{ph}})^2 \int_{-\infty}^{\infty} \mathrm{d}t\, e^{i\xi t} e^{\chi(t)}, \tag{5.10}$$

$$\chi(t) = -2\sum_{pp'} \frac{|\Delta V_{pp'}|^2}{(\varepsilon_p - \varepsilon_{p'})^2 + 1/\tau^2}\, n_p(1 - n_{p'})(1 - e^{-i(\varepsilon_{p'} - \varepsilon_p)t}). \tag{5.11}$$

In this expression, just as in eq. (3.20), the characteristic energy of electron–hole pairs is of the order of $(\Omega, \xi)_{\max}$; by virtue of this, we may omit the quantity $1/\tau^2$ from the denominator.

The expression for χ can conveniently be rewritten in the universal form [cf. eq. (3.21)]

$$\chi(t) = -\int_0^{\omega_0} \frac{\mathrm{d}\omega}{\omega} f(\omega)\,[(1 - \cos(\omega t))\coth(\omega/2T) + i\sin(\omega t)]. \tag{5.12}$$

Here

$$f(\omega) = 2\sum_{pp'} \frac{|\Delta V_{pp'}|^2}{(\varepsilon_p - \varepsilon_{p'})}\,(n_{p'} - n_p)\,\delta(\varepsilon_p - \varepsilon_{p'} - \omega). \tag{5.13}$$

Direct calculation of eq. (5.13), using the definition (2.57), yields the following result:

$$f(\omega) = \mathrm{const} = 2b. \tag{5.14}$$

As we have already discussed (see section 3.3), such a behavior of the function $f(\omega)$ reflects the tunneling process in the ohmic dissipation regime and leads to the characteristic asymptotics of eq. (3.32) at long times.

Using a standard transformation analogous to the transition from eq. (3.21) to eq. (3.27), we can rewrite eq. (5.12) as follows:

$$\chi = -2b\left\{\int_0^{\omega_0}\frac{d\omega}{\omega}\tanh(\omega/4T) + \int_0^{\omega_0}\frac{d\omega}{\omega}\frac{1-\cos(\omega t)}{\sinh(\omega/2T)}\right\}.$$

Taking into account the condition $T \ll \omega_0$, we have

$$\chi(t) = -2b\left\{\ln\left(\frac{\gamma\omega_0}{2\pi T}\right) + \ln(\cosh(\pi T t))\right\}. \qquad (5.15)$$

The first term describes the polaron renormalization of the tunneling amplitude [eq. (5.8)] and leads in eq. (5.10) to the replacement $J_0 \to \tilde{\Delta}_0$. The second term in eq. (5.15) has the same structure as in the case of two-phonon coupling [eq. (3.29)]. Calculation of the transition probability, therefore, yields an expression analogous to eq. (3.49):

$$W(\xi, T) = 2\frac{\tilde{\Delta}_0^2(T)\Omega}{\xi^2+\Omega^2}\,e^{\xi/2T}\,\frac{|\Gamma(1+b+i\xi/2\pi T)|^2}{\Gamma(1+2b)}. \qquad (5.16)$$

This expression has been derived with account taken of the explicit form of eq. (5.8) by Kagan and Prokof'ev (1986b). An analogous result for the interaction of a particle with a model boson thermostat has been obtained in the works of Grabert and Weiss (1985) and Fisher and Dorsey (1985).

5.3. *Quantum diffusion in a metal*

The results that have been obtained in the previous sections permit us to analyze quantum diffusion in a metal. In the absence of defects, when $\xi = 0$,

$$W = 2\frac{\tilde{\Delta}_0^2(T)}{\Omega}\frac{\Gamma^2(1+b)}{\Gamma(1+2b)}. \qquad (5.17)$$

For crystals of cubic symmetry eq. (5.17) gives the diffusion coefficient as

$$D(T) = \frac{za^2}{3}\frac{\tilde{\Delta}_0^2(T)}{\Omega}\frac{\Gamma^2(1+b)}{\Gamma(1+2b)}. \qquad (5.18)$$

The structure of this expression coincides with the value found for $D(T)$ [eq. (3.36)] in the case of the phonon coupling with the appropriate replacement of $\tilde{\Delta}_0$ and Ω by the values in eqs. (5.8) and (5.9). The temperature dependence now proves to be weak:

$$D \propto T^{-(1-2b)}. \qquad (5.19)$$

This result was first established by Kondo (1984) and Yamada (1984). In a typical situation, $b < \frac{1}{2}$, the diffusion coefficient increases with decreasing temperature.

It should be noted that even at T of the order of $\omega_0/2\pi$ the electronic polaron effect, in fact, disappears. It is essential that as the particle mass increases the value of ω_0 falls off and so does the polaron effect. In particular, here the problem of the limit $M \to \infty$ is cancelled.

In an irregular crystal, when $\xi \neq 0$, it is necessary to use the general expression (5.16) for the transition probability and a kinetic equation in the form given by eq. (3.57) to solve the diffusion problem. The entire analysis of the solution of this equation given in section 3.4 for phonon coupling may be extended substantially to the case of electron coupling. Emphasis should, however, be made on certain features inherent in tunneling diffusion in a metal. They are associated with the fact that in the general case the dimensionless interaction parameter b is not small and the phase relaxation frequency [eq. (5.9)] depends on temperature only linearly. It is easy to see that under these conditions the inequality $\xi \gg \Omega$, in fact, becomes valid when the condition $\xi \gg T$ is simultaneously satisfied. As a result, the contribution of low-frequency excitations to the polaron effect is cut off at the scale of ξ and

$$\tilde{\Delta}_0(T) \to \tilde{\Delta}_0(\xi) = \tilde{\Delta}_0^{\text{ph}}(\xi/\gamma\omega_0)^b. \tag{5.20}$$

On the other hand, Ω changes effectively to $\Omega(\xi) = \pi b \xi$. Equation (5.16) then reduces to

$$W(\xi) = 2\pi \frac{\tilde{\Delta}_0^2(\xi)}{\xi \Gamma(2b)} \frac{1}{1 - \exp(-\xi/T)}. \tag{5.21}$$

The probability of a transition with energy transfer to the electron subsystem ($\xi > 0$) does not depend on temperature at all and

$$W \propto \xi^{-(1-2b)}. \tag{5.22}$$

If for any reason b is anomalously small, then W assumes the value given by eq. (3.50) when $\tilde{\Delta}_0 \approx \tilde{\Delta}_0^{\text{ph}}$. In such a case there appears a temperature region $\Omega < \xi < T$ with the inverse temperature course

$$W \propto \Omega/\xi^2 \sim T/\xi^2. \tag{5.23}$$

5.4. Band tunneling diffusion

In considering the band motion ($\xi = 0, \Omega < \mu\Delta_c$) of a tunneling particle in a metal, we again use the formalism of the kinetic equation for the density matrix given in section 3.5. All intermediate computations are to be performed; it is only necessary in going from eq. (3.72) to eq. (3.81) to introduce the interaction [eqs.

(2.16) and (2.17)] instead of $\mathscr{V}_2$. In this case, for the transition probability in the collision integral we have

$$W_{p,p-q} = 2\pi \sum_{k\sigma} |V(q)|^2 n_k(1 - n_{k+q})\delta(\varepsilon_k - \varepsilon_{k+q} + \delta\mathscr{E}(p, q)), \tag{5.24}$$

where $\delta\mathscr{E}(p, q) = \mathscr{E}_p - \mathscr{E}_{p-q}$ is the change of the particle energy upon scattering. Since $\Delta_c \ll \varepsilon_F$, this expression readily simplifies to the form

$$W_{p,p-q} = \theta(2k_F - q)\frac{4\pi\rho_F|V(q)|^2\delta\mathscr{E}(p, q)}{v_F q(1 - \exp(-\delta\mathscr{E}(p, q)/T))}. \tag{5.25}$$

Using the representation [eq. (3.83)] for the correction to the equilibrium distribution function $\chi_p(k)$, we obtain a kinetic equation in the form shown in eq. (3.84).

With T greater than the particle bandwidth and, hence, with $T > \delta\mathscr{E}$, the transition probability [eq. (5.25)] ceases to be dependent on p. The solution of eq. (3.84) will be sought for in the form $\chi_p(k) = -ik \cdot v_p A_p$. It is not difficult to see, taking into account the relation [see eq. (3.80)]

$$v_{p-q} = i\Delta_c \sum_g a_g e^{ip \cdot a_g} e^{-iq \cdot a_g}, \tag{5.26}$$

that $A_p = 1/\Omega$ (here Ω has the value given in eq. (5.9)). Using eq. (3.87), we have

$$D(T) = \frac{za^2}{3}\frac{\Delta_c^2}{\Omega}. \tag{5.27}$$

Taking cognizance of eq. (5.8) and comparing eq. (5.27) with eq. (5.18), we see that in the general case, with $\Omega\tau \approx 1$, there occurs a smooth transition from the band regime to the incoherent diffusion regime.

With T much smaller than the bandwidth, the particle spectrum near the bottom of the band may be considered quadratic. In an ordinary metal, with $k_F a \sim 1$, the solution of eq. (3.84) in the τ-approximation or within the framework of the method of moments leads to the following result:

$$D(T) = \text{const}\,\frac{a^2\Delta_c^2}{\Omega}. \tag{5.28}$$

The structure of this expression is associated with the fact that $\langle v_p^2 \rangle \sim T/M_*$ and the momentum transferred to the electrons is limited by the scale of the particle momentum $p \sim \sqrt{TM_*} \ll k_F$. This leads effectively to the value $1/\tau \sim \Omega(p/k_F)^2$. Considering $M_* = 3/(z\Delta_c a^2)$, we arrive at the relation (5.28).

Thus, instead of the temperature dependence [eq. (5.19)], with $T < \Delta_c$ there appears the following dependence:

$$D \propto 1/T. \tag{5.29}$$

This result has been obtained by Jäckle and Kehr (1983).

In dealing with a semimetal or a degenerate semiconductor we have a low-temperature range, within which the inequality $k_F \ll p(T)$ can be satisfied. Here, on the one hand, the factor $(p/k_F)^2$ is eliminated in $1/\tau$ and, on the other, by virtue of the transport effect, there appears an inverse factor, $(k_F/p)^2$. In this case, the kinetic equation can be reduced to the Fokker–Planck equation (3.85), which, as we have seen in section 3.5, leads to the exact solution of eq. (3.89). Here

$$D \propto T.$$

This result has been obtained by Morosov and Sigov (1985) and Kagan and Prokof'ev (1986b). Of course, at very low T, the fall of the diffusion coefficient will be replaced by an increase according to the law [eq. (5.28)] and eq. (5.29) with an additional factor $(k_F a)^4$ appearing in eq. (5.28).

5.5. *Tunneling diffusion in a superconductor*

In this section we will briefly discuss the variation of $D(T)$ with transition of a metal to the superconducting state. In this case, two points are of fundamental importance to quantum diffusion: the appearance of a gap Δ_s in the spectrum of elementary excitations and an exponential decrease of the number of normal excitations with decreasing temperature when $T < T_c$. The former leads to a change in the electronic polaron effect when $\tau^{-1} < \Delta_s$, and the latter to a drastic decrease in Ω.

To obtain quantitative relations, we have to construct an electronic polaron operator for the superconducting state. To do this, we can use, restricting ourselves to the BCS model, a standard (u, v) transformation of the second-quantization operators in eq. (2.16) (e.g., De Gennes 1966). Here, to the same approximation as in the derivation of eqs. (2.49) and (2.52), the operator Λ assumes the form

$$\Lambda = \prod_{p>p'} \Lambda_{pp'}, \quad \Lambda_{pp'} = \exp\left\{ -\left[\Delta V_{pp'} \left(\frac{u_p u_{p'} - v_p v_{p'}}{E_p - E_{p'} - \mathrm{i}/\tau} a^+_{p\sigma} a_{p'\sigma} + \frac{u_p v_{-p'} + v_p u_{-p'}}{E_p + E_{p'} - \mathrm{i}/\tau} \sigma_z a^+_{p\sigma} a^+_{-p'-\sigma} \right) - \text{h.c.} \right] \right\}, \tag{5.30}$$

where $u_p^2, v_p^2 = (1/2 \pm \varepsilon_p/2E_p)$, $\varepsilon_p = v_F(p - k_F)$ and $E_p = (\varepsilon_p^2 + \Delta_s^2)^{1/2}$. To determine the coherent amplitude Δ_c^s [eq. (5.1)], we have to find the diagonal matrix element of Λ [eq. (5.30)]. Calculations analogous to the determination of eq. (2.53) yield a polaron exponential in the form shown in eq. (2.55) with the function $\Omega(u)$ being defined by

$$\Omega_s(u) = \tfrac{1}{2}\pi u f_s(u) \coth(u/2T), \tag{5.31}$$

$$f_s(u) = \frac{2b}{u}\left\{\int_{\Delta_s}^{\omega_0} \mathrm{d}E\, g_E g_{E+u}(E(E+u) - \Delta_s^2)(n_E - n_{E+u}) + \theta(u - 2\Delta_s)\int_{\Delta_s}^{u-2\Delta_s} \mathrm{d}E\, g_E g_{u-E}(E(u-E) + \Delta_s^2)(\tfrac{1}{2} - n_E)\right\}.$$

Here we have introduced the function $g_E = (E^2 - \Delta_s^2)^{-1/2}$.

When $T \ll T_c$, the occupation numbers $n_E \to 0$ and straightforward evaluation of the integral in eq. (5.31) gives

$$f_s(u) = \Theta(u - 2\Delta_s) 2b\, \boldsymbol{E}(\sqrt{1 - (2\Delta_s/u)^2}), \tag{5.32}$$

where $\boldsymbol{E}(x)$ is the full elliptic integral of the second kind. Substituting this expression into $\Omega(u)$ and using eq. (2.55), we find under the assumption that $\Delta_s \tau \gg 1$

$$Z_s = b \ln\left(\frac{2\omega_0}{\mathrm{e}\Delta_s}\right). \tag{5.33}$$

This result leads for Δ_c^s to eq. (5.6) introduced earlier.

When $u \gg 2\Delta_s$, the function $f_s(u)$ takes on the value $2b$ characteristic of a normal metal [eq. (5.14)]. The same result is deduced from eq. (5.31) when $T > \Delta_s$. Both these results have an obvious physical meaning.

Of prime importance for quantum diffusion in coherent and incoherent regimes is the phase correlation damping frequency Ω_s at neighboring lattice sites. This quantity is found from the static limit (5.31):

$$\Omega_s(0) = \frac{4\pi b T}{1 + \exp(\Delta_s/T)} \equiv 2\pi b^{\mathrm{eff}} T. \tag{5.34}$$

This expression coincides with that derived by Black and Fulde (1979) and Morosov (1979). From eq. (5.34) it immediately follows that a decrease in the number of normal excitations in a superconductor is effectively equivalent to a decrease in the interaction.

Particular expressions for the diffusion coefficient in a superconductor can be derived through the simple replacement $\Omega \to \Omega_s$ and of Δ_c in eq. (5.7) by Δ_c^s [eq. (5.6)]. But if $\Delta_s < 1/\tau$, then Δ_c retains the value given by eq. (5.7). From this we see at once that in an ideal superconductor, after it has passed over to the superconducting state, the diffusion coefficient increases exponentially with decreasing T. In a crystal with defects, when $\mu\Delta_c^s \ll \xi < 2\Delta_s$, as the temperature falls off, the condition $\Omega_s < \xi$ is rapidly attained. Then, according to eq. (5.16) or eq. (3.50) (with $b^{\mathrm{eff}} \ll 1$ we may use this expression) $W \sim \Omega_s$ and, hence, the delocalization that takes place in the region considered only due to induced transitions, is exponentially suppressed. When $T \to 0$, complete localization occurs in a situation where no localization is observed in a normal metal.

5.6. *The role of barrier fluctuations due to electron coupling*

So far, while considering electron coupling, we have ignored the effect of barrier fluctuations on tunneling motion. At first glance this is natural since, as we have pointed out in section 2.4, $\mathscr{B} \lesssim 1$ because of the limited volume of the nonadiabatic excitation band. However, in a discussion of inelastic transitions induced by barrier fluctuations we deal with an appreciable enhancement of the transition probability due to an infrared divergence in rescattering of electrons. This circumstance was first pointed out by Kondo (1976b). Vladar and and Zawadowski (1983a, b, c) and Vladar et al. (1988a, b) have carried out a detailed analysis of the problem using the multicomponent renormalization-group method for the partition function of the system and have revealed an appreciable enhancement of this interaction channel. However, the important question of whether the scattering on barrier fluctuations in the real tunneling problem in a metal could be comparable to intrawell scattering has remained open. The analysis given below is based on the results obtained by Kagan and Prokof'ev (1989a).

Note that the problem under consideration has one more aspect. Since the infrared enhancement due to rescattering occurs only for fermions, the positive answer to the question would imply that in the tunneling problem the Fermi and Bose thermostats are fundamentally different. At the same time, most of the works on tunneling in metallic systems have been based on their equivalence.

Let us again consider the transition matrix element (2.45). The polaron operator Λ, which links, according to eq. (2.44), the eigenfunctions of the Hamiltonians $\mathscr{H}^{(1)}$ and $\mathscr{H}^{(2)}$, is defined on the basis of the interaction [eqs. (2.47) and (2.48)]. In this section we abandon perturbation theory in $\mathscr{V}$ and construct, instead of eqs. (2.49) and (2.50), the general exact expression for the polaron operator.

The wave function $\Psi_n^{(2)}$ in eq. (2.44) will be sought in the form of an expansion in the states of the Hamiltonian $\mathscr{H}^{(1)}$. Here, $\Psi_n^{(2)}$ differs from $\Psi_n^{(1)}$ by an arbitrary number of electron–hole pairs. To macroscopic accuracy the amplitude of the creation of two and more pairs decomposes into the product of pair amplitudes. In this case we may write

$$\Lambda = S\exp\left\{-\sum_{pp'}^{n} c_{pp'}a_p^+ a_{p'}\right\} = S\prod_{pp'}^{n}(1 - C_{pp'}a_p^+ a_{p'}), \tag{5.35}$$

where S is a normalization factor. In the product in eq. (5.35) the index p enumerates only the empty single-particle states in $|n\rangle$, while the index p' enumerates only the occupied states, this being indicated by the symbol n. It is this circumstance that allows us to pass over in eq. (5.35) to the last equality, since $\{p\}$ and $\{p'\}$ do not overlap and all the terms in the exponential are mutually commuting.

The coefficients $C_{pp'}$ have the meaning of the probability amplitudes of finding in $\Psi_n^{(2)}$ a state with a single electron–hole pair $a_p^+ a_{p'}|n\rangle$. If we now write for $\Psi_n^{(2)}$ the Schrödinger equation and make use of eqs. (2.44) and (5.35), then for these amplitudes we find the following integral equation (Kagan and Prokof'ev 1989a)

$$C_{pp'} = \frac{1}{\varepsilon_p - \varepsilon_{p'}} \Big\{ \Delta V_{pp'} - \sum_k (1 - n_k) \Delta V_{pk} C_{kp'} + \sum_{k'} n_{k'} \Delta V_{k'p'} C_{pk'} - \sum_{kk'} (1 - n_k) n_{k'} \Delta V_{k'k} C_{pk'} C_{kp'} \Big\}. \tag{5.36}$$

The structure of this expression has an obvious meaning: a state with one pair can be produced by direct creation of a pair from the state $|n\rangle$, by rescattering an electron or hole or by annihilation of the extra pair in a state with two pairs.

Let us expand $\Delta V_{pp'}$ in a certain complete system of functions defined on a unit sphere for the arguments $\boldsymbol{p}/p$ and $\boldsymbol{p}'/p'$ separately. The Hamiltonian $\mathscr{H}^{(2)}$ with $\boldsymbol{p}, \boldsymbol{p}'$ being near the Fermi surface may be reduced to a diagonal form. So,

$$\Delta V_{pp'} = \sum_j \Delta V_j(\varepsilon, \varepsilon') \Omega_j(\boldsymbol{p}) \Omega_j^*(\boldsymbol{p}'). \tag{5.37}$$

In order to make the treatment more transparent, we adopt the separability of ΔV_j:

$$\Delta V_j(\varepsilon, \varepsilon') = \Delta V_j \alpha_j(\varepsilon) \alpha_j^*(\varepsilon') \quad (\alpha_j(0) = 1). \tag{5.38}$$

Then, the solution of eq. (5.36) may be sought in the form

$$C_{pp'} = \sum_j \Omega_j(\boldsymbol{p}) \Omega_j^*(\boldsymbol{p}') C_j(\varepsilon, \varepsilon'), \tag{5.39}$$

$$C_j(\varepsilon, \varepsilon') = \frac{\Delta V_j(\varepsilon, \varepsilon')}{\varepsilon - \varepsilon'} \xi_j(\varepsilon) \eta_j(\varepsilon'). \tag{5.40}$$

By solving the resultant system of equations for ξ and η, we arrive at the final result in the form [for more detail, see Kagan and Prokof'ev (1989a)]

$$C_j(\varepsilon, \varepsilon') = \frac{\Delta V_j(\varepsilon, \varepsilon')}{\varepsilon - \varepsilon'} \left| \frac{\varepsilon'}{\varepsilon} \right|^{\delta_j/\pi} \frac{\sin \delta_j}{\pi g_j} \quad (T = 0), \tag{5.41}$$

where $g_j = \rho(\varepsilon_F) \Delta V_j$ is a dimensionless interaction constant. This expression contains the electron scattering phase on the Fermi surface in the potential ΔV_j:

$$\delta_j = \tan^{-1} \pi G_j(0), \tag{5.42}$$

$$G_j(\varepsilon) = g_j \left(1 + \Delta V_j \int \frac{|\alpha_j(x)|^2}{x - \varepsilon} \rho(x) \, dx \right)^{-1}. \tag{5.43}$$

We give here an approximate expression for the amplitudes C_j corresponding to finite temperatures:

$$C_j(\varepsilon, \varepsilon') \approx \frac{\Delta V_j(\varepsilon, \varepsilon')}{\varepsilon - \varepsilon'} \left| \frac{(\varepsilon', T)_{\max}}{(\varepsilon, T)_{\max}} \right|^{\delta_j/\pi} \frac{\sin \delta_j}{\pi g_j}. \tag{5.44}$$

In what follows, it will be important to know $\xi_j(\varepsilon)$ and $\eta_j(\varepsilon')$ separately. The explicit form of the solutions for these functions may be written as

$$\xi_j(\varepsilon) \approx A_j \left| \frac{\omega_0}{(\varepsilon, T)_{\max}} \right|^{\delta_j/\pi} \qquad \eta_j(\varepsilon') = \frac{G_j}{g_j \xi_j(\varepsilon')}. \tag{5.45}$$

The normalization factor in eq. (5.35), which is equal to the overlap integral $\langle \Psi_n^{(2)} | \Psi_n^{(1)} \rangle$, can be found from the condition $\langle n | \Lambda^+ \Lambda | n \rangle = 1$. The answer coincides with the result known from the literature (Nozieres and de Dominicis 1969, Yamada and Yosida 1982)

$$S = \exp\left(-b \int_{\varepsilon_{\min}}^{\omega_0} \frac{d\varepsilon}{\varepsilon} \right), \quad b = \sum_j \left(\frac{\delta_j}{\pi} \right)^2. \tag{5.46}$$

This is just the generalized expression for b in eq. (5.2). In a real tunneling problem the cutoff of the logarithmic divergence in the exponential occurs at the value τ^{-1}.

The result obtained for the polaron operator enables a straightforward estimation of the effect of the fluctuational preparation of the barrier on coherent and incoherent tunneling processes. If we expand the exponential $e^{-\mathscr{B}}$ in eq. (2.45), retaining the first two terms, the correction to the coherent amplitude Δ_c [eqs. (5.1) and (5.2)] will be

$$- J_0 (n | \mathscr{B} \Lambda | n) \tag{5.47}$$

The expression for $\mathscr{B}$ in eq. (2.42) may be represented in second quantization as

$$\mathscr{B} = \sum_{pp'} \frac{B_{pp'}}{\omega_0} a_p^+ a_{p'}. \tag{5.48}$$

We now expand $B_{pp'}$ in the same system of functions Ω_j:

$$B_{pp'} = \sum_{jj'} B_{jj'} \Omega_j(p) \Omega_{j'}^*(p'). \tag{5.49}$$

The commutability of all the terms in the exponent in eq. (5.35) predetermines the simplicity of calculation of the matrix elements containing the operator Λ. Substituting eq. (5.49) into eq. (5.47) and using the explicit form (5.39) and (5.41), we directly find

$$2\Delta_c \sum_j \frac{B_{jj} \Delta V_j \sin \delta_j}{\pi g_j} \rho^2(\varepsilon_F) \iint_{-\omega_0}^{+\omega_0} \frac{(1 - n_\varepsilon) n_{\varepsilon'} \, d\varepsilon \, d\varepsilon'}{\varepsilon - \varepsilon'} \left| \frac{\varepsilon'}{\varepsilon} \right|^{\delta_j/\pi}. \tag{5.50}$$

This integral is connected with the upper limits of integration, and with account taken of the estimate (2.42), it yields the following result:

$$J_0\langle n|\mathscr{B}\Lambda|n\rangle \sim \Delta_c \rho(\varepsilon_F) V \sin\delta. \tag{5.51}$$

Bearing in mind that $\rho(\varepsilon_F)V \lesssim 1$, we see that the Δ_c renormalization is not important. This result is also valid at finite temperatures $T \ll \omega_0$ since in the integral (5.50) ε, $\varepsilon' \sim \omega_0$ are important. It is easy to see that the corrections from the higher terms of the expansion of $e^{-\mathscr{B}}$ leave the resultant estimate unchanged.

Thus, the barrier fluctuations due to nonadiabatic electronic excitations, whose phase volume has the smallness of ω_0/ε_F, do not lead to a substantial renormalization of Δ_c over the entire temperature range $T < \omega_0$. In this lies the fundamental distinction from the case of the interaction with the lattice, where the entire phonon spectrum may be nonadiabatic.

We will now consider the incoherent tunneling regime and elucidate the role of particle scattering on barrier fluctuations. To this end, we employ the general expression (3.14) for the transition probability. We begin with perturbation theory with respect to $\mathscr{B}$ and at first we neglect the rescattering of electron–hole pairs. For the probability of a transition induced by barrier fluctuations we then have

$$\begin{aligned} W_b &= J_0^2 \sum_{pp'} \left|\frac{B_{pp'}}{\omega_0}\right|^2 (1-n_p)n_{p'} \int_{-\infty}^{+\infty} dt\, e^{i(\xi+\varepsilon_{p'}-\varepsilon_p)t+\chi(t)} \\ &= \sum_{pp'} \left|\frac{B_{pp'}}{\omega_0}\right|^2 (1-n_p)n_{p'} W(\xi+\varepsilon_{p'}-\varepsilon_p), \end{aligned} \tag{5.52}$$

where W has the value given by eq. (5.16). The energy of electron–hole pairs is bounded in eq. (5.52) by the scale of $(\xi, \Omega)_{max}$.

This immediately yields the estimate

$$W_b \sim W\left(\frac{(\Omega,\xi)_{max}}{\omega_0}\right)^2. \tag{5.53}$$

This result is known to be also valid even if we go beyond first-order perturbation theory with respect to $\mathscr{B}$. This can easily be seen to be the case if we take into account that the components of the operator $\mathscr{B}$ have in the denominator the frequency ω_0 instead of $|\varepsilon_{p'}-\varepsilon_p| \lesssim (\xi,\Omega)_{max}$ in the amplitude $C_{pp'}$ in eq. (5.35).

In contrast to the polaron operator [eq. (5.35)], the operator $\mathscr{B}$ leads to rescattering of excitations. In order to take the rescattering into account, we evaluate the electron–hole pair creation amplitude:

$$\tilde{B}_{nm} = S^{-1}\langle n|\mathscr{B}\Lambda|m\rangle, \tag{5.54}$$

where the state $|n\rangle = a_p^+ a_{p'}|m\rangle$. Taking cognizance of the structure of the operator Λ, the explicit expression for $\tilde{B}_{pp'}$ reads

$$\tilde{B}_{pp'} = B_{pp'} - \sum_k (1 - n_k) B_{pk} C_{kp'} + \sum_{k'} n_{k'} B_{k'p'} C_{pk'} - \sum_{kk'} (1 - n_k) n_{k'} B_{k'k} C_{kp'} C_{pk'}. \tag{5.55}$$

This expression is quite analogous to eq. (5.36) and we can immediately write an exact solution:

$$\tilde{B}_{jj'}(\varepsilon, \varepsilon') = B_{jj'} \xi_j(\varepsilon) \eta_{j'}(\varepsilon'). \tag{5.56}$$

Using eq. (5.45), we obtain

$$\tilde{B}_{jj'}(\varepsilon, \varepsilon') \sim B_{jj'} \left| \frac{\omega_0}{(\varepsilon, T)_{\max}} \right|^{\delta_j/\pi} \left| \frac{\omega_0}{(\varepsilon', T)_{\max}} \right|^{-(\delta_{j'}/\pi)} \tag{5.57}$$

This expression allows for the possibility of an appreciable increase of the incoherent transition amplitude due to barrier fluctuations if the electron and hole are scattered in channels with different j. This circumstance was first pointed out by Kondo (1976b). Vladar and Zawadowski (1983a, b, c) and Vladar et al. (1988a, b) undertook a detailed analysis of the renormalization of this amplitude. A result analogous to eq. (5.57) was first obtained with the use of a multicomponent renormalization-group method by Vladar et al. (1988a, b) for a model with a commutator $[\mathscr{V}(\boldsymbol{R}_1), \mathscr{V}(\boldsymbol{R}_2)] = 0$.

Let us now substitute the renormalized amplitudes $\tilde{B}_{jj'}$ into eq. (5.52). If $\tilde{B}_{jj'}$ [eq. (5.57)] increases compared to $B_{jj'}$, the probability of incoherent tunneling induced by barrier fluctuations will be defined by the following relation instead of eq. (5.53):

$$W_b \sim W \left| \frac{(\xi, \Omega)_{\max}}{\omega_0} \right|^{2(1-\theta)}, \quad \theta = \left\{ \frac{\delta_j - \delta_{j'}}{\pi} \right\}_{\max} \tag{5.58}$$

A detailed analysis carried out by Kagan and Prokof'ev (1989a) has shown that with the very general nature of the interaction of a tunneling particle with electrons the phases δ_j characterizing the scattering in a nonspherical potential $\mathscr{V}(\boldsymbol{r}, \boldsymbol{R}_2) - \mathscr{V}(\boldsymbol{r}, \boldsymbol{R}_1)$ cannot exceed $|\pi/2|$. For a particular case when the electron–particle coupling in an individual well is described by a single scattering phase, an analogous result has been obtained by Yamada et al. (1986) and by Tanabe and Ohtaka (1986). This limitation on the phases leads to the condition $\theta < 1$. Thus, we are led to the general statement that in the case of the tunneling motion of the particle the interaction with barrier fluctuations is always less than the intrawell interaction, despite the infrared singularity.

From the above treatment it becomes clear that, strictly speaking, the fermion and boson thermostats are nonequivalent. This may be ignored whenever all contributions associated with barrier fluctuations may be omitted.

5.7. *A general picture of quantum diffusion in a metal*

A general picture of quantum diffusion in a metal qualitatively has much in common with the picture outlined in section 3.7 for the tunneling motion in an insulator (see fig. 4). In ordinary metals with $k_F a \sim 1$, at low temperatures (in fact, up to T comparable to θ_D), the decisive role in particle scattering is played by the interaction with electrons. The interaction with phonons enters into the answer only via the polaron renormalization of the tunneling amplitude $J_0 \to \tilde{\Delta}_0^{ph}$ [see also eqs. (5.7) and (5.8)]. By virtue of this, the branch of $D(T)$, which increases with decreasing temperature and which at first is associated with incoherent diffusion, obeys the law $D(T) \propto T^{-(1-2b)}$ which is replaced by the law $D(T) \propto 1/T$ when $T < \mu\Delta_c$. A small defect concentration limits the rise of $D(T)$ when $T \to 0$. With an appreciable defect concentration we must use eq. (5.16) for an elementary tunneling act, and a general analysis of long-scale diffusion and localization is carried out in the same way as it was done in sections 3.4 and 3.7, of course, with account taken of the results of section 5.3. It is interesting that the localization picture is drastically changed upon transition to the superconducting state (see section 5.5).

The high-temperature behavior of the diffusion coefficient for the interaction with electrons sharply differs from the case of phonon coupling. Indeed, the electronic polaron effect disappears even when $T \sim \omega_0$ and incoherent diffusion, in principle, continues to decrease with the rise of temperature. As we have seen in section 5.6, the role of barrier fluctuations and transitions stimulated by them remains to be limited. Therefore, the $D(T)$ branch increasing with temperature (fig. 4) appears because the dominant scattering mechanism changes from the electronic to the phononic one. As a result, with $T > T_{min}$ the behavior is again dictated either by phonon-stimulated diffusion or by classical overbarrier diffusion. When $T \lesssim \Theta_D$, the phononic and electronic scattering mechanisms are balanced and in order to evaluate the probability of an incoherent transition we should employ the general relations (3.23) and (3.24) with the generalized value

$$\Omega(\omega) = \Omega^{ph}(\omega) + 2\pi bT,$$

in which $\Omega^{ph}(\omega)$ is defined according to eq. (3.31). For a superconducting metal the quantity $2\pi bT$ should be replaced by the more general expression (5.31).

5.8. *Heavy electrons*

The physical picture corresponding to the generation of the electronic polaron effect in tunneling of a heavy particle (see sections 5.1 and 2.3) may be extended to a case where, apart from ordinary electrons, a metal has a subsystem of electrons with a small group velocity v_g (Kagan and Prokof'ev 1987b). Indeed, in this case the inverse lifetime of the electron in an individual unit cell is given by

$$\tau^{-1} \sim v_g/a \ll \varepsilon_0,$$

where ε_0 is the characteristic energy scale for broad-band electrons. As long as the slow electron is present in the unit cell, a rearranged cloud of fast electrons is formed around it. It is important that the subbarrier motion is always governed by the bare particle mass, i.e., occurs with a frequency $\omega_0 \sim \varepsilon_0$. Therefore, the entire electron–hole excitation spectrum appears to be nonadiabatic. As a result, the Coulomb interaction of the electrons of this subsystem with other electrons gives rise to a strong polaron effect, which causes an effective decrease of v_g.

The complete picture is revealed in a spectacular way in the case of intersecting narrow and broad bands with a common Fermi level. The electronic polaron effect inevitably leads to a narrowing of the initial band according to eq. (5.3), with ω_0 replaced in this expression by ε_0, to the logarithmic accuracy adopted. Since in the case under discussion there are no grounds for considering the parameter b small, the narrowing may be very strong, especially if the bare ratio Δ/ε_0 is sufficiently small for the narrow band. If the value of τ^{-1} becomes less than Θ_D as a result of such narrowing, the phonon polaron effect will also be operative and the bandwidth will be determined by eq. (5.7). It is important that the Fermi level will lie, as before, in the narrow band after an arbitrary-scale narrowing if it was in the original band.

Qualitatively the above results remain valid when the spectrum of single-particle electronic states exhibits a density peak associated with the hybridization of the f- and d-levels of ions with broad-band electrons (Kagan and Prokof'ev 1987b, Liu 1987, 1988). An unimportant feature in this case is the fact that a polarization "cloud" composed of the electron–hole pairs of the broad band is formed only when the electron is present at a quasi-localized level on the ion and is not formed when the electron is transferred to the continuous spectrum. The resultant polaron effect leads to a decrease in the vertex in the standard hybridization Hamiltonian and, along with this, to a drastic narrowing of the density-of-state peak.

Thus, if within the solution of the one-electron problem there is a peak in the density of energy states in an interval appreciably smaller than ε_0 due to the presence of a narrow band or hybridization, then the interelectronic interaction will invariably lead to a narrowing of that peak. With all the factors taken into account, the narrowing may be very strong and the renormalized width may reach a value of 100 K or even 10 K. We, thereby, find ourselves in the region of heavy electrons or heavy fermions. The general nature of the results obtained allows us to believe that we have here an alternative concept of the formation of heavy electrons, which is not associated with the Kondo effect.

Of particular importance in this connection is the fact that, according to the results of section 5.1, at a temperature of the order of Δ there must occur a dynamic destruction of the band with a slow diffusion of the carriers, which begin to play the role of independent (spin-carrying) scatterers. This involves a continuous natural transition at low T from the coherent to the incoherent picture, this being the most characteristic feature of heavy-electron systems. The thermodynamic and kinetic properties calculated within the present treatment reproduce qualitatively the experimentally observed picture (Kagan and Prokof'ev 1987b, Liu 1987, 1988).

5.9. *Dynamics of tunneling of two-level systems*

There exists a large class of systems, whose properties are determined by the tunneling dynamics of a set of two-well configurations. The most striking example is furnished by insulating and metallic glasses, the description of which in most cases reduces to an ensemble of two-level systems (TLS) with a continuous distribution of the parameters. Based on the line of reasoning adopted in the previous sections, we shall consider here only the tunneling dynamics of an individual two-level system and the linear response function of such a system. In doing so, we shall be interested first of all in the evolution of dynamic properties upon transition from the coherent to the incoherent regime of motion, which, as can be seen from the above treatment, reflects the variation of the ratio Δ_c/Ω. This problem is particularly crucial for metallic systems, in which this parameter becomes less than unity even at T comparable to Δ_c.

The kinetic equation for the density matrix of a two-level system in the absence of an external field has the following form if use is made of the general relations (3.67) and (3.70):

$$\omega f_{cd} - [\mathscr{H}_0, f]_{cd} - \sum_{ab} \Omega_{ab}^{cd} f_{ab} = 0. \tag{5.59}$$

Here the indices run over two values and the superoperator Ω [eq. (3.72)] represents a 3×3 matrix in the space of the independent density matrix elements [$\mathrm{Sp}\, f(\omega) = 0$].

Let $T \gg \Delta_0, \xi$. As has been mentioned earlier, in going over to eqs. (3.73) and (3.74) it is convenient to use the site representation, in which only the following elements are different from zero:

$$\Omega_{12}^{12} = \Omega_{21}^{21} = -\mathrm{i}\Omega(\omega), \quad \Omega(\omega) = \tfrac{1}{2}\pi\omega f(\omega)\coth(\omega/2T) \approx \pi T f(\omega). \tag{5.60}$$

Here $f(\omega)$ has the value given by eq. (3.24) for phonons and that given by eqs. (5.14) and (5.31) for electrons. Accordingly, the system of equations (5.59) will be rewritten as

$$\begin{aligned}
&\omega f_{11}(\omega) + \tilde{\Delta}_0(f_{12}(\omega) - f_{21}(\omega)) = 0,\\
&(\omega - \xi + \mathrm{i}\Omega(\omega)) f_{12}(\omega) + 2\tilde{\Delta}_0 f_{11}(\omega) = 0,\\
&(\omega + \xi + \mathrm{i}\Omega(\omega)) f_{21}(\omega) - 2\tilde{\Delta}_0 f_{11}(\omega) = 0,
\end{aligned} \tag{5.61}$$

in which $\tilde{\Delta}_0$ has the value given by eq. (3.11) or by eq. (5.8).

Note that the notation $\frac{1}{2}J_0$ generally adopted in the theory of glasses for the tunneling transition amplitude differs by a factor of 2 from the one introduced earlier [see eqs. (2.5) and (2.43)].

The determinant of the system (5.61) defines the equation for the eigenvalues of the TLS, which has the form (Kagan and Maksimov 1980)

$$\omega((\omega + \mathrm{i}\Omega)^2 - \tilde{\varepsilon}^2) - 4\mathrm{i}\tilde{\Delta}_0^2\Omega = 0, \tag{5.62}$$

$$\tilde{\varepsilon}^2 = \xi^2 + 4\tilde{\Delta}_0^2. \tag{5.63}$$

In the case of electron or phonon coupling, where the determining factor is the two-phonon scattering mechanism, we may put $\omega = 0$ in the argument of the function Ω. Here, for a symmetric TLS the solution of eq. (5.62) takes the form

$$\omega^{(1)} = -\mathrm{i}\Omega, \quad \omega^{(2,3)} = \pm\tfrac{1}{2}\sqrt{16\tilde{\Delta}_0^2 - \Omega^2} - \mathrm{i}\tfrac{1}{2}\Omega. \tag{5.64}$$

If $\Omega \ll \tilde{\Delta}_0$, then the poles of eq. (5.64) give a result characteristic of perturbation theory: the level splitting is equal to $2\tilde{\Delta}_0$ and the inverse relaxation time $\gamma_2 = \frac{1}{2}\gamma_1 = \frac{1}{2}\Omega$. The picture changes drastically when $\Omega > 4\tilde{\Delta}_0$. In this case, all the roots become imaginary. In the limit $\Omega \gg \tilde{\Delta}_0$

$$\omega^{(1)} \approx -4\mathrm{i}\tilde{\Delta}_0^2/\Omega, \qquad \omega^{(2)} = -\mathrm{i}\Omega, \qquad \omega^{(3)} = -\mathrm{i}\Omega + 4\mathrm{i}\tilde{\Delta}_0^2/\Omega. \tag{5.65}$$

We have changed the nomenclature of the roots, retaining the index (1) for the slowest root responsible, in the general case, for the relaxation of the level populations. The disappearance of the real part of the spectrum is connected with the transition from the coherent to the incoherent regime, and it is natural that this takes place in the same parametric region as does the suppression of the coherent transition amplitude [see eqs. (3.10) and (5.8)].

Suppose now that $\xi \neq 0$. We take advantage of the fact that with $\Omega \gg \tilde{\Delta}_0$ or $\xi \gg \tilde{\Delta}_0$ and also with $\Omega \ll \tilde{\Delta}_0$ one of the roots of eq. (5.62) is small compared to the others and is approximately equal to (cf. Kagan and Maksimov 1980)

$$\omega^{(1)} \equiv -\mathrm{i}\gamma_1 \approx -4\mathrm{i}\frac{\tilde{\Delta}_0^2\Omega}{\tilde{\varepsilon}^2 + \Omega^2}. \tag{5.66}$$

The remaining two roots are given by

$$\begin{aligned} &\omega^{(2,3)} \approx \mp\tilde{\varepsilon} - \mathrm{i}\Omega + \mathrm{i}\tfrac{1}{2}\gamma_1 \quad (\Omega \ll \tilde{\varepsilon}), \\ &\omega^{(2,3)} \approx \mp\xi - \mathrm{i}\Omega \qquad\qquad (\Omega \gg \tilde{\varepsilon}). \end{aligned} \tag{5.67}$$

The latter result, in which the real part does not depend on $\tilde{\Delta}_0$ at all, again reflects the suppression of the coherent tunneling state. Perhaps, the most important result here is that in a wide region of values of the parameters of the TLS the relaxation of the phase, i.e., $T_2^{-1} = \gamma_2$, is much larger than $T_1^{-1} = \gamma_1$. Contrary to the generally accepted viewpoint, this is characteristic of an individual TLS (Kagan and Maksimov 1980, Kagan and Prokof'ev 1990a).

At low temperatures, when $\Omega \ll \tilde{\varepsilon}$, we can solve the kinetic equation using in eqs. (5.59) and (3.72) the representation of coherent tunneling states and perturbation theory with respect to $\Omega/\tilde{\varepsilon}$ (Kagan and Maksimov 1980, Kagan and Prokof'ev 1990a). Referring the reader to these works for details, we give here only the results for the poles of the density matrix for this particular case:

$$\gamma_1 \approx \frac{4\tilde{\Delta}_0^2}{\tilde{\varepsilon}^2}\,\Omega(\tilde{\varepsilon}),$$

$$\omega^{(2,3)} \approx \mp \tilde{\varepsilon} - \mathrm{i}\gamma_2, \quad \gamma_2 \approx \frac{\xi^2}{\tilde{\varepsilon}^2}\,\Omega(\omega = 0) + \frac{2\tilde{\Delta}_0^2}{\tilde{\varepsilon}^2}\,\Omega(\tilde{\varepsilon}). \tag{5.68}$$

For electron coupling the amplitude $\tilde{\Delta}_0$, which appears in eq. (5.63), must be determined from the self-consistent equation

$$\tilde{\Delta}_0(\tilde{\varepsilon}) = \tilde{\Delta}_0^{\mathrm{ph}}\left(\frac{2\pi T}{\omega_0}\right)^b \exp\left[-b\,\mathrm{Re}\,\psi\left(1 + \mathrm{i}\frac{\tilde{\varepsilon}}{2\pi T}\right)\right], \tag{5.69}$$

where ψ is the digamma function. It should be emphasized that eq. (5.68) includes Ω at a finite frequency $\omega = \tilde{\varepsilon}$ and that, with $\tilde{\varepsilon} > T$, we must use the general expression (5.60). When $T \gg \tilde{\varepsilon}$ and $\Omega < \tilde{\varepsilon}$, the roots of eq. (5.68) coincide with eqs. (5.66) and (5.67). For phonon coupling and with $b \ll 1$ in the case of electron coupling the regions of applicability of eq. (5.68) and eqs. (5.66) and (5.67) overlap appreciably.

Note that, with $T \ll \tilde{\varepsilon}$, relation (5.69) yields a simple expression for the tunneling state spectrum:

$$\tilde{\varepsilon}^2 = \xi^2 + 4\Delta_{\mathrm{c}}^2\left(\frac{\tilde{\varepsilon}}{2\Delta_{\mathrm{c}}}\right)^{2b}, \quad \Delta_{\mathrm{c}} = \tilde{\Delta}_0^{(\mathrm{ph})}(2\tilde{\Delta}_0^{(\mathrm{ph})}/\omega_0)^{b/(1-b)}. \tag{5.70}$$

The expression for Δ_{c} coincides with eq. (5.7) if we take into account that the coefficient $\mu = 2$ for the two-well problem. The structure of the expression for the spectrum reflects the change of the polaron effect with increasing ξ, as we have already discussed in the analysis of incoherent transitions [see eq. (5.20)]. When $T > \tilde{\varepsilon}$, the amplitude $\tilde{\Delta}_0(\tilde{\varepsilon}) = \tilde{\Delta}_0(T)$ [eq. (5.8)].

The results given in this section have been obtained under the assumption that in evaluating the collision integral the interaction with excitations may be considered to be weak. As regards the polaron effect, which remains arbitrary in magnitude in a metal, it enters only into the renormalization of the tunneling transition amplitude. This has made it possible to use the results originally obtained by Kagan and Maksimov (1980) for phonon coupling. The dynamic properties of the TLS upon arbitrary interaction with an ohmic thermostat [$f(\omega) = \mathrm{const}$] have been explored using the functional integration technique in a number of works [see the reviews by Leggett et al. (1987), Grabert et al. (1986), Grabert (1987) and by Weiss and Wollensak (1989)].

The spectrum and relaxation of a two-level system manifest themselves in an experiment via its interaction with an external field, be it deformation, electromagnetic radiation or neutron scattering. The interaction with an external field leads to the appearance of an additional term in the Hamiltonian:

$$\mathscr{H}_{\text{int}} = \varepsilon_0 \eta \sigma_z \cos(\omega t), \tag{5.71}$$

where η is the characteristic energy scale and ε_0 is the dimensionless field amplitude. The term proportional to σ_x is small compared to eq. (5.71) and may be neglected. For interaction with an acoustic wave, e.g., η has the meaning of the deformation potential and ε_0 has the meaning of one of the components of the deformation tensor. Within the linear response ($\varepsilon_0 \to 0$), to which we will restrict ourselves here, the properties of the system are described by the susceptibility (the modulus of elasticity):

$$M_\omega = 2\eta \frac{\partial}{\partial \varepsilon_0} \{\operatorname{Tr} \rho(\omega)\sigma_z\} \equiv 4\eta \frac{\partial}{\partial \varepsilon_0} f_{11}(\omega), \tag{5.72}$$

where $f_{11}(\omega)$ is the diagonal element of the particle density matrix [eq. (3.66)] in the site representation.

In an approximation linear in ε_0, in the original equation for the density matrix [eq. (3.63)] we must add to the right-hand side the term $[\mathscr{H}_{\text{int}}(t), \rho_{\text{t}}]$, where ρ_{t} is the equilibrium density matrix of the system. A detailed analysis of the problem of the linear response of two-level systems over the entire range of variation of the parameters and temperature can be found in Kagan and Prokof'ev (1990a) [see also Weiss and Wollensak (1989)]. We shall give here two limiting results, which will be sufficient for the analysis of the experimental data given below.

For the region of $T \gg \tilde{\varepsilon}$ the result is simple in structure. This result clearly reflects the nature of the poles of eq. (5.62):

$$M_\omega = 4\eta^2 \frac{\tilde{\Delta}_0^2}{T} \frac{\omega + \mathrm{i}\Omega}{\omega[(\omega + \mathrm{i}\Omega)^2 - \tilde{\varepsilon}^2] - \mathrm{i}4\tilde{\Delta}_0^2 \Omega}. \tag{5.73}$$

In most acoustic experiments the condition $\omega \ll T$ is realized. In this case, when $\Omega \ll \tilde{\varepsilon}$,

$$M_\omega = 2\eta^2 \tilde{\varepsilon}^2 \frac{\mathrm{i}\gamma_1 \left(\dfrac{\partial \tilde{\varepsilon}}{\partial \xi}\right) \dfrac{\partial \tanh(\tilde{\varepsilon}/2T)}{\partial \xi} + (\omega + \mathrm{i}\gamma_1) \dfrac{\partial^2 \tilde{\varepsilon}}{\partial \xi^2} \tanh(\tilde{\varepsilon}/2T)}{(\omega + \mathrm{i}\gamma_1)[(\omega + \mathrm{i}\gamma_2)^2 - \tilde{\varepsilon}^2]}. \tag{5.74}$$

In the region of $\tilde{\varepsilon} < T$, with $\Omega < \tilde{\varepsilon}$, eqs. (5.73) and (5.74) coincide. It is essential that for weak interaction with medium excitations there is an appreciable region in which the two expressions are simultaneously valid. In this case, we can construct a common solution.

In spite of the seeming simplicity, the expression given for M_ω [eq. (5.73)] describes a rather nontrivial evolution of the linear response with change of temperature. In order to see that this is really the case, it will suffice to consider the spectral density described by the imaginary part of M_ω. Figure 17 presents

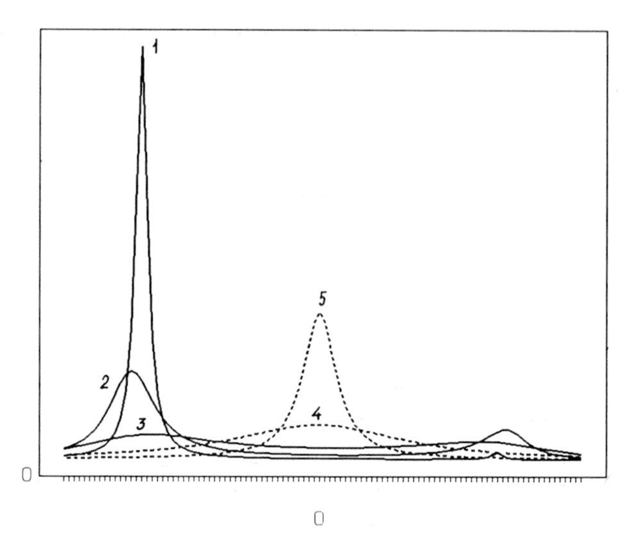

Fig. 17. Spectral function of absorption $S(\omega) \sim \operatorname{Im} M_\omega/(e^{\omega/T} - 1)$ [eq. (5.81)] with $\xi = 0$ and $b = 0.05$: (1) $T/\Delta_c = 0.25$; (2) $T/\Delta_c = 1$; (3) $T/\Delta_c = 3$; (4) $T/\Delta_c = 10$; (5) $T/\Delta_c = 50$.

spectral curves for the symmetric case $\xi = 0$ corresponding to successive increase of temperature. At low T, two distinct peaks appear, corresponding to the splitting $2\tilde{\Delta}_0(T)$ with a finite width $\gamma_2 = \Omega/2$ [see eq. (5.64)]. As the temperature increases there is first observed an increase in $\tilde{\Delta}_0(T)$ [eq. (5.8)], which is rapidly compensated for by the approach of the poles $\operatorname{Re}\omega^{(2,3)}$ [eq. (5.64)] due to an increase in $\Omega(T)$. At the same time the resonance picture becomes blurred. However, with further rise of temperature it changes to a narrow central peak, which increases with increasing T (see Weiss and Wollensak 1989, Kagan and Prokof'ev 1990a). The function M_ω then assumes a purely relaxation structure:

$$M_\omega = -\frac{\eta^2}{T}\frac{i\gamma_1}{\omega + i\gamma_1}. \tag{5.75}$$

It is interesting that with $\xi \gg \tilde{\Delta}_0$ the resonance structure is practically absent and the linear response function is described at all T by eq. (5.75). This is because the transition probability between the wells contains a small factor $(\tilde{\Delta}_0/\xi)^2$. Note that for the relaxation regime we may write a general expression for γ_1, encompassing the entire region of the parameters, including the quantity b (Kagan and Prokof'ev 1987, 1990a):

$$\gamma_1 \cong 4\frac{\tilde{\Delta}_0^2(T)\Omega}{\tilde{\varepsilon}^2 + \Omega^2}\frac{|\Gamma(1 + b + i\tilde{\varepsilon}/2\pi T)|^2}{\Gamma(1 + 2b)}\cosh(\tilde{\varepsilon}/2T). \tag{5.76}$$

6. Experimental study of quantum diffusion in a metal

6.1. Quantum diffusion of μ^+-muon

The measurement of the depolarization of μ^+-muons in a metal has opened up the interesting possibility of studying the diffusion of this light particle. The method that was used for the first time in the works of Gurevich et al. (1972) and Grebinnik et al. (1975) is based on the difference in depolarization rate, depending upon whether the particle is at rest or is moving. In a nonmagnetic metal the depolarization itself is associated with the scatter, from unit cell to unit cell, of the local hyperfine interaction between the μ^+-muon and the surrounding nuclei. During the motion of the particle, the random fields average out and the temporary decay of polarization proves to be dependent on the particle lifetime τ in the unit cell and, hence, on the diffusion coefficient. In rapid motion the time law for polarization $P(t)$ coincides with the result well-known from the theory of the so-called dynamic narrowing in NMR (e.g., Abragam 1961):

$$P(t) = P_0 e^{-\Lambda t}, \quad \Lambda = \delta^2 \tau. \tag{6.1}$$

Here δ^2 is the second moment of the local field distribution with a numerical coefficient dependent on the geometry of the experiment. The usual scale of δ is $\delta \lesssim 10^{-5}$ K, and the measurement time is limited by the lifetime of the μ^+-muon $\tau^{(0)} \approx 2 \times 10^{-6}$ s and, as a rule, does not exceed $10\tau^{(0)}$. In a perfect crystal the coherent bandwidth is known to exceed δ by many orders of magnitude. Therefore, practically no depolarization occurs during a purely band motion.

Actually, depolarization can take place only in regions of strong static or dynamic destruction of the band when ξ or $\Omega \gg \Delta$. Therefore, at low temperatures, where electron coupling predominates, for τ to be determined we must apply relation (5.16) and at high temperatures, where phonon coupling becomes significant, use must be made of the general expressions (3.23) and (3.31). If the dynamic destruction of the band prevails, the problem continues to be homogeneous and relation (6.1) retains its form. The inhomogeneity of the problem in a case where the level shifts predominate requires a solution of the kinetic equation which takes into account both local diffusion and local depolarization of particles (Kagan and Prokof'ev 1987c).

If at a low defect concentration diffusion is still rapid at distances of the order of the average distance $\bar{R}$ between the defects, then the strong depolarization regions occupy only a small fraction of space. At low T, from distances of the order of R_t [eq. (3.122)], there begins an irreversible capture into the defect region. (The interaction through a deformation field or with a spatially oscillating Kohn–Friedel potential always provides regions of attraction to the defect region.) In this case the law $P(t)$ is described by an expression analogous to eq. (6.1), with the replacement $\Lambda \to K$, where the trapping rate K is defined in accordance with eqs. (3.120) and (3.121). In the general case, in particular when the region of rapid motion is large, the two depolarization mechanisms are operating and

$$P(t) \approx P_0 \mathrm{e}^{-\sigma(T)t}, \quad \sigma(T) = \Lambda + K. \tag{6.2}$$

An example of a quasihomogeneous system in which impurities play an insignificant role is copper. The occurrence of quantum diffusion in this metal was established at an earlier stage when it was experimentally found that the activation course of $\tau^{-1}(T)$ at high $T > 100\,\mathrm{K}$ has an anomalously small pre-exponential factor, of the order of $10^{7.5}\,\mathrm{s}^{-1}$ (Gurevich et al. 1972, Grebinnik et al. 1975). As has been discussed in section 3.6, this is an indication of the tunneling nature of the motion. Here, the activation energy was nearly one order of magnitude less than for the diffusion of a proton in copper. The entire behavior of the $\tau^{-1}(T)$ curve for very pure copper samples has been obtained by Kadono et al. (1984–1986, 1989). It is shown in fig. 18.

First of all, mention should be made of the increase of D below 50 K with decreasing temperature, which is just characteristic of quantum diffusion. With $T < 10\,\mathrm{K}$, when electron coupling is known to predominate, the diffusion coefficient obeys the law given by eqs. (5.18) and (5.19). It has been found that $b \approx 0.16$. The increase of the diffusion coefficient in copper with decreasing temperature was first observed by Hartmann et al. (1980, 1981) (see also Clawson et al. 1982, 1983, Welter et al. 1983, 1984). At higher temperatures, phonon coupling begins to predominate. The sharp fall-off of $D(T)$ is tied up with the dynamic destruction of the band due to a two-phonon coupling, which is characterized by a strong dependence on T (see section 3.3). Just as in the case of the diffusion of muonium in KCl and NaCl (see section 4.1), the passage through a minimum and the appearance of a branch, which increases with T, is associated with activation subbarrier diffusion.

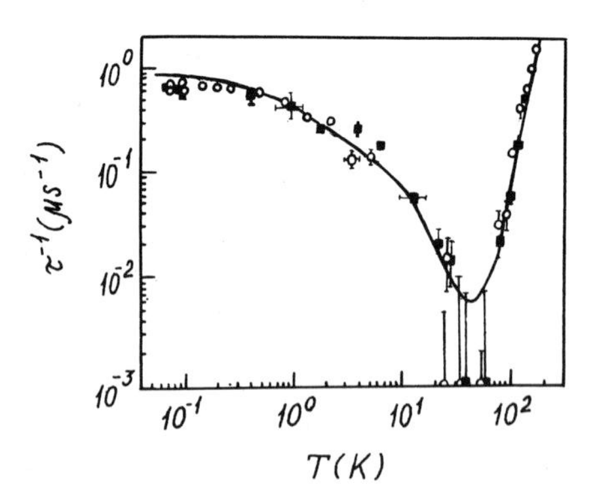

Fig. 18. Jump rate of μ^+-muon in copper: ○ sample 1 with residual resistivity ratio RRR = 18 000; ■ sample 2 (RRR = 7350) (Kadono et al. 1989).

The value found for the activation energy, $E_a \sim 700\,\mathrm{K}$, together with the value of the pre-exponential factor, makes it possible, through comparison with relation (3.104), to find J_0 and the exponent in eq. (3.11) by separating out the polaron effect or the fluctuational barrier preparation effect (see section 4.1). On the other hand, at low T, apart from b, we can also independently find $\tilde{\Delta}_0^{(\mathrm{ph})}$ through quantitative comparison with eq. (5.18). It becomes clear that the decisive factor in this case is the polaron effect and that

$$J_0 \approx 0.3\,\mathrm{K}, \quad \tilde{\Delta}_0^{(\mathrm{ph})} \approx 2 \times 10^{-3}\,\mathrm{K} \tag{6.3}$$

(in the estimates use was made of the value $\omega_0 \approx 10^3\,\mathrm{K}$). The coherent tunneling amplitude [eq. (5.7)], which determines the band, appears to be equal to

$$\Delta_c \approx 10^{-4}\,\mathrm{K}. \tag{6.4}$$

With such a value of Δ_c, the entire experimental temperature range corresponds to the dynamic destruction of the band and the diffusion coefficient is described by eq. (5.18) rather than by the band expression [eq. (5.28)].

The evolution to a constant value when $T \to 0$ is most likely associated with the transition to a regime under which the characteristic level shifts ξ become larger than T [see eq. (5.21)].

Another interesting material is aluminum. It is with this metal that the delocalization of the μ^+-muon was first experimentally detected in very pure samples, the process being characterized by the complete absence of depolarization (Hartmann et al. 1977, 1980). Since this was observed over a wide temperature range, it is evident that in aluminum Δ_c is substantially higher than in copper and that the dynamic destruction of the band does not lead to a quasilocalization of the particles. This has made it possible to study the depolarization with trapping into the impurity regions. The basic results of the investigations carried out are summarized by Hartmann et al. (1988). Figure 19 shows the low-temperature behavior of the depolarization rate $\sigma(T)$, which coincides in this case with K in eq. (6.2).

The temperature course was traced at nearly two orders of variation of T, from 0.03 to 2 K, and led to the dependence $K \sim T^{-0.7}$ (Kehr et al. 1982, Hartmann et al. 1986), which is insensitive to impurity species. In treating the theoretical dependence [eq. (3.121)] we must take into account not only the dependence $D(T)$ [eqs. (5.18) and (5.19)] but also the variation of the temperature radius of the trap, R_t, in accordance with relation (3.122). Within the T range considered the radius R_t inevitably increases with temperature according to the law $R_t \propto a(\mathscr{U}_0/T)^{1/3}$ if $\mathscr{U}(R)$ falls off with distance as R^{-3}. This yields the following relationship:

$$K \propto T^{-(1-2b+1/3)}. \tag{6.5}$$

The experimental data were analyzed without taking into account the variation of R_t with temperature, which led to the estimate $b \lesssim 0.15$. From the

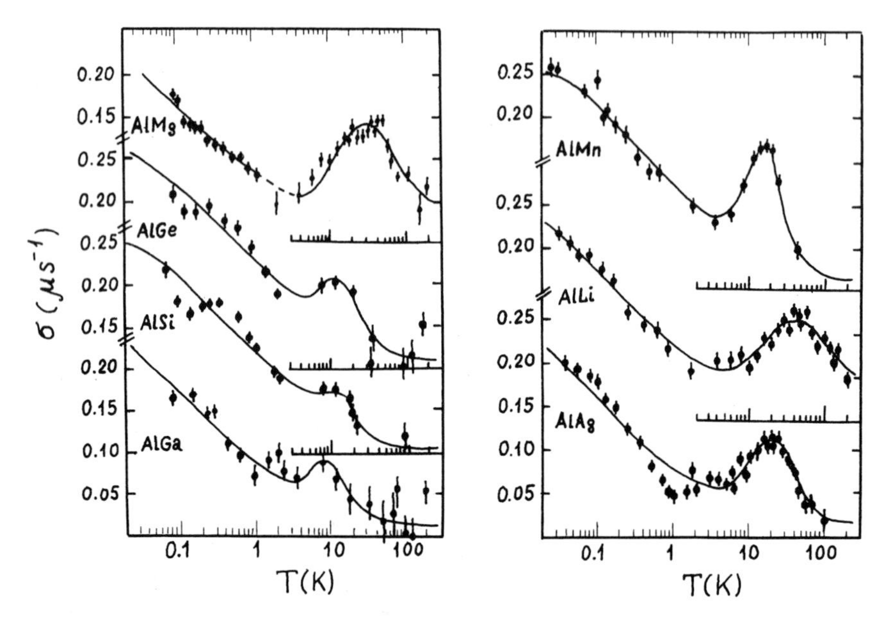

Fig. 19. Depolarization rate: (a) AlMg (92 ppm), AlGe (117 ppm), AlSi (108 ppm), and AlGa (163 ppm) (Hartmann et al. 1988); (b) AlMn (70 ppm), AlLi (70 ppm), AlAg (117 ppm) (Kehr et al. 1982).

law indicated above for K, it follows that the value of b must be close to 0.3, this being very close to a theoretical estimate (Hartmann et al. 1988).

In experiments, a slight deviation of the concentration dependence of K from the linear plot was observed. It is not clear whether this circumstance should be considered important.

A comparison of the experimental results with eq. (5.18) yields the estimate

$$\tilde{\Delta}_0^{(\mathrm{ph})} \sim 5 \times 10^{-2}\,\mathrm{K}, \qquad \Delta_c \sim 2 \times 10^{-3}\,\mathrm{K}. \tag{6.6}$$

It can be seen that the regime of dynamic band destruction persists down to very low temperatures.

An interesting picture of depolarization of the μ^+-muon must arise upon transition of the metal to the superconducting state. Let $\bar{R} \gg R_t$. The fall-off of Ω_s [eq. (5.34)] with decreasing T near T_c results in a certain increase of $D(R_t)$ since in a normal metal at the trapping radius, $\Omega > \xi(R_t)$. This leads to an increase of K [eq. (3.121)] and, hence, to an enhancement of the particle depolarization. This increase, however, is rapidly replaced by an exponential decay of $D(R_t)$ [cf. eq. (5.16)]:

$$D(R_t) \approx \frac{za^2}{3} \frac{(\Delta_c^s)^2 \Omega_s(T)}{\xi^2(R_t)}. \tag{6.7}$$

The depolarization rate, which is connected with K in the exponent of eq. (6.2), falls off rapidly. The slowing-down of the particle motion rapidly covers the entire volume of the crystal if $\xi(\bar{R})$ is larger than the bandwidth, which now is associated with Δ_c^s [eq. (5.6)]. In this case the depolarization is governed by the quantity Λ in the exponent of eq. (6.2), which is proportional to τ according to eq. (6.1). When $\xi(\bar{R}) \ll \xi(R_t)$, the increase of τ and $\sigma(T)$ begins with a noticeable shift towards low temperatures from T_c. After this, the depolarization increases very sharply.

If $\xi(\bar{R})$ is less than the bandwidth, then in a superconductor "lakes" of band motion are left, in which the particles will not undergo depolarization. The particles do not leave the "lakes" since the probability of transition into the localization region is proportional to Ω_s and is, therefore, exponentially low. As a result, with $T \ll T_c$ two independent ensembles are created, in which the particles are either completely depolarized or retain their polarization.

It should be noted that in a real case the region of increase near T_c may be either smeared or absent if $\xi(R_t)|_{T=T_c}$ is close to $\Omega(T_c)$.

This picture predicted by Kagan and Prokof'ev (1987c) has recently been detected experimentally in a study of the depolarization of the μ^+-muon upon transition of aluminum to the superconducting state ($x \sim 10^{-4}$) (Hartmann et al. 1989, Kiefl et al. 1990). The results obtained by these authors are shown in fig. 20. What attracts attention here is a large shift, from T_c, of the temperature at which a sharp increase of $\sigma(T)$ begins. This is a striking example of how the transition to the superconducting state leads, instead of the enhancement of diffusion, to a localization of the particles in a crystal with defects. The authors cited have also revealed other aspects of the picture described above, in particular the appearance of "lakes" where no depolarization takes place when they used very pure samples.

An example illustrating the characteristic change of the depolarization picture with temperature is provided by the results obtained for vanadium. The

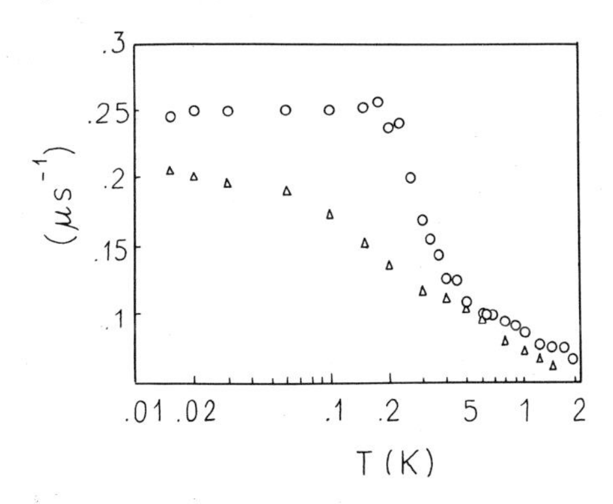

Fig. 20. Depolarization rate in AlLi (76 ppm): △ normal; ○ superconducting ($T_c = 1.15$ K) (Kiefl et al. 1991).

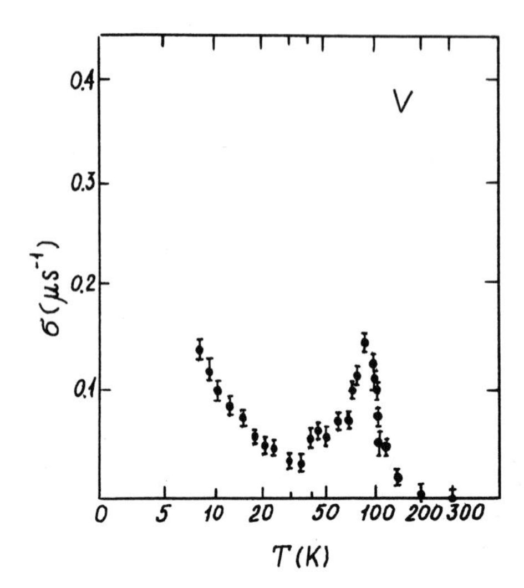

Fig. 21. Depolarization rate in vanadium with 15 ppm O (Heffner et al. 1979).

experimental data for a relatively pure sample (Heffner et al. 1979) are shown in fig. 21. We will give here a qualitative explanation of the picture. The high-temperature peak, whose position $T_{\max}$ and height are practically independent of the impurity concentration, is an indication of a change of the diffusion regime. The branch that falls off with $T > T_{\max}$ corresponds to activation diffusion, which is replaced by quantum diffusion at $T < T_{\max}$ under the conditions of dynamic destruction of the band due to phonon and electron coupling. As the temperature drops, $D(T)$ increases and the particle depolarization regime is replaced by trapping into the impurity regions. The increase of $\sigma(T)$ in eq. (6.2) with $T \to 0$ is associated with an increase of K [eqs. (6.2) and (3.121)]. The relatively weak dependence on T close to T^{-1} is evidence for the determining role of electron coupling.

Of substantial interest is the study of the change of the depolarization picture upon transition of V to the superconducting state ($T_c \approx 4.5$).

We shall limit ourselves to these examples, although there exist a number of other metals, say Nb (e.g., Hartmann et al. 1977, 1984, Borghini et al. 1978) and Bi (e.g., Barsov et al. 1984, Gugax et al. 1988), where the quantum diffusion of the μ^+-muon is very distinct.

6.2. *Tunneling of hydrogen in niobium*

Hydrogen in a metal is a very intriguing object for the study of tunneling dynamics at low temperatures. However, in a pure metallic matrix the clustering of particles (or a phase separation) hinders the observation of such motion. At

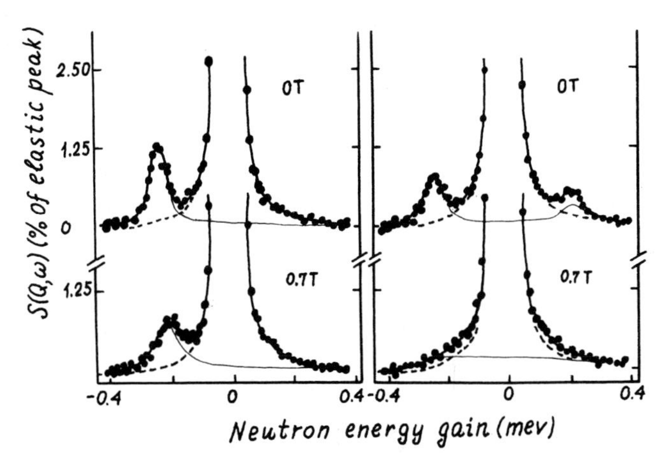

Fig. 22. Neutron spectra of $NbO_{0.0002}H_{0.0002}$ sample at 0.2 K (a) and 4.3 K (b) in the superconducting (0 T) and normal (0.7 T) states. The total, inelastic and elastic intensities are represented by bold, thin and broken lines, respectively (Wipf et al. 1987).

the same time, even at an early stage (Sellers et al. 1974) it has been found that the presence of gas impurities of N or O leads to the formation of N–H or O–H pairs, whose configuration induces the degenerate two-well system for hydrogen. Evidence for the appearance of tunnel states in this case has been furnished in experiments on heat capacity (Sellers et al. 1974, Morkel et al. 1978, Wipf and Neumaier 1984), thermal conductivity (Locatelli et al. 1978) and sound absorption (Andronikashvili et al. 1976, Poker et al. 1979). However, the most spectacular results have been obtained in studies of the inelastic scattering of neutrons on the system under consideration (Wipf et al. 1981, 1987, Magerl et al. 1986, Steinbinder et al. 1988). Comparative experiments on a single sample in the normal and superconducting states are undoubtedly of prime importance.

Figure 22 shows the neutron spectra obtained at temperatures 0.2 K (a) and 4.2 K (b). The spectra clearly exhibit a satellite peak corresponding to the transition between tunneling states with a level splitting [eqs. (5.63) and (5.70)], which at $T = 4.2\,\mathrm{K}$ is retained in the superconducting phase and actually disappears in the normal state. The latter is associated with a large magnitude of damping Ω [see eq. (5.64)] with a simultaneous decrease of the real part of the spectrum, $\omega^{(2,3)}$. These results accurately reflect the picture sketched in fig. 17. The experimental data given in fig. 22 clearly demonstrate that the tunneling splitting in the superconducting state is larger than that in the normal state, as has been pointed out by Yu and Granato (1985). This is a spectacular manifestation of the role of the electronic polaron effect and its sensitivity to a low-frequency cutoff of the infrared divergence. We have established earlier that upon transition to the superconducting state, with $2\Delta_s \gg \Delta_c$, the expression for Δ_c [eq. (5.78)] is replaced by eq. (5.6). The ratio of these quantities,

$$\left.\frac{\Delta_c^s}{\Delta_c}\right|_{T\to 0} = \left(\frac{e\Delta_s}{2\Delta_c}\right)^b > 1, \tag{6.8}$$

permits the determination of the value of the electron coupling parameter b.

Figure 23 gives the results of a treatment of experimental data for the temperature dependence of the quantity $\tilde{\Delta}_0$ and damping $\Omega(T)$ in the normal and superconducting phases. The results refer to the case of a very pure sample, where the two-level system is close to a symmetric one. It is interesting that in this case ($\xi = 0$) the linear response function (5.73) has a frequency dependence characteristic of a damped harmonic oscillator, the model of which was used for treating the experimental data.

In analyzing the results obtained, we note first the temperature independence of the quantity Δ_c^s when $T < \Delta_s(T)$ in a superconductor and the initial rise of $\tilde{\Delta}_0$ in the normal state, which corresponds to relation (5.69) or in the limiting form to eq. (5.8). Of no less interest is the relative behavior of $\Omega(T)$. In the case of a normal metal we have a linear dependence $\Omega(T)$ in accordance with eq. (5.9). In the superconducting case there is clearly observed an

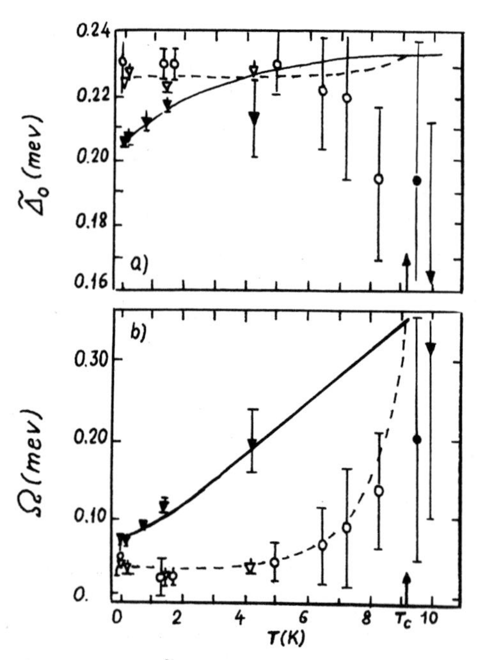

Fig. 23. Effective tunneling amplitude $\tilde{\Delta}_0$ (a) and phase damping Ω (b) versus temperature: (▼) normal and (▽) superconducting $NbO_{0.0002}H_{0.0002}$ (Wipf et al. 1987); (●) normal and (○) superconducting $NbO_{0.002}H_{0.002}$ (Magerl et al. 1986). The results of theoretical calculation with $b = 0.05$ in normal and superconducting states are shown by solid and broken lines, respectively (Wipf et al. 1987).

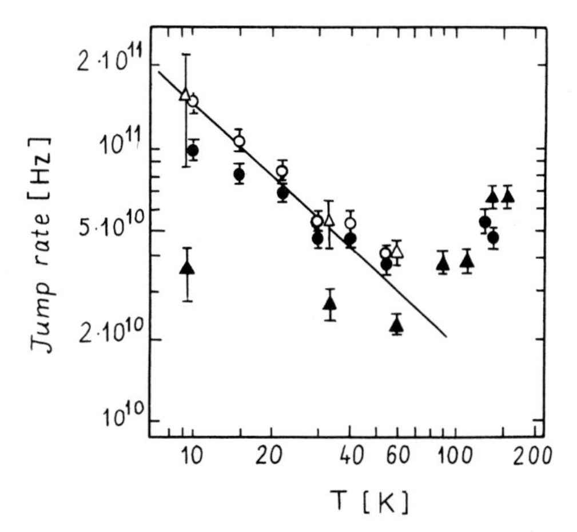

Fig. 24. Jump rates of H in $Nb(OH)_x$: ● and ○ $x = 0.002$; ▲ and △ $x = 0.011$ (Steinbinder et al. 1988). The open circles are recalculated values for a perfect crystal ($\varepsilon_i = 0$). The solid line is the $\tau^{-1} \propto T^{-(1-2b)}$ law.

exponential decrease of $\Omega(T)$ with decreasing temperature, corresponding to the result (5.34). The most remarkable point is that the entire set of data is described quantitatively by a single value of the constant b (≈ 0.05–0.06).

At high temperatures when the tunneling motion is purely incoherent and the linear response function is described by a quasielastic peak (5.75), the neutron experiments carried out by Steinbinder et al. (1988) on the same system have made it possible to determine the dependence $\gamma_1(T)$. The results of the measurements are given in fig. 24. The filled circles and triangles represent the direct measurement results. It can be seen that in a purer sample (●) the value of γ_1 falls off with temperature up to $T \sim 60\,K$. In a less pure sample (■) the temperature dependence is weak. This outcome is a direct consequence of the behavior of the hopping probability [eqs. (5.16) and (5.21)] (the quantity γ_1 defined by eq. (5.76) is the half-sum of the probabilities of direct and reverse transitions). The decrease of τ^{-1}, which is characteristic of quantum diffusion, is replaced with rise of temperature by an increase at high T due to phonon-induced transitions or classical overbarrier diffusion.

The authors cited above have made an attempt to recalculate the low-temperature branch for a perfect crystal ($\xi = 0$) by specifying the distribution function for level shifts. The results obtained are given in fig. 24 (open circles and triangles). The temperature behavior of $\tau^{-1}(T)$ found is close to the law described by eq. (5.18) (Kondo 1984, Yamada 1984, with the same value of the constant b.

Thus, the neutron experiments have made it possible to reveal the basic regularities characteristic of the tunneling in a metal when the decisive role is played by electron coupling.

6.3. Acoustic properties of metallic and superconducting glasses

The low-temperature properties of amorphous glasses are nicely described within the framework of the concept of the ensemble of two-level systems (TLS) (Anderson et al. 1972, Phillips 1972). A distribution function $P(\ln J_0, \xi)$ is introduced for the parameters of a TLS and all results obtained for an individual TLS must be averaged using this distribution function. In dealing with the acoustic properties we see that, at least under the assumption of a weak interaction of the TLS with excitations of a crystal, all the results are rather universal. The difference between phonons and electrons is associated in this case only with the difference in the function $\Omega(\varepsilon)$ [eq. (5.60)] in the expression for γ_1 [eq. (5.68)] (e.g., Black 1981, Hunklinger and Raychaudhuri 1986).

In the general case, however, as we have seen in the previous sections, electron coupling is strong, which leads, on the one hand, to the appearance of an electronic polaron effect and a self-consistent renormalization of the spectrum and, on the other, to a dynamic destruction of the coherent two-well states. A large number of experiments have recently been carried out, in which the acoustic properties of metallic glasses in the normal and superconducting states have been compared. The qualitative features that have been revealed go beyond the framework of the standard concepts. We will focus our attention only on this series of experiments. A description of the ordinary acoustic and thermodynamic properties of insulating and metallic glasses can be found in detailed reviews (e.g., Hunklinger and Arnold 1976, Black 1981, Hunklinger and Raychaudhuri 1986).

As a result of the interaction of sound with TLSs, two contributions arise to both the renormalization of sound velocity, $\Delta v/v$, and the absorption of sound α: a resonance and a relaxation contribution. For frequencies $\omega \ll T$ they are distinctly separated. Below we shall see that upon averaging over the parameters J_0 and ξ, the major contribution comes from TLSs with energies higher than or of the order of temperature. In this case eq. (5.74) assumes an especially simple form ($\Omega \ll \tilde{\varepsilon}$):

$$M_\omega = -2\eta^2 \left\{ \left(\frac{\partial \tilde{\varepsilon}}{\partial \xi} \right) \frac{\partial \tanh(\tilde{\varepsilon}/2T)}{\partial \xi} \frac{\mathrm{i}\gamma_1}{(\omega + \mathrm{i}\gamma_1)} + \frac{\partial^2 \tilde{\varepsilon}}{\partial \xi^2} \tanh(\tilde{\varepsilon}/2T) \right\}. \tag{6.9}$$

The renormalization of sound velocity and the sound absorption are determined, respectively, by the real and the imaginary part of eq. (6.9):

$$\frac{\Delta v}{v} = \frac{1}{2\rho v^2} \langle \operatorname{Re} M_\omega \rangle_{\mathrm{TLS}}, \qquad \alpha = -\frac{\omega}{2\rho v^3} \langle \operatorname{Im} M_\omega \rangle_{\mathrm{TLS}}, \tag{6.10}$$

where ρ is the density of a metallic glass and the notation $\langle \ldots \rangle_{\mathrm{TLS}}$ signifies the averaging of the expression with a weight $P(\ln J_0, \xi)\, \mathrm{d}\xi\, \mathrm{d}J_0/J_0$. Usually, the distribution function P is taken to be equal to a certain constant $\bar{P}$.

The resonance renormalization of sound, $(\Delta v/v)_{\text{res}}$, is associated with the modulation of the energy splitting $\tilde{\varepsilon}$ for TLSs with the occupation numbers remaining unchanged. This mechanism is efficient when $\tilde{\varepsilon} > T$ and is described by the second term in eq. (6.9):

$$\left.\frac{\Delta v}{v}\right|_{\text{res}} = -C\int_0^{J_0^{\max}} \frac{\mathrm{d}J_0}{J_0}\int_0^{\xi^{\max}} \mathrm{d}\xi\, \frac{\partial^2\tilde{\varepsilon}}{\partial\xi^2}\tanh(\tilde{\varepsilon}/2T), \quad C = \frac{\bar{P}\eta^2}{\rho v^2}. \tag{6.11}$$

The integral (6.11) may be written to logarithmic accuracy as follows:

$$\left.\frac{\Delta v}{v}\right|_{\text{res}} \approx -C\int_{\Delta_c \sim T}^{J_0^{\max}} \frac{\mathrm{d}J_0}{J_0}. \tag{6.12}$$

The important point here is that the lower limit of integration is determined by the renormalized tunneling amplitude [eq. (5.70)], which, in contrast to J_0, is a true physical parameter of TLSs.

The relaxation contribution to $\Delta v/v$ is associated with the change of the level population in the field of a sound wave and is described by the first term in eq. (6.9):

$$\left.\frac{\Delta v}{v}\right|_{\text{rel}} = -C\int_0^{\infty} \frac{\mathrm{d}J_0}{J_0}\int_0^{\infty} \mathrm{d}\xi \left(\frac{\partial\tilde{\varepsilon}}{\partial\xi}\right)^2 \frac{1}{2T\cosh^2(\tilde{\varepsilon}/2T)}\,\frac{1}{1+(\omega/\gamma_1)^2}. \tag{6.13}$$

It can be seen that this mechanism is efficient only for thermal TLSs with $\tilde{\varepsilon} \sim T$, whose inverse longitudinal relaxation time is sufficiently large ($\omega/\gamma_1 < 1$). Since $\omega \ll T$, in a normal metal there is, in fact, always an appreciable relaxation contribution. To the same logarithmic accuracy,

$$\left.\frac{\Delta v}{v}\right|_{\text{rel}} = -C\int_{\Delta_c \sim \Delta_c'}^{\Delta_c \sim T} \frac{\mathrm{d}J_0}{J_0}, \tag{6.14}$$

where the quantity Δ_c' is determined by the condition $\omega/\gamma_1 \approx 1$. Using the expression for γ_1 [eq. (5.76)] we obtain the estimate

$$J_0' \approx T\left(\frac{\omega}{\Omega}\right)^{1/2}\left(\frac{\omega}{T}\right)^{b}, \qquad \Delta_c' \approx T\left(\frac{\omega}{\Omega}\right)^{1/[2(1-b)]}. \tag{6.15}$$

From eqs. (6.12) and (6.14) it follows that the two contributions to $\Delta v/v$ have a negative sign but opposite temperature dependences. Upon transition of a metal to the superconducting state, the relaxation times sharply increase due to the exponential fall-off of Ω_s [eq. (5.34)] and the relaxation contribution rapidly disappears. One would think that the overall renormalization of the sound velocity in a superconductor must always be less in absolute value than $|\Delta v/v|$ in the normal phase. This is not the case in reality. It has been experimentally established (Raychaudhuri and Hunklinger 1984, Esquinazi et

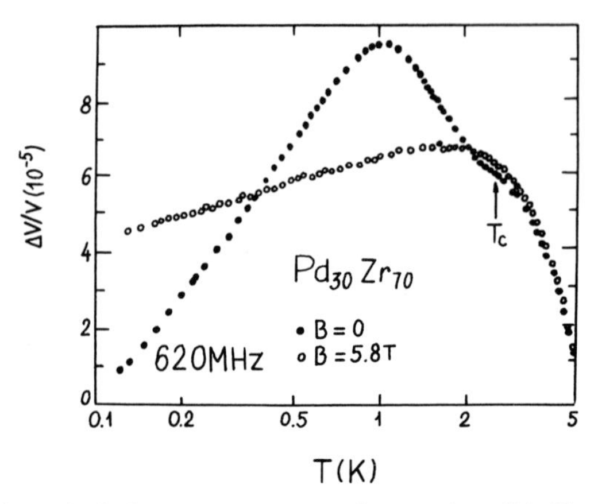

Fig. 25. Variation of sound velocity versus temperature in amorphous $Pd_{30}Zr_{70}$ in normal (○) and superconducting (●) states (Esquinazi et al. 1986).

al. 1986, Weiss 1990) that the $\Delta v/v$ curves intersect in the normal and superconducting phases of a $Pd_{30}Zr_{70}$ alloy and that at low T the sound velocity has a lower value in a superconductor (see fig. 25).

The seeming contradiction is easily removed if we take into account the nonlinear nature of the polaron effect in metallic glasses (Kagan and Prokof'ev 1988, 1990a). Indeed, the change of the variable J_0 (or $\tilde{\Delta}_0^{\mathrm{ph}}$) by the renormalized parameter Δ_c [eq. (5.70)] gives

$$\frac{\mathrm{d}J_0}{J_0} = (1 - b)\,\frac{\mathrm{d}\Delta_c}{\Delta_c}. \tag{6.16}$$

The appearance of the factor $(1 - b)$ may be regarded as a change in the distribution function P in going over to the physical parameters of the system. It is essential that relation (6.16) explicitly depends on the interaction of two-level systems with electrons. In a superconductor, for two-level systems with $\Delta_c < 2\Delta_s$ there is a linear relationship $\Delta_c^s \propto J_0$ [eq. (5.6)] and eq. (6.16) does not contain the quantity b. On the basis of these considerations, we have

$$\left.\frac{\Delta v}{v}\right|_{\mathrm{res}} \approx \begin{cases} -\,C(1-b)\ln(T_0/T), \\ -\,C\ln(2\Delta_s/T) - C(1-b)\ln(T_0/2\Delta_s), \quad T < T_c. \end{cases} \tag{6.17}$$

Performing an analogous replacement of the variables of eq. (6.16) in relation (6.14), we get

$$\left.\frac{\Delta v}{v}\right|_{\mathrm{rel}} \approx -\,C/2\ln(\Omega/\omega) \tag{6.18}$$

or, for the sum in the normal state,

$$\frac{\Delta v}{v} \approx -C(1/2-b)\ln(T_0/T) - \text{const} \tag{6.19}$$

In a superconducting metal the relaxation contribution is important only as long as $\omega \ll \Omega_s$ (or $\ll \Omega^{ph}$ if $T \ll T_c$) and formally has the same structure [eqs. (6.14) and (6.18)], with the replacement $\Omega \to \Omega_s$. Here, the value of the minimum amplitude in the logarithmic integral (6.14) is given by

$$(\Delta_c^s)' \approx T\left(\frac{\omega}{\Omega_s}\right)^{1/2} \tag{6.20}$$

From a comparison of eq. (6.17) in the normal and superconducting phases it follows that the amplitude with the $\ln T$ resonance renormalization of sound is larger in the superconducting state. This is just the explanation for the intersection of the curves in fig. 25. If we use relations (6.17) and (6.19), then from the intersection of the curves we can estimate the value of the parameter b. The value of this parameter for the $Pd_{30}Zr_{70}$ alloy has been found to be ≈ 0.2 (Weiss 1990).

Thus, the explanation of the observed experimental results does not require the abandoning of the model of tunnel states in an amorphous metal, as has been discussed. Quite the contrary, they demonstrate the role of the self-consistent nonlinear renormalization of the spectrum, which is characteristic for a strong interaction of tunnel states with electrons in a metal. It should be stressed that such an interaction leads to the nonuniversality of the behavior of the sound velocity in metallic glasses.

The absorption of sound [eq. (6.10)] in metallic glasses is completely governed by relaxation processes ($\omega \ll T$). It is well known [in agreement with eq. (6.9)] that under this regime the result is determined by two-level systems with $\varepsilon \sim T$, to which there simultaneously corresponds the condition $\omega/\gamma_1 \approx 1$. In such a case we find straightforwardly

$$\alpha|_{\text{rel}} \approx C\frac{\pi\omega}{2v} P(\ln J_0')/\bar{P} \tag{6.21}$$

[J_0' is defined from the condition (6.15)]. In the superconducting state, after the condition $\Omega_s \sim \omega$ is satisfied, the absorption of sound begins to fall sharply since $\alpha \sim \gamma_1 \sim \Omega_s$ (or $\alpha \sim \gamma_1^{ph}$ if the phonon scattering mechanism prevails). It is exactly this picture, which provides an evidence for the role of electron coupling in tunneling dynamics in a metal, that has been detected by Weiss et al. (1980) and Arnold et al. (1982).

However, in later experiments (Raychaudhuri and Hunklinger 1984, Esquinazi et al. 1986, Neckel et al. 1986, Esquinazi and Luzuriaga 1988) a different picture was observed for the case of extremely low sound frequencies

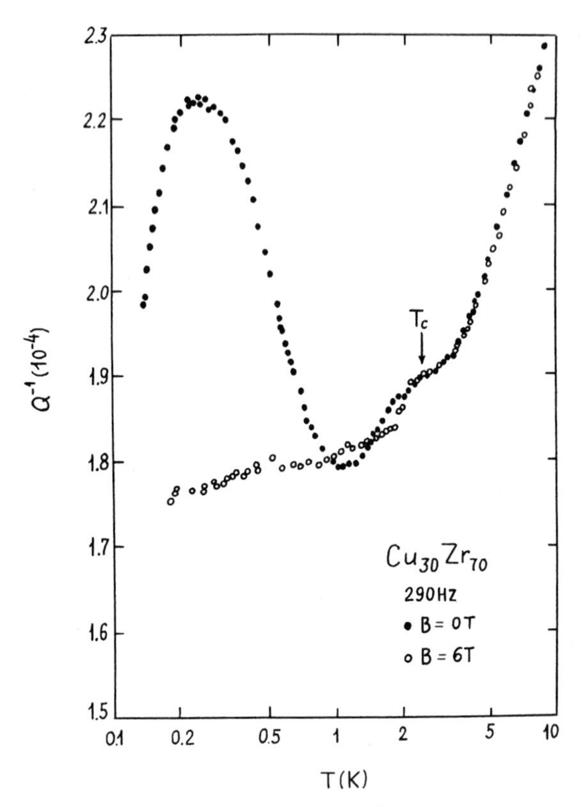

Fig. 26. Temperature dependence of internal friction $Q^{-1} = \alpha v/\omega$ in amorphous $Cu_{30}Zr_{70}$ in normal (○) and superconducting (●) states (Esquinazi et al. 1986).

$\omega \sim 1$–10^3 Hz in $Pd_{30}Zr_{70}$ and $Cu_{30}Zr_{70}$. The sound absorption, which usually varies little in a normal metal at low T, increased sharply upon transition to the superconducting state (see fig. 26). In order to account for this result (Kagan and Prokof'ev 1988), we go back to eq. (6.21) and analyze the characteristic values of the quantity J'_0 [eq. (6.15)]. For the range of low frequencies indicated we have a direct estimate, $J'_0 \sim 10^{-3}$–10^{-4} K. It seems reasonable to presume that for such low values of the tunneling amplitude we can no longer regard the distribution function to be constant. Physically, it is evident that in the limit $J_0 \to 0$ the distribution function must fall off. In a normal metal $J'_0 \propto T^{1/2-b}$ and we have, with decrease of T, a gradual fall-off of J'_0 and, hence, of $P(\ln J'_0)$ in relation (6.21). In the superconducting phase the exponential decay of Ω_s [eq. (5.34)] leads, on the contrary, to a drastic increase of $(\Delta^s_c)'$ and J'_0 [eq. (6.20)] in accordance with the law $T^{1/2}e^{\Delta_s/2T}$. As a result, the fall of α with decreasing T is replaced by an increase in sound absorption. The passage of α through a maximum and its subsequent decrease are associated, as before, with the fulfillment of the condition $\gamma_1^{max} \approx \Omega_s \sim \omega$. Thus, the low-frequency picture of sound absorption is tied up in many respects with the distribution of a two-level system over the tunneling parameter in the region of low J_0. Here it is necessary to abandon the traditional P = const concept.

The validity of the assertions made above can be supported independently by considering the low-frequency renormalization of sound velocity. Indeed, if the characteristic value of J_0^{min} (or Δ_c^{min}) at which the distribution function begins to fall off appreciably is found to be higher than the value specified by eq. (6.15), then the total value of $\Delta v/v$ in a normal metal ceases to be temperature dependent:

$$\frac{\Delta v}{v} = -C \int_{J_0^{min}}^{J_0^{max}} \frac{dJ_0}{J_0} = \text{const.} \tag{6.22}$$

Here the transition to the superconducting state must not alter, at first, the universal result (6.22) as long as $(\Delta_c^s)' < (\Delta_c^s)^{min}$. Such a behavior of $\Delta v/v$ has been distinctly revealed for $Pd_{30}Zr_{70}$ and $Cu_{30}Zr_{70}$ by Raychaudhuri and Hunklinger (1984), Esquinazi et al. (1986) and Neckel et al. (1986). In these experiments the authors have detected a noticeable latent interval below T_c, in which the sound velocity was, in fact, the same in the normal and superconducting states.

References

Abragam, A., 1961, The Principles of Nuclear Magnetism (Clarendon Press, Oxford).

Alefeld, G., and J. Völkl, eds, 1978, Hydrogen in Metals, Vol. 1 (Springer, Berlin).

Allen, A.R., and M.G. Richards, 1978, Phys. Lett. A **65**, 36.

Allen, A.R., M.G. Richards and J. Schratter, 1982, J. Low Temp. Phys. **47**, 289.

Anderson, P.W., 1967, Phys. Rev. Lett. **18**, 1049; Phys. Rev. **164**, 352.

Anderson, P.W., B.I. Halperin and C. Varma, 1972, Philos. Mag. **25**, 1.

Andreev, A.F., 1975, Zh. Eksp. & Teor. Fiz. **68**, 2341 [1976, Sov. Phys.-JETP **41**, 1170].

Andreev, A.F., and I.M. Lifshitz, 1969, Zh. Eksp. & Teor. Fiz. **52**, 2057 [1969, Sov. Phys.-JETP **29**, 1107].

Andronikashvili, E.I., V.A. Melik-Shakhnazarov and I.A. Naskidashvili, 1976, J. Low Temp. Phys. **23**, 1.

Appel, J., 1968, Solid State Phys. **21**, 193.

Arnold, W., A. Billman, P. Doussineau and A. Levelut, 1982, J. Phys. Colloq. (Paris) **43**, C9-537.

Barsov, S.G., A.L. Getalov, V.G. Grebinnik, V.A. Gordeev, I.I. Gurevich, Yu.M. Kagan, A.I. Klimov, S.P. Kruglov, L.A. Kuzmin, A.B. Lazarev, S.M. Mikirtychyants, B.A. Nikolsky, A.V. Pirogov, A.N. Ponomarev, V.I. Selivanov, V.A. Suetin and V.A. Zhukov, 1984, Hyperfine Interact. **17–19**, 145.

Bilz, H., and W. Kress, 1979, Phonon Dispersion Relations in Insulators (Springer, Berlin).

Black, J.L., 1981, Glassy Metals, in: Topics in Applied Physics, Vol. 46, eds H.J. Guntherodt and H. Beck (Springer, Berlin).

Black, J.L., and P. Fulde, 1979, Phys. Rev. Lett. **43**, 453.

Borghini, M., T.O. Niinikoski, J.C. Soulie, O. Hartmann, E. Karlsson, L.O. Norlin, K. Pernestal, K.W. Kehr, D. Richter and E. Walker, 1978, Phys. Rev. Lett. **40**, 1723.

Brovman, E.G., and Yu.M. Kagan, 1967, Zh. Eksp. & Teor. Fiz. **52**, 557 [1967, Sov. Phys.-JETP **25**, 362].

Burin, A.L., L.A. Maksimov and I.Ya. Polishchuk, 1989, Zh. Eksp. & Teor. Fiz. Pis'ma **49**, 680.

Caldeira, A.O., and A.J. Leggett, 1981, Phys. Rev. Lett. **46**, 211.

Clawson, C.W., K.M. Crowe, S.E. Kohn, S.S. Rosenblum, C.Y. Huang, J.L. Smith and J.H. Brewer, 1982, Physica B+C **109–110**, 2164.

Clawson, C.W., K.M. Crowe, S.E. Kohn, S.S. Rosenblum, C.Y. Huang, J.L. Smith and J.H. Brewer, 1983, Phys. Rev. Lett. **51**, 114.

De Gennes, P.G., 1966, Conductivity of Metals and Alloys (Benjamin, New York).

Eselson, B.N., V.N. Grigoriev, V.N. Ivantsov, E.Ya. Rudavsky, D.G. Sanikidze and I.A. Serbin, 1973, Solutions of Quantum He^3–He^4 Liquids (in Russian) (Nauka, Moscow).

Esquinazi, P., and J. Luzuriaga, 1988, Phys. Rev. B **37**, 7819.

Esquinazi, P., H.M. Ritter, H. Neckel and S. Hunklinger, 1986, Z. Phys. B **64**, 81.

Fisher, M.P.A., and A.T. Dorsey, 1985, Phys. Rev. Lett. **54**, 1609.

Fleurov, V.N., and L.I. Trakhtenberg, 1983, Solid State Commun. **46**, 755.

Flynn, C.P., and A.M. Stoneham, 1970, Phys. Rev. B **1**, 3966.

Fujii, S., 1979, J. Phys. Soc. Jpn. **46**, 1833.

Gogolin, A.A., 1980, Zh. Eksp. & Teor. Fiz. Pis'ma **32**, 30 [1980, JETP Lett. **32**, 281].

Goldanski, V.I., L.I. Trakhtenberg and V.N. Fleurov, 1986, Tunneling Phenomena in Chemical Physics (Nauka, Moscow).

Grabert, H., 1987, in: Quantum Aspects of Molecular Motions in Solids, Springer Proceedings in Physics, Vol. 17, eds A. Heidemann et al. (Springer, Berlin) p. 130.

Grabert, H., and U. Weiss, 1985, Phys. Rev. Lett. **54**, 1605.

Grabert, H., S. Linkwitz, S. Dattagupta and U. Weiss, 1986, Europhys. Lett. **2**, 631.

Gradshteyn, I.S., and I.M. Ryzhik, 1965, Tables of Integrals, Series and Products (Academic Press, New York).

Grebinnik, V.G., I.I. Gurevich, V.A. Zhukov, A.P. Manich, E.A. Meleshko, I.A. Muratova, B.A. Nikolsky, V.I. Selivanov and V.A. Suetin, 1975, Zh. Eksp. & Teor. Fiz. **68**, 1548 [1975, Sov. Phys.-JETP **41**, 777].

Grigoriev, V.N., B.N. Eselson, V.A. Mikheev and Yu.E. Shul'man, 1973, Pis'ma Zh. Eksp. & Teor. Fiz. **17**, 25 [1973, JETP Lett. **17**, 161].

Gugax, F.N., B. Hitti, E. Lippelt, A. Schenck and S. Barth, 1988, Z. Phys. B **71**, 473.

Gurevich, I.I., E.A. Meleshko, I.A. Muratova, B.A. Nikolsky, V.S. Roganov, V.I. Selivanov and B.V. Sokolov, 1972, Phys. Lett. A **40**, 143.

Guyer, R.A., and L.I. Zane, 1969, Phys. Rev. **188**, 445.

Guyer, R.A., R.C. Richardson and L.I. Zane, 1971, Rev. Mod. Phys. **43**, 532.

Hartmann, O., E. Karlsson, K. Pernestal, M. Borghini, T.O. Niinikoski and L.O. Norlin, 1977, Phys. Lett. A **61**, 141.

Hartmann, O., E. Karlsson, L.O. Norlin, T.O. Niinikoski, K.W. Kehr, D. Richter, J.M. Welter, A. Yaouanc and J. Le Hericy, 1980, Phys. Rev. Lett. **44**, 337.

Hartmann, O., L.O. Norlin, A. Yaouanc, J. Le Hericy, E. Karlsson and T.O. Niinikoski, 1981, Hyperfine Interact. **8**, 533.

Hartmann, O., E. Karlsson, R. Wäppling, D. Richter, R. Hempelmann, B. Patterson, E. Holzshuh, W. Kündig, K. Schulze and S.F.J. Cox, 1984, Hyperfine Interact. **17–19**, 183.

Hartmann, O., E. Karlsson, E. Wäckelgärd, R. Wäppling, D. Richter, R. Hempelmann and T.O. Niinikoski, 1986, Hyperfine Interact. **31**, 223.

Hartmann, O., E. Karlsson, E. Wäckelgärd, R. Wäppling, D. Richter, R. Hempelmann and T.O. Niinikoski, 1988, Phys. Rev. B **37**, 4425.

Hartmann, O., E. Karlsson, S. Harris, R. Wäppling and T.O. Niinikoski, 1989, Phys. Lett. A **142**, 504.

Heffner, R.H., J.A. Brown, R.L. Hutson, M. Leon, D.M. Parkin, M.E. Schillaci, W.B. Gauster, O.N. Carlson, D.K. Renbein and A.T. Fiory, 1979, Hyperfine Interact. **6**, 237.

Holstein, T., 1959, Ann. Phys. (New York) **28**, 343.

Hunklinger, S., and W. Arnold, 1976, in: Physical Acoustics, Vol. XII, eds W.P. Mason and R.N. Thurston (Academic Press, New York) p. 155.

Hunklinger, S., and A.K. Raychaudhuri, 1986, in: Progress in Low Temperature Physics, Vol. X, ed. D.F. Brewer (North-Holland, Amsterdam) ch. 3.

Ikeya, M., L.O. Schwan and T. Miki, 1978, Solid State Commun. **27**, 891.

Iskovskih, A.S., A.Ya. Katunin, I.I. Lukashevich, V.V. Skyarevski, V.V. Suraev, V.V. Filippov and V.A. Shevtsov, 1986, Zh. Eksp. & Teor. Fiz. **91**, 1832 [Sov. Phys.-JETP **64**, 1085].

Ivliev, A.V., A.Ya. Katunin, I.I. Lukashevich, V.V. Skyarevski, V.V. Suraev, V.V. Filippov, N.I. Filippov and V.A. Shevtsov, 1982, Pis'ma Zh. Eksp. & Teor. Fiz. **36**, 391 [JETP Lett. **36**, 4721].

Ivliev, A.V., A.S. Iskovskih, A.Ya. Katunin, I.I. Lukashevich, V.V. Skyarevski, V.V. Suraev, V.V. Filippov, N.I. Filippov and V.A. Shevtsov, 1983, Pis'ma Zh. Eksp. & Teor. Fiz. **38**, 317 [JETP Lett. **38**, 3791].

Jäckle, J., and K.W. Kehr, 1983, J. Phys. F **13**, 753.

Jäckle, J., L. Piché, W. Arnold and S. Hunklinger, 1976, J. Non-Cryst. Solids **20**, 365.

Kadono, R., J. Imazato, K. Nishiyama, K. Nagamine, T. Yamazaki, D. Richter and J.M. Welter, 1984, Hyperfine Interact. **17–19**, 109.

Kadono, R., J. Imazato, K. Nishiyama, K. Nagamine, T. Yamazaki, D. Richter and J.M. Welter, 1985, Phys. Lett. A **109**, 61.

Kadono, R., T. Matsuzaki, K. Nagamine, T. Yamazaki, D. Richter and J.M. Welter, 1986, Hyperfine Interact. **31**, 205.

Kadono, R., J. Imazato, T. Matsuzaki, K. Nishiyama, K. Nagamine, T. Yamazaki, D. Richter and J.M. Welter, 1989, Phys. Rev. B **39**, 23.

Kadono, R., R.F. Kiefl, E.J. Ansaldo, J.H. Brewer, M. Celio, S.R. Kreitzman and G.M. Luke, 1990, Phys. Rev. Lett. **64**, 665.

Kagan, Yu., 1981, Defects in Insulating Crystals, Proc. of Intern. Conf., Riga (Springer, Berlin).

Kagan, Yu., and M.I. Klinger, 1974, J. Phys. C **7**, 2791.

Kagan, Yu., and M.I. Klinger, 1976, Zh. Eksp. & Teor. Fiz. **70**, 255 [Sov. Phys.-JETP **43**, 1321].

Kagan, Yu., and L.A. Maksimov, 1973, Zh. Eksp. & Teor. Fiz. **65**, 622 [1974, Sov. Phys.-JETP **38**, 307].

Kagan, Yu., and L.A. Maksimov, 1980, Zh. Eksp. & Teor. Fiz. **79**, 1363 [Sov. Phys.-JETP **52**, 688].

Kagan, Yu., and L.A. Maksimov, 1983a, Zh. Eksp. & Teor. Fiz. **84**, 792 [Sov. Phys.-JETP **57**, 459].

Kagan, Yu., and L.A. Maksimov, 1983b, Phys. Lett. A **95**, 242.

Kagan, Yu., and L.A. Maksimov, 1984, Zh. Eksp. & Teor. Fiz. **87**, 348 [Sov. Phys.-JETP **60**, 201].

Kagan, Yu., and N.V. Prokof'ev, 1986a, Pis'ma Zh. Eksp. & Teor. Fiz. **43**, 434 [JETP Lett. **43**, 558].

Kagan, Yu., and N.V. Prokof'ev, 1986b, Zh. Eksp. & Teor. Fiz. **90**, 2176 [Sov. Phys.-JETP **63**, 1276].

Kagan, Yu., and N.V. Prokof'ev, 1987a, Pis'ma Zh. Eksp. & Teor. Fiz. **45**, 91 [JETP Lett. **45**, 115].

Kagan, Yu., and N.V. Prokof'ev, 1987b, Zh. Eksp. & Teor. Fiz. **93**, 366 [Sov. Phys.-JETP **66**, 211].

Kagan, Yu., and N.V. Prokof'ev, 1987c, Preprint IAE-446/9.

Kagan, Yu., and N.V. Prokof'ev, 1988, Solid State Commun. **65**, 1385.

Kagan, Yu., and N.V. Prokof'ev, 1989a, Zh. Eksp. & Teor. Fiz. **96**, 1473 [Sov. Phys.-JETP **69**, 836].

Kagan, Yu., and N.V. Prokof'ev, 1989b, Zh. Eksp. & Teor. Fiz. **96**, 2209 [Sov. Phys.-JETP **69**, 1350].

Kagan, Yu., and N.V. Prokof'ev, 1990a, Zh. Eksp. & Teor. Fiz. **97**, 1698 [Sov. Phys.-JETP **70**, 990].

Kagan, Yu., and N.V. Prokof'ev, 1990b, Phys. Lett. A **150**, 320.

Kagan, Yu., L.A. Maksimov and N.V. Prokof'ev, 1982, Pis'ma Zh. Eksp. & Teor. Fiz. **36**, 204 [JETP Lett. **36**, 2951].

Katunin, A.Ya., I.I. Lukashevich, S.T. Orosmamatov, V.V. Skyarevski, V.V. Suraev, V.V. Filippov, N.I. Filippov and V.A. Shevtsov, 1981, Pis'ma Zh. Eksp. & Teor. Fiz. **34**, 375 [JETP Lett. **34**, 357].

Katunin, A.Ya., I.I. Lukashevich, S.T. Orosmamatov, V.V. Skyarevski, V.V. Suraev, V.V. Filippov, N.I. Filippov and V.A. Shevtsov, 1982, Phys. Lett. A. **87**, 483.

Kehr, K.W., D. Richter, J.M. Welter, O. Hartmann, E. Karlsson, L.O. Norlin, T.O. Niinikoski and A. Yaouanc, 1982, Phys. Rev. B **26**, 567.

Kiefl, R.F., R. Kadono, L.H. Brewer, G.M. Luke, H.K. Yen, M. Celio and E.J. Ansaldo, 1989, Phys. Rev. Lett. **62**, 792.

Kiefl, R.F., R. Kadono, S.R. Kreitzman, Q. Li, T. Pfiz, T.M. Riseman, H. Zhou, R. Wäppling, S. Harris, O. Hartmann, E. Karlsson, R. Hempelmann, D. Richter, T.O. Niinikoski, L.P. Le, G.M. Luke, B. Sternlieb and E.J. Ansaldo, 1990, Hyperfine Interact. **64**, 737.

Klinger, M.I., 1968, Rep. Prog. Phys. **31**, 225.

Kondo, J., 1976a, Physica B+C **84**, 40.

Kondo, J., 1976b, Physica B+C **84**, 207.

Kondo, J., 1984, Physica B+C **126**, 377.

Kondo, J., 1988, Fermi Surface Effects, in: Springer Series in Solid State Science, Vol. 77, ed. P. Fulde (Springer, Berlin) p. 1.

Kramers, H.A., 1940, Physica **7**, 284.

Landesman, A., 1975, Phys. Lett. A **54**, 137.

Leggett, A.J., S. Chakravarty, A.T. Dorsey, M.P.A. Fisher, A. Garg and W. Zwerger, 1987, Rev. Mod. Phys. **59**, 1.

Lifshitz, I.M., and Yu.M. Kagan, 1972, Zh. Eksp. & Teor. Fiz. **62**, 385 [Sov. Phys.-JETP **35**, 206].

Liu, S.H., 1987, Phys. Rev. Lett. **58**, 2706.

Liu, S.H., 1988, Phys. Rev. B **37**, 3542.

Locatelli, M., K. Neumaier and H. Wipf, 1978, J. Phys. Colloq. (Paris) **39**, C6-995.

Magerl, A., A.-J. Dianoux, H. Wipf, K. Neumaier and I.S. Anderson, 1986, Phys. Rev. Lett. **56**, 159.

Mahan, G.D., 1981, Many-Particle Physics (Plenum Press, New York).

Mel'nikov, V.I., 1984, Zh. Eksp. & Teor. Fiz. **87**, 663 [Sov. Phys.-JETP **60**, 383].

Mikheev, V.A., B.N. Eselson, V.N. Grigoriev and N.P. Mikhin, 1977, Fiz. Nizk. Temp. **3**, 386 [Sov. J. Low Temp. Phys. **3**, 186].

Mikheev, V.A., V.A. Maydanov and N.P. Mikhin, 1982, Fiz. Nizk. Temp. **8**, 1000 [Sov. J. Low Temp. Phys. **8**, 505].

Mikheev, V.A., N.P. Mikhin and V.A. Maydanov, 1983a, Fiz. Nizk. Temp. **9**, 901 [Sov. J. Low Temp. Phys. **9**, 4615].

Mikheev, V.A., V.A. Maydanov and N.P. Mikhin, 1983b, Solid State Commun. **48**, 361.

Mikheev, V.A., V.A. Maydanov and N.P. Mikhin, 1984, Proc. 17th Int. Conf. on Low Temperature Physics (North-Holland, Amsterdam) Part 1, p. 533.

Miyazaki, T., K.-P. Lee, K. Fueki and A. Takeuchi, 1984, J. Phys. Chem. **88**, 4959.

Morkel, C., H. Wipf and K. Neumaier, 1978, Phys. Rev. Lett. **40**, 947.

Morosov, A.I., 1979, Zh. Eksp. & Teor. Fiz. **77**, 147 [Sov. Phys.-JETP **50**, 738].

Morosov, A.I., and A.S. Sigov, 1985, Solid State Commun. **53**, 31.

Neckel, H., P. Esquinazi, G. Weiss and S. Hunklinger, 1986, Solid State Commun. **57**, 151.

Nozières, P., and C.T. de Dominicis, 1969, Phys. Rev. **178**, 1097.

Petzinger, K.G., 1982, Phys. Rev. **26**, 6530.

Phillips, W.A., 1972, J. Low Temp. Phys. **7**, 351.

Poker, D.B., G.C. Setser, A.V. Granato and H.K. Birnbaum, 1979, Z. Phys. Chem. **116**, 39.

Raychaudhuri, A.K., and S. Hunklinger, 1984, Z. Phys. B **57**, 113.

Richards, M.G., J. Pope and A. Widom, 1972, Phys. Rev. Lett. **29**, 708.

Richards, M.G., J. Pope, P.S. Tofts and J.H. Smith, 1976, J. Low Temp. Phys. **24**, 1.

Schmid, A., 1983, Phys. Rev. Lett. **51**, 1506.

Sellers, G.J., A.C. Anderson and H.K. Birnbaum, 1974, Phys. Rev. B **10**, 2771.

Shklovsky, B.I., and A.L. Efros, 1979, Elektronnye Svoistva Legirovannykh Poluprovodnikov (Nauka, Moscow); English Edition: 1984, Electronic Properties of Doped Semiconductors (Springer, Berlin).
Slyusarev, V.A., M.A. Strzhemechny and I.A. Burakhovich, 1977, Fiz. Nizk. Temp. **3**, 1229 [Sov. J. Low Temp. Phys. **3**, 591].
Steinbinder, D., H. Wipf, A. Magerl, D. Richter, A.-J. Dianoux and K. Neumaier, 1988, Europhys. Lett. **6**, 535.
Tanabe, Y., and K. Ohtaka, 1986, Phys. Rev. B **34**, 3763.
Teichler, H., and A. Seeger, 1981, Phys. Lett. A **82**, 91.
Vladár, K., and A. Zawadowski, 1983a, Phys. Rev. B **28**, 1564.
Vladár, K., and A. Zawadowski, 1983b, Phys. Rev. B **28**, 1582.
Vladár, K., and A. Zawadowski, 1983c, Phys. Rev. B **28**, 1596.
Vladár, K., A. Zawadowski and G.T. Zimányi, 1988a, Phys. Rev. B **37**, 2001.
Vladár, K., A. Zawadowski and G.T. Zimányi, 1988b, Phys. Rev. B **37**, 2015.
Weiss, G., 1991, Mater. Sci. & Eng. A **133**, 45.
Weiss, G., W. Arnold, K. Dransfeld and H.J. Guntherodt, 1980, Solid State Commun. **33**, 111.
Weiss, U., and M. Wollensak, 1989, Phys. Rev. Lett. **62**, 1663.
Welter, J.M., D. Richter, R. Hempelmann, O. Hartmann, E. Karlsson, L.O. Norlin, T.O. Niinikoski and D. Lenz, 1983, Z. Phys. B **52**, 303.
Welter, J.M., D. Richter, R. Hempelmann, O. Hartmann, E. Karlsson, L.O. Norlin and T.O. Niinikoski, 1984, Hyperfine Interact. **17–19**, 117.
Wipf, H., and K. Neumaier, 1984, Phys. Rev. Lett. **52**, 1308.
Wipf, H., A. Magerl, S.M. Shapiro, S.K. Satija and W. Thomlinson, 1981, Phys. Rev. Lett. **46**, 947.
Wipf, H., D. Steinbinder, K. Neumaier, P. Gutsmiedl, A. Magerl and A.-J. Dianoux, 1987, Europhys. Lett. **4**, 1379.
Yamada, K., 1984, Prog. Theor. Phys. **72**, 195.
Yamada, K., and K. Yosida, 1982, Prog. Theor. Phys. **68**, 1504.
Yamada, K., A. Sakurai, S. Miyazima and H.S. Hwang, 1986, Prog. Theor. Phys. **75**, 1030.
Yu, Clare C., and A.V. Granato, 1985, Phys. Rev. B **32**, 4793.

CHAPTER 3

The Decay of a Metastable State in a Multidimensional Configuration Space

Ulrich ECKERN

Kernforschungszentrum Karlsruhe
Institut für Nukleare Festkörperphysik
Postfach 3640
W-7500 Karlsruhe 1, Germany

and

Albert SCHMID

Institut für Theorie der Kondensierten Materie
Universität Karlsruhe
Postfach 6980
W-7500 Karlsruhe 1, Germany

Quantum Tunnelling in Condensed Media
Edited by
Yu. Kagan and A.J. Leggett

Contents

1. Introduction

Metastable states with large, though finite, lifetimes occur rather frequently in nature. As a model of a metastable state, one takes a particle in a potential well which is separated from states of lower energy by a potential barrier. Such a metastable state may decay by thermal activation and in many systems as they occur in condensed matter physics, this is presumably the most frequent decay mode. However, as one succeeds in reaching lower and lower temperatures, there appears also the possibility that quantum tunneling may play a role.

It was Leggett (1978, 1980) who drew attention to the quantum decay mode in large systems. In particular, he had argued that it may serve as a demonstration of peculiarities connected with macroscopic quantum phenomena. Consider, e.g., a system which features a metastable state in connection with a collective coordinate of macroscopic dimensions. In general, this collective coordinate will be coupled more or less strongly to the remaining degrees of freedom of the system; thus, a situation emerges which may be best described by a macroscopic object coupled to a dissipative environment.

Subsequently, Caldeira and Leggett (1981, 1983, 1984) have modelled such a system as follows: The macroscopic object is represented by a particle in a potential with a metastable minimum as shown in fig. 1, and the environment is taken to be a set of harmonic oscillators (see section 3.6) which are coupled to the object.*

Let us consider first the case of zero temperature, $T = 0$. In this case, and for their model, Caldeira and Leggett have calculated the quantum decay rate Γ using a field-theoretic technique (instanton technique), which was invented by Langer (1967) and independently somewhat later by Lifshitz and Kagan (1972). The instanton technique was subsequently discussed by Coleman (1979, 1985) in his most illuminating and elegant paper. Note that this technique is of quasi-classical accuracy, in the sense that Planck's constant $\hbar$ is considered there to be a small quantity.

The instanton technique has sometimes met with skepticism. The seemingly heuristic treatment of those quantum fluctuation modes that produce zero, or even negative contributions to the action, has been questioned. More

* Frequently, we will call this ensemble a dissipative object.

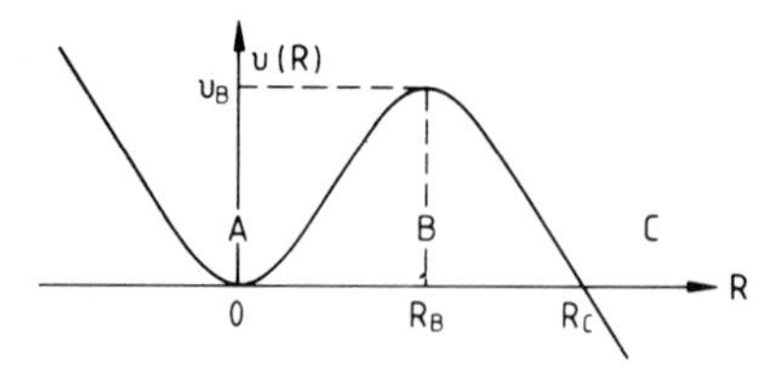

Fig. 1. Potential energy of an object featuring a metastable minimum. Regions A, B, and C represent, respectively, the metastable well, the barrier, and the outside region, where the object is found after the decay.

specifically, it has been questioned whether the instantons of the Caldeira–Leggett model, which feature a long-range interaction, allow the dilute-gas approximation.*

Such questions motivated an investigation (Schmid 1986) by one of the authors. As an alternative to the instanton technique, it has been proposed there to calculate the wave function of the decaying mode in quasiclassical approximation, which is an approximation where the phase of the wave function is expanded in powers of $\hbar$. The quasiclassical approximation has been used frequently in one-dimensional tunneling problems, but difficulties arise in systems of several degrees of freedom. In the last case, it seems that Kapur and Peierls (1937) have been the first ones to investigate the properties of quantum decay in systems with several degrees of freedom. Later, specific calculations of wave functions in quasiclassical accuracy have been carried out by Banks et al. (1973), Banks and Bender (1973) and de Vega et al. (1978).

However, the investigations of Banks et al. (1973) and Banks and Bender (1973) on quantum decay were concerned mostly with systems of a rather specific symmetry, whereas de Vega et al. (1978) were interested predominantly in tunneling between degenerate minima of the potential energy. Since the wave function of a decaying state exhibits some unique features, it seemed to be worthwhile to carry out a quasiclassical calculation (Schmid 1986) for a general model in a multidimensional configuration space.

From a technical point of view, differences in method seem to separate the decay at zero temperature and that at finite temperatures. At $T = 0$, e.g., Coleman (1979, 1985) has been able to demonstrate explicitly that the probability of the object to remain in the metastable well decreases exponentially with time, i.e., is $\propto \exp(-\Gamma t)$ (dilute gas of instantons). For finite temperatures $T > 0$, on the other hand, the calculation of the decay rate Γ proceeds by a more or less direct identification of Γ with an imaginary part of the free energy, as was first suggested by Langer (1967) and later also by Lifshitz and Kagan (1974).

* We cannot disperse completely this suspicion. See the last paragraph of section 3.6.

This identification has also been emphasized by Affleck in his lucid and transparent article (Affleck 1981).

A generalization of Affleck's work to a dissipative object was formulated by Waxman and Leggett (1985), but their work has been left uncompleted. Later, this program was taken up by Ludviksson (1989) in his thesis, where it has been carried out successfully.

This chapter consists mostly of a review of the papers by Schmid (1986) and Ludviksson (1989), with an emphasis on a comprehensive and unified presentation. The formal procedure is based on the quasiclassical approximation either of wave functions (Landau and Lifshitz 1975) or of Feynman's path integrals (Feynman and Hibbs 1965, Schulman 1981).

At this point, we wish to acknowledge the work of Miller (1975) on semiclassical methods in chemical physics. We should also not fail to recall the pioneering work of Kramers (1940) on the classical decay problem 50 years ago, which started and motivated many other investigations on the decay problem. In this context, we mention recent review articles on the decay problem by Hänggi et al. (1990) and Mel'nikov (1991).

The chapter is organized as follows. Section 2 is meant to introduce the fundamental concepts, which are illustrated explicitly for a one-dimensional system. The multidimensional quantum decay for a wave function, i.e., at zero temperature, will be discussed in section 3. Finally, section 4 is devoted to the discussion of the statistical matrix and to the multidimensional decay at finite temperatures, and a conclusion is given in section 5.

A detailed discussion of a one-dimensional system is given in sections 2.1–2.6. In section 2.1, the concept of a quasistationary state is introduced. This concept is the basis of all the following calculations; in fact, it seems difficult to work out a general theory without this assumption. The matching of the quasiclassical wave function is discussed in section 2.2; and in section 2.3, it is shown how this matching procedure can be reduced to comparatively simple operations. The decay at finite temperatures as put forward by Affleck (1981) is reviewed in section 2.4. The statistical matrix appropriate to a decaying state is introduced in section 2.5 and it is also shown how it can be obtained in quasiclassical approximation from a Feynman path integral. This approximation is based on complex extremal paths and section 2.6 provides a demonstration that complex orbits of arbitrary multiplicity are required for a unified theory. It is also shown that the generalization to multiple orbits leads to quantitative, if not qualitative, changes as compared to the Affleck ansatz.

Sections 3.1–3.6 discuss the multidimensional quasiclassical wave function. The direct approach of section 2.3 for calculating wave functions decaying from a metastable well is applied in section 3.1 to quantum decay for a N-dimensional system. The quasiclassical approximation there leads to nonlinear partial differential equations of first order which can be solved by the method of characteristics. This method, however, requires the evaluation of a classical equation of motion in N dimensions; it is shown in section 3.2 how this rather complicated equation can be solved in the small-fluctuation approximation provided that the escape path is known. In section 3.3, the analytical properties of the principal wave are investigated near the caustic (which is the generalization of a turning point to N dimensions) and it is shown how the transmitted wave can be obtained by analytical continuation. There, an unpublished work of Landauer (1950) has been most helpful to the authors in clarifying important concepts. The decay rate is then calculated in section 3.4; and complete agreement is found with the result of the instanton technique. Section 3.5 discusses the construction of the reflected wave and the validity of the applied procedure in N dimensions. For illustration, the Caldeira–Leggett model is presented in section 3.6 as an example of quantum decay in a multidimensional system.

The theory of the multidimensional statistical matrix is explained through sections 4.1–4.3. As pointed out in section 4.1, the quasiclassical approximation is built on a set of extremal and complex paths which incorporates periodic orbits which can be seen as a generalization of the escape path referred to above. In section 4.2, we include the Gaussian fluctuations about these paths and calculate their contribution to the decay current in the outside region; thus, we obtain a generalization to N dimensions of Afflecks result for the decay rate. However, the summation over multiple orbits is a nontrivial problem. This will be demonstrated in section 4.3 and we will find qualitative changes as compared to the standard theory. For the Caldeira–Leggett model a comparison with the results of the standard theory is carried out at the end.

A discussion in section 4.4 summarizes the main results.

There are various appendices which are meant to remove detailed calculations from the main body. In Appendix A we calculate the quantum-mechanical transmission of a smooth barrier in the quasiclassical approximation; the result confirms the idea of summing up the contribution of multiple orbits. Appendix B contains some results for the one-dimensional decay. In Appendix C, the nontrivial problem of Gaussian fluctuations about complex paths is investigated. In Appendix D we extend the work of Larkin and Ovchinnikov (1983 a, b, 1984) on the pre-factor for a heavily damped object to multiple orbits. Finally, conceptual and computational details on the decay rate of a heavily damped object are contained in Appendix E.

2. *Decay in one dimension*

2.1. *Quasistationary state*

The concept of a quasistationary state is essential to all considerations which will follow in this chapter. It is only for the sake of an easy demonstration that we discuss this concept here for the one-dimensional decay. Let us recall the standard model of a metastable state, which consists of a particle ("object") in a potential well (region A in fig. 1), separated from the decayed state (region C) by a potential barrier (region B). Characteristic parameters of this model are the mass m of the object; the height v_B; and width R_C of the barrier. Of interest is also the oscillation frequency ω_A at A and ω_B at B for the inverted potential, respectively. Later (section 3.6), we will also specify the interaction of the object with its environment but for the moment we will do without it.

In extreme cases, the decay may occur entirely either by quantum tunneling (quantum decay; QD) or by thermal activation (classical decay; CD). If the condition for a quasistationary state applies, the theory leads to an exponential decay law

$$P(t) = \exp(-\Gamma t), \qquad (2.1)$$

where $P(t)$ is the probability for the object to remain in the well. Furthermore, it is commonly found that the decay rate Γ is of the form

$$\Gamma = \mathscr{A} \exp(-\mathscr{B}). \qquad (2.2)$$

For orientation, we give some simple relations

$$\mathscr{B} \propto v_B/\hbar\omega_B \quad \text{(QD)},$$

$$\mathscr{B} = v_B/kT \quad \text{(CD)},$$

$$\mathscr{A} \propto \omega_A. \qquad (2.3)$$

Within the limits of this simple picture, the condition for a quasistationary decaying state* is given by

$$\Gamma \ll \omega_A. \qquad (2.4)$$

Physically, this condition means that the metastable state may be considered to be internally in equilibrium. Using eqs. (2.2) and (2.3), we may rewrite this condition as

$$\mathscr{B} \gg 1. \qquad (2.5)$$

2.2. *Matching of the quasiclassical wave function*

At zero temperature, the concept of a quasistationary decaying state allows us to look for a wave function $\psi(R)$ which is a solution of the time-independent Schrödinger** equation

$$\mathscr{H}\psi = E\psi, \qquad (2.6)$$

* See, e.g., Landau and Lifschitz (1975) for a discussion of quasistationary states and the Breit–Wigner formula.

** The time-dependent Schrödinger equation has been solved repeatedly in the past for simple models of decay. See Ludviksson (1987) for a recent publication.

where the Hamiltonian

$$\mathscr{H} = -\frac{\hbar^2}{2m}\frac{\partial^2}{\partial R^2} + v(R) \tag{2.7}$$

includes a potential energy $v(R)$ of a type shown in fig. 1. The ansatz above also requires that the wave function has to satisfy the following conditions:

(i) Far to the right, ψ must carry an outgoing probability current.
(ii) ψ must vanish far to the left.
(iii) Near the metastable minimum, ψ must resemble the ground state wave function of a harmonic oscillator.

For such a wave function, the Schrödinger equation (2.6) can be solved only for a complex energy $E = \frac{1}{2}\hbar(\omega_A - i\Gamma)$; it follows that in a time representation, the probability decays exponentially in time at a rate Γ, namely $|\psi(t)|^2 \propto \exp(-\Gamma t)$. Obviously,

$$\Gamma = -2\,\mathrm{Im}\,E/\hbar \tag{2.8}$$

has to be identified with the decay rate.

In a region near the origin, $R = 0$, as well as in a region near to the classical turning point $R = R_C$, the general solution of the differential equation (2.6) may be expressed by known functions. This follows from the fact that there $v(R)$ can be approximated by a harmonic and a linear potential, respectively. Accordingly,

$$v(R) = \begin{cases} \frac{1}{2}m\omega_A^2 R^2, & |R| \ll R_B, \\ -F_C(R - R_C), & |R - R_C| \ll R_C, \end{cases} \tag{2.9}$$

where we have also indicated the regions of validity of the approximation for a smooth potential. The simplest of such a smooth potential is

$$v(R) = \frac{1}{2}m\omega_0^2 R^2(1 - R/R_C), \tag{2.10a}$$

where

$$\omega_A = \omega_B = \omega_0, \qquad R_B = \frac{2}{3}R_C,$$
$$v_B = \frac{1}{6}m\omega_0^2 R_B^2, \qquad F_C = \frac{1}{2}m\omega_0 R_C. \tag{2.10b}$$

Concerning the region near the classical turning point R_C, we recognize that a finite energy in eq. (2.6) means only a redefinition of R_C. Therefore, we may put $E = 0$ (i.e., $\mathrm{Re}\,E = \mathrm{Im}\,E = 0$). Then the solution which satisfies boundary condition (i) is of the form

$$\psi(R) = \mathscr{N}\left[\mathrm{Bi}\left(\frac{R_C - R}{a_C}\right) + i\mathrm{Ai}\left(\frac{R_C - R}{a_C}\right)\right], \quad a_C = (\hbar^2/2mF_C)^{1/3}, \tag{2.11a}$$

where Ai and Bi are Airy functions (Abramowitz and Stegun 1968). To the right of R_C and for $a_C \ll |R - R_C|$, the asymptotic form of $\psi(R)$ is

$$\tilde{\psi}(R) = \mathcal{N}\pi^{-1/2}\left(\frac{R - R_C}{a_C}\right)^{-1/4} \exp\left\{ i\left[\frac{2}{3}\left(\frac{R - R_C}{a_C}\right)^{3/2} + \frac{\pi}{4}\right]\right\}, \tag{2.11b}$$

which represents clearly an outgoing wave. The asymptotic form of $\psi(R)$ to the left of R_C may be written as $\psi_0(R) + \psi_1(R)$, which is a sum of what we wish to call a principal and a reflected wave. Asymptotically, their form is*

$$\psi_0(R) = \mathcal{N}\pi^{-1/2}\left(\frac{R_C - R}{a_C}\right)^{-1/4} \exp\left[\frac{2}{3}\left(\frac{R_C - R}{a_C}\right)^{3/2}\right],$$

$$\psi_1(R) = \frac{i}{2}\mathcal{N}\pi^{-1/2}\left(\frac{R_C - R}{a_C}\right)^{-1/4} \exp\left[\frac{-2}{3}\left(\frac{R_C - R}{a_C}\right)^{3/2}\right]. \tag{2.11c}$$

Considering the differential equation (2.6) in the region near the origin, we find that a solution satisfying the boundary conditions (ii) and (iii) is

$$\psi(R) = D_\nu(-R/a_0), \quad a_0 = (\hbar/2m\omega_0)^{1/2}, \ \nu = -i\Gamma/2\omega_0, \tag{2.12a}$$

where D_ν is a parabolic cylinder function (Abramowitz and Stegun (1981)). The asymptotic form of eq. (2.12a) to the right of the origin and for $a_0 \ll R$ is again of the form $\psi_0(R) + \psi_1(R)$, where

$$\psi_0(R) = |R/a_0|^\nu \exp(-R^2/4a_0^2),$$

$$\psi_1(R) = -\nu(2\pi)^{1/2}(a_0/R)^{1+\nu} \exp(R^2/4a_0^2). \tag{2.12b}$$

The asymptotic form of $\psi(R)$ to the left of the origin is identical to $\psi_0(R)$ in eq. (2.12b). [Note that $a_C/R_C \equiv (2^{1/2}a_0/R_C)^{4/3}$.]

Outside the regions $R \sim 0$ and $R \sim R_C$ considered above, the wave function can be calculated in the quasiclassical approximation.** In fact, we will find that the condition for quasistationarity $\mathcal{B} \gg 1$ of eq. (2.5) guarantees the validity of this approximation.

In this approximation, the asymptotic waves of eq. (2.11) and of eq. (2.12) can be connected as follows. We consider first the principal wave $\psi_0(R)$. In the classically inaccessible region we expect it to be a real quantity; therefore, we put

$$\psi_0(R) = \exp\left\{-\frac{1}{\hbar}[\mathcal{W}(R) + \hbar\mathcal{W}^{(1)}(R) + O(\hbar^2)]\right\}, \tag{2.13}$$

* Since ψ_1 is purely imaginary, it is legitimate to retain this subdominant contribution next to the real dominant part ψ_0. See also the interesting discussion in Coleman (1979, 1985).

** An excellent discussion of the quasiclassical approximation for one-dimensional systems can be found in Landau and Lifschitz (1975). Frequently, this approximation is called the WKB or WKBJ approximation.

where the expression $\mathscr{W}(R)$ – the first term in the square bracket – may be called the (abbreviated) Euclidean action. We insert the ansatz (2.13) into the Schrödinger equation (2.6) and then equate separately the terms of equal power in $\hbar$, observing that $E = \hbar\omega_0(\frac{1}{2} + \nu)$ is of order $\hbar$. Thus, we obtain
*in leading order the eikonal equation**

$$\frac{1}{2m}\left(\frac{\partial \mathscr{W}}{\partial R}\right)^2 - v(R) = 0, \tag{2.14}$$

in next order the transport equation

$$\frac{1}{m}\frac{\partial \mathscr{W}^{(1)}}{\partial R}\frac{\partial \mathscr{W}}{\partial R} - \frac{1}{2m}\frac{\partial^2 \mathscr{W}}{\partial R^2} + \omega_0(\tfrac{1}{2} + \nu) = 0. \tag{2.15}$$

The solution of the eikonal equation (2.14) is

$$\mathscr{W}(R) = \int_0^R \mathrm{d}R'[2mv(R')]^{1/2}, \tag{2.16a}$$

where the sign of the square root has been chosen such that ψ_0 decreases exponentially with increasing R. For convenience, an arbitrary integration constant has been fixed by the condition $\mathscr{W}(0) = 0$. In the case where $v(R)$ is given by eq. (2.10), we obtain

$$\frac{1}{\hbar}\mathscr{W}(R) = \frac{1}{2}\mathscr{B}\left[1 - \left(1 + \frac{3R}{2R_\mathrm{C}}\right)\left(1 - \frac{R}{R_\mathrm{C}}\right)^{3/2}\right] \tag{2.16b}$$

where

$$\mathscr{B} = \frac{8}{15}\frac{m\omega_0^2 R_\mathrm{C}^2}{\hbar\omega_0} = \frac{36}{5}\frac{v_\mathrm{B}}{\hbar\omega_0}. \tag{2.17}$$

Limiting expressions for $\mathscr{W}(R)$ are as follows:

$$\frac{1}{\hbar}\mathscr{W}(R) = \begin{cases} \dfrac{R^2}{4a_0^2}, & |R| \ll R_\mathrm{C}, \\ \dfrac{1}{2}\mathscr{B} - \dfrac{2}{3}\left(\dfrac{R_\mathrm{C} - R}{a_\mathrm{C}}\right)^{3/2}, & |R_\mathrm{C} - R| \ll R_\mathrm{C}, \end{cases} \tag{2.16c}$$

where a_C and a_0 are given by eqs. (2.11) and (2.12a).

The transport equation (2.15) can be solved by a straightforward integration; choosing a suitable integration constant, one obtains

$$\mathscr{W}^{(1)}(R) = \frac{1}{2}\ln\left[\frac{R}{4R_\mathrm{C}}\left(1 - \frac{R}{R_\mathrm{C}}\right)^{1/2}\right] - \left(\frac{1}{2} + \nu\right)\ln\frac{1 - \sqrt{1 - R/R_\mathrm{C}}}{1 + \sqrt{1 - R/R_\mathrm{C}}} - \nu\ln\frac{4R_\mathrm{C}}{a_0}. \tag{2.18a}$$

* This terminology has been used, e.g., in Banks and Bender (1973) and Banks et al. (1973).

Limiting expressions for $\mathscr{W}^{(1)}(R)$ are as follows:

$$\exp[-\mathscr{W}^{(1)}(R)] = \begin{cases} \left(\dfrac{R}{a_0}\right)^{\nu}, & |R| \ll R_C, \\ \left(\dfrac{4R_C}{a_0}\right)^{\nu}\left(\dfrac{R_C - R}{16\,R_C}\right)^{1/4}, & |R_C - R| \ll R_C. \end{cases} \tag{2.18b}$$

Comparing the limiting expressions, we recognize that $\psi_0 = \exp[-(\mathscr{W} + \hbar\mathscr{W}^{(1)})/\hbar]$ agrees with ψ_0 of eq. (2.12) and also that it agrees with ψ_0 of eq. (2.11c) if we choose

$$\mathscr{N} = 2\pi^{1/2}\left(\frac{R_C}{a_C}\right)^{1/4}\left(\frac{4R_C}{a_0}\right)^{\nu}\exp\left(\tfrac{1}{2}\mathscr{B}\right). \tag{2.19}$$

The quasiclassical ansatz for the reflected wave $\psi_1(R)$ is chosen similar to eq. (2.13) with an Euclidean action $\mathscr{W}_1(R) + \hbar\mathscr{W}_1^{(1)}(R)$. Clearly, the structure of the eikonal equation and of the transport equation remains unchanged. The different boundary condition, however, requires the opposite sign of the square root; hence,

$$\mathscr{W}_1(R) = -\mathscr{W}(R). \tag{2.20}$$

As a consequence, $\mathscr{W}_1^{(1)}(R)$ is similar to eq. (2.18), with, however, a change in the sign of $(\frac{1}{2} + \nu)$; therefore,

$$\mathscr{W}_1^{(1)}(R) = \frac{1}{2}\ln\frac{R}{R_C}\left(1 - \frac{R}{R_C}\right)^{1/2} + \left(\frac{1}{2} + \nu\right)\ln\frac{1 - \sqrt{1 - R/R_C}}{1 + \sqrt{1 + R/R_C}}$$

$$+ (1 + \nu)\ln\frac{4R_C}{a_0} - \ln(2\pi)^{1/2}(-\nu), \tag{2.21}$$

where a suitable change in the integration constant has been included. Thus, the limiting expressions are

$$\exp[-\mathscr{W}_1^{(1)}(R)]$$

$$= \begin{cases} -(2\pi)^{1/2}\,\nu\left(\dfrac{a_0}{R}\right)^{1+\nu}, & |R| \ll R_C, \\ -(2\pi)^{1/2}\,\nu\left(\dfrac{a_0}{4R_C}\right)^{1+\nu}\left(\dfrac{R_C - R}{16\,R_C}\right)^{-1/4}, & |R_C - R| \ll R_C, \end{cases} \tag{2.22}$$

Comparing again the limiting expressions, we recognize that $\psi_1 = \exp[-(\mathscr{W}_1 + \hbar\mathscr{W}_1^{(1)})]/\hbar$ agrees with ψ_1 of eq. (2.12b). We also recognize that it agrees with ψ_1 of eq. (2.11c) if we choose

$$\nu = -\mathrm{i}\frac{\Gamma}{2\omega_0} = -\mathrm{i}\frac{1}{2(2\pi)^{1/2}}\left(\frac{4R_C}{a_0}\right)^{1+2\nu}\exp(-\mathscr{B}), \tag{2.23}$$

and $\mathscr{N}$ according to eq. (2.19).

If $|\nu| \ll 1$, the ν-dependence on the right-hand side can be neglected. Then, eq. (2.23) can be rewritten in the standard form [eq. (2.2)], $\Gamma = \mathscr{A}\exp(-\mathscr{B})$, where $\mathscr{B}$ is given by eq. (2.17) and where

$$\mathscr{A} = 60^{1/2}\left(\frac{\mathscr{B}}{2\pi}\right)^{1/2}\omega_0. \tag{2.24}$$

This expression for the decay rate is similar to eq. (2.3) except for numerical constants and a factor $\mathscr{B}^{1/2}$ contributing to $\mathscr{A}$.

It is known (Landau and Lifschitz 1975) that the quasiclassical approximation relies on the inequality

$$\hbar\left|\frac{\partial^2 \mathscr{W}}{\partial R^2}\right| \ll \left|\frac{\partial \mathscr{W}}{\partial R}\right|^2. \tag{2.25}$$

In the present case, where $\partial\mathscr{W}/\partial R = \pm m\omega_0 R(1 - R/R_C)^{1/2}$, this inequality reduces either to $a_0 \ll R_C$ or to $a_C \ll |R - R_C|$. Therefore, we conclude that the present method of solving the Schrödinger equation by matching exact solutions for an approximate potential with approximate solutions for the exact potential requires the inequalities

$$a_0 \ll R_C, \qquad a_C \ll R_C. \tag{2.26}$$

Inserting the appropriate definitions, we find that inequalities (2.26) are satisfied provided that eq. (2.5), i.e., $\mathscr{B} \gg 1$, is satisfied. Therefore, the concept of a quasistationary state and the quasiclassical approximation are valid in the quasiclassical limit.

2.3. A simple approach to quantum decay

We have put the discussion of the one-dimensional quantum decay on a broad basis in order to demonstrate how so many intricate details conspire to bring forth the simplicity of the final result [eqs. (2.17) and (2.24)]. One might wonder whether there is a much simpler scheme leading to the same end. Indeed, there is an alternative method which ultimately relies on the fact that the principal wave ψ_0 can be found, to a sufficient degree of accuracy, without prior knowledge of Γ.

Considering the Schrödinger equation (2.6), we conclude that $\operatorname{Im}\psi^*[\mathscr{H} - E]\psi = 0$. Integrating this expression from $-\infty$ to R, and recalling eq. (2.8), we obtain

$$\Gamma = \left[\int_{-\infty}^{R} \mathrm{d}R'\,|\psi(R')|^2\right]^{-1} J(R), \tag{2.27}$$

where the probability current

$$J(R) = (\hbar/m)\operatorname{Im}\psi^*\,\partial\psi/\partial R. \tag{2.28}$$

Later we will take $R \sim R_C$ for the current $J(R)$. On the other hand, the choice $R' \ll R_B$ in the denominator of eq. (2.27) is sufficient since it includes most of the normalization of a decaying state.

One might quite easily argue that the ratio (2.27) of probability current divided by the probability of the object to be found in the region of the well may a priori be considered as the definition of the decay rate and that no reference to eq. (2.8) is required. We will make use of this idea later.

Consider the limit $\Gamma = \nu = 0$, where eq. (2.12b) assumes the simple form $\psi_0(R) = \exp(-R^2/4a_0^2)$. Note also that in the same limit the quasiclassical approximation to the wave function as given implicitly by eqs. (2.16) and (2.18) agrees with $\psi(R)$ in the region $|R| \ll R_C$. Therefore, we conclude that the following statement is true:

The quasiclassical solution can be found directly from the eikonal equation (2.14) *and the transport equation* (2.15) *if we put* $\Gamma = \nu = 0$ *and if we extend these equations to the origin* $R = 0$, *supplementing them by the boundary condition*

$$\mathscr{W}(0) = \mathscr{W}^{(1)}(0) = 0, \qquad \mathscr{W}(R) \geqslant 0. \tag{2.29}$$

Thus, the quasiclassical method leads us directly from the harmonic oscillator wave function near the origin to the asymptotic form of the principle wave $\psi_0(R)$ near the classical turning point R_C. According to eqs. (2.16) and (2.18), we obtain for $R \approx R_C$,

$$\psi_0(R) = 2\left(\frac{R_C - R}{R_C}\right)^{-1/4} \exp\left[\frac{2}{3}\left(\frac{R_C - R}{a_C}\right)^{3/2} - \frac{\mathscr{B}}{2}\right]. \tag{2.30}$$

In principle, we could use the connection formula contained in eq. (2.11) in order to construct the outgoing wave $\tilde{\psi}(R)$. However, it has been found* that the breakdown of the quasiclassical approximation near the turning point R_C and the matching of the wave function can be avoided if one generalizes the eikonal equation and the transport equation to complex R. Then, it becomes possible to pass from one side of the turning point to the other side on a semicircle in the complex plane which remains at a sufficiently large distance from R_C.

In order to obtain a wave with outgoing probability current, we have to perform an analytical continuation via the upper R-half plane. Thus, we are led to substitute

$$(R_C - R) \to e^{-i\pi}(R - R_C) \tag{2.31}$$

* In a recent edition of Landau and Lifschitz (1975), the method of analytical continuation is attributed to Zwaan (1929). There are doubts whether it is valid in general. Presently, one might say that it depends on some analytical properties of the Airy functions. See also the discussion in section 3.8 of Bender and Orszag (1978).

in the principal wave of eq. (2.30), whence we obtain the outgoing, i.e., the transmitted wave

$$\tilde{\psi}(R) = 2\left(\frac{R - R_C}{R_C}\right)^{-1/4} \exp\left[\frac{2i}{3}\left(\frac{R - R_C}{a_C}\right)^{3/2} - \frac{\mathscr{B}}{2} + \frac{i\pi}{4}\right] \tag{2.32}$$

which agrees with eq. (2.11b) together with eq. (2.19).

Calculating the probability current (2.28) from $\tilde{\psi}(R)$, we need to differentiate only with respect to the argument of the exponential. Thus, we obtain $J(R) = 4\omega_0 R_C \exp(-\mathscr{B})$ which is independent of R. Concerning the normalization, we note that it is given by $\int dR|\psi_0|^2 = (2\pi)^{1/2} a_0$ by a high degree of accuracy. Inserting these expressions in eq. (2.27), we recover the same result for Γ as given in eqs. (2.2), (2.17), and (2.24).

As a final comment on the possibilities and ambiguities enclosed in the concept of analytical continuation, we remark that a substitution

$$(R_C - R) \to e^{-2i\pi}(R_C - R) \tag{2.33}$$

in the principal wave of eq. (2.30) does lead to the reflected wave $\psi_1(R)$ as given by eq. (2.11c) except for a factor of $\frac{1}{2}$. It is possible to implement this factor by the additional requirement of current conservation.

2.4. *Decay at finite temperatures*

In this section, we review the theory which has long roots but which Affleck (1981) has presented lucidly in his paper. Consider again the metastable potential as shown in the pictogram of fig. 2. In quasiclassical approximation, the energies E_n of the levels in the well are implicitly given by (Bohr–Sommerfeld rule)

$$W(E_n) = 2\pi\hbar(n + \tfrac{1}{2}), \tag{2.34}$$

where the abbreviated action W for a closed orbit in the well is calculated according to

$$W(E) = \oint P\, dR,$$

$$P(R) = \{2m[E - v(R)]\}^{1/2}. \tag{2.35}$$

In the simplest quasiclassical approximation, the transmission probability $D(E)$ through the barrier is given by

$$D(E) = \exp\left[-\frac{1}{\hbar}\mathscr{W}(E)\right], \tag{2.36}$$

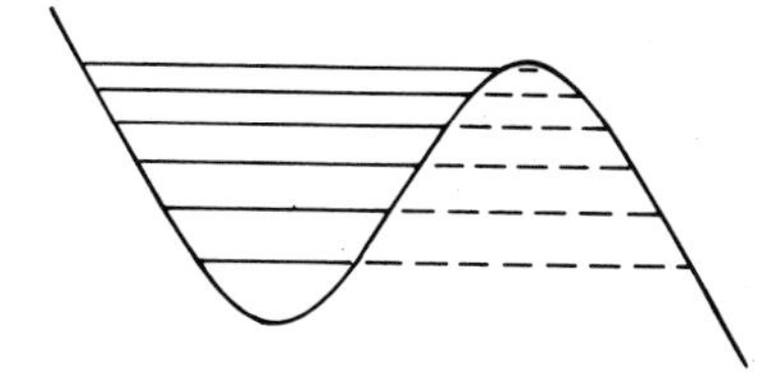

Fig. 2. Quasistationary levels in the well (full lines) and the tunneling distances (broken lines).

where $\mathscr{W}(E)$ is the abbreviated action for a closed orbit in the inverted potential $-v(R)$ of the barrier. Hence,

$$\mathscr{W}(E) = \oint \mathscr{P}\,\mathrm{d}R,$$

$$\mathscr{P}(R) = \{2m[v(R) - E]\}^{1/2}. \tag{2.37}$$

Recall that, classically, $\partial W/\partial E$ is the time the object needs to complete the closed orbit in the metastable well. Therefore,

$$\Gamma_n = \left(\frac{\partial W}{\partial E}\right)^{-1} D(E)\bigg|_{E=E_n} \tag{2.38}$$

is the rate by which the object leaves the nth level of the well.

Let us now assume that the levels are populated as being in thermal equilibrium, i.e., with probability $\propto \exp(-E_n/kT)$. Then, the average decay rate is given by

$$\bar{\Gamma} = Z_0^{-1} \sum_n \Gamma_n \exp(-\tau_1 E_n/\hbar), \tag{2.39}$$

where Z_0 is the normalization and where, with respect to further developments, we have introduced the Euclidean time

$$\tau_1 = \hbar/kT \tag{2.40}$$

as a measure of the inverse temperature.

We emphasize that the ansatz (2.39) may have only a restricted validity since it ignores changes in the population of the levels which result from a level-dependent decay rate. On the other hand, we wish to emphasize that such an ansatz is completely in accordance with the concept of a quasistationary state introduced in section 2.

Concerning the normalization

$$Z_0 = \sum_n \exp(-\tau_1 E_n/\hbar), \tag{2.41}$$

one should observe that, effectively, only the lowest levels contribute. There, the well may be approximated by a harmonic potential; hence ($\omega_A = \omega_B = \omega_0$),

$$Z_0 = (2 \sinh \omega_0 \tau_1 /2)^{-1}. \tag{2.42}$$

We consider now the case where the level spacing is so small that we may replace the summation in eq. (2.39) by an energy integration. Since the level density is $(\partial W/\partial E)/2\pi\hbar$, we recognize a cancellation of a corresponding term in the expression (2.38) for Γ_n. Thus, we arrive at

$$\bar{\Gamma} = Z_0^{-1} \int \frac{\mathrm{d}E}{2\pi\hbar} \exp\left(-\frac{1}{\hbar}[\mathscr{W}(E) + \tau_1 E]\right). \tag{2.43}$$

In the quasiclassical limit, it is possible to evaluate the integral (2.43) by steepest descent. Thus, we obtain for the average decay rate a relation of the form [eq. (2.2)], which is $\bar{\Gamma} = \mathscr{A} \exp(-\mathscr{B})$, where the prefactor is given by

$$\mathscr{A} = Z_0^{-1}[2\pi\hbar\partial^2 \mathscr{W}(E)/\partial E^2]^{-1/2} \tag{2.44}$$

and where the argument of the exponential (except for a factor $\hbar$),

$$\hbar\mathscr{B} = \mathscr{S}(\tau_1) = \mathscr{W}(E) + \tau_1 E,$$

$$\tau_1 = -\partial \mathscr{W}(E)/\partial E, \tag{2.45}$$

is the Legendre transform of $\mathscr{W}$. Alternatively, we may write the prefactor as follows:

$$\mathscr{A} = Z_0^{-1/2}[-2\pi\hbar\partial\tau_1(E)/\partial E]^{-1/2} = Z_0^{-1}(2\pi\hbar)^{-1/2}[\partial^2 \mathscr{S}(\tau_1)/\partial^2 \tau_1]^{1/2}. \tag{2.46}$$

An important point in Affleck's paper has been his demonstration that $\bar{\Gamma}$ is related with the imaginary part of the free energy [compare eq. (2.39)]

$$\bar{\Gamma} = -\frac{2}{\hbar} \operatorname{Im} \mathscr{F}, \tag{2.47}$$

which means a generalization of eq. (2.8) to finite temperatures. In the first step, Affleck proposes to calculate the full partition function Z by a path integral*

$$Z = \int \mathrm{d}(R_\tau) \exp\left(-\frac{1}{\hbar}\mathscr{S}([R_\tau];\tau_1)\right), \tag{2.48}$$

* For a general introduction to path integrals, see Feynman and Hibbs (1965) and Schulman (1981).

where the Euclidean action is given by*

$$\mathscr{S}([R_\tau]; \tau_1) = \int_0^{\tau_1} d\tau \left[\frac{m}{2} (\partial_\tau R_\tau)^2 + v(R_\tau) \right], \tag{2.49}$$

and where the integration includes all closed paths $R_{\tau_1} = R_{\tau=0}$.

In quasiclassical approximation (see the discussion by Dashen et al. (1974)), the predominant contribution to the path integral is from extremal paths r_τ that obey the equation

$$m\ddot{r}_\tau - \frac{\partial v(r_\tau)}{\partial r_\tau} = 0. \tag{2.50}$$

Specifically, the relevant extremal paths that contribute to Z are closed orbits that are completed in time τ_1. There is the trivial orbit $r_\tau = 0$ (metastable minimum) which provides, together with its Gaussian fluctuations (periodic in time τ_1), the normalization Z_0 of eq. (2.42). The nontrivial orbit $r_\tau \neq 0$ produces, first of all, the exponential

$$\hbar \mathscr{B} = \mathscr{S}([r_\tau]; \tau_1), \tag{2.51}$$

which can easily be shown to agree in its value with eq. (2.45).

Concerning the Gaussian fluctuations about r_τ, one should observe that they are periodic in time with period τ_1. Furthermore, one should note that there is a mode $\propto \dot{r}_\tau$ with eigenvalue $\lambda_1 = 0$ and also one mode with a negative eigenvalue λ_0. In contrast to the Minkowski case, the negative eigenvalue poses a delicate problem in the Euclidean case. However, one can argue (Caldeira and Leggett 1981, 1983, 1984) that for the present case, it contributes with a factor $\frac{1}{2}(\lambda_0)^{-1/2} = -\frac{1}{2}(\mathrm{i})|\lambda_0|^{-1/2}$. For the remaining factors we refer to Dashen et al. (1974); accordingly, the overall contribution of the nontrivial orbit is

$$Z_1 = Z_0 \frac{\mathrm{i}\tau_1}{2} \mathscr{A} \exp(-\mathscr{B}), \tag{2.52}$$

where for the present case

$$\mathscr{A} = \left(\frac{\mathscr{S}_0([r_\tau]; \tau_1)}{2\pi\hbar} \right)^{1/2} \left| \frac{\det[-\partial_\tau^2 + \omega_0^2]}{\det'[-\partial_\tau^2 + v''(r_\tau)/m]} \right|_{\mathrm{PB}}^{1/2}. \tag{2.53}$$

In the above relation

$$\mathscr{S}_0([r_\tau]; \tau_1) = m \int_0^{\tau_1} d\tau \, (\partial_\tau r_\tau)^2 \tag{2.54}$$

and det means a determinant in τ-space (det′: zero eigenvalue omitted). The subscript PB means that the eigenvalues are calculated for periodic boundary

* For convenience, we mark at appropriate places the time dependence by a subscript, e.g., $R_\tau = R(\tau)$.

conditions. If one makes use of relations given in Dashen et al. (1974), one finds that the prefactor of eq. (2.53) agrees exactly with the one given by eq. (2.44). Since the exponent $\mathscr{B}$ in eq. (2.52) is clearly identical to that given by (2.45), this completes the justification of eq. (2.47).

2.5. *Statistical matrix in quasiclassical approximation*

Again, we assume that the system is in a quasistationary state corresponding to thermal equilibrium. This means that we should look for a statistical matrix $\rho(R_1, R; \tau_1)$ which has to satisfy the following conditions:

(i) Far to the right, ρ must supply an outgoing probability current.
(ii) ρ must vanish far to the left.
(iii) ρ should contain a large part which represents the object localized in the metastable well.

Let us represent the statistical matrix by the path integral (Feynman and Hibbs 1965):

$$\rho(R_1, R; \tau_1) = \int \mathrm{d}[R_\tau] \exp\left(-\frac{1}{\hbar}\mathscr{S}([R_\tau]; \tau_1)\right), \tag{2.55}$$

where $\mathscr{S}$ is given by eq. (2.49) and where the integration above includes all paths R_τ which start from R at $\tau = 0$ and lead to R_1 at $\tau = \tau_1$. In the quasiclassical approximation, the path integral is dominated by extremal paths r_τ which obey eq. (2.50). There are different classes of paths (see fig. 3) but the final choice has to be in accordance with conditions (i)–(iii). In the discussion that follows, we will realize that in fig. 3, the class (a) provides the large part which is important for the normalization* whereas class (b) contributes to the outgoing probability current.** At first sight, there seem to be two different paths of class (b) which differ for $R_1 = R$ in the sense of circulation. However, the requirement of an outgoing current eliminates one of the possibilities.

The outgoing current is calculated from the statistical matrix according to the prescription

$$J(R) = \frac{\hbar}{2mi}\left(\frac{\partial}{\partial R_1} - \frac{\partial}{\partial R}\right)\rho(R_1, R; \tau_1)\bigg|_{R_1 = R}. \tag{2.56}$$

The average decay rate then is given by

$$\bar{\Gamma} = Z_0^{-1} J(R), \tag{2.57}$$

* Note that in the present case the Gaussian fluctuations about extremal paths have to be calculated for zero boundary conditions. Alternatively, class (a) is equivalent to the trivial orbit including Gaussian fluctuations for periodic boundary conditions.

** It is this selection of paths by which we disagree with the work of Waxman and Leggett (1985).

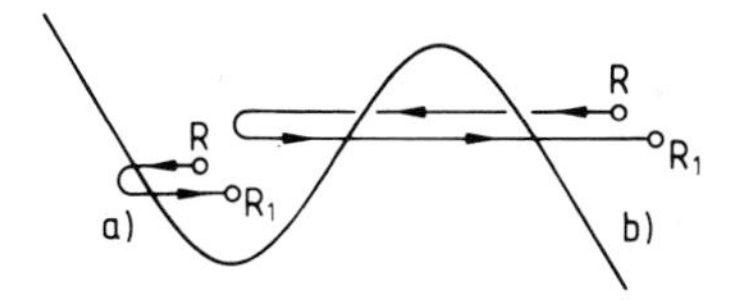

Fig. 3. Extremal path contributing to the statistical matrix $\rho(R_1, R; \tau_1)$ in quasiclassical approximation: (a) Contribution to the normalization. (b) Contribution to the outgoing probability current. At first sight, there seem to be two different paths (for $R_1 = R$, they differ in the sense of circulation). However, the requirement of an outgoing current eliminates one of the two possibilities.

where

$$Z_0 = \int_{-\infty}^{R} \mathrm{d}R' \rho(R', R'; \tau_1). \tag{2.58}$$

As already discussed following eq. (2.28), we take $R \gtrsim R_C$ in the expression for the current $J(R)$, and $R' \lesssim R_B$ for the normalization (see fig. 3).

In quasiclassical accuracy, the decay current can be calculated by taking the derivatives only with respect to the argument of the exponential, i.e., with respect to the action. Considering this action as a function of the coordinates, $\mathscr{S}([r_\tau]; \tau_1) = \mathscr{S}(R_1, R; \tau_1)$, we find that

$$\begin{aligned} J(R) &= -\frac{1}{2mi}\left[\frac{\partial \mathscr{S}}{\partial R_1} - \frac{\partial \mathscr{S}}{\partial R}\right]_{R_1 = R} \rho(R, R; \tau_1) \\ &= -\frac{1}{2mi}[\mathscr{P}_1 - (-\mathscr{P})]_{R_1 = R}\rho(R, R; \tau_1), \end{aligned} \tag{2.59}$$

where $\mathscr{P}_i = \mathscr{P}(R_i)$ is given by eq. (2.37) with R_i in the outside region (C). Clearly, the requirement of an outgoing probability current is met if we choose the sign of the square root such that $\operatorname{Im} \mathscr{P}_i < 0$. Observing that $m\dot{r}_{\tau_1} = m\dot{r}_0 = \mathscr{P}_1 = \mathscr{P}$, we may write

$$J(R) = |\dot{r}_0| \rho(R, R; \tau_1). \tag{2.60}$$

In retrospect, we realize that we need to calculate only extremal paths r_τ that are closed. For the one-dimensional case we are presently considering, these paths are also periodic since $\dot{r}_{\tau_1} = \dot{r}_0$. It is worth noting that it will also be a periodic path which will play a prominent role in multidimensional systems.

A further property of a periodic path can easily be demonstrated in one dimension and for the cubic potential [eq. (2.10)]. There, the extremal paths r_τ are found to be doubly periodic functions in the complex τ-plane; more precisely, they are found to be Weierstrass functions (Abramowitz and Stegun 1968). Specifically, the (real) period τ_1 fixes the energy as well as the second period which is purely imaginary (and which is the period of the periodic motion in the classically accessible region). Now, if one allows also complex extremal paths $r_\tau^* \neq r_\tau$, where τ is real and $0 \leqslant \tau \leqslant \tau_1$, then one can see that for any τ_1, there exists a complex path where $r_0 = r_{\tau_1} = R$ is real and larger than R_C. In the complex R-plane such a path is completely characterized by its topology with respect to the branch points (classical turning points; see fig. 4). Also, one recognizes at once that any deformation of the path in the complex plane leaves the action $S([r_\tau]; \tau_1)$ and the final momentum $\mathscr{P}_i = -\mathrm{i}(2m[E - v(R_i)])^{1/2}$ unchanged.

In order to elucidate a basic feature of the complex path, let us compare it with a real path which will, however, require a complex time. Considering fig. 5, we recognize that the real paths $R_0 \to R_2$ and $R_3 \to R_1$ in the classical accessible region of fig. 5a run along the imaginary τ-axis in fig. 5b. Of importance is that at the classical turning points R_2 and R_3, the velocities vanish; and it is this very fact which allows us to introduce corners in the contour of the complex τ-plane.

We anticipate the fact that in multidimensional problems the turning points are replaced by caustics where the velocity does not vanish; in this case, only the concept of complex paths carries through without modifications.

Of importance is also the fact that complex paths cannot be avoided in the case $E > v_B$ (high temperatures); see fig. 4c for illustration. Concerning corrections due to Gaussian fluctuations about the extremal paths, useful relations can

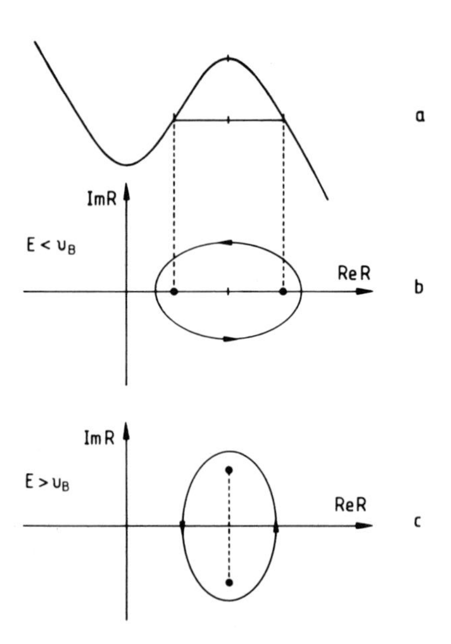

Fig. 4. Extremal orbits completed in time $\tau_1 = \hbar/kT$ under the barrier: (a) Potential energy and turning points; (b) complex extremal path which encloses the two branch points for $E < v_B$ ($T < T_B$); (c) the same for $E > v_B$ ($T > T_B$).

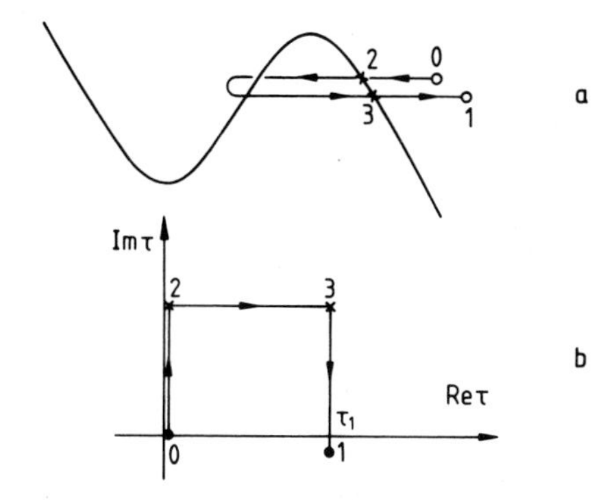

Fig. 5. (a) Real path under the barrier (2 → 3) in time τ_1 and real paths in the outside region (0 → 2, 3 → 1) for imaginary τ. (b) The corresponding contour in the complex τ-plane. (For $R_1 = R_0 \equiv R$, the endpoint of the contour is on the real axis.) Note the corners at the classical turning points.

be found in Dashen et al. (1974). There, they are derived for a real path but we will show in Appendix C that they are also valid for complex paths with appropriate interpretation. Accordingly,

$$\rho(R, R; \tau_1) = \sum_{\text{CP}} [2\pi\hbar\,|\dot{r}_0|^2(-\,\partial\tau_1/\partial E)]^{-1/2} \exp\left(-\frac{1}{\hbar}\mathscr{S}([r_\tau]; \tau_1)\right), \quad (2.61)$$

where the sum includes all closed paths. Eventually, we arrive at the relation

$$J(R) = \sum_{\text{PP}} [2\pi\hbar(-\,\partial\tau_1/\partial E)]^{-1/2} \exp\left(-\frac{1}{\hbar}\mathscr{S}([r_\tau]; \tau_1)\right), \quad (2.62)$$

where now only periodic paths have to be included.

We recognize that eqs. (2.62) and (2.57) are equivalent to eqs. (2.45) and (2.46), except for contributions of all periodic paths that satisfy the requirements.

2.6. Unified theory by multiple orbits

As pointed out by Affleck (1981), that part of the theory which we have presented in section 2.4 is not applicable to higher temperatures. In what follows, we present a unified theory which comprises the separate parts of Affleck's theory which are meant to cover different temperature ranges. In order to emphasize at least one of the differences between our theory and Affleck's one, we note that the trivial path $r_\tau = R_B$ cannot contribute to an outgoing probability current; hence, it does not play a role in the foundation of our theory.

As already mentioned, there are nontrivial complex orbits (Weierstrass functions) even for small values of τ_1, i.e., for large temperatures (see fig. 4c). Therefore, we may also look for periodic orbits that are completed in time

(primitive period)

$$\tau_p = \frac{1}{p}\tau_1, \quad p = 1, 2, \ldots, \tag{2.63}$$

and which are traversed p times. Recalling the summation of eq. (2.62), we write

$$\bar{\Gamma} = \sum_{p=1}^{\infty} \bar{\Gamma}_p, \tag{2.64}$$

where according to eqs. (2.57) and (2.62) we have

$$\bar{\Gamma}_p = Z_0^{-1}(-1)^{p+1}[2\pi\hbar p(-\partial\tau_p/\partial E)]^{-1/2} \exp\left(-\frac{p}{\hbar}\mathscr{S}([r_\tau]; \tau_p)\right). \tag{2.65}$$

The factor $(-1)^{p+1}$ above deserves a comment. In a saddle point approximation, the overall sign depends on the direction in which the "mountain pass" is being traversed. Quite generally, this direction has to be found from a proper deformation of the original integration contour. For an integration in function space, however, it seems to be difficult to ascertain the details of the proper deformation. At least, this appears to be the case with Euclidean functional integrals and we have been unable to find general sign rules in the literature. On the other hand, the factor $(-1)^{p+1}$ above does lead to correct results in cases where we have been able to check it.*

Since $\mathscr{S}([r_\tau]; \tau_p) \to -\infty$ for $\tau_p \to 0$, the sum given by eq. (2.64) is divergent.** However, it is possible to interpret this summation in a meaningful way as follows. In a first step, we calculate the Laplace transform in quasiclassical approximation

$$\bar{\Gamma}_p(E) = \int_0^{\infty} \mathrm{d}\tau_1\, \bar{\Gamma}_p(\tau_1) \exp\left(-\frac{1}{\hbar}E\tau_1\right)$$
$$= Z_0^{-1}(-1)^{p+1} \exp\left[-\frac{1}{\hbar}p\mathscr{W}(E)\right], \tag{2.66}$$

where $\mathscr{W}(E)$ is defined by eq. (2.37) – or by the inverse of the Laplace transform (2.45). (Note that τ_p is the same function of E independent of p.) Summation of the series leads to

$$\bar{\Gamma}(E) = \sum_{p=1}^{\infty} \bar{\Gamma}_p(E) = Z_0^{-1} D(E), \tag{2.67}$$

* For an illustration see Appendix B.

** This is a general property of asymptotic expansions (see section 3.8 of Bender and Orszag 1978).

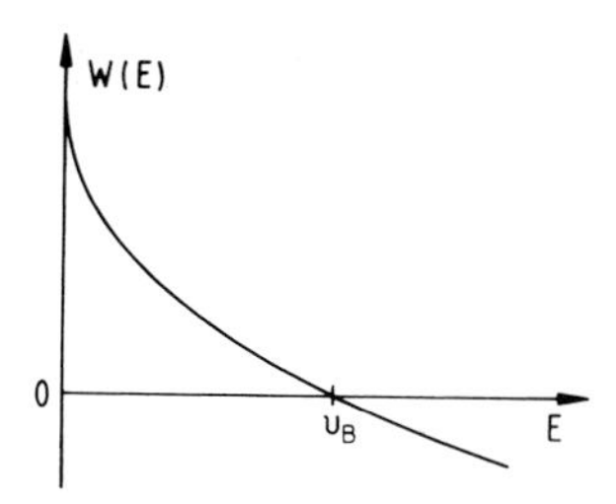

Fig. 6. Abbreviated Euclidean action $\mathscr{W}(E)$ for a closed orbit under the barrier. We have for $\tau = -\partial\mathscr{W}(E)/\partial E$ the values ∞ and $\tau_B = \hbar/kT_B$ at $E = 0$ and $E = v_B$, respectively.

where

$$D(E) = \left[1 + \exp\frac{1}{\hbar}\mathscr{W}(E)\right]^{-1} \tag{2.68}$$

means a generalization of the barrier transmission probability of eq. (2.36). Next, we invert the Laplace transformation and obtain ($E\tau_1/\hbar = E/kT$)

$$\bar{\Gamma} = (2\pi\hbar Z_0)^{-1}\int \mathrm{d}E\, D(E)\mathrm{e}^{-E/kT}. \tag{2.69}$$

In his paper, Affleck (1981) has discussed three different regions of temperature. They correspond, to some extent, to the different types of approximations* which are suggested by some properties of eqs. (2.68) and (2.69). For an orientation, consider fig. 6 for a typical energy dependence of $\mathscr{W}(E)$. Note that

$$\mathscr{W}(v_B) = 0,$$
$$-\left.\frac{\partial\mathscr{W}(E)}{\partial E}\right|_{E=v_B} = \tau_B = \frac{\hbar}{kT_B}, \tag{2.70}$$

where T_B is called the crossover temperature. (In the one-dimensional case we are considering here $kT_B = \hbar\omega_0/2\pi$.)

(i) In the case of very low temperatures $T \ll T_B$, only small energies contribute in eq. (2.69) and we may approximate $D(E) \approx \exp[-(1/\hbar)\mathscr{W}(E)]$ as shown in eq. (2.36). Thus, we recover eqs. (2.44) and (2.45).

(ii) Next, we consider temperatures where T is distinctly larger than T_B. Then, we expect that energies $|E - v_B| \lesssim kT$ contribute mostly in eq. (2.69). This justifies the approximation $\mathscr{W} \approx \mathscr{W}_1$, where the first-order expression $\mathscr{W}_1$ is given [cf. eq. (2.70)] by

$$\frac{1}{\hbar}\mathscr{W}_1(E) = -(E - v_B)/kT_B. \tag{2.71}$$

* See also Hänggi and Hontscha (1988).

In this case, we obtain

$$\bar{\Gamma} = (2\hbar Z_0)^{-1} kT_B (\sin \pi T_B/T)^{-1} \exp(-v_B/kT). \tag{2.72}$$

For temperatures $kT \gg kT_B, \hbar\omega_0$, we may expand Z_0 of eq. (2.42) as well as the trigonometric function above. Thus, we obtain in leading order

$$\bar{\Gamma} = (\omega_0/2\pi) \exp(-v_B/kT), \tag{2.73}$$

which should be compared with eq. (2.3).

At this point we wish to draw attention to an alternative way to obtain eq. (2.72), which can already be found in Affleck (1981) and which has been worked out in some detail in Wolynes (1981) and Grabert and Weiss (1984). Starting from a modification of eq. (2.47), namely

$$\bar{\Gamma} = -\frac{2}{\hbar}\frac{T_B}{T} \operatorname{Im} \mathscr{F}, \tag{2.74}$$

one calculates the free energy $\mathscr{F} = -kT \ln Z$ from $Z = Z_0 + Z_1$, where Z_1 is the contribution of the trivial orbit $r_\tau = R_B$, including its Gaussian fluctuations. Thus,

$$Z_1 = \tfrac{1}{2}\mathrm{i}\,[2 \sin(\hbar\omega_0/2kT)]^{-1} \exp(-v_B/kT), \tag{2.75}$$

where the factor i is a consequence of the unstable mode at R_B and where the factor $\frac{1}{2}$ needs a separate justification.

(iii) Close to the crossover temperature $T \approx T_B$, Affleck proposes the expression

$$\bar{\Gamma} = (2\pi\hbar Z_0)^{-1} \int_{-\infty}^{v_B} \mathrm{d}E \exp\left\{-\left[\frac{1}{\hbar}\mathscr{W}_2(E) + E/kT\right]\right\}, \tag{2.76}$$

where $\mathscr{W}_2(E)$ is the second-order expansion given by

$$\frac{1}{\hbar}\mathscr{W}_2(E) = \frac{1}{\hbar}\mathscr{W}_1(E) + \frac{1}{4}\left(\frac{E - v_B}{kT_2}\right)^2, \tag{2.77}$$

with

$$\left(\frac{1}{kT_2}\right)^2 = \frac{2}{\hbar}\frac{\partial^2 \mathscr{W}(E)}{\partial E^2}\bigg|_{E=v_B}. \tag{2.78}$$

[In the one-dimensional case, $(kT_2)^2 \sim kT_B v_B$.] This ansatz leads to*

$$\bar{\Gamma} = Z_0^{-1} \frac{kT_2}{2\pi\hbar} \pi^{1/2} \operatorname{erfc}\left(\frac{T_2}{T_B} - \frac{T_2}{T}\right) \exp\left[-\frac{v_B}{kT} + \frac{1}{2}\left(\frac{T_2}{T_B} - \frac{T_2}{T}\right)^2\right], \tag{2.79}$$

* See also eq. (63) of Larkin and Ovchinnikov (1992).

where $\operatorname{erfc} z = 2\pi^{-1/2}\int_z^\infty \mathrm{d}t \exp(-t^2)$ is the (complementary) error function (Abramowitz and Stegun 1968).

For large values of $T_2/T_B - T_2/T$, we obtain from eq. (2.79)

$$\bar{\Gamma} = Z_0^{-1} \frac{1}{2\pi\hbar} \left[\frac{1}{kT_B} - \frac{1}{kT} \right]^{-1} e^{-v_B/kT}, \tag{2.80}$$

which reduces to expression (2.72) for temperatures $T \gtrsim T_B$ above the crossover temperature.

In particular, we have for $T \to T_B$, where erfc $(T_2/T_B - T_2/T) \to 1$,

$$\bar{\Gamma} = Z_0^{-1} \frac{kT_2}{2\pi^{1/2}\hbar} e^{-v_B/T_B}. \tag{2.81}$$

A detailed discussion of the procedures above for a one-dimensional system is given in Appendix B.

3. *Quasiclassical wave function in multidimensional quantum decay*

3.1. *Quantum decay in N dimensions*

The following considerations concern a system with N degrees of freedom. Without loss of generality, one may assume all masses to be equal. Then the Hamiltonian may be written as

$$\mathscr{H} = -\frac{\hbar^2}{2m} \sum_{k=1}^{N} \frac{\partial^2}{\partial R_k^2} + V(\{R_k\}), \tag{3.1}$$

where the potential energy V is supposed to have one metastable minimum. We take this minimum to be at the origin such that

$$V = \tfrac{1}{2} m\, U_{kk'}^{(0)} R_k R_{k'} + 0(R^3), \qquad U_{kk'}^{(0)} = \frac{1}{m} \frac{\partial^2 V}{\partial R_k \partial R_{k'}}\bigg|_{R=0}. \tag{3.2}$$

Here and in what follows, summation over repeated indices is implied; we will also use vector notation $\boldsymbol{R} = \{R_k\} = (R_1, \ldots, R_N)$, if convenient.

Although the potential energy V is positive in a region which includes the origin, it should become negative in a range of directions and at larger distances. Consequently, there is a surface Σ_0, defined by $V(\boldsymbol{R}) = 0$, which separates the outside region C (classically accessible) from the well and barrier regions A + B (classically inaccessible), i.e., the regions where $V < 0$ and $V > 0$, respectively. See fig. 7 for illustration.

As pointed out in section 2.2, the wave function $\psi(\boldsymbol{R})$ of a quasistationary state has to be a solution of the Schrödinger equation [eq. (2.6)], where the Hamiltonian $\mathscr{H}$ now is given by eq. (3.1). However, conditions (i) and (ii) on the

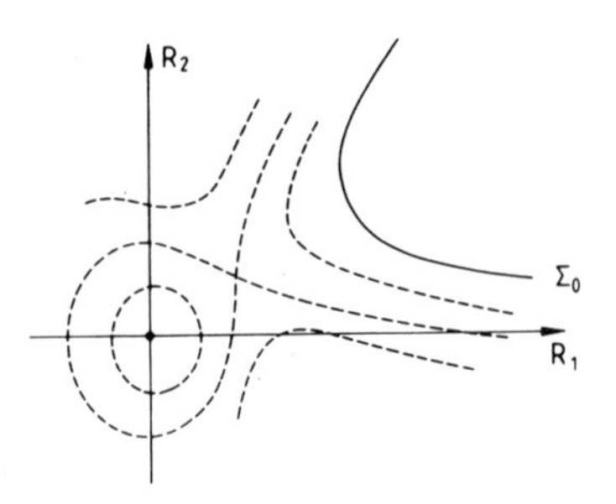

Fig. 7. Surfaces of constant potential energy $V(\boldsymbol{R})$ for a two-dimensional system with a metastable minimum (potential well A) at $\boldsymbol{R} = 0$. The surface Σ_0 defined by $V(\boldsymbol{R}) = V(0) = 0$ separates the well and barrier region (A + B) from the outside region (C).

behavior of the wave function at large distances are of little value on account of the missing information for $N - 1$ degrees of freedom. Therefore, we have to rely entirely on the simple approach of section 2.3.

We start from the quasiclassical ansatz (2.13) for the principal wave $\psi_0(\boldsymbol{R})$, where now the action is a function of $\boldsymbol{R}$. Inserting this ansatz into the Schrödinger equation, where $E = \frac{1}{2}\hbar\omega_T$ is real, we obtain
the eikonal equation

$$H = \frac{1}{2m} P_k^2 - V(\{R_k\}) = 0,$$

$$P_k = \frac{\partial \mathscr{W}(\{R_k\})}{\partial R_k}, \tag{3.3}$$

the transport equation

$$H^{(1)} = \frac{1}{m} P_k^{(1)} \frac{\partial \mathscr{W}}{\partial R_k} - \frac{1}{2m} \frac{\partial^2 \mathscr{W}}{\partial R_k^2} + \frac{1}{2}\omega_T = 0,$$

$$P_k^{(1)} = \frac{\partial \mathscr{W}^{(1)}(\{R_k\})}{\partial R_k}. \tag{3.4}$$

The equations above are nonlinear partial differential equations of first order which have to be solved for the boundary condition (2.29). Note also the special notation for the partial derivatives introduced above.

The solution to the eikonal equation (3.3) can be found by the method of characteristics.* The characteristics can be presented in parametric form as trajectories in the $2N$-dimensional $\{\boldsymbol{R}, \boldsymbol{P}\}$-space

$$R_k = R_k(\tau), \qquad P_k = P_k(\tau) \tag{3.5}$$

* A concise presentation of the method of characteristics can be found in section 2.13 of Whitham (1974). See also section II of Courant and Hilbert (1962) for a broad discussion.

such that they obey $H = 0$, together with the canonical equations of motion

$$\dot{R}_k = \frac{\partial H}{\partial P_k} = \frac{1}{m} P_k, \qquad \dot{P}_k = -\frac{\partial H}{\partial R_k} = \frac{\partial V}{\partial R_k}, \tag{3.6}$$

where the dot means differentiation with respect to τ.

In addition, we also introduce the concept of paths which are defined as projections of the trajectories on the N-dimensional $\{\boldsymbol{R}\}$-space. These paths obey the equation of motion

$$m\ddot{R}_k = \frac{\partial V}{\partial R_k}, \qquad \tfrac{1}{2}m\dot{R}_k^2 - V = 0, \tag{3.7}$$

where the condition $H = 0$ has been added for completeness. Clearly, eq. (3.7) corresponds to the classical equation of motion in the inverted potential for zero energy.

The change of $\mathscr{W}$ along a characteristic is given by

$$\dot{\mathscr{W}} = P_k \frac{\partial H}{\partial P_k} = m\dot{R}_k^2. \tag{3.8}$$

At this point, it is appropriate to recall that the relation $\partial \mathscr{W}/\partial \boldsymbol{R} = \boldsymbol{P} = m\dot{\boldsymbol{R}}$ means that the paths follow the lines of steepest descent of $\mathscr{W}(\boldsymbol{R})$.

Solving the transport equation (3.4) by the same method, one recognizes that the paths are the same as above. Furthermore,

$$\dot{\mathscr{W}}^{(1)} = P_k^{(1)} \frac{\partial H^{(1)}}{\partial P_k^{(1)}} = \frac{1}{2m} \frac{\partial^2 \mathscr{W}}{\partial R_k^2} - \frac{1}{2}\omega_{\mathrm{T}}, \tag{3.9}$$

where the condition $H^{(1)} = 0$ has been used.

Since $\mathscr{W} = \mathscr{W}^{(1)} = 0$ at the origin is given, we obtain the action $\mathscr{W}(\boldsymbol{R})$ and $\mathscr{W}^{(1)}(\boldsymbol{R})$ for other points in space by calculating the paths which connect the origin with these particular points. Being paths in the inverted potential, they will be confined to the classically inaccessible regions, A + B, acquiring large distances for sufficiently large τ. As an illustration, such paths are shown in fig. 8 for a two-dimensional system. The figure suggests that there is also a collection of paths which are reflected at some distance from Σ_0, having curvatures either to the left or to the right. Following this suggestion, we conclude that there exists, as a case on the border line, a single path which approaches Σ_0 perpendicularly and with zero velocity.

The particular role of such a singular path in quantum decay has been stressed by Banks et al. (1978). We follow their suggestion and assume that in the most general quantum decay problem, there always exists such a singular path – possibly a discrete number of such paths* – which connect the metastable

* One might suspect symmetries to be responsible for the appearance of several of such singular paths.

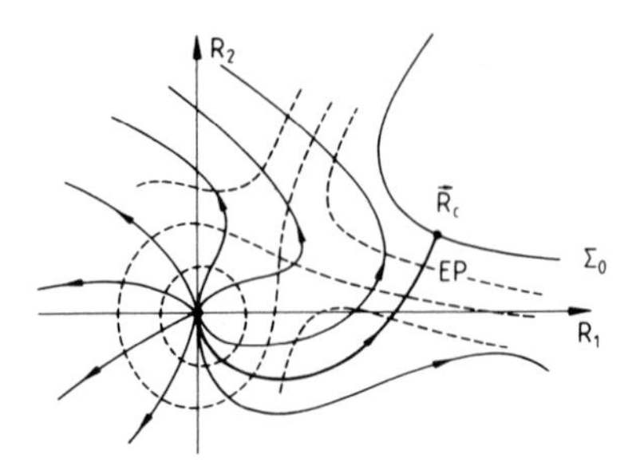

Fig. 8. The continuous lines (with arrow) represent paths that are solutions of the classical equation of motion in the *inverted* potential. These paths have left the unstable point at $\boldsymbol{R} = 0$ in the infinite past. There is a unique path, the escape path (EP), which connects $\boldsymbol{R} = 0$ with $\boldsymbol{R}_{\mathrm{C}}$ on Σ_0.

minimum with the boundary between inaccessible and accessible regions. Investigations on specific models support this assumption. In fact, one has found one such path for the Caldeira–Leggett model (see section 3.6).

For simplicity, we consider the case where there is just one such path, say $\boldsymbol{R} = \boldsymbol{r}(\tau) = \boldsymbol{r}_\tau$. It obeys the equation of motion (3.7) together with the appropriate boundary conditions; explicitly,

$$m\ddot{\boldsymbol{r}}_\tau = \frac{\partial V}{\partial \boldsymbol{r}_\tau}, \qquad \frac{m}{2}\dot{\boldsymbol{r}}_\tau^2 - V = 0,$$

$$\boldsymbol{r}_{-\infty} = 0, \qquad \boldsymbol{r}_0 = \boldsymbol{R}_{\mathrm{C}}, \tag{3.10}$$

where $\boldsymbol{R}_{\mathrm{C}}$ is on Σ_0. We will call $\boldsymbol{r}_\tau$ the escape path* (EP) and $\boldsymbol{R}_{\mathrm{C}}$ the escape point (see fig. 8).

Clearly, $\dot{\boldsymbol{r}}_\tau = 0$ for $\tau \to 0$; but there exists a unit vector

$$(\dot{\boldsymbol{r}}_\tau/|\dot{\boldsymbol{r}}_\tau|)_{\tau\to 0} = \boldsymbol{n}_{\mathrm{C}}, \tag{3.11}$$

which is perpendicular to Σ_0 at $\boldsymbol{R}_{\mathrm{C}}$.

If one follows the consequences of this assumption, a picture as shown in fig. 9 emerges. There is a dense bundle of paths which start (at $\tau = -\infty$) at the origin and which follow the escape path closely for a long distance. Eventually, they enter a region close to Σ_0 and then, it becomes evident that their velocity component parallel to Σ_0 remains finite preventing these paths from reaching Σ_0. Observe that in this region, the form of these paths is almost parabolic, and that there is an envelope Σ_{C} to these paths which may be called a caustic (Courant and Hilbert 1962). We conclude that, beyond Σ_{C}, there is no solution $\mathscr{W}(\boldsymbol{R})$ of the eikonal equation (3.3) which is real and which satisfies the boundary conditions at the origin.

* This is an abbreviation of the term "most probable escape path" found in Banks et al. (1978).

3.2. *Small-fluctuation approximation*

Let us assume that the wave function falls off quickly if one moves away perpendicular from the escape path.* In this case, we may restrict our attention to paths in the vicinity of the escape path. Thus, we are led to introduce the small-fluctuation approximation where $\boldsymbol{R}_\tau = \boldsymbol{r}_\tau + \boldsymbol{v}_\tau$, such that $\boldsymbol{v}_\tau$ is small. It follows from eq. (3.7) and from the initial condition discussed before, that, through first order in $\boldsymbol{v}$,

$$\ddot{\boldsymbol{v}}_\tau = \hat{U}_\tau \cdot \boldsymbol{v}_\tau, \qquad \boldsymbol{v}_{-\infty} = 0, \tag{3.12}$$

where the operator $\hat{U}_\tau$ is defined by**

$$U_{kk'}(\tau) = \frac{1}{m} \frac{\partial^2 V}{\partial R_k \partial R_{k'}}\bigg|_{\boldsymbol{R}=\boldsymbol{r}_\tau} \tag{3.13}$$

A quantity of most importance is the action along the escape path,

$$\mathscr{W}_{\mathrm{EP}}(\tau) = m \int_{-\infty}^{\tau} \mathrm{d}\tau' \, \dot{\boldsymbol{r}}_{\tau'}^2. \tag{3.14}$$

Concerning the action in the vicinity of the escape path, one may calculate it as follows, using eq. (3.8) and the condition $H = 0$:

$$\begin{aligned} \mathscr{W}(\tau) &\equiv \int_{-\infty}^{\tau} \mathrm{d}\tau' \, m\dot{\boldsymbol{R}}_{\tau'}^2 \\ &= \int_{-\infty}^{\tau} \mathrm{d}\tau' \left\{ \tfrac{1}{2} m\dot{\boldsymbol{R}}_{\tau'}^2 - V(\boldsymbol{R}_{\tau'}) + \tfrac{1}{2} m\dot{\boldsymbol{R}}_{\tau'}^2 + V(\boldsymbol{R}_{\tau'}) \right\} \\ &= \int_{-\infty}^{\tau} \mathrm{d}\tau' \left\{ \tfrac{1}{2} m\dot{\boldsymbol{r}}_{\tau'}^2 + m\dot{\boldsymbol{r}}_{\tau'} \cdot \dot{\boldsymbol{v}}_{\tau'} + \tfrac{1}{2} m\dot{\boldsymbol{v}}_{\tau'}^2 + V(\boldsymbol{r}_{\tau'} + \boldsymbol{v}_{\tau'}) \right\}. \end{aligned} \tag{3.15}$$

One expands $V(\boldsymbol{r} + \boldsymbol{v}) = V(\boldsymbol{r}) + (\boldsymbol{v} \cdot \partial/\partial \boldsymbol{r}) V + (1/2)(\boldsymbol{v} \cdot \partial/\partial \boldsymbol{r})^2 V$, then one performs partial integrations and observes that $\boldsymbol{v}_\tau \cdot m\ddot{\boldsymbol{r}}_\tau - (\boldsymbol{v}_\tau \cdot \partial/\partial \boldsymbol{r}_\tau) V = 0$ [cf. eqs. (3.12) and (3.13)] as well as $\boldsymbol{v}_\tau \cdot m\ddot{\boldsymbol{v}}_\tau - (\boldsymbol{v}_\tau \cdot \partial/\partial \boldsymbol{r}_\tau)^2 V = 0$. Thus, one obtains in quadratic accuracy

$$\mathscr{W}(\tau) = \mathscr{W}_{\mathrm{EP}}(\tau) + m\dot{\boldsymbol{r}}_\tau \cdot \boldsymbol{v}_\tau + \tfrac{1}{2} m (\dot{\boldsymbol{v}}_\tau \cdot \boldsymbol{v}_\tau). \tag{3.16}$$

We express $\mathscr{W}(\tau)$ in coordinate space as follows. By choosing τ, we select a point $\boldsymbol{r}_\tau$ on the escape path. A vicinity of this point will be represented by

$$\boldsymbol{R} = \boldsymbol{r}_\tau + \boldsymbol{\eta}, \tag{3.17}$$

*In section 3.3, we will show that this assumption is not really independent but fits in the framework of the quasiclassical approximation.

**In cases where the quantities carry Cartesian subscripts, we note the time dependence in the usual way, e.g., $(\hat{U}_\tau)_{kk'} = U_{kk'}(\tau)$.

where $|\boldsymbol{\eta}|$ is small. Of course, this representation is not unique. It could be made unique by requiring that $\boldsymbol{\eta}$ be locally perpendicular to the escape path (Banks and Bender 1973, Banks et al. 1973, Gervais and Sakita 1977) or by another convention, as will be shown in section 3.3. For the time being, however, it is not necessary to do so.

Next, we define a small-fluctuation operator $\hat{K}_\tau$ such that it obeys the equation of motion

$$[-\hat{1}\partial_\tau^2 + \hat{U}_\tau]\hat{K}_\tau = 0 \tag{3.18a}$$

and such that it vanishes for $\tau \to -\infty$. Note that in this limit $\hat{U}_\tau$ approaches the constant $\hat{U}^{(0)}$ introduced in eq. (3.2). Therefore, if

$$\hat{\omega}_0 = [\hat{U}^{(0)}]^{1/2} \tag{3.19}$$

is the positive square root, we have

$$\hat{K}_\tau \to \exp \hat{\omega}_0 \tau, \quad \tau \to -\infty. \tag{3.18b}$$

Since, $\hat{U}^{\mathrm{T}} = \hat{U}$, we may derive the relation

$$\dot{\hat{K}}_\tau^{\mathrm{T}} \hat{K} - \hat{K}_\tau^{\mathrm{T}} \dot{\hat{K}}_\tau = 0, \tag{3.20}$$

where the superscript T denotes the transpose of a matrix.

It is now easy to associate with each τ and $\boldsymbol{\eta}$ a path $\boldsymbol{v}_{\tau'}$ such that $\boldsymbol{v}_\tau = \eta$ and

$$\boldsymbol{v}_{\tau'} = \hat{K}_{\tau'} \hat{K}_\tau^{-1} \cdot \boldsymbol{\eta}. \tag{3.21}$$

Therefore, in the small-fluctuation approximation, the action assumes the form

$$\mathscr{W}(\boldsymbol{\eta}, \tau) = \mathscr{W}_{\mathrm{EP}}(\tau) + m\dot{\boldsymbol{r}}_\tau \cdot \boldsymbol{\eta} + \tfrac{1}{2} m\boldsymbol{\eta} \cdot \hat{\Omega}_\tau \cdot \boldsymbol{\eta}, \tag{3.22}$$

where

$$\hat{\Omega}_\tau = \dot{\hat{K}}_\tau \hat{K}_\tau^{-1}. \tag{3.23}$$

We conclude from eq. (3.20) that $\hat{\Omega}_\tau^{\mathrm{T}} = \hat{\Omega}_\tau$. Furthermore, $\hat{\Omega}_{-\infty} = \hat{\omega}_0$. Considering eq. (3.18), one may derive for $\hat{\Omega}_\tau$ the following Riccati-type equation

$$\dot{\hat{\Omega}}_\tau = \hat{U}_\tau - \hat{\Omega}_\tau^2. \tag{3.24}$$

The redundancy of representation (3.17) can be used to calculate the derivative in the direction of the escape path in two ways. Accordingly, $\partial/\partial\tau = \dot{\boldsymbol{r}} \cdot \partial/\partial\boldsymbol{\eta}$, where the left-hand side is taken at $\boldsymbol{\eta} \equiv 0$ and the right-hand side requires the limit $\boldsymbol{\eta} \to 0$ to be taken only at the end. We apply this relation to $\partial\mathscr{W}/\partial\eta_k = \dot{r}_k + \Omega_{kk'}\eta_{k'}$, and obtain the useful relation

$$\ddot{\boldsymbol{r}}_\tau = \hat{\Omega}_\tau \cdot \dot{\boldsymbol{r}}_\tau \tag{3.25}$$

According to eq. (3.9), the calculation of the first-order correction $\mathscr{W}^{(1)}$ requires the evaluation of $\partial^2\mathscr{W}/\partial R_k^2$. In the small-fluctuation approximation, this is equal to $\partial^2 \mathscr{W}/\partial\eta_k^2$; therefore, from eqs. (3.9) and (3.22),

$$\dot{\mathscr{W}}^{(1)}(\tau) = \tfrac{1}{2}\mathrm{Tr}\,\hat{\Omega}_\tau - \tfrac{1}{2}\omega_\mathrm{T} \tag{3.26}$$

For $\mathscr{W}^{(1)}$ to be finite near the origin (where $\tau \to -\infty$), we require that

$$\omega_\mathrm{T} = \mathrm{Tr}\,\hat{\Omega}_{-\infty} = \mathrm{Tr}\,\hat{\omega}_0. \tag{3.27}$$

Then it follows that

$$\mathscr{W}^{(1)}(\tau) = \tfrac{1}{2}\mathrm{Tr}\ln\{\hat{K}_\tau \exp(-\hat{\omega}_0\tau)\}. \tag{3.28}$$

Considering eqs. (2.13), (3.17), (3.22) and (3.28) and using the identity $\mathrm{Tr}\ln\hat{A} = \ln\det\hat{A}$, the principal wave function may be written as

$$\psi_0(\boldsymbol{\eta}, \tau) = \frac{\exp[-\{\mathscr{W}_\mathrm{EP}(\tau) + m\dot{\boldsymbol{r}}_\tau\cdot\boldsymbol{\eta} + \frac{1}{2}m\boldsymbol{\eta}\cdot\hat{\Omega}_\tau\cdot\boldsymbol{\eta}\}/\hbar]}{[\det\{\hat{K}_\tau\exp(-\hat{\omega}_0\tau)\}]^{1/2}}. \tag{3.29}$$

Concerning the wave function near the origin, we assume that there is a sufficiently large environment where the potential is purely harmonic. This means that

$$\boldsymbol{r}_\tau = \hat{\omega}_0^{-1}\hat{K}_\tau\cdot\boldsymbol{\chi}_1^- = \hat{\omega}_0^{-1}\cdot\dot{\boldsymbol{r}}_\tau, \tag{3.30}$$

provided that $\boldsymbol{r}_\tau$ is sufficiently small ($\tau \to -\infty$). One may understand this relation by noting that $\boldsymbol{\chi}_1^- = \hat{\omega}_0^{-1}\hat{K}_\tau^{-1}\cdot\dot{\boldsymbol{r}}_\tau$ is a time-independent vector. It also follows that in the above limit, $\mathscr{W}_\mathrm{EP}(\tau) = (m/2)\boldsymbol{r}_\tau\cdot\hat{\omega}_0\cdot\boldsymbol{r}_\tau$ and that

$$\mathscr{W}(\boldsymbol{\eta}, \tau) = \tfrac{1}{2}m[\boldsymbol{r}_\tau + \boldsymbol{\eta}]\cdot\hat{\omega}_0\cdot[\boldsymbol{r}_\tau + \boldsymbol{\eta}] = \tfrac{1}{2}m\boldsymbol{R}\cdot\hat{\omega}_0\cdot\boldsymbol{R}, \tag{3.31}$$

which means that eq. (3.29) reduces to the correct harmonic oscillator wave function $\psi_0 = \exp[-(m/2\hbar)\boldsymbol{R}\cdot\hat{\omega}_0\cdot\boldsymbol{R}]$. The normalization of the wave function is determined mostly by contributions from the environment near the origin; therefore,

$$\int \mathrm{d}^N R\,\psi_0^2 = \left[\frac{(\pi\hbar/m)^N}{\det\hat{\omega}_0}\right]^{1/2}. \tag{3.32}$$

3.3. Principal and transmitted wave near the caustic

For convenience, we introduce a coordinate system $\boldsymbol{\zeta} = \boldsymbol{R} - \boldsymbol{R}_\mathrm{C} = (\zeta_1, \ldots, \zeta_N)$ such that the ζ_1-axis is parallel to $\boldsymbol{n}_\mathrm{C}$, which is the tangent vector to the escape path at its end as given by eq. (3.11). Note that at the escape point, the force $\boldsymbol{F}_\mathrm{C} = -(\partial V/\partial\boldsymbol{R})_{\boldsymbol{R}=\boldsymbol{R}_\mathrm{C}} = (F_\mathrm{C}, 0, \ldots, 0)$ points in the same direction.

Let us also shift the coordinates of the escape path according to $\boldsymbol{z}_\tau = \boldsymbol{r}_\tau - \boldsymbol{R}_\mathrm{C}$. Observe now that the solution to eq. (3.10) can be represented in the form of a power series in τ^2:

$$z_k(\tau) = -\frac{1}{2m}\tau^2 F_{Ck} - \frac{1}{4!m}\tau^4 U^C_{kk'} F_{Ck'} + \cdots, \tag{3.33}$$

where* $\hat{U}^C = \hat{U}_{\tau=0}$. Note that in leading order, $z_1 = -\tau^2 F_C/2m$, and that $z_\perp \propto \tau^4 \propto z_1^2$, where $z_\perp$ is the component perpendicular to $\boldsymbol{n}_C$.

Consider now $\mathscr{W}(\boldsymbol{\eta}, \tau)$ as given by eq. (3.22), and observe that $\boldsymbol{\eta} = \boldsymbol{\zeta} - \boldsymbol{z}_\tau$. The redundancy in the representation (3.17) is used to choose $\tau = \tau_1$ such that

$$z_1(\tau_1) = \zeta_1, \quad \tau_1 = -(-2m\zeta_1/F_C)^{1/2} + \cdots, \tag{3.34}$$

where the second expression holds in leading order. Then,

$$\mathscr{W}(\zeta_1, \ldots, \zeta_N) = \mathscr{W}_{EP}(\tau_1) + m\dot{r}_k(\tau_1)[\zeta_k - z_k(\tau_1)] + \tfrac{1}{2}m\Omega_{kk'}(\tau_1)[\zeta_k - z_k(\tau_1)][\zeta_{k'} - z_{k'}(\tau_1)]. \tag{3.35}$$

Since $\zeta_1 - z_1(\tau_1) = 0$, the summation above includes only indices $k, k' > 1$. In particular, $\mathscr{W}$ depends on ζ_1 only through its dependence on τ_1.

One expects that $\mathscr{W}(\boldsymbol{\zeta})$ satisfies the eikonal equation (3.3). However, due to the small-fluctuation approximation, this will be true only through terms of order $(\zeta - z)^2 \sim \zeta_\perp^2, \zeta_\perp \zeta_1^2, \zeta_1^4$. This can be checked by differentiating eq. (3.35) as follows. We introduce indices $j, j' \geqslant 2$, e.g., $\boldsymbol{\zeta} = (\zeta_1, \{\zeta_j\})$. Then we observe eqs. (3.14) and (3.25), whence we obtain

$$\frac{1}{m}\frac{\partial \mathscr{W}}{\partial \zeta_1} = \frac{d\tau_1}{d\zeta_1}\{\dot{r}_1^2 + \dot{r}_1\Omega_{1j}[\zeta - z]_j + \tfrac{1}{2}\dot{\Omega}_{jj'}[\zeta - z]_j[\zeta - z]_{j'}\},$$

$$\frac{1}{m}\frac{\partial \mathscr{W}}{\partial \zeta_j} = \dot{r}_j + \Omega_{jj'}[\zeta - z]_{j'}. \tag{3.36}$$

As an abbreviation, we introduce again $\eta_j = \zeta_j - z_j(\tau_1)$. Then one obtains through $O(\eta_j^2)$

$$\frac{1}{m^2}\left(\frac{\partial \mathscr{W}}{\partial \zeta_k}\right)^2 = \dot{r}_k^2 + 2\dot{r}_k\Omega_{kj}\eta_j + \dot{\Omega}_{jj'}\eta_j\eta_{j'} + \Omega_{kj}\Omega_{kj'}\eta_j\eta_{j'}$$

$$= \dot{r}_k^2 + 2\ddot{r}_j\eta_j + U_{jj'}\eta_j\eta_{j'}. \tag{3.37}$$

Note that the last line follows from eqs. (3.24) and (3.25). Using eq. (3.10), we find through order η^2 that $(\partial\mathscr{W}/\partial\zeta_k)^2 - 2mV = 0$, as expected.

Of central importance is the fact that $\boldsymbol{r}_{\tau_1}$ and $\hat{\Omega}_{\tau_1}$ can be expanded in powers of τ_1. This follows from the equations of motion (3.10) and (3.24); eq. (3.33) may serve as an example. On the other hand, it is also possible to express τ_1 in a series of powers in $(-\zeta_1)^{1/2}$ the leading term of which is shown in eq. (3.34). Hence, it follows that all quantities can be represented by power series in $(-\zeta_1)^{1/2}$. Thus,

* Note that $\hat{U}^C$ has (at least) one negative eigenvalue.

we may write eq. (3.35) as follows:

$$\begin{aligned}\mathscr{W}(\zeta) &= \mathscr{W}_{\text{EP}}(0) - \tfrac{2}{3}(2mF_{\text{C}})^{1/2}(-\zeta_1)^{3/2} + \cdots \\ &\quad + \frac{m}{2}\left(\frac{2m}{F_{\text{C}}}\right)^{1/2}(-\zeta_1)^{3/2}\, U^{\text{C}}_{1j}[\zeta_j - z_j(\zeta_1)] + \cdots \\ &\quad + \frac{m}{2}\,\Omega_{jj'}(0)[\zeta_j - z_j(\zeta_1)][\zeta_{j'} - z_{j'}(\zeta_1)] + \cdots,\end{aligned} \tag{3.38}$$

where only terms of leading order are shown explicitly.

Note that $\mathscr{W}^{(1)}$, the first-order correction to the action, does not depend on ζ_j in the small-fluctuation approximation. Hence, its expansion is of a simple form except for one peculiarity which is connected with a logarithmic divergence for $\zeta_1 \to 0$. Therefore, we write

$$\mathscr{W}^{(1)}(\zeta) = \tfrac{1}{4}\ln(-2mF_{\text{C}}\zeta_1)^{1/2} + \cdots + \tfrac{1}{2}\{\operatorname{Tr}\ln \hat{K}_\tau - \ln \dot{r}_1(\tau)\}_{\tau\to 0} + \cdots. \tag{3.39}$$

The unique role of ζ_1 clearly demonstrates that in the small-fluctuation approximation the caustic Σ_{C} is approximated by the hypersurface $\zeta_1 = 0$. It follows from eq. (3.38) that the surfaces $\mathscr{W}(\zeta) = \text{const} < \mathscr{W}_{\text{EP}}(0)$ are of the form $-\zeta_1 \propto (\zeta_\perp - \zeta_0)^{2/3}$, whereas the surface $\mathscr{W}(\zeta) = \mathscr{W}_{\text{EP}}(0)$ is characterized by $-\zeta_1 \propto \zeta_\perp^{4/3}$. See fig. 9 for an illustration.

Concerning the validity of the quasiclassical approximation we take an appropriate generalization of the criterion [eq. (2.25)] which is given by

$$\hbar\left|\frac{\partial \mathscr{W}^{(1)}}{\partial \boldsymbol{R}}\right| \ll \left|\frac{\partial \mathscr{W}}{\partial \boldsymbol{R}}\right|. \tag{3.40}$$

In the present case, no problem arises with this condition for $\boldsymbol{R}$ near the origin, provided that $\hbar$ is sufficiently small. The situation is different near the caustic. Taking the derivatives in eq. (3.40) in the normal direction, we obtain from eqs.

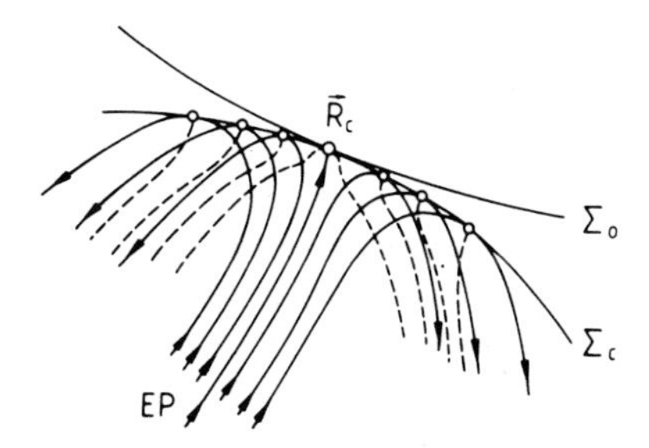

Fig. 9. Paths (→→) in the inverted potential near the escape point. The caustic (——) is marked by Σ_{C}. The surfaces for $\mathscr{W}(\boldsymbol{R}) = \text{const}$ (– – –) are seen to intersect the incoming paths orthogonally.

(3.38) and (3.39), and for $\zeta_\perp = 0$, the condition $|\zeta_1|^3 \gg \hbar^2/mF_C$ which corresponds to one of the inequalities (2.26). Therefore, the quasiclassical approximation breaks down near a caustic the same way it does near a turning point in a one-dimensional problem.

In accordance with the principle of analytical continuation outlined in section 2.3, it is possible to bypass the caustic (which is given by $\zeta_1 = 0$ in the small-fluctuation approximation) if we allow ζ_1 to assume complex values. At this point, we observe that $\mathscr{W}(\zeta)$ of eq. (3.38) is a solution of the eikonal equation (3.3) also for complex ζ_1. A similar statement can also be made with regard to $\mathscr{W}^{(1)}(\zeta)$. Therefore, and in view of eq. (2.31), we are led to substitute

$$-\zeta_1 \to \mathrm{e}^{-\mathrm{i}\pi}\zeta_1 \tag{3.41}$$

in both expressions (3.38) and (3.39). Let us now mark all quantities obtained by this analytical continuation by a tilde, e.g., $\tilde{\mathscr{W}}(\zeta) = \mathscr{W}[-\zeta_1 \to \mathrm{e}^{-\mathrm{i}\pi}\zeta_1]$. Then, the transmitted wave may be written as

$$\tilde{\psi}(\zeta) = \exp\left\{-\frac{1}{\hbar}[\tilde{\mathscr{W}}(\zeta) + \hbar\tilde{\mathscr{W}}^{(1)}(\zeta)]\right\}, \tag{3.42}$$

where $\zeta_1 > 0$.

For a general orientation, let us remark that, beyond the caustic, $\tilde{\mathscr{W}}(\zeta)$ is a complex function. This complex function may be represented by surfaces $\mathrm{Re}\,\tilde{\mathscr{W}} = \text{const}$ and $\mathrm{Im}\,\tilde{\mathscr{W}} = \text{const}$. It follows from the eikonal equation (3.3) that at each point, the two normal directions to these surfaces are perpendicular (Landauer 1950). Furthermore, the surface $\mathrm{Im}\,\tilde{\mathscr{W}} = 0$ coincides with the caustic. Schematically, this is shown in fig. 10 for a two-dimensional system.

It is possible to obtain the wave function [eq. (3.42)] directly by modifying appropriately the method of characteristics introduced in section 3.1. For the time being, we are interested only in the small-fluctuation approximation. Then one may proceed as follows.* We allow the parameter τ to assume purely imaginary values. In view of eqs. (3.41) and (3.34) we substitute

$$\tau \to \mathrm{i}t, \tag{3.43}$$

where t is real and positive. Note that this procedure is justified by the very fact already made use of earlier, namely that $\hat{\Omega}_\tau$ and $\boldsymbol{r}_\tau$ allow a representation as power series in τ and that

$$\dot{\hat{\Omega}}_0 = 0 \quad \text{as well as} \quad \dot{\boldsymbol{r}}_0 = 0.$$

The extension of the escape path $\tilde{\boldsymbol{r}} = \tilde{\boldsymbol{r}}_t$ is defined by

$$m\ddot{\tilde{\boldsymbol{r}}}_t = -\frac{\partial V}{\partial \tilde{\boldsymbol{r}}_t}, \quad \tilde{\boldsymbol{r}}_0 = \boldsymbol{R}_C, \quad \dot{\tilde{\boldsymbol{r}}}_0 = 0, \tag{3.44}$$

where the dot means here differentiation with respect to t. (Note that the t-differentiation operates always on quantities marked by a tilde; therefore, no

* Recall the discussion in the second part in section 2.5 and fig. 5.

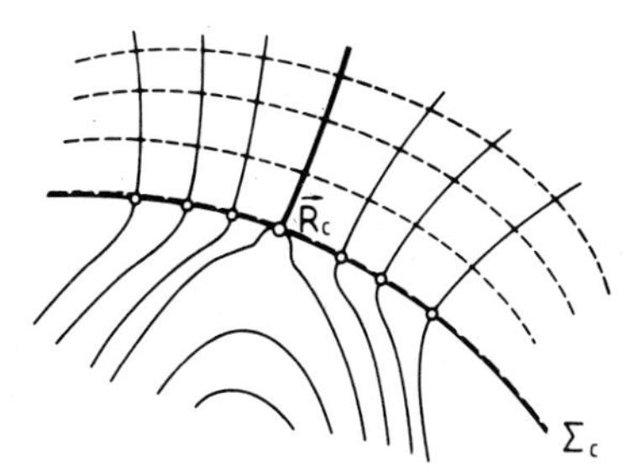

Fig. 10. Surfaces of constant Re $\tilde{\mathscr{W}}(\boldsymbol{R})$ (——) and of constant Im $\tilde{\mathscr{W}}(\boldsymbol{R})$ (– – –) near the caustic. The surfaces of constant $\mathscr{W}(\boldsymbol{R})$ (——) in the classically forbidden region complete the picture in a consistent way.

ambiguity arises with the two types of dot-symbols.*) This path describes a classical motion in the potential V in contrast to eq. (3.10), where the potential is inverted.

In the small-fluctuation approximation, eq. (3.17) is replaced by $\boldsymbol{R} = \tilde{\boldsymbol{r}}(t) + \boldsymbol{\eta}$. Further changes are as follows:

(a) Equation (3.18) is replaced by

$$[\hat{1}\partial_t^2 + \hat{\tilde{U}}_t]\hat{\tilde{K}}_t = 0, \qquad \hat{\tilde{K}}_0 = \hat{K}_0, \qquad -\mathrm{i}\dot{\hat{\tilde{K}}}_0 = \dot{\hat{K}}_0, \tag{3.45}$$

where $\hat{\tilde{U}}_t$ is obtained from eq. (3.13) if $\tilde{\boldsymbol{r}}_t$ is substituted in place of $\boldsymbol{r}_\tau$.

(b) Equations (3.14), (3.22) and (3.23) are replaced by

$$\tilde{\mathscr{W}}_{\mathrm{EP}}(t) = \mathscr{W}_{\mathrm{EP}}(0) - im\int_0^t \mathrm{d}t'\,\dot{\tilde{\boldsymbol{r}}}_{t'}^2,$$

$$\tilde{\mathscr{W}}(\boldsymbol{\eta}, t) = \tilde{\mathscr{W}}_{\mathrm{EP}}(t) - im\dot{\tilde{\boldsymbol{r}}}_t\cdot\boldsymbol{\eta} + \tfrac{1}{2}m\boldsymbol{\eta}\cdot\hat{\tilde{\Omega}}_t\cdot\boldsymbol{\eta},$$

$$\hat{\tilde{\Omega}}_t = -\,\mathrm{i}\dot{\hat{\tilde{K}}}_t\hat{\tilde{K}}_t^{-1}, \tag{3.46}$$

(c) Equation (3.28) is replaced by

$$\mathscr{W}^{(1)}(t) = \tfrac{1}{2}\mathrm{Tr}\ln\{\hat{\tilde{K}}_t\cdot\exp(-\mathrm{i}\hat{\omega}_0 t)\}. \tag{3.47}$$

It is advantageous to use again the ζ-coordinate system. Specifically, we put $\tilde{z}_t = \tilde{\boldsymbol{r}}_t - \boldsymbol{R}_{\mathrm{C}}$. In place of eq. (3.34), we now choose $t = t_1$ such that $\tilde{z}_1(t_1) = \zeta_1 > 0$. Considering eq. (3.46), and introducing $\eta_j = \zeta_j - \tilde{z}_j(t_1)$ as an abbreviation, we obtain analogously to eq. (3.35) the expression

$$\tilde{\mathscr{W}}(\zeta) = \tilde{\mathscr{W}}_{\mathrm{EP}}(t_1) - im\dot{\tilde{\boldsymbol{r}}}_{t_1}\cdot\boldsymbol{\eta} + \tfrac{1}{2}m\boldsymbol{\eta}\cdot\hat{\tilde{\Omega}}_{t_1}\cdot\boldsymbol{\eta}. \tag{3.48}$$

* Following eq. (3.41), we have introduced the convention to mark the analytically continued quantities with a tilde. Note that this notation might be ambiguous by a factor $\pm\mathrm{i}$ for time-integrated or time-differentiated quantities. In what follows, however, the explicit definition applies.

Note specifically

$$|\tilde{\psi}(\boldsymbol{\eta})|^2 = \exp\left\{-2\mathscr{W}_{\mathrm{EP}}(0)/\hbar - \mathrm{Re}\left[\frac{m}{\hbar}\boldsymbol{\eta}\cdot\hat{\hat{\Omega}}_{t_1}\cdot\boldsymbol{\eta} + 2\tilde{\mathscr{W}}^{(1)}(t_1)\right]\right\}. \tag{3.49}$$

We assume that $\mathrm{Re}\,\hat{\hat{\Omega}}_{t_1}$ is a positive definite operator in the subspace perpendicular to $\boldsymbol{n}_{\mathrm{C}}$.

Let us now calculate the probability current through the plane $\zeta_1 = \text{const}$:

$$\begin{aligned} J(\zeta_1) &= \int \mathrm{d}^{N-1}\{\zeta_j\}\boldsymbol{n}_{\mathrm{C}}\cdot\frac{\hbar}{m}\,\mathrm{Im}\,\tilde{\psi}^*\frac{\partial\tilde{\psi}}{\partial\boldsymbol{R}} \\ &= \frac{-1}{m}\int \mathrm{d}^{N-1}\{\zeta_j\}|\tilde{\psi}|^2\,\mathrm{Im}\left[\frac{\partial\tilde{\mathscr{W}}}{\partial\zeta_1} + \hbar\frac{\partial\tilde{\mathscr{W}}^{(1)}}{\partial\zeta_1}\right]. \end{aligned} \tag{3.50}$$

Calculating $\partial\tilde{\mathscr{W}}/\partial\zeta_1$, we obtain a result similar to eq. (3.36). Due to the form of $|\tilde{\psi}|^2$, the terms linear in η_j vanish upon integration. We also observe that terms contained in $\partial\tilde{\mathscr{W}}/\partial\zeta_1$ which are bilinear in η_j may be neglected since their contribution to J is smaller than the leading term by a factor $\hbar$. Within the same approximation, $\partial\tilde{\mathscr{W}}^{(1)}/\partial\zeta_1$ may also be neglected. Thus, we obtain in leading order

$$J(\zeta_1) = \left[\frac{(\pi\hbar/m)^{N-1}}{\det\mathrm{Re}\,\hat{\hat{\Omega}}_\perp(t)}\right]^{1/2}\left.\frac{\dot{\bar{r}}_1(t)\exp(-2\mathscr{W}_{\mathrm{EP}}(0)/\hbar)}{\det\mathrm{Re}\{\hat{\hat{K}}_t\exp(-\mathrm{i}\hat{\omega}t)\}}\right|_{t=t_1}, \tag{3.51}$$

where $\hat{\hat{\Omega}}_\perp$ is the projection of $\hat{\hat{\Omega}}$ onto the subspace perpendicular to $\boldsymbol{n}_{\mathrm{C}}$.

It is worth noting that the leading contribution to the current as given in eq. (3.51) does not suffer from the small-fluctuation approximation. In fact, systematic corrections to this approximation are of higher order in η_j than the order of the leading terms in $\tilde{\mathscr{W}}$ and $\tilde{\mathscr{W}}^{(1)}$. Consequently, these corrections will give rise to changes in J only by relative order of $\hbar$.

The current $J(\zeta_1)$ as given above does not depend on the value of ζ_1. One way* to understand this is as follows. Since eqs. (3.3) and (3.4) are also valid for $\tilde{\mathscr{W}}$ and $\tilde{\mathscr{W}}^{(1)}$, we find that

$$\frac{\partial j_k}{\partial R_k} = \frac{\hbar}{m}\,\mathrm{Im}\left[\left(\frac{\partial\tilde{\mathscr{W}}^{(1)}}{\partial R_k}\right)^2 - \frac{\partial^2\tilde{\mathscr{W}}^{(1)}}{\partial R_k^2}\right]|\tilde{\psi}|^2. \tag{3.52}$$

Integrating this expression over a volume enclosed between two planes $\zeta_1 = \zeta_1'$ and $\zeta_1 = \zeta_1''$, we find that the difference $J(\zeta_1') - J(\zeta_1'')$ is smaller than, say, $J(\zeta_1')$ by a factor $\hbar$.

*It is also possible to show that this property follows directly from eqs. (3.44–3.47). See also a similar discussion in section 3.5.

3.4. *Decay rate in quasiclassical approximation and in instanton technique*

In analogy with the procedure leading to eq. (2.27), we integrate $\mathrm{Im}\,\psi^*(\mathscr{H} - E)\psi$ with respect to $\mathrm{d}^N R$ in the region to the "left" side of the plane $\zeta_1 = \text{const}$. Thus, we obtain Γ as the ratio of the current $J(\zeta_1)$ as given in eq. (3.51) and a normalization which is given by eq. (3.32). Since $J(\zeta_1)$ does not depend on ζ_1, one may take as well the limit $\zeta_1 \to 0$, i.e., $t_1 \to 0$.

However, this limit has to be taken with care since both $\dot{\tilde{r}}_1(t)$ as well as $\det \mathrm{Re}\, \hat{\tilde{K}}_t$ vanish for $t \to 0$, although their ratio remains finite. On the other hand, the analytical properties ensure that

$$\left(\frac{\dot{\tilde{r}}_1(t)}{\det \mathrm{Re}\,\hat{\tilde{K}}_t}\right)_{t\to 0} = \left(\frac{\dot{r}_1(\tau)}{\det \hat{K}_\tau}\right)_{\tau\to 0} \tag{3.53a}$$

Thus, we may write $\Gamma = \mathscr{A}\exp(-\mathscr{B})$ in accordance with eq. (2.2) where the exponent is

$$\mathscr{B} = 2\mathscr{W}_{\mathrm{EP}}(0)/\hbar \tag{3.54}$$

and the prefactor is

$$\mathscr{A} = \left[\frac{m \det \hat{\omega}_0}{\pi\hbar \det \hat{\Omega}_\perp(0)}\right]^{1/2} \left(\frac{\dot{r}_1(\tau)}{\det \hat{K}_\tau}\right)_{\tau\to 0}. \tag{3.55a}$$

The restriction of $\hat{\Omega}$ to a subspace in the determinant above is of some inconvenience. It follows from eq. (3.25) that in leading order, $\Omega_{11}(\tau) \to \ddot{r}_1(\tau)/\dot{r}_1(\tau)$ for $\tau \to 0$. Consequently, we have

$$\det \hat{\Omega}_\perp(0) = \left(\frac{\dot{r}_1(\tau)}{\ddot{r}_1(\tau)} \det \hat{\Omega}_\tau\right)_{\tau\to 0}, \tag{3.53b}$$

since the contributions of the finite elements Ω_{1j} and Ω_{j1} to $\det \hat{\Omega}$ drop out in the limit taken above. Using this relation and the definition (3.23), we may write eq. (3.55a) in the form

$$\mathscr{A} = \left[\frac{m\dot{r}_1(\tau)\ddot{r}_1(\tau)\det \hat{\omega}_0}{\pi\hbar \det\{\dot{\hat{K}}_\tau \hat{K}_\tau\}}\right]^{1/2}_{\tau\to 0}. \tag{3.55b}$$

As a technical device, we select a number z, and change the equation of motion (3.18) for the small-fluctuation operator $\hat{K}_\tau^z$ as follows:

$$\begin{aligned}
&[-\hat{1}\partial_\tau^2 + \hat{U}_\tau + z\hat{1}]\hat{K}_\tau^z = 0,\\
&\hat{K}_\tau^z \to \exp\{[\hat{\omega}_0^2 + z\hat{1}]^{1/2}\tau\}, \quad \tau \to -\infty,\\
&\mathrm{Re}[\hat{\omega}_0^2 + z\hat{1}]^{1/2} \geqslant 0.
\end{aligned} \tag{3.56}$$

In the following, we choose $z = \varepsilon$, where ε is a small quantity, $\varepsilon > 0$. Considering eq. (3.30), we introduce

$$\dot{\boldsymbol{r}}_\tau^\varepsilon = \hat{K}_\tau^\varepsilon \cdot \chi_1^- . \tag{3.57a}$$

Note that $\dot{\boldsymbol{r}}_\tau^\varepsilon$ satisfies the small-fluctuation equation (3.12) if $\hat{U}$ is replaced by $\hat{U} + \varepsilon\hat{1}$ and that there exists a Wronskian type of relation

$$\ddot{\boldsymbol{r}}_\tau^\varepsilon \cdot \dot{\boldsymbol{r}}_\tau - \dot{\boldsymbol{r}}_\tau^\varepsilon \cdot \ddot{\boldsymbol{r}}_\tau = \varepsilon \int_{-\infty}^{\tau} \mathrm{d}\tau' \, \dot{\boldsymbol{r}}_{\tau'}^\varepsilon \cdot \dot{\boldsymbol{r}}_{\tau'} . \tag{3.57b}$$

In leading order, one may replace $\dot{\boldsymbol{r}}^\varepsilon$ by $\dot{\boldsymbol{r}}$ on the right-hand side. Therefore,

$$- \dot{\boldsymbol{r}}_0^\varepsilon \cdot \ddot{\boldsymbol{r}}_0 = \frac{\varepsilon}{2m} \mathscr{B} . \tag{3.57c}$$

We proceed now on the assumption that the limits $\tau \to 0$ and $\varepsilon \to 0$ can be interchanged. Then, we obtain from eqs. (3.55b) and (3.57c)

$$\mathscr{A} = \left[\frac{m \dot{r}_1^\varepsilon(0) \ddot{r}_1^\varepsilon(0) \det \hat{\omega}_0}{\pi\hbar \det \hat{K}_0^\varepsilon \dot{\hat{K}}_0^\varepsilon} \right]_{\varepsilon\to 0}^{1/2} = \left[\frac{(-\varepsilon)\mathscr{B} \det \hat{\omega}_0}{2\pi\hbar \det \hat{K}_0^\varepsilon \dot{\hat{K}}_0^\varepsilon} \right]_{\varepsilon\to 0}^{1/2} , \tag{3.58}$$

where we have made use of the fact that $\ddot{\boldsymbol{r}}_0^\varepsilon$ is parallel to $\boldsymbol{n}_\mathrm{C}$ up to corrections of $\mathrm{O}(\varepsilon)$.

We compare now the present result with the one obtained by the instanton technique (Coleman 1979, 1985). The escape path here corresponds to the bounce there if the escape path is supplied by a path of return, which is done by putting $\boldsymbol{r}_{-\tau} = \boldsymbol{r}_\tau$. Clearly, the action of a bounce is then $2\mathscr{W}_{\mathrm{EP}}(0) = \hbar\mathscr{B}$.

In the instanton technique, the prefactor contains products of eigenvalues λ of the small-fluctuation equation. These eigenvalues are defined by

$$[-\hat{1}\partial_\tau^2 + \hat{U}_\tau] \cdot \boldsymbol{\varphi}_\tau = \lambda \boldsymbol{\varphi}_\tau , \tag{3.59}$$

where $\boldsymbol{\varphi}_\tau$ is required to remain bounded for $|\tau| \to \infty$. Since $\hat{U}_{-\tau} = \hat{U}_\tau$, the eigenfunctions can be chosen to be either odd or even functions of τ. Therefore, we must have either $\boldsymbol{\varphi}_0 = 0$ or $\dot{\boldsymbol{\varphi}}_0 = 0$.

Consider now an arbitrary complex number z, and let us construct the small-fluctuation operator $\hat{K}_\tau^z$ according to eq. (3.56). Then it is necessary and sufficient for $-z$ to be an eigenvalue λ to an odd eigenfunction that there exists a nontrivial vector $\boldsymbol{c}$ such that

$$\hat{K}_0^z \cdot \boldsymbol{c} = 0. \tag{3.60a}$$

Furthermore, $-z$ is an eigenvalue λ to an even eigenfunction only if

$$\dot{\hat{K}}_0^z \cdot \boldsymbol{d} = 0. \tag{3.60b}$$

The two relations above can be solved for $\boldsymbol{c}$ or for $\boldsymbol{d}$ nontrivially only if the appropriate determinants vanish. Therefore, $-z = \lambda$ only if

$$\det \hat{K}_0^z \dot{\hat{K}}_0^z = 0. \tag{3.60c}$$

Consider now the case where $\hat{U}_\tau$ is replaced by its asymptotic form $\hat{U}^{(0)}$. Correspondingly, we may construct the small-fluctuation operator $\hat{K}_\tau^{(0)z}$. Then we assert that

$$\frac{\det\{-\hat{1}(\partial_\tau^2 - z) + \hat{U}^{(0)}\}}{\det\{-\hat{1}(\partial_\tau^2 - z) + \hat{U}_\tau\}} = \frac{\det \hat{K}_0^{(0)z} \dot{\hat{K}}_0^{(0)z}}{\det \hat{K}_0^z \dot{\hat{K}}_0^z}, \tag{3.61}$$

where det means a determinant in continuous τ and N-dimensional configuration space.

The proof of the assertion is the same as the one given by Coleman (1979, 1985) for the case $N = 1$; it is based on a comparison of the analytical properties of both sides of eq. (3.61) as functions of z. In particular, one finds that the zeros and the poles agree, as well as the asymptotic behavior for $|z| \to \infty$.

Since $\hat{K}_\tau^{(0)z} = \exp([\hat{\omega}_0^2 + z\hat{1}]^{1/2}\tau)$, we have

$$\det \hat{K}_0^{(0)z} \dot{\hat{K}}_0^{(0)z} = \det[\hat{\omega}_0^2 + z\hat{1}]^{1/2}. \tag{3.62}$$

Note that in all cases of interest, $\det([\hat{\omega}_0^2 + \varepsilon\hat{1}]/\hat{\omega}_0^2) \to 1$ for $\varepsilon \to 0$. Therefore, we may write eq. (3.58) as

$$\mathscr{A} = \left(\frac{\mathscr{B}}{2\pi}\right)^{1/2} \left[-\varepsilon \frac{\det\{-\hat{1}\partial_\tau^2 + \hat{U}^{(0)}\}}{\det\{-\hat{1}\partial_\tau^2 + \hat{U}_\tau + \varepsilon\hat{1}\}}\right]_{\varepsilon \to 0}^{1/2}. \tag{3.63}$$

Recall that $\dot{\boldsymbol{r}}_\tau = -\dot{\boldsymbol{r}}_{-\tau}$ is an odd eigenfunction of eq. (3.59) with eigenvalue 0. This means that we may write eq. (3.63) in the form best known in instanton technique (Coleman 1979, 1985), namely,

$$\mathscr{A} = \left(\frac{\mathscr{B}}{2\pi}\right)^{1/2} \left[-\frac{\det\{-\hat{1}\partial_\tau^2 + \hat{U}^{(0)}\}}{\det'\{-\hat{1}\partial_\tau^2 + \hat{U}_\tau\}}\right]^{1/2}, \tag{3.64}$$

where det′ means that the eigenvalue zero has to be omitted. Thus, the prefactor [eq. (3.58)] obtained from our many-dimensional WKB technique is identical to the instanton expression (3.64).

So far we have assumed that there is only one escape path. If there are several such paths, the total decay rate is the sum of contributions of the type (3.54) and (3.64) from each escape path.

3.5. *Reflected wave*

Essentially, the wave function of a metastable state consists of a principle wave ψ_0, which resembles near the origin the ground state of a set of harmonic oscillators and which extends towards the caustic in a narrow tube surrounding the escape path. There, the wave is partially transmitted and it emerges as the transmitted wave $\tilde{\psi}$ in the classically accessible region. It remains now to calculate the reflected wave ψ_1. It has been shown in the one-dimensional case that ψ_1 can be obtained by analytical continuation. Considering eq. (2.33), we perform the substitution

$$-\zeta_1 \to -e^{-2\pi i}\zeta_1 \tag{3.65}$$

in eqs. (3.38) and (3.39). Let us mark all quantities obtained by this analytical continuation with a prime, e.g., $\mathscr{W}'(\zeta) = \mathscr{W}([-\zeta_1 \to -e^{-2\pi i}\zeta_1])$. Clearly, $\mathscr{W}'$ and $\mathscr{W}'^{(1)}$ satisfy the eikonal equation (3.3) and the transport equation (3.4) to the same degree of accuracy as the original expressions do. Therefore, we conclude that the reflected wave is given by

$$\psi_1(\zeta) = \frac{1}{2}\exp\left(-\frac{1}{\hbar}[\mathscr{W}' + \mathscr{W}'^{(1)}]\right), \tag{3.66}$$

where we have inserted an additional factor* of $\frac{1}{2}$ in order to obtain from $\psi_0 + \psi_1$ the same probability current density normal to the plane $\zeta_1 \to -0$, as we do from the wave function $\tilde{\psi}(\zeta)$ at $\zeta_1 \to +0$.

Again, it is possible to obtain the wave function [eq. (3.66)] directly; in the present case, this is done by extending the method of characteristics to $\tau > 0$. Thus, the extension of the escape path is just the return path $\boldsymbol{r}_\tau = \boldsymbol{r}_{-\tau}$, and, in the small-fluctuation approximation, most of the relations of section 3.2 apply to the present case as well. For instance, $\mathscr{W}'(\boldsymbol{\eta}, \tau)$ is equal to expression (3.22) for $\mathscr{W}(\boldsymbol{\eta}, \tau)$ in the case $\tau > 0$. However, care has to be taken with expression (3.28) for $\mathscr{W}'^{(1)}(\tau)$ on account of the logarithmic branch point at $\tau = 0$. Specifically, we have for $\tau > 0$

$$\mathscr{W}'^{(1)}(\tau) = \mathrm{Re}\{\tfrac{1}{2}\mathrm{Tr}[\ln\{\hat{K}_\tau \exp(-\hat{\omega}_0\tau)\}]\} + i\pi . \tag{3.67}$$

Consider now the equation of motion (3.18) of the small-fluctuation operator $\hat{K}_\tau$. Since $\hat{U}_\tau = \hat{U}_{-\tau}$, we may derive the following Wronskian-type of relation

$$\dot{\hat{K}}^{\mathrm{T}}_{-\tau}\hat{K}_\tau + \hat{K}^{\mathrm{T}}_{-\tau}\dot{\hat{K}}_\tau = \mathrm{const}, \tag{3.68}$$

which will be found useful in a moment.

Next we aim to calculate the current $J(\tau)$ from $\psi = \psi_0 + \psi_1$ through a plane perpendicular to the escape path, i.e., through a plane perpendicular to $\dot{\boldsymbol{r}}_\tau$ at $\boldsymbol{r}_\tau$. We obtain the result

$$J(\tau) = \left[\frac{(\pi\hbar/m)^{N-1}}{\det\{\frac{1}{2}[\hat{\Omega}_\perp(\tau) + \hat{\Omega}_\perp(-\tau)]\}}\right]^{1/2}\left[\frac{\dot{\boldsymbol{r}}_\tau^2}{\det(\hat{K}_\tau\hat{K}_{-\tau})}\right]^{1/2}\exp(-\tfrac{1}{2}\mathscr{B}), \tag{3.69}$$

which may be compared with $J(\zeta_1)$ of eq. (3.51). Using a local basis with one axis pointing in the direction of $\dot{\boldsymbol{r}}_\tau$ and using eq. (3.68), we find after some lengthy arguments that $J(\tau)$ is independent of τ. In view of eq. (3.53a), we conclude that $J(\tau) = J(\zeta_1 = 0)$.

* See also the comment below eq. (2.33).

With increasing τ, the return path approaches the origin. Again, we assume that there is a sufficiently large environment of the origin where the potential is purely harmonic. Then, for τ large and positive, the small-fluctuation operator assumes the form

$$\hat{K}_\tau = e^{\hat{\omega}_0 \tau} \hat{A} + e^{-\hat{\omega}_0 \tau} \hat{B}, \tag{3.70}$$

where $\hat{A}$ and $\hat{B}$ are constant matrices. According to eq. (3.20), they have to obey the relation

$$\hat{A}^{\mathrm{T}} \hat{\omega}_0 \hat{B} = \hat{B}^{\mathrm{T}} \hat{\omega}_0 \hat{A}. \tag{3.71}$$

It has already been remarked in connection with eq. (3.20) that there is a vector χ_1^- such that $\hat{K}_\tau \cdot \chi_1^- = \dot{\boldsymbol{r}}_\tau$. Since $\dot{\boldsymbol{r}}_\tau \to 0$ for $\tau \to \pm\infty$, we find that $\hat{A} \cdot \chi_1^- = 0$. This means that $\det \hat{A} = 0$. On the other hand, there are no arguments which tell us that $\det \hat{B}$ should vanish. Therefore, we assume $\det \hat{B} \neq 0$ and write

$$\hat{\Omega}_\tau = \hat{\omega}_0 - 2\hat{\omega}_0 e^{-\hat{\omega}_0 \tau} [\hat{C} + \hat{\omega}_0 e^{-2\hat{\omega}_0 \tau}]^{-1} \hat{\omega}_0 e^{-\hat{\omega}_0 \tau}, \tag{3.72}$$

where the matrix

$$\hat{C} = \hat{\omega}_0 \hat{A} \hat{B}^{-1} = \hat{C}^{\mathrm{T}} \tag{3.73}$$

is symmetric on account of eq. (3.71). Furthermore, its determinant is zero. Let χ_1^+ be the normalized eigenvector of $\hat{C}$ to the eigenvalue zero. Then, for sufficiently* large τ, we obtain

$$\hat{\Omega}_\tau - \hat{\omega}_0 = 2 \frac{\hat{\omega}_0 e^{-\hat{\omega}_0 \tau} \hat{P}_1^+ \hat{\omega}_0 e^{-\hat{\omega}_0 \tau}}{\chi_1^+ \cdot \hat{\omega}_0 e^{-2\hat{\omega}_0 \tau} \cdot \chi^+}, \tag{3.74}$$

where $\hat{P}_1^+$ is the projector on χ_1^+. We conclude that, in general, $\mathscr{W}'(\boldsymbol{\eta}, \tau)$ cannot be put in a simple form comparable with eq. (3.31).

Consider now the case where the lowest eigenvalue ω_{01} of $\hat{\omega}_0$ is smaller than the remaining ones, $\omega_{0n}, n = 2, \ldots, N$, such that $\exp(-\omega_{01}\tau) \gg \exp(-\omega_{0n}\tau)$. Taking the matrix elements of eq. (3.74) with respect to the normal coordinates, we then obtain

$$\Omega_{nn'} = \delta_{nn'} \begin{cases} -\omega_{01}, & n = 1, \\ \omega_{0n}, & n = 2, \ldots, N. \end{cases} \tag{3.75}$$

Clearly, we will find in this case that $\dot{r}_n(\tau) \propto \delta_{n1} \exp(-\omega_{01}\tau)$ which means that the return path (escape path) leads to (leaves) the origin on a straight line in the direction of the low-frequency mode. Furthermore, it follows from eq. (3.26) that $\dot{\mathscr{W}}^{(1)} = -\omega_{01}$, i.e.,

$$\mathscr{W}'^{(1)} = -\omega_{01}\tau + \text{const} = \ln \bar{R}_1 + \text{const}. \tag{3.76}$$

* Precisely, the condition is $\chi_k^+ \cdot \hat{\omega}_0 \exp(-2\hat{\omega}_0 \tau) \cdot \chi_k^+ \ll C_k$, where $\hat{C} \cdot \chi_k^+ = C_k \chi_k^+$ for $k = 2, \ldots, N$. Note that eq. (3.74) may also be written as $\hat{\Omega}_\tau = \hat{\omega}_0 - 2\hat{P}_\tau(\ddot{\boldsymbol{r}}_\tau \cdot \ddot{\boldsymbol{r}}_\tau)/(\dot{\boldsymbol{r}} \cdot \ddot{\boldsymbol{r}})$, where $\hat{P}_\tau$ is the projector on $\dot{\boldsymbol{r}}_\tau$.

In writing down the last part of eq. (3.76), we have chosen a coordinate system $(\bar{R}_1, \ldots, \bar{R}_N)$ in the direction of the normal modes, and we have also chosen $\boldsymbol{\eta}$ in eq. (3.17) such that $\boldsymbol{\eta}\cdot\boldsymbol{r}_\tau = 0$. As the final result, we obtain

$$\psi_1(\boldsymbol{R}) = \frac{\mathrm{i}\Gamma}{2\omega_{01}}(2\pi)^{1/2}\frac{a_{01}}{\bar{R}_1}\exp\left(\frac{\bar{R}_1^2}{2a_{01}^2} - \sum_{n=2}^{N}\frac{\bar{R}_n^2}{2a_{0n}^2}\right), \tag{3.77}$$

where $a_{0n} = (\hbar/2m\omega_{0n})^{1/2}$ and Γ is given by eq. (3.54). The overall constant in eq. (3.77) has been obtained from the condition that the probability current calculated from $\psi_0 + \psi_1$ through a plane normal to the $\bar{R}_1$-direction be equal to the result of eq. (3.69).

The reflected wave of eq. (3.77) is an appropriate generalization of the one-dimensional form (2.12b) to N-dimensional systems.* The above presentation suggests that this result can be obtained only if restrictions are imposed on the frequencies of the metastable state. In particular, it appears that one slow mode has to enslave the remaining fast-moving ones. Alternatively, one may require that the escape path leaves the origin on a straight line. At present, we do not know whether there are significant corrections – if any at all – to the decay rate (presumably only to the prefactor $\mathscr{A}$), if this condition is not satisfied.**

At the end of this section, we wish to comment on the extension of the method of characteristics which we have introduced so far only in the small-fluctuation approximation. We recall that this extension has been meant to provide solutions to the eikonal equation and to the transport equation for complex coordinates. We will give arguments which show that this can be done quite generally by solving the equations of motion eqs. (3.7)–(3.9) for complex coordinates.

As an introduction, we consider first the case of a harmonic potential. There, the solution satisfying the boundary conditions of the present problem is

$$\begin{aligned} \boldsymbol{R}_\tau &= \mathrm{e}^{\hat{\omega}_0\tau}\cdot\boldsymbol{\chi}, \\ \mathscr{W} &= \tfrac{1}{2}m\boldsymbol{\chi}\cdot\hat{\omega}_0\mathrm{e}^{2\hat{\omega}_0\tau}\cdot\boldsymbol{\chi}, \end{aligned} \tag{3.78}$$

where the vector $\boldsymbol{\chi}$ represents an initial direction which is arbitrary. Clearly, complex coordinates are obtained if the times τ and the initial direction $\boldsymbol{\chi}$ are chosen to be complex quantities.*** Although the mapping $(\tau; \chi_1, \ldots, \chi_N) \to (R_1, \ldots, R_N)$ is a projection which does not allow inversion, in general, it is possible to eliminate $(\tau; \chi_1, \ldots, \chi_N)$ in $\mathscr{W}$ in favor of $(R_1, \ldots, R_N)$ since there is a relation between the time and the initial directions which expresses energy conservation. For the present case this elimination is trivial on account of the

* The exponent $1 + \nu$ is now replaced by 1 since the imaginary part of the energy has been omitted in the Schrödinger equation.

** One may wonder whether this condition is necessary to guarantee a quasistationary state.

*** Actually, τ could have been chosen to be real. However, one may use the redundancy of the (τ, χ)-representation in the more complicated case as discussed below to ones advantage.

linear nature of this mapping. For instance, we may write $\boldsymbol{\chi} = \exp(-\hat{\omega}_0\tau)\cdot \boldsymbol{R}$, insert this in the expression for the action and then we obtain $\mathscr{W} = \frac{1}{2}m\boldsymbol{R}\cdot\hat{\omega}_0\cdot\boldsymbol{R}$.

Consider now the case where the potential energy $V(\boldsymbol{R})$ is an analytical function of the coordinates; e.g., a polynomial of third degree is of sufficient complexity in the problem of quantum decay. The solution of the equations of motion (3.7) and (3.8),

$$R_k = R_k(\tau; \chi_1, \ldots, \chi_N),$$
$$\mathscr{W} = \mathscr{W}(\tau; \chi_1, \ldots, \chi_N), \tag{3.79}$$

is known to be unique if $R_k(\tau)$ is bounded. Therefore, we expect that the expressions on the right-hand side of eq. (3.79) are analytical functions of τ and also of $(\chi_1, \ldots, \chi_N)$, with the possible exception of poles. Clearly, this property allows us to conclude that relation (3.79) does not depend on the path of integration in the complex τ-plane.

The elimination procedure described above may now be applied to $\mathscr{W}$ of eq. (3.79). There, we may consider one variable, e.g., χ_N, to be fixed for reasons discussed above (energy conservation). Note, however, that this inversion fails if the corresponding Jacobian vanishes; this defines a hypersurface in $\{R_k\}$-space which should be identified with the caustic. In the space of complex coordinates, this hypersurface is of dimension $2(N-1)$, i.e., two less than the dimensions of the space itself. Therefore, the complex coordinate space will not be disconnected by the caustic.* We have made use of this property in the analytical continuation procedure where we obtain an expression for $\mathscr{W}(\boldsymbol{R})$ which should agree with eq. (3.79) in the sense explained above. Note that the caustic appears as a branching surface of $\mathscr{W}(\boldsymbol{R})$. Similar considerations apply to $\mathscr{W}^{(1)}(\boldsymbol{R})$.

3.6. The Caldeira–Leggett model

This model consists (Caldeira and Leggett 1981, 1983, 1984) of one particle (which we will call the object in the following) in a metastable potential as shown in fig. 1, which interacts with a dissipative environment. Specifically, the cubic form of eq. (2.10a, b) for a potential is considered. The environment is represented by a set of harmonic oscillators, say, with coordinates x_j, interacting with the particle by linear forces. Specifically, we take from Caldeira and Leggett (1983) the following Lagrangian:

$$L = \frac{1}{2}M\dot{q}^2 - v(q) + \frac{1}{2}\sum_{j=2}^{N} m_j\left[\dot{x}_j^2 - \omega_j^2\left(x_j - \frac{C_j}{m_j\omega_j^2}q\right)^2\right]. \tag{3.80}$$

Eventually, one takes the limit $N \to \infty$ such that the frequencies ω_j are distributed continuously. Of importance is the quantity [see eqs. (4.8) and (4.9) of

* See also the discussion on complex orbits in sections 2.5 and 4.1.

Caldeira and Leggett (1983)]

$$J(\omega) = \tfrac{1}{2}\pi \sum_j \frac{C_j^2}{m_j \omega_j} \delta(\omega - \omega_j) \to \eta\omega. \tag{3.81}$$

The equation above reveals a redundancy in the description of the environment. Therefore, we may put without loss of generality $M = m_j = m$, $C_j = m\omega_j^2$. Furthermore, we introduce a uniform notation for the coordinates $q = R_1$, $x_j = R_j, j = 2, \ldots, N$. Then eqs. (3.80) and (3.81) assume the form

$$L = \sum_{k=1}^{N} \tfrac{1}{2} m \dot{R}_k^2 - V(\{R_k\}),$$

$$V = v(R_1) + \sum_{j=2}^{N} \tfrac{1}{2} m\omega_j^2 (R_j - R_1)^2,$$

$$J(\omega) = \tfrac{1}{2}\pi m \sum_j \omega_j^3 \delta(\omega - \omega_j) \to \eta\omega. \tag{3.82}$$

In the above representation, the environment appears to consist of a cloud of springs attached to the particle carrying masses at the other end.

By now it is clear that the Hamiltonian of the Caldeira–Leggett model is just a special case of eq. (3.1). In order to make progress in the calculation of the decay rate of the metastable state in the well at $\boldsymbol{R} \sim 0$, we have to find $\boldsymbol{r}_\tau$ of the escape path. According to eqs. (3.10) and (3.82), it obeys the equation

$$m\ddot{r}_1 = v'(r_1) - \sum_{j=2}^{N} m\omega_j^2 (r_j - r_1), \tag{3.83a}$$

$$m\ddot{r}_j = m\omega_j^2 (r_j - r_1). \tag{3.83b}$$

As already found previously, it is advantageous to supplement a return path by putting $\boldsymbol{r}_{-\tau} = \boldsymbol{r}_\tau$, $\boldsymbol{r}_{\pm\infty} = 0$. Then, one may introduce Fourier transforms,

$$\boldsymbol{r}_\tau = \int \frac{\mathrm{d}\omega}{2\pi} \mathrm{e}^{-\mathrm{i}\omega\tau} \boldsymbol{r}_\omega, \tag{3.84}$$

and eq. (3.83b) is solved by

$$r_j(\omega) = \frac{\omega_j^2}{\omega^2 + \omega_j^2} r_1(\omega). \tag{3.85}$$

Substituting this result in eq. (3.83a), one obtains the following equation of motion for the coordinate r_1 of the object

$$-m\ddot{r}_1(\tau) + H * r_1 + v'(r_1(\tau)) = 0,$$

$$H * r_1 = \int \mathrm{d}\tau' \, H(\tau - \tau') r_1(\tau'), \tag{3.86}$$

where the linear operator H is defined most conveniently by its Fourier transform

$$H_\omega = m \sum_{j=2}^{N} \frac{\omega_j^2 \omega^2}{\omega^2 + \omega_j^2} \to \eta |\omega| . \tag{3.87}$$

In the limit indicated above, H represents a friction linear in the velocity.

Caldeira and Leggett (1981, 1983, 1984) have succeeded in solving the equation $H * r_1 + v' = 0$, which is eq. (3.86) without acceleration term. It can be considered as the heavy damping limit of eq. (3.86), which is realized when $\gamma = \eta/m \gg \omega_0$. The solution is given by

$$r_1^{\mathrm{CL}}(\tau) = \frac{4R_{\mathrm{C}}}{3} \frac{1}{\tau^2 \omega_{\mathrm{B}}^2 + 1}, \quad \omega_{\mathrm{B}} = \frac{2\pi}{\tau_{\mathrm{B}}} = \frac{m\omega_0^2}{\eta} = \frac{\omega_0^2}{\gamma},$$

$$r_1^{\mathrm{CL}}(\omega) = \frac{4\pi R_{\mathrm{C}}}{3\omega_{\mathrm{B}}} \exp(-|\omega|/\omega_{\mathrm{B}}). \tag{3.88}$$

The quantities $r_j^{\mathrm{CL}}(\omega)$ follow from eq. (3.85). We wish to add that numerical calculations (Chang and Chakravarty 1984) have shown that the above heavy damping limit is approached very smoothly.

Interesting properties of the escape path (3.88) can be calculated as follows. We define

$$\begin{aligned} \Delta(\boldsymbol{r}_\tau \cdot \boldsymbol{r}_{\tau'}) &= \boldsymbol{r}_\tau \cdot \boldsymbol{r}_{\tau'} - r_1(\tau) r_1(\tau') \\ &= \int \frac{\mathrm{d}\omega \, \mathrm{d}\omega'}{(2\pi)^2} \mathrm{e}^{-\omega\tau - \mathrm{i}\omega\tau'} \sum_{j=2}^{N} \frac{\omega_j^4}{(\omega^2 + \omega_j^2)(\omega'^2 + \omega_j^2)} r_1^{\mathrm{CL}}(\omega) r_1^{\mathrm{CL}}(\omega') \\ &= \frac{4\gamma R_{\mathrm{C}}^2}{9\omega_{\mathrm{B}}^2} \int \mathrm{d}\omega \, \mathrm{d}\omega' \frac{\exp[-\mathrm{i}\omega\tau - \mathrm{i}\omega'\tau' - (|\omega| + |\omega'|)/\omega_{\mathrm{B}}]}{|\omega| + |\omega'|}, \end{aligned} \tag{3.89}$$

where we have obtained the last line of this equation by making use of eq. (3.88). One recognizes that eq. (3.89) can also be written as

$$\begin{aligned} \Delta(\boldsymbol{r}_\tau \cdot \boldsymbol{r}_{\tau'}) &= \frac{4\gamma R_{\mathrm{C}}^2}{9\omega_{\mathrm{B}}^2} \int_0^{\omega_{\mathrm{B}}} \mathrm{d}\omega \frac{4}{(\tau^2\omega^2 + 1)(\tau'^2\omega^2 + 1)} \\ &= \frac{4\gamma R_{\mathrm{C}}^2}{9\omega_{\mathrm{B}}^2} \begin{cases} \omega_{\mathrm{B}}[4 - 3\omega_{\mathrm{B}}^2(\tau^2 + \tau'^2)], & |\omega_{\mathrm{B}}\tau|, |\omega_{\mathrm{B}}\tau'| \ll 1, \\ 2\pi/(|\tau| + |\tau'|), & |\omega_{\mathrm{B}}\tau|, |\omega_{\mathrm{B}}\tau'| \gg 1. \end{cases} \end{aligned} \tag{3.90}$$

In particular, we have near the origin, $\Delta(\boldsymbol{r}_\tau^2) \gg r_1^2(\tau)$; this means that the escape path starts perpendicular to the R_1-axis.

Using eq. (3.90), we may also calculate the curvature κ of the escape path

$$\kappa_\tau^2 = \frac{(\dot{\boldsymbol{r}}_\tau)^2(\ddot{\boldsymbol{r}}_\tau)^2 - (\dot{\boldsymbol{r}}_\tau \cdot \ddot{\boldsymbol{r}}_\tau)^2}{[(\dot{\boldsymbol{r}}_\tau)^2]^3} = \begin{cases} \text{const}, & |\omega_{\mathrm{B}}\tau| \ll 1, \\ |9\tau\omega_{\mathrm{B}}^3|/(8\pi R_{\mathrm{C}}^2\gamma), & |\omega_{\mathrm{B}}\tau| \gg 1. \end{cases} \tag{3.91}$$

We recognize that this path leaves the origin with infinite curvature.

Very schematically, the escape path of fig. 8 may be thought of describing the present situation if we identify $-r_2(\tau)$ there with $[\Delta(\boldsymbol{r}_\tau^2)]^{1/2}$ here. It is interesting to note that at the escape point* where $R_{C1} = r_1^{CL}(\tau = 0) = 4R_C/3 > R_C$, the particle has lost the potential energy $v(4R_C/3) = -4v_B$; this energy is now stored in the springs.

Considering definitions (3.14) and (3.54), we calculate

$$\hbar\mathscr{B}^{CL} = 2\int_{-\infty}^{0} d\tau\, m(\dot{\boldsymbol{r}}_\tau^{CL})^2 = \int \frac{d\omega}{2\pi} m\omega^2 |\boldsymbol{r}_\omega^{CL}|^2 = \frac{2\pi\eta R_C^2}{9} + \frac{4\pi m\omega_B R_C^2}{9}. \qquad (3.92)$$

In the limit $m \to 0$, this agrees with the result of Caldeira and Leggett (1981, 1983, 1984).

Concerning the prefactor $\mathscr{A}$, note that the square root of the ratio of determinants in eq. (3.63) can be expressed as the ratio of two Gaussian functional integrals with respect to N-dimensional paths.** Observe now the simple structure of $\hat{U}_\tau$ in the Caldeira–Leggett model, where all matrix elements but one are time-independent. Therefore, it is possible to perform the functional integrals with respect to $N - 1$ components of the paths ("integrating out the coordinates of the environment"). The result of this operation is expressed by the relation

$$\frac{\det\{-\hat{1}\partial_\tau^2 + \hat{U}^{(0)}\}}{\det\{-\hat{1}(\partial_\tau^2 - \varepsilon) + \hat{U}_\tau\}} = \frac{\det\{-m\partial_\tau^2 + H + m\omega_0^2\}}{\det\{-m(\partial_\tau^2 - \varepsilon) + H^\varepsilon + v''[r_1(\tau)]\}}, \qquad (3.93)$$

where the operator H is defined by eq. (3.86) and H^ε is obtained by replacing $\omega^2 \to \omega^2 + \varepsilon$ in eq. (3.87). Therefore,

$$H_\omega^\varepsilon = H_\omega + \varepsilon \sum_j \frac{m\omega_j^4}{(\omega^2 + \omega_j^2)^2} + O(\varepsilon^2). \qquad (3.94)$$

Concerning the operator $-m(\partial_\tau^2 - \varepsilon) + H^\varepsilon + v''$, note that in the case of $\varepsilon = 0$, it has an eigenvalue zero which belongs to the eigenfunction $\dot{r}_1(\tau)$. Therefore, we may calculate for finite ε, the eigenvalue λ^ε closest to zero by perturbation theory:

$$\lambda^\varepsilon = \frac{\varepsilon}{\mathscr{B}_0} \int \frac{d\omega}{2\pi} m\left[1 + \sum_j \frac{\omega_j^4}{(\omega^2 + \omega_j^2)^2}\right] \omega^2 |r_1(\omega)|^2 = \varepsilon \frac{\mathscr{B}}{\mathscr{B}_0}, \qquad (3.95)$$

where $\hbar\mathscr{B} = 2\mathscr{W}_{EP}(0)$ and the normalization is given by

$$\hbar\mathscr{B}_0 = \hbar\mathscr{B}_0\{[r_1(\tau)]\} = 2m \int_{-\infty}^{0} d\tau [\dot{r}_1(\tau)]^2. \qquad (3.96)$$

* Note that in the present context, R_C of eq. (2.10) is just a parameter which has nothing to do with the one coordinate $R_{C1} = 4R_C/3$ of the escape point.

** The problem of a negative eigenvalue in the denominator determinant can be handled by deforming the contour of integration properly.

Thus, we conclude that the prefactor can be expressed alternatively as

$$\mathscr{A} = \left(\frac{\mathscr{B}_0}{2\pi m}\right)^{1/2} \left[-\frac{\det\{-m\partial_\tau^2 + H + m\omega_0^2\}}{\det'\{-m\partial_\tau^2 + H + v''\}} \right]^{1/2}. \tag{3.97}$$

This form is well-established in the literature (Caldeira and Leggett 1981, 1983, 1984, Chang and Chakravarty 1984, Grabert and Weiss 1984).

Although the present formalism of calculating decay rates does not raise any problems, we should observe that the Caldeira–Leggett model does not conform so well with the concept of a quasistationary decaying state on account of the presence of arbitrarily small frequency modes. For the same reasons, the escape path leaves the origin with infinite curvature. This seems to leave some questions open with respect to the concept of a quasistationary state.

4. *Statistical matrix and multidimensional decay at finite temperatures*

4.1. *Decay at finite temperatures in N dimensions*

In the discussion of section 2.5 we have argued that our theory requires closed extremal orbits which connect the metastable region (A) with the outside region (C) and which are traversed in time $\tau_1 = \hbar/kT$. We have also pointed out that only an extension to complex extremal paths provides sufficient flexibility to meet these requirements in the multidimensional problem.

An attempt to describe the properties of such a complex orbit in a multi-dimensional space should be guided by the experience we have gained in the zero temperature ($\tau_1 = \infty$) case of section 3. There, an essential part of the theory is the escape path which connects the bottom of the well with the outside region at the escape point $\boldsymbol{R}_C$ (fig. 8). A closed orbit (bounce) is obtained by adding a return path which retraces the escape path. Obviously, the total time needed to complete such an orbit is $\tau_1 = \infty$. We have also learned in section 3.3 that the escape path can be extended in region C by letting $\tau \to it$; see, e.g., eq. (3.43). This procedure reminds us of the discussion in the second part of section 2.5 which has been illustrated by fig. 5.

It is fairly obvious that at finite temperatures $kT = \hbar/\tau_1$, we need an escape path (see fig. 11) which connects in the time $\tau_1/2$ the turning points on two surfaces of the same potential energy, but surfaces pertaining to the well region A and outside region C, respectively. Again, this escape path together with its return path forms a closed orbit which will be completed in time τ_1. We may also argue that two external legs can be attached at $\boldsymbol{R}_{2,3}$ by letting $\tau \to it$. The corresponding diagram in space and complex time is shown in fig. 12a, b; it should be compared with fig. 5. Evidently, for $\boldsymbol{R}_1 = \boldsymbol{R}$ we have $\boldsymbol{P}_1 = \boldsymbol{P}$; the

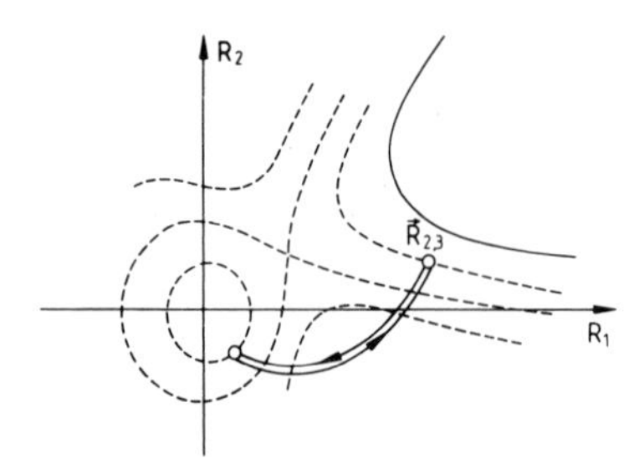

Fig. 11. Periodic orbit under the barrier completed in time τ_1. The endpoints are turning points on surfaces of equal potential energy.

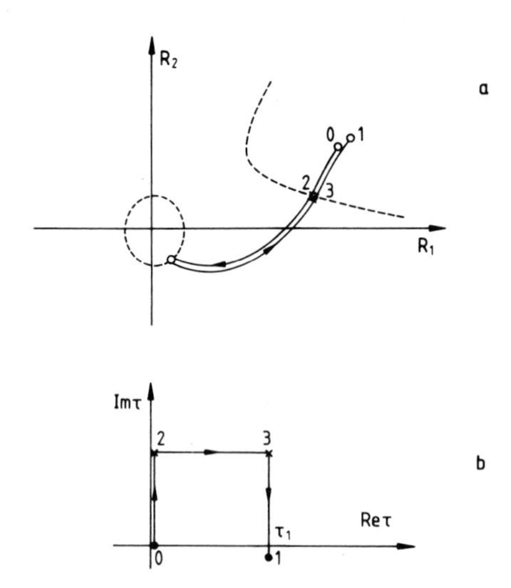

Fig. 12. (a) Real path under the barrier ($2 \to 3$) in time τ_1 and real paths in the outside world ($0 \to 2$, $3 \to 1$) for imaginary τ. (b) The corresponding contour in the complex τ-plane. (For $\boldsymbol{R}_1 = \boldsymbol{R}_0 \equiv \boldsymbol{R}$, the endpoint of the contour is on the real axis.) Note the corners at the classical turning points.

trajectory is closed and also periodic in time τ_1. Now, we assume that this combination of escape and return paths augmented by external legs can be deformed continuously in the multidimensional complex planes $\boldsymbol{R} = \{R_k\}$ such that for real τ and $0 \leqslant \tau \leqslant \tau_1$ it is the solution of the equation*

$$m\ddot{\boldsymbol{r}}_\tau = \frac{\partial V}{\partial \boldsymbol{r}_\tau}. \tag{4.1}$$

* It differs from eq. (3.10) by different boundary conditions.

We expect that this solution is an analytic function of τ; moreover, we expect that such an orbit exists for any choice of the starting point $\boldsymbol{R}$ and that the action $\mathscr{S}([\boldsymbol{r}_\tau];\tau_1)$ depends only on the topology of the orbit. We will give some arguments in favor of this assumption at the end of Appendix C; moreover, it can be verified for the Caldeira–Leggett model by analytical (see section 4.3) and numerical (Ludviksson 1989) calculations.

With only a moderate increase in the action $\mathscr{S}$, there are also closed extremal paths $\bar{\boldsymbol{r}}_\tau$ in the multidimensional vicinity of the periodic orbit which contribute also to the outgoing current. To evaluate their contribution, we put

$$\bar{\boldsymbol{r}}_\tau = \boldsymbol{r}_\tau + \boldsymbol{x}_\tau, \quad \boldsymbol{x}_0 = \boldsymbol{x}_{\tau_1} = \boldsymbol{\rho}, \tag{4.2}$$

where $\boldsymbol{x}_\tau$ is small. We now expand [cf. eq. (3.15) and below]

$$V(\bar{\boldsymbol{r}}_\tau) = V(\boldsymbol{r}_\tau) + (\boldsymbol{x}_\tau \cdot \partial_{\boldsymbol{r}_\tau}) V(\boldsymbol{r}_\tau) + \tfrac{1}{2}(\boldsymbol{x}_\tau \cdot \partial_{\boldsymbol{r}_\tau})^2 V(\boldsymbol{r}_\tau) + \cdots, \tag{4.3}$$

and obtain the action through second order in $\boldsymbol{x}_\tau$:

$$\begin{aligned} \mathscr{S}([\bar{\boldsymbol{r}}_\tau];\tau_1) &= \mathscr{S}([\boldsymbol{r}_\tau];\tau_1) \\ &\quad + \int_0^{\tau_1} \mathrm{d}\tau \{ m\dot{\boldsymbol{r}}_\tau \cdot \dot{\boldsymbol{x}}_\tau + \tfrac{1}{2} m \dot{\boldsymbol{x}}_\tau^2 + (\boldsymbol{x}_\tau \cdot \partial_{\boldsymbol{r}_\tau}) V(\boldsymbol{r}_\tau) + \tfrac{1}{2}(\boldsymbol{x}_\tau \cdot \partial_{\boldsymbol{r}_\tau})^2 V(\boldsymbol{r}_\tau) \}. \end{aligned} \tag{4.4}$$

Taking into account the periodicity of the orbit $\boldsymbol{r}_\tau$ as well as its equation of motion (4.1), we obtain after integration by parts

$$\mathscr{S}([\bar{\boldsymbol{r}}_\tau];\tau_1) - \mathscr{S}([\boldsymbol{r}_\tau];\tau_1) = \tfrac{1}{2} m \int_0^{\tau_1} \mathrm{d}\tau (\dot{\boldsymbol{v}}_\tau^2 + \boldsymbol{v}_\tau \cdot \hat{U}_\tau \cdot \boldsymbol{v}_\tau), \tag{4.5}$$

where $\boldsymbol{v}_\tau$ is the first-order approximation to $\boldsymbol{x}_\tau$, which satisfies the small-fluctuation equation

$$[-\hat{1}\partial_\tau^2 + \hat{U}_\tau] \cdot \boldsymbol{v}_\tau = 0 \tag{4.6}$$

and the boundary condition

$$\boldsymbol{v}_0 = \boldsymbol{v}_{\tau_1} = \boldsymbol{\rho}. \tag{4.7}$$

Above, we have introduced the operator $\hat{U}_\tau$ which is defined analogous to eq. (3.13)

$$(\hat{U}_\tau)_{kk'} = \frac{1}{m} \left. \frac{\partial^2 V}{\partial R_k \partial R_{k'}} \right|_{\boldsymbol{R}=\boldsymbol{r}_\tau} \tag{4.8}$$

but for a different extremal path.

Let us choose a linearly independent set of solutions to the small-fluctuation equation as follows:

$$[-\hat{1}\partial_\tau^2 + \hat{U}_\tau]\begin{Bmatrix}\hat{K}_\tau \\ \hat{M}_\tau\end{Bmatrix} = 0, \tag{4.9}$$

where $\hat{K}_\tau$ and $\hat{M}_\tau$ satisfy the initial conditions

$$\hat{K}_0 = 0, \qquad \dot{\hat{K}}_0 = \hat{1}, \qquad \hat{M}_0 = \hat{1}, \qquad \dot{\hat{M}}_0 = 0. \tag{4.10}$$

In terms of $\hat{K}_\tau$ and $\hat{M}_\tau$ we may write $\boldsymbol{v}_\tau$ [cf. eq. (3.21)] as follows:

$$\boldsymbol{v}_\tau = \{\hat{K}_\tau \hat{K}_1^{-1}[\hat{1} - \hat{M}_1] + \hat{M}_\tau\}\cdot\boldsymbol{\rho}. \tag{4.11}$$

Above, we have introduced

$$\hat{K}_1 = \hat{K}_{\tau_1}, \qquad \hat{M}_1 = \hat{M}_{\tau_1} \tag{4.12}$$

and have assumed that $\det \hat{K}_1 \neq 0$.

In eq. (4.5) we integrate by parts again and observe that $\boldsymbol{v}_\tau$ satisfies eqs. (4.6) and (4.7). Then, we obtain

$$\mathscr{S}([\bar{\boldsymbol{r}}_\tau]; \tau_1) - \mathscr{S}([\boldsymbol{r}_\tau]; \tau_1) = \tfrac{1}{2} m\, \boldsymbol{v}_0 \cdot (\dot{\boldsymbol{v}}_{\tau_1} - \dot{\boldsymbol{v}}_0). \tag{4.13}$$

Considering eqs. (4.7) and (4.11), we may write

$$\mathscr{S}([\bar{\boldsymbol{r}}_\tau]; \tau_1) = \mathscr{S}([\boldsymbol{r}_\tau]; \tau_1) + \tfrac{1}{2} m \boldsymbol{\rho}\cdot\hat{\Omega}\cdot\boldsymbol{\rho}, \tag{4.14}$$

where

$$\hat{\Omega} = -[\hat{1} - \dot{\hat{K}}_1]\hat{K}_1^{-1}[\hat{1} - \hat{M}_1] + \dot{\hat{M}}_1. \tag{4.15}$$

Obviously, $\hat{\Omega}$ is a symmetric matrix. Furthermore, since $\dot{\boldsymbol{r}}_\tau$ is a periodic solution of the small-fluctuation equation (4.6), the vector $\dot{\boldsymbol{r}}_0$ $(=\dot{\boldsymbol{r}}_{\tau_1})$ is an eigenvector of $\hat{\Omega}$ with eigenvalue zero. For convenience, let us choose a representation where the first basis vector $\boldsymbol{e}_1$ is parallel to $\dot{\boldsymbol{r}}_0$. Then

$$\hat{\Omega}\cdot\boldsymbol{e}_1 = 0, \quad \boldsymbol{e}_1 \parallel \dot{\boldsymbol{r}}_0. \tag{4.16}$$

This means that we have

$$\Omega_{1k} = \Omega_{k1} = 0, \quad k = 1, 2, \ldots, N. \tag{4.17}$$

4.2. *Gaussian fluctuations about periodic orbits*

To complete the quasiclassical approximation to the statistical matrix $\rho(\boldsymbol{R} + \boldsymbol{\rho}, \boldsymbol{R} + \boldsymbol{\rho}; \tau_1)$ we must calculate the prefactor due to Gaussian fluctuations about $\bar{\boldsymbol{r}}_\tau = \boldsymbol{r}_\tau + \boldsymbol{x}_\tau$. In view of the approximation $\boldsymbol{x}_\tau \to \boldsymbol{v}_\tau$, this prefactor will be, to lowest order, the same as for the periodic orbit $\boldsymbol{r}_\tau$.

Since $\boldsymbol{r}_\tau$ and, hence, the operator $\hat{U}_\tau$ is complex-valued, the calculation of the Gaussian fluctuations follows somewhat unusual lines. A complete calculation has been carried through for the one-dimensional case in Appendix C. The outcome is fairly simple: the final result is the same as for real-valued operators $\hat{U}_\tau$ provided that we replace the determinant det, which means the product of all eigenvalues, by $\det^+$, which now includes only positive eigenvalues as explained by eqs. (C.47), and (C.48).

For the sake of simplicity, therefore, we will consider $\hat{U}_\tau$ to be real in the following discussions, which allows us to disregard the details mentioned above. Generalization to complex $\hat{U}_\tau$ follows the arguments of Appendix C and is straightforward though lengthy.

We introduce $\hat{K}_\tau^z$ and $\hat{M}_\tau^z$ that are solutions of

$$\{-\hat{1}\partial_\tau^2 + \hat{U}_\tau + z\,\hat{1}\}\begin{Bmatrix}\hat{K}_\tau^z\\ \hat{M}_\tau^z\end{Bmatrix} = 0 \tag{4.18}$$

subject to the initial conditions

$$\hat{K}_0^z = 0, \qquad \dot{\hat{K}}_0^z = \hat{1}, \qquad \hat{M}_0^z = \hat{1}, \qquad \dot{\hat{M}}_0^z = 0. \tag{4.19}$$

First let us consider the eigenvalue problem

$$[-\hat{1}\partial_\tau^2 + \hat{U}_\tau]\,\varphi_\tau = \lambda\varphi_\tau, \tag{4.20}$$

where the eigenfunction φ_τ satisfies zero boundary conditions

$$\varphi_0 = \varphi_{\tau_1} = 0. \tag{4.21}$$

For $\lambda = -z$, we may write any solution of the differential equation (4.20) which vanishes at $\tau = 0$ as follows:

$$\varphi_\tau = \hat{K}_\tau^z \cdot \boldsymbol{a}. \tag{4.22}$$

For $-z$ to be an eigenvalue λ we must demand ($\hat{K}_1^z = \hat{K}_{\tau_1}^z$, etc.)

$$\hat{K}_1^z \cdot \boldsymbol{a} = 0. \tag{4.23}$$

This equation has a nontrivial solution only if

$$\det \hat{K}_1^z = 0. \tag{4.24}$$

Similar relations may be derived in the case where $\hat{U}_\tau$ is replaced by $\hat{U}^{(0)}$ as defined by [see also eq. (3.2)]

$$\hat{U}_{kk'}^{(0)} = \frac{1}{m}\left.\frac{\partial^2 V}{\partial R_k \partial R_{k'}}\right|_{R=0} \tag{4.25}$$

and where the matrices $\hat{K}_\tau^{(0)}$ and $\hat{M}_\tau^{(0)}$ are constructed according to eqs. (4.9) and (4.10) with $\hat{U}^{(0)}$ replacing $\hat{U}_\tau$.

We assert that

$$\frac{\prod_n(\lambda_n^{(0)} + z)}{\prod_n(\lambda_n + z)} = \frac{\det\{-\hat{1}\partial_\tau^2 + \hat{U}^{(0)} + \hat{1}z\}}{\det\{-\hat{1}\partial_\tau^2 + \hat{U}_\tau + \hat{1}z\}} = \frac{\det \hat{K}_1^{(0)z}}{\det \hat{K}_1^z}. \tag{4.26}$$

The proof of the assertion is based on the fact that both sides have the same zeros and poles and approach unity as $|z| \to \infty$. Thus, we have for zero boundary (ZB) conditions, taking $z = 0$ in the above equation,

$$R_{\mathrm{ZB}} = \left.\frac{\det\{-\hat{1}\partial_\tau^2 + \hat{U}^{(0)}\}}{\det\{-\hat{1}\partial_\tau^2 + \hat{U}_\tau\}}\right|_{\mathrm{ZB}} = \frac{\det \hat{K}_1^{(0)}}{\det K_1}. \tag{4.27}$$

Next we consider the same eigenvalue problem [eq. (4.20)] but this time with periodic boundary conditions. Although the eigenvalues and eigenfunctions are different now, we retain the labeling λ and $\boldsymbol{\varphi}$.

The eigenfunctions now satisfy

$$\boldsymbol{\varphi}_0 = \boldsymbol{\varphi}_{\tau_1}, \qquad \dot{\boldsymbol{\varphi}}_0 = \dot{\boldsymbol{\varphi}}_{\tau_1}. \tag{4.28}$$

Now we must write an arbitrary solution of eq. (4.20) as follows:

$$\boldsymbol{\varphi}_\tau = \hat{K}_\tau^z \cdot \boldsymbol{c} + \hat{M}_\tau^z \cdot \boldsymbol{d}. \tag{4.29}$$

For $-z$ to be an eigenvalue λ we must require that the boundary conditions (4.28) are satisfied, i.e.,

$$\begin{aligned} \hat{K}_1^z \cdot \boldsymbol{c} + \hat{M}_1^z \cdot \boldsymbol{d} &= \boldsymbol{d}, \\ \dot{\hat{K}}_1^z \cdot \boldsymbol{c} + \dot{\hat{M}}_1^z \cdot \boldsymbol{d} &= \boldsymbol{c}, \end{aligned} \tag{4.30}$$

which is equivalent to

$$\det\begin{pmatrix} \hat{1} - \hat{M}_1^z & -\hat{K}_1^z \\ -\dot{\hat{M}}_1^z & \hat{1} - \dot{\hat{K}}_1^z \end{pmatrix} = 0. \tag{4.31}$$

We deduce by the usual argument that for periodic boundary conditions

$$\left.\frac{\det\{-\hat{1}\partial_\tau^2 + \hat{U}^{(0)} + \hat{1}z\}}{\det\{-\hat{1}\partial_\tau^2 + \hat{U}_\tau + \hat{1}\hat{z}\}}\right|_{\mathrm{PB}} = \frac{\det\begin{pmatrix} \hat{1} - \hat{M}_1^{(0)z} & -\hat{K}_1^{(0)z} \\ -\dot{\hat{M}}_1^{(0)z} & \hat{1} - \dot{\hat{K}}_1^{(0)z} \end{pmatrix}}{\det\begin{pmatrix} \hat{1} - \hat{M}_1^z & -\hat{K}_1^z \\ -\dot{\hat{M}}_1^z & \hat{1} - \dot{\hat{K}}_1^z \end{pmatrix}}. \tag{4.32}$$

Let us now assume that $\hat{K}_1^z$ is nonsingular. In that case the system (4.30) is equivalent to the existence of a nontrivial solution of

$$\{-[\hat{1} - \dot{\hat{K}}_1^z][\hat{K}_1^z]^{-1}[\hat{1} - \hat{M}_1^z] + \dot{\hat{M}}_1^z\} \cdot \boldsymbol{d} = 0 \tag{4.33}$$

Consider now the function of the complex variable w

$$\frac{(\det \hat{K}_1^z)\det\{[w\hat{1} - \dot{\hat{K}}_1^z][\hat{K}_1^z]^{-1}[w\hat{1} - \hat{M}_1^z] - \dot{\hat{M}}_1^z\}}{\det\begin{pmatrix} w\hat{1} - \hat{M}_1^z & -\hat{K}_1^z \\ -\dot{\hat{M}}_1^z & w\hat{1} - \dot{\hat{K}}_1^z \end{pmatrix}}. \tag{4.34}$$

The zeros of numerator and denominator coincide and for $|w| \to \infty$ their ratio approaches unity. Hence, it is equal to one everywhere and we may write

$$\left.\frac{\det\{-\hat{1}\partial_\tau^2 + \hat{U}^{(0)} + z\hat{1}\}}{\det\{-\hat{1}\partial_\tau^2 + \hat{U}_\tau + z\hat{1}\}}\right|_{\text{PB}} = \frac{(\det \hat{K}_1^{(0)z}) \det \hat{\Omega}^{(0)z}}{(\det \hat{K}_1^z) \det \hat{\Omega}^z}, \tag{4.35}$$

where, in accordance with eq. (4.15) we have put

$$\hat{\Omega}^z = -[\hat{1} - \dot{\hat{K}}_1^z][\hat{K}_1^z]^{-1}[\hat{1} - \hat{M}_1^z] + \dot{\hat{M}}_1^z. \tag{4.36}$$

We know that the small-fluctuation-operator with periodic boundary conditions has an eigenvalue zero (corresponding to the eigenfunction $\dot{\boldsymbol{r}}_\tau$), which has to be omitted. To take account of this, we consider

$$R_{\text{PB}} = \left.\frac{\det\{-\hat{1}\partial_\tau^2 + \hat{U}^{(0)}\}}{\det'\{-\hat{1}\partial_\tau^2 + \hat{U}_\tau\}}\right|_{\text{PB}} = \lim_{\varepsilon \to 0} \left.\frac{\varepsilon \det\{-\hat{1}\partial_\tau^2 + \hat{U}^{(0)}\}}{\det\{-\hat{1}\partial_\tau^2 + \hat{U}_\tau + \varepsilon\hat{1}\}}\right|_{\text{PB}}, \tag{4.37}$$

where, again, the determinants are calculated for periodic boundary conditions and the prime denotes omission of the zero eigenvalue.

Due to the property (4.16) we may write

$$R_{\text{PB}} = \frac{(\det \hat{K}_1^{(0)}) \det \hat{\Omega}^{(0)}}{\alpha (\det \hat{K}_1) \det_\perp \hat{\Omega}}, \tag{4.38}$$

where the subscript $\perp$ denotes restriction to the subspace perpendicular to $\dot{\boldsymbol{r}}_0$. Furthermore,

$$\alpha = \lim_{\varepsilon \to 0} \frac{1}{\varepsilon} \{-[\hat{1} - \dot{\hat{K}}_1^\varepsilon][\hat{K}_1^\varepsilon]^{-1}[\hat{1} - \hat{M}_1^\varepsilon] + \dot{\hat{M}}_1^\varepsilon\}_{11}. \tag{4.39}$$

To evaluate α, we consider the vector

$$\boldsymbol{v}_\tau^\varepsilon = \{\hat{K}_\tau^\varepsilon[\hat{K}_1^\varepsilon]^{-1}[\hat{1} - \hat{M}_1^\varepsilon] + \hat{M}_\tau^\varepsilon\} \cdot \boldsymbol{e}_1, \tag{4.40}$$

which, according to eqs. (4.6) and (4.11) is a closed solution of

$$[-\partial_\tau^2 \hat{1} + \hat{U}_\tau + \varepsilon\hat{1}] \cdot \boldsymbol{v}_\tau^\varepsilon = 0. \tag{4.41}$$

In terms of $\boldsymbol{v}_\tau^\varepsilon$ we may express α as follows:

$$\alpha = \lim_{\varepsilon \to 0} \frac{1}{\varepsilon} \boldsymbol{e}_1 \cdot [\dot{\boldsymbol{v}}_{\tau_1}^\varepsilon - \dot{\boldsymbol{v}}_0^\varepsilon]. \tag{4.42}$$

On the other hand, we have

$$\partial_\tau \{\boldsymbol{v}_\tau^\varepsilon \cdot \dot{\boldsymbol{v}}_\tau - \dot{\boldsymbol{v}}_\tau^\varepsilon \cdot \boldsymbol{v}_\tau\} = -\varepsilon \boldsymbol{v}_\tau^\varepsilon \cdot \boldsymbol{v}_\tau, \tag{4.43}$$

where $\boldsymbol{v}_\tau$ is defined as in eq. (4.40) with $\varepsilon = 0$. Integrating eq. (4.43) and observing that $\boldsymbol{v}_0 = \boldsymbol{v}_{\tau_1}$, we find to lowest order

$$(\dot{\boldsymbol{v}}_{\tau_1}^\varepsilon - \dot{\boldsymbol{v}}_0^\varepsilon) \cdot \boldsymbol{v}_0 = \varepsilon \int_0^{\tau_1} d\tau\, \boldsymbol{v}_\tau \cdot \boldsymbol{v}_\tau. \tag{4.44}$$

Eventually, we conclude that

$$\alpha = \int_0^{\tau_1} \mathrm{d}\tau (\boldsymbol{v}_\tau \cdot \boldsymbol{v}_\tau) = \frac{1}{|\dot{\boldsymbol{r}}_0|^2} \int_0^{\tau_1} \mathrm{d}\tau \, |\dot{\boldsymbol{r}}_\tau|^2. \tag{4.45}$$

We are now in a position to write down the quasiclassical approximation to the statistical matrix at $\boldsymbol{R}_1 = \boldsymbol{R} + \boldsymbol{\rho}$, $\boldsymbol{R} = \boldsymbol{r}_0 = \boldsymbol{r}_1$:

$$\rho(\boldsymbol{R}+\boldsymbol{\rho}, \boldsymbol{R}+\boldsymbol{\rho}; \tau_1) = \left(\det \frac{m\hat{\omega}_0}{2\pi\hbar \sinh \hat{\omega}_0 \tau_1} \right)^{1/2} \left[\frac{\det \hat{K}_1^{(0)}}{\det \hat{K}_1} \right]^{1/2} \times \exp\left(-\frac{1}{\hbar} \mathscr{S}([r_\tau]; \tau_1) - \frac{m}{2\hbar} \boldsymbol{\rho} \cdot \hat{\Omega} \cdot \boldsymbol{\rho} \right), \tag{4.46}$$

where again $\hat{\omega}_0 = (\hat{U}^{(0)})^{1/2}$. We have chosen a representation where a ratio of determinants appears since it is easier to handle. That the overall constant is chosen properly can be seen by inserting the trivial orbit $\boldsymbol{r}_\tau = 0$, where $\hat{U}_\tau = \hat{U}^{(0)}$. Then one obtains $\rho(0, 0; \tau_1) = [\det(m\hat{\omega}_0/2\pi\hbar \sinh \hat{\omega}_0 \tau_1)]^{1/2}$, which is known to be valid for harmonic oscillators.

Eventually, we need the total current flowing through a plane perpendicular to $\boldsymbol{e}_1$ which is defined in eq. (4.16). Since $\hat{\Omega} \cdot \boldsymbol{e}_1 = 0$, we need to take the derivative in the exponent only with respect to $\mathscr{S}([\boldsymbol{r}_\tau]; \tau_1)$. This operation produces the factor

$$(\partial_{\boldsymbol{R}} - \partial_{\boldsymbol{R}_1}) \mathscr{S}([\boldsymbol{r}_\tau]; \tau_1)|_{\boldsymbol{R}_1 = \boldsymbol{R}} = 2\boldsymbol{P} = 2m\dot{\boldsymbol{r}}_0. \tag{4.47}$$

Thus, we obtain the decay rate

$$\Gamma = \frac{1}{Z_0} \int \mathrm{d}^{N-1} \rho_\perp (\boldsymbol{e}_1 \cdot \boldsymbol{j}) = \frac{1}{Z_0} \left[\frac{m}{2\pi\hbar} \det\left(\frac{\hat{\omega}_0}{\sinh \hat{\omega}_0 \tau_1} \right) \right]^{1/2} \left[\frac{\det \hat{K}_1^{(0)}}{\det \hat{K}_1} \right]^{1/2} \frac{|\dot{\boldsymbol{r}}_0|}{(\det_\perp \hat{\Omega})^{1/2}} \times \exp\left(-\frac{1}{\hbar} \mathscr{S}[(\boldsymbol{r}_\tau]; \tau_1) \right). \tag{4.48}$$

Here, in analogy with the one-dimensional case, Z_0 is the partition function corresponding to the harmonic approximation of $V(\boldsymbol{R})$ at the origin. Hence,

$$Z_0^{-1} = \det(2 \sinh \hat{\omega}_0 \tau_1 / 2). \tag{4.49}$$

Furthermore, we note that

$$\hat{K}_\tau^{(0)} = \frac{1}{\hat{\omega}_0} \sinh \hat{\omega}_0 \tau_1, \qquad \hat{M}_\tau^{(0)} = \cosh \omega_0 \tau_1 . \tag{4.50}$$

Considering eq. (4.36), we obtain

$$\hat{\Omega}^{(0)} = 2\hat{\omega}_0 \tanh \hat{\omega}_0 \tau_1 / 2 . \tag{4.51}$$

Combining suitably eqs. (4.37), (4.38), (4.45), (4.49) and (4.51) and inserting it into eq. (4.48), we obtain a decay rate which can be written in the standard form $\bar{\Gamma} = \mathscr{A} e^{-\mathscr{B}}$ of eq. (2.2), where the exponent is given by

$$\mathscr{B} = \frac{1}{\hbar} \mathscr{S}([\boldsymbol{r}_\tau]; \tau_1), \tag{4.52}$$

whereas the prefactor

$$\mathscr{A} = \left(\frac{\mathscr{S}_0([\boldsymbol{r}_\tau]; \tau_1)}{2\pi\hbar} \right)^{1/2} \left| \frac{\det(-\partial_\tau^2 \hat{1} + \hat{U}^{(0)})}{\det'(-\partial_\tau^2 \hat{1} + \hat{U}_\tau)} \right|_{\mathrm{PB}}^{1/2}. \tag{4.53}$$

In addition, we have introduced the quantity

$$\mathscr{S}_0([\boldsymbol{R}_\tau]; \tau_1) = m \int_0^{\tau_1} \mathrm{d}\tau |\dot{\boldsymbol{R}}_\tau|^2. \tag{4.54}$$

Note one peculiarity of this result. At first, we had to calculate the Gaussian fluctuations for zero boundary condition (ZB). Now, it has been found that the contribution to the current from the closed paths in the vicinity of the periodic orbit has, in effect, changed the requirement of zero boundary condition to the one of periodic boundary condition (PB). This allows us to interpret the above quantities as being part of a contribution $Z_1 = (\mathrm{i}\tau_1/2) Z_0 \mathscr{A} \exp(-\mathscr{B})$ to the partition function that leads to an imaginary part of the free energy $\mathscr{F}$ and from there – see eq. (2.47) – to the decay rate $\bar{\Gamma}$.

Thus, the relations above confirm and generalize Affleck's theory and interpretation to the multidimensional case. Specifically, they mean a generalization of eqs. (2.51) and (2.53). In the limit $T \to 0$ ($\tau_1 \to \infty$, $\mathscr{S}_0 \to \mathscr{S} \to 2\mathscr{W}_{\mathrm{EP}}(0)$), this relation agrees also with eqs. (3.54) and (3.64). It is this form (4.53) which emerges naturally (although less rigorously) from instanton-type calculations and forms the starting-point of many finite-temperature calculations in the literature (see, e.g., Larkin and Ovchinnikov 1992).

4.3. *The Caldeira–Leggett model at finite temperatures*

In what follows, we wish to apply the theory developed in sections 2.5, 2.6, 4.1 and 4.2 to the Caldeira–Leggett model as presented in section 3.6. Most of the motivation for such an application originates from our desire to demonstrate the multiple orbit theory of section 2.6 for a nontrivial model.

First of all, note that the equation of motion [eqs. (3.83a, b)] is still valid, but now it has to be solved for a periodic orbit. Consequently, we introduce the discrete frequencies

$$\omega_n = \omega(n) = \frac{2\pi}{\tau_1} n \tag{4.55}$$

and define Fourier transforms as follows:

$$\boldsymbol{r}_\tau = \frac{1}{\tau_1} \sum_n \mathrm{e}^{-\mathrm{i}\omega_n \tau} \boldsymbol{r}_{\omega_n},$$

$$\boldsymbol{r}_{\omega_n} = \int_0^{\tau_1} \mathrm{d}\tau \, \mathrm{e}^{\mathrm{i}\omega_n \tau} \boldsymbol{r}_\tau. \tag{4.56}$$

Elimination of the environmental coordinates r_k, $k = 2, \ldots, N$, leads to the same type [eq. (3.86)] of equation of motion for the coordinate r_1 of the object except for the redefinitions (4.55) and (4.56) of the Fourier transform; specifically, the dissipative part [cf. eqs. (3.86) and (3.87)] is given by

$$(H * r_1)_{\omega_n} = \eta |\omega_n| r_1(\omega_n). \tag{4.57}$$

Therefore,

$$\{m\omega_n^2 + \eta|\omega_n| + m\omega_0^2\} \xi_n - \tfrac{3}{2} m\omega_0^2 \frac{1}{\tau_1} \sum_{n'} \xi_{n-n'} \xi_{n'} = 0, \tag{4.58}$$

where the coordinate of the object has been rescaled according to

$$\xi_n = \frac{1}{R_\mathrm{C}} r_1(\omega_n). \tag{4.59}$$

In the limit of heavy damping

$$\gamma = \frac{\eta}{m} \gg \omega_0, \tag{4.60}$$

the acceleration term $m\omega_n^2$ in the curly brackets of eq. (4.58) can be neglected; Larkin and Ovchinnikov (1983a, b, 1984) have shown how to solve eq. (4.58) in this limit. Specifically, for an orbit of multiplicity p, where the primitive period is

$$\tau_p = \frac{\tau_1}{p}, \tag{4.61}$$

we have

$$\xi_{pn} = \tfrac{2}{3} \tau_1 (\tanh b) \exp[-b|n| + \mathrm{i}\omega_{pn}\tau_a],$$

$$\xi_\tau = \tfrac{2}{3} (\tanh b) \sum_n \exp[-b|n| - \mathrm{i}\omega_{pn}(\tau - \tau_a)], \tag{4.62}$$

where

$$\tanh b = \frac{\tau_\mathrm{B}}{\tau_p}, \qquad \tau_\mathrm{B} = \frac{2\pi\gamma}{\omega_0^2}. \tag{4.63}$$

Note that $T_\mathrm{B} = \hbar/k\tau_\mathrm{B}$ is the crossover temperature in the above mentioned limit. Also, we wish to draw attention to the fact that $r_1(\tau = 0) = R_\mathrm{C}\xi_{\tau=0}$ can be given any value by an appropriate choice of the integration constant τ_a (e.g., $\operatorname{Im}\tau_a \neq 0$). Note also that we have orbits for $\tau_p < \tau_B$; in this case $\operatorname{Im} b = \pi/2$.

Calculating the action for a single orbit, we find that

$$\mathscr{S}^{(1)} = \mathscr{S}([\boldsymbol{r}_\tau]; \tau_1) = \tau_B v_B \left[\frac{3}{2} - \frac{1}{2}\left(\frac{\tau_B}{\tau_1}\right)^2\right], \tag{4.64}$$

and that its Legendre transform is given by

$$\mathscr{W}^{(1)}(E) = \tfrac{3}{2}\,\tau_B v_B \left[1 - \left(\frac{E}{v_B}\right)^{2/3}\right], \tag{4.65}$$

which should be compared with fig. 6.

The corresponding quantities for an orbit of multiplicity p are

$$\mathscr{S}^{(p)} = p\mathscr{S}([\boldsymbol{r}_\tau]; \tau_p) = p\tau_B v_B \left[\frac{3}{2} - \frac{1}{2}\left(\frac{\tau_B}{\tau_p}\right)^2\right],$$

$$\mathscr{W}^{(p)} = p\mathscr{W}^{(1)}(E). \tag{4.66}$$

Of interest will also be the quantity

$$\frac{\partial^2 \mathscr{S}^{(p)}}{\partial \tau_1^2} = -3\,\frac{p^3 \tau_B^3 v_B}{\tau_1^4}. \tag{4.67}$$

The computational procedure for the fluctuation determinants which appear in the expression (4.53) for the prefactor, follows the same reasoning as in section 3.6, that has led us to eq. (3.97). Accordingly, we have

$$\mathscr{A} = \left(\frac{\mathscr{S}_0([r_1(\tau)]; \tau_1)}{2\pi\hbar m}\right)^{1/2} \left|\frac{\det(-m\partial_\tau^2 + H + m\omega_0^2)}{\det'(-m\partial_\tau^2 + H + v''(r_1(\tau)))}\right|^{1/2}. \tag{4.68}$$

The calculation of the prefactor is carried out in Appendix D. There it is shown that it depends on the multiplicity p in a nontrivial way. Accordingly, $\mathscr{A} \to \mathscr{A}^{(p)}$, which is given by

$$\mathscr{A}^{(p)}(\tau_1) = \left(\frac{9\pi v_B}{\hbar \tau_B}\right)^{1/2} \left(\frac{\tau_B}{\tau_1}\right)^2 \frac{p^{1/2}}{\Gamma^2(p)} \frac{\Gamma(2p + 2\sigma)}{\Gamma(2\sigma)}, \tag{4.69}$$

where σ is defined by

$$\sigma = \frac{1}{2}\left[1 + \frac{\tau_1}{\tau_B}\left(\frac{\gamma^2}{\omega_0^2} - 1\right)\right]. \tag{4.70}$$

According to what has been said in section 2.6, we should sum the contributions of all orbits with arbitrary multiplicity p. Hence,

$$\bar{\Gamma} = \sum_{p=1}^{\infty} (-1)^{p+1} \mathscr{A}^{(p)}(\tau_1) \exp\left(-\frac{1}{\hbar}\mathscr{S}^{(p)}\right). \tag{4.71}$$

As in the dissipation-free case, we solve the summation problem by a Laplace transform, evaluated by the method of steepest descent. Thus, we obtain*

$$\bar{\Gamma} = \sum_{p=1}^{\infty} (-1)^{p+1} \int \mathrm{d}E\, \tilde{\mathscr{A}}^{(p)} \exp\left\{ -\frac{1}{\hbar} [\mathscr{W}^{(p)}(E) + E\tau_1] \right\}, \tag{4.72}$$

where, by virtue of eq. (4.67),

$$\tilde{\mathscr{A}}^{(p)} = \mathscr{A}^{(p)} \left| \frac{1}{2\pi\hbar} \frac{\partial^2 \mathscr{S}^{(p)}}{\partial \tau_1^2} \right|^{-1/2} = \frac{1}{\hbar} \frac{\Gamma(2p + 2\sigma)}{p\Gamma^2(p)\Gamma(2\sigma)}. \tag{4.73}$$

Interchanging the order of summation and integration in eq. (4.72), the decay rate can finally be put into the form

$$\bar{\Gamma} = \frac{1}{\hbar} \int_0^{\infty} \mathrm{d}E\, \mathrm{e}^{-E/kT} \phi(z), \quad z = \mathrm{e}^{-\mathscr{W}(E)/\hbar}, \tag{4.74}$$

where $\phi(z)$ can be expressed in terms of the Gauss hypergeometric function (Abramowitz and Stegun 1986) as

$$\begin{aligned} \phi(z) &= \sum_{p=1}^{\infty} (-1)^{p+1} \frac{p\Gamma(2p + 2\sigma)}{\Gamma^2(p + 1)\Gamma(2\sigma)} z^p \\ &= 4z\sigma(\sigma + \tfrac{1}{2}) F(\sigma + 1, \sigma + \tfrac{3}{2}; 2; -4z). \end{aligned} \tag{4.75}$$

At this point we would like to mention that in Larkin and Ovchinnikov (1984), and subsequently also in Grabert and Weiss (1984) and Grabert et al. (1987), a calculation of $\bar{\Gamma}$ close to the crossover temperature T_{B} has been carried through according to Affleck's prescription (2.76) listed in (iii) of section 2.6:

$$\bar{\Gamma}^{\mathrm{A}} = \int_{-\infty}^{v_{\mathrm{B}}} \mathrm{d}E\, \tilde{\mathscr{A}}^{(1)}(\tau_1) \exp\left\{ -\frac{1}{\hbar} [\mathscr{W}_2^{(1)}(E) + E\tau_1] \right\}, \tag{4.76}$$

where $\mathscr{W}_2^{(1)}$ is the second-order expansion of $\mathscr{W}^{(1)} = \mathscr{W}$ in $E - v_{\mathrm{B}}$ as shown in eq. (2.77). Some arguments concerning this prescription are found in Appendix E, where also the explicit form of $\bar{\Gamma}^{\mathrm{A}}$ is given in eq. (E.11).

Ludviksson (1989) has calculated the decay rate [eq. (4.72)] numerically. (For details see also Appendix E.) In Table 1, we have listed some of his results for the case $\gamma = 4\omega_0$, $v_{\mathrm{B}} = \hbar\omega_0$ in the form of the ratio $Q = \bar{\Gamma}/\Gamma_{\mathrm{R}}(T)$, where

$$\Gamma_{\mathrm{R}}(T) = \begin{cases} \dfrac{\omega_0^2}{2\pi\gamma} \mathrm{e}^{-v_{\mathrm{B}}/kT}, & T > T_{\mathrm{B}}, \\ \dfrac{\omega_0^2}{2\pi\gamma} \mathrm{e}^{-v_{\mathrm{B}}/kT_{\mathrm{B}}}, & T < T_{\mathrm{B}}. \end{cases} \tag{4.77}$$

* See also Appendix E.

Table 1
Decay rate divided by $\Gamma_{\mathrm{R}}(T)$, which is the classical Kramers rate truncated for $T < T_{\mathrm{B}}$. The parameters are $\gamma = 4\omega_0$ and $v_{\mathrm{B}} = \hbar\omega_0$. Present theory: Q; Affleck's prescription: Q^{A}

T/T_{B}	Q	Q^{A}
0.4	0.0161	0.0177
0.5	0.0545	0.0585
0.6	0.241	0.253
0.7	1.35	1.43
0.8	8.58	10.5
0.9	52.0	101
1.0	268	1010
2.0	6.13	9.28
4.0	1.70	2.03
6.0	1.24	1.39
8.0	1.09	1.18
10.0	1.02	1.09

Note that for $T > T_{\mathrm{B}}$, $\Gamma_{\mathrm{R}}(T)$ is the classical Kramers rate in the heavy damping limit. For the sake of comparison, we have also listed numerical data $Q^{\mathrm{A}} = \bar{\Gamma}^{\mathrm{A}}/\Gamma_{\mathrm{R}}(T)$, which have been calculated in Grabert et al. (1987) according to Affleck's prescription. In view of the very different analytical expressions, one may call the agreement reasonable* above and below the crossover temperature, with some reservation in the crossover region. Fair agreement is obtained also for other values of γ/ω_0 and v_{B}, with a tendency of improvement for large values of $v_{\mathrm{B}}/\hbar\omega_0$.

The agreement for temperatures far above the crossover temperatures is not a coincidence. In fact, as we have shown in Appendix E, our theory is, in the high-temperature limit, equivalent to the ansatz of a quadratic expansion of $V(\boldsymbol{R})$ at $\boldsymbol{R}_{\mathrm{B}}$. Such an expression has been discussed by Affleck (1981) and others.**

For temperatures far below the crossover temperature, only small energies contribute to the decay rate [eq. (4.72)]. Consequently, only the simple orbit $p = 1$ is of importance and we agree with the standard theory.

* The agreement here relates only to the comparison of different methods. Arguments for the validity of semiclassical approximations, even for the quoted parameter values, are given in Larkin and Ovchinnikov (1992).

** See, e.g., the paper by Wolynes (1981).

5. Conclusion

This chapter has been concerned with the decay of a metastable state which may exist in a system where the potential energy $V(\boldsymbol{R})$ of an object features a relative minimum at some point of the multidimensional configuration space $\boldsymbol{R}$. In the present paper, such a minimum has been assumed to be at $\boldsymbol{R} = 0$, where we have put $V(0) = 0$. Clearly, there must be an environment of $\boldsymbol{R} = 0$ where $V(\boldsymbol{R}) \geqslant 0$, i.e., which is classically inaccessible; but there must also be an extensive outside, i.e., an accessible region $V(\boldsymbol{R}) < 0$, where the object is found in the decayed state.

There are two main parts: one corresponding to the decay at zero temperature, where a wave function ψ suffices for a description of the state, and the other corresponding to the decay at finite temperatures, where a statistical matrix ρ is required.

Let us first recall the case $T = 0$, where the wave function ψ is calculated in quasiclassical accuracy. Specifically, this calculation is based entirely on the standard quasiclassical ansatz, where the wave function is put equal to $\exp(-[\text{action}]/\hbar)$ and the (Euclidean) action is expanded in powers of $\hbar$. Generally, the action is a complex function of position. In case of a metastable state, however, it is real in a very large part of the inaccessible region.

This ansatz substituted in the Schrödinger equation leads to a coupled set of nonlinear first-order partial differential equations (eikonal equation and transport equation) which can be solved by the methods of characteristics. These characteristics are trajectories in a phase space (configuration and momentum space), which obey classical equations of motion.

Of central importance is one single trajectory called the escape path. This is a classical trajectory of zero energy in the "inverted" potential $-V(\boldsymbol{R})$, which connects the metastable minimum at $\boldsymbol{R} = 0$ with an escape point $\boldsymbol{R}_{\mathrm{C}}$ on the border of the outside region. There are also other trajectories in the inverted potential which start with zero energy at $\boldsymbol{R} = 0$ and which follow the escape path for some distance. Eventually, however, they will be reflected off at more or less close distances to the boundary of the accessible region. There is an envelope to these paths which may be called a caustic, and which lies entirely in the inaccessible region except for the escape point.

One can show that the wave function falls off rapidly at distances which scale with $\hbar^{1/2}$ from the escape path. Therefore, it is possible to introduce the small-fluctuation approximation, where only paths which are sufficiently close to the escape path are taken into consideration.

The problem now arises how the wave function can be obtained in the region beyond the caustic, which is by and large classically accessible. It is shown that a process of analytical continuation in the configuration space extended to complex coordinates resolves the problem. Alternatively, it is possible to extend the trajectories (essentially, as they touch the caustic) to complex phase space and to complex times. Note as a special case that the extension of the escape path, which starts at $\boldsymbol{R}_C$ into the accessible region, is a real classical trajectory in the true potential $V(\boldsymbol{R})$.

We should keep in mind that this complexity merely reflects the fact that in multidimensional systems cases are rarely found where the action is either purely real or purely imaginary.

Concerning the wave function, the following picture emerges. Essentially, it consists of a principal wave ψ_0 ($= \psi_0^*$) which resembles, near the origin, the ground state of a set of harmonic oscillators and which extends toward the caustic in a narrow tube surrounding the escape path. There, it is partially transmitted and reflected. The transmitted wave $\tilde{\psi}$ ($\neq \tilde{\psi}^*$) represents a wave propagating in the direction of the extension of the escape path. The reflected wave ψ_1 ($= -\psi_1^*$) returns to the origin.

By general arguments, the decay rate Γ of the metastable state is equal to the probability current which penetrates into the outside region. It has been shown that the decay rate thus obtained is exactly the same as that calculated in the instanton technique. This equivalence can partially be understood by noting that the escape path here corresponds to the bounce there.

At finite temperatures, we represent the statistical matrix in the form of a Feynman path integral. In quasiclassical approximation, the integration with respect to the paths is dominated by extremal paths, that are paths for which the action is stationary. Two types of such paths are of importance. There are some that contribute to the normalization and others that lead to a finite probability current in the outside, i.e., the classically accessible region. The extremal paths are solutions of classical equations of motion in the inverted potential. Of importance are periodic orbits completed in time $\tau_1 = \hbar/kT$ and which connect the metastable well with the outside region. One can easily see that for $\tau_1 \to \infty$ ($T \to 0$) this periodic orbit degenerates into the bounce discussed above.

For the same reasons that has led us above to the conclusion that, in general, the action is a complex quantity, we have to extend the class of extremal orbits to include also complex orbits. Also, we have found it necessary to include multiple orbits (see, e.g., Hänggi and Hontscha 1988), where a primitive orbit completed in time $\tau_p = \tau_1/p$ is traversed p times. This generalization constitutes the main message of our theory as compared with the one put forward, e.g., by Affleck.

In the vicinity of the periodic orbit there are closed extremal paths which also contribute to the outgoing current. These neighboring paths can be taken into account by a technique similar to what we have referred to above as the small-fluctuation approximation.

One result is worthy of note. As is quite generally the case in the quasiclassical approximation to path integrals, the Gaussian fluctuations have to be calculated for zero boundary conditions. Now, it has been found that the integration of the contribution to the current from those neighboring paths mentioned above changes the expression for Gaussian fluctuations at zero boundary condition exactly to an expression for Gaussian fluctuations at periodic boundary conditions. This allows us to interpret the overall current contribution as a contribution to the partition function where the paths are known to obey periodic boundary conditions. Thus, we have been able to confirm strictly the connection between the decay rate and the imaginary part of the free energy.

However, our theory means a generalization since it requires the inclusion of complex periodic orbit of arbitrary multiplicity. The summation with respect to the multiplicity is simple in a one-dimensional system but it might become a formidable task in the multidimensional case. We have carried through such a program for the Caldeira–Leggett model in the case of heavy damping and found that, numerically, the agreement between our theory and the standard one may be called reasonable in view of the fact that the corresponding analytical expressions are quite different. On the other hand, we have been able to show that for temperatures far above and far below the crossover temperature, our theory does lead to the standard limiting expression.

Although progress has been made, we have not found quantitative criteria for the assumption on the quasistationarity of the decaying state.

Appendix A. Transmission through a smooth barrier in quasiclassical approximation

For an illustration of the quasiclassical expression for the transmission coefficient [eq. (2.68)] let us compare the quasiclassical result with the analytical one for a barrier of the shape

$$v(R) = \frac{v_B}{\cosh^2 \alpha R}. \tag{A.1}$$

Let us calculate the abbreviated action $\mathscr{W}(E)$ defined by eq. (2.37) for a closed orbit and for $E < v_B$. Then ($v(R_0) = E$)

$$\begin{aligned}\mathscr{W}(E) &= 4 \int_0^{R_0} \mathrm{d}R \sqrt{2m(v(R) - E)} \\ &= \frac{\sqrt{2mv_B}}{\alpha}(1-\varepsilon) \int_0^1 (1-s)^{1/2}(1-(1-\varepsilon)s)^{-1} s^{-1/2}\, \mathrm{d}s,\end{aligned} \tag{A.2}$$

where we have made the substitution

$$\varepsilon = \frac{E}{v_B}, \quad s = \frac{1}{1-\varepsilon}\tanh^2 \alpha R. \tag{A.3}$$

After some calculations, we obtain

$$\mathscr{W}(E) = \frac{2\pi}{\alpha}(1-\varepsilon^{1/2})(2mv_B)^{1/2}. \tag{A.4}$$

Although we initially assumed $E < V_B(\varepsilon < 1)$, the above result holds for all values of E (> 0). We thus obtain within quasiclassical accuracy

$$\frac{1}{D(E)} = 1 + \exp\left[\frac{2\pi}{\hbar\alpha}(2mv_B)^{1/2}(1 - \varepsilon^{1/2})\right], \tag{A.5}$$

whereas the exact expression (Landau and Lifschitz 1975) is

$$\frac{1}{D(E)} = \frac{\sinh^2(\pi k/\alpha) + \cosh^2[\pi(2mv_B/\hbar^2\alpha^2 - \frac{1}{4})^{1/2}]}{\sinh^2(\pi k/\alpha)}, \tag{A.6}$$

where $\hbar k = (2mE)^{1/2}$. We realize that eqs. (A.5) and (A.6) agree in the quasiclassical limit $\hbar \to 0$, i.e., in the range

$$(\hbar\alpha)^2/2m \ll E, v_B. \tag{A.7}$$

In contrast, the expression for the transmission coefficient in harmonic approximation, i.e., replacing the barrier by a parabolic one with the same curvature at the top, is given by

$$\frac{1}{D(E)} = 1 + \exp\left[\frac{\pi}{\hbar\alpha}(2mv_B)^{1/2}(1 - \varepsilon)\right], \tag{A.8}$$

which differs from eq. (A.5). Thus, the agreement of expressions (A.5) and (A.6) in the quasiclassical limit is not related to the fact that the smooth potential (A.1) can be approximated by a parabola close to its maximum.

Appendix B. Some results for the one-dimensional decay

In the case of the cubic potential (2.10) one may calculate $\mathscr{W}(E)$ explicitly for all E in terms of the Gauss hypergeometric function (Abramowitz and Stegun 1968), with the result

$$\mathscr{W}(E) = m\omega_0 R_C^2 \frac{1}{\sqrt{2}}(-3c)^{5/2}\frac{\pi}{4} z F\left(-\frac{1}{4}, \frac{1}{4}; 2, z\right). \tag{B.1}$$

The quantities in the expression above are explained as follows: c is the smallest root of

$$c^3 - \frac{1}{3}c = \frac{2}{27}\left(1 - \frac{2E}{v_B}\right), \tag{B.2}$$

and

$$z = \frac{1}{3}\left[\left(\frac{2}{3c}\right)^2 - 1\right]. \tag{B.3}$$

Note that $-\infty < c < -\frac{1}{3}$ and $-\frac{1}{3} < z < 1$ for $\infty > E/v_B > 0$. Consequently, $\mathscr{W}(E)$ is an analytical function for $\mathrm{Re}\, E > 0$.

Consider now the expressions (2.68) and (2.69) for $D(E)$ and $\bar{\Gamma}$, respectively. According to the concluding comments of section 2.2, the approximation (i), where $D(E) = \exp[-(1/\hbar)\mathscr{W}(E)]$, may be used for low temperatures. There, we require $\mathscr{W}(E)$ only for small values of E. An appropriate expansion* of eq. (B.1) leads to

$$\frac{1}{\hbar}\mathscr{W}(E) = \frac{1}{\hbar\omega_0}\left[\frac{36v_B}{5} - E\ln\frac{432v_B}{E} - E\right]. \tag{B.4}$$

Thus, we may write

$$\bar{\Gamma} = \frac{1}{2\pi\hbar Z_0}\int_0^\infty dE\exp\left(-\frac{E}{kT} - \frac{1}{\hbar\omega_0}\left[\frac{36v_B}{5} - E\ln\frac{432v_B}{E} - E\right]\right). \tag{B.5}$$

If v_B is large compared to kT, $\hbar\omega_0$, we may evaluate the integral by steepest descent. In the limit $T \to 0$, we recover indeed the decay rate as given by eqs. (2.17) and (2.24). For the sake of completeness, we remark that at finite temperatures, we obtain nominally an exponent

$$\mathscr{B}(T) = \frac{v_B}{\hbar\omega_0}\left[\frac{36}{5} - 432\exp\left(-\frac{\hbar\omega_0}{kT}\right)\right]. \tag{B.6}$$

However, the finite temperature correction is irrelevant, particularly in comparison with the correction of the normalization $Z_0(\tau_1)$.

For large temperatures, we approximate $\mathscr{W}(E) \approx \mathscr{W}_1(E)$ according to eq. (2.71) whence we find that

$$kT_B = \hbar\omega_0/2\pi. \tag{B.7}$$

Thus, eq. (2.72) may be written in the form

$$\bar{\Gamma} = \frac{\omega_0}{2\pi}\frac{1}{2Z_0\sin(\hbar\omega_0/2kT)}e^{-v_B/kT}. \tag{B.8}$$

Note that $2Z_0\sin(\hbar\omega_0/2kT) \to 1$ for $kT \gg \hbar\omega_0$; this confirms eqs. (2.2) and (2.3).

The divergence of eq. (B.8) for $T \to T_B+$ is a consequence of grossly overestimating the small energy contribution in the replacement $\mathscr{W} \to \mathscr{W}_1$ (see fig. 6). Therefore, it seems reasonable to make use of the following approximation

$$\bar{\Gamma} = \frac{1}{2\pi\hbar Z_0}\int_0^\infty dE\, e^{-E/kT}\{[1 + e^{(1/\hbar)\mathscr{W}_1(E)}]^{-1} + \theta(v_B - E)[e^{-(1/\hbar)\mathscr{W}_2(E)} - e^{-(1/\hbar)\mathscr{W}_1(E)}]\}, \tag{B.9}$$

*One should make use of the appropriate linear transformation formulas (Abramowitz and Stegun 1968).

where $\mathscr{W}_2(E)$ is given by eq. (2.77). For the present case, we have

$$\frac{1}{kT_2} = \left(\frac{5}{18\,kT_\mathrm{B}v_\mathrm{B}}\right)^{1/2}. \tag{B.10}$$

In the form (B.9), we may extend the lower limit of integration to $-\infty$. Thus, we obtain

$$\bar{\Gamma} = \frac{kT_\mathrm{B}}{2\pi\hbar Z_0}\,\mathrm{e}^{-v_\mathrm{B}/kT}\left\{\frac{\pi}{\sin(\pi T_\mathrm{B}/T)} - \frac{T}{T-T_\mathrm{B}} + \frac{T_2}{T_\mathrm{B}}\pi^{1/2}\,\mathrm{erfc}\left(\frac{T_2}{T_\mathrm{B}} - \frac{T_2}{T}\right)\exp\left(\frac{T_2}{T_\mathrm{B}} - \frac{T_2}{T}\right)^2\right\}. \tag{B.11}$$

Clearly, in the region of the crossover temperature $T \sim T_\mathrm{B}$, the first two terms in the curly brackets vanish and one recovers Affleck's prescription (iii), which has led us to eq. (2.79). It appears that we have to require that $T_2/T_\mathrm{B} \sim (v_\mathrm{B}/\hbar\omega_0)^{1/2} \gg 1$ in order to have this approximation to be meaningful.

In the limit of high temperatures, we obtain the classical result including quantum correction:

$$\bar{\Gamma} = \frac{\omega_0}{2\pi}\,\mathrm{e}^{-v_\mathrm{B}/kT}\left[1 + \frac{1}{12}\left(\frac{\hbar\omega_0}{kT}\right)^2 - \frac{5\pi^2}{9}\frac{(\hbar\omega_0)^2}{kTv_\mathrm{B}}\right]. \tag{B.12}$$

The second term in the brackets represents quantum corrections already found by Wigner (1932). The third term is a much smaller correction in the limit $v_\mathrm{B} \gg kT$.

Appendix C. Gaussian fluctuations about complex paths

In a quasiclassical approximation to the path integral (2.55) it is necessary to calculate the contributions of Gaussian fluctuations about the extremal path $\boldsymbol{r}_\tau$, which is presently a complex path.

In this section, we are concerned with a one-dimensional system. Accordingly, we have

$$R_\tau = r_\tau + z_\tau, \quad z = x + \mathrm{i}y,$$
$$W_\tau = \frac{1}{m}v''(r_\tau) = U_\tau + \mathrm{i}V_\tau, \tag{C.1}$$

the second-order contribution to the action is given by

$$\mathscr{S}_2 = \tfrac{1}{2}m\int_0^{\tau_1}\mathrm{d}\tau\{\dot{z}_\tau^2 + W_\tau z_\tau^2\} = \mathscr{S}'_2 + \mathrm{i}\mathscr{S}''_2. \tag{C.2}$$

Since Gaussian fluctuations have to be calculated for zero boundary (ZB) conditions $z_{\tau_1} = z_0$, we obtain by partial integrations

$$\mathscr{S}'_2 = \tfrac{1}{2} m \int_0^{\tau_1} \mathrm{d}\tau \{x_\tau(-\ddot{x}_\tau + U_\tau x_\tau - V_\tau y_\tau) + y_\tau(\ddot{y}_\tau - U_\tau y_\tau - V_\tau x_\tau)\}$$

$$\mathscr{S}''_2 = \tfrac{1}{2} m \int_0^{\tau_1} \mathrm{d}\tau \{x_\tau(-\ddot{y}_\tau + U_\tau y_\tau - V_\tau x_\tau) + y_\tau(-\ddot{x}_\tau + U_\tau x_\tau - V_\tau y_\tau)\}. \quad \text{(C.3)}$$

Next, we introduce the two-component set of functions (x_n, y_n) which satisfies the equations

$$\left\{(-\partial_\tau^2 + U_\tau)\begin{pmatrix} 1 & 0 \\ 0 & -1 \end{pmatrix} - V_\tau \begin{pmatrix} 0 & 1 \\ 1 & 0 \end{pmatrix}\right\}\begin{pmatrix} x_n \\ y_n \end{pmatrix} = \lambda_n \begin{pmatrix} x_n \\ y_n \end{pmatrix},$$

$$\int_0^{\tau_1} \mathrm{d}\tau (x_n x_{n'} + y_n y_{n'}) = \delta_{nn'}. \quad \text{(C.4)}$$

One may convince oneself easily that $(x_n, y_n) = (y_{\bar{n}}, -x_{\bar{n}})$ is also an eigenfunction and that its eigenvalue is $\lambda_n = -\lambda_{\bar{n}}$. It is important to note that this set with positive and negative eigenvalues is complete. Therefore, we may expand

$$\begin{pmatrix} x \\ y \end{pmatrix} = \sum_n c_n \begin{pmatrix} x_n \\ y_n \end{pmatrix}, \quad \text{(C.5)}$$

and inserting eq. (C.5) into eq. (C.3), we arrive at

$$\mathscr{S}'_2 = \tfrac{1}{2} m \sum_n \lambda_n c_n^2,$$

$$\mathscr{S}''_2 = -\tfrac{1}{2} m \sum_n c_{\bar{n}} c_n \lambda_n, \quad \text{(C.6)}$$

where $\bar{n}$ is so defined that $\lambda_{\bar{n}} = -\lambda_n$.

The path integral in eq. (2.55) is now replaced by Gaussian integrals with respect to the expansion coefficients c_n. For these integrals to converge we must restrict the expansion (C.5) to such values of n for which $\lambda_n > 0$. In that case we have $\mathscr{S}''_2 = 0$. We may interpret the restriction as follows: For each degree of freedom n there is only a one-dimensional integration in the complex c_n-plane; furthermore, the integration contour is in the direction of steepest descent from the saddle point;* the direction of steepest ascent – corresponding to a rotation by $\frac{1}{2}\pi$ in the complex plane – is excluded.

Let us introduce the Jacobi fields

$$z_j(\tau; \xi) = x_j(\tau; \xi) + \mathrm{i} y_j(\tau; \xi), \quad j = 1, 2, \quad \text{(C.7)}$$

* Of course, the sign problem ("which way to cross the mountain pass") mentioned in section 2.6 [eq. (2.65)] remains.

which are solutions of the equation

$$\left\{(-\partial_\tau^2 + U)\begin{pmatrix}1 & 0\\ 0 & -1\end{pmatrix} - V\begin{pmatrix}0 & 1\\ 1 & 0\end{pmatrix} - \xi\begin{pmatrix}1 & 0\\ 0 & 1\end{pmatrix}\right\}\begin{pmatrix}x_j\\ y_j\end{pmatrix} = 0 \tag{C.8}$$

subject to the initial conditions

$$z_j(0;\xi) = 0, \qquad \dot{z}_1(0;\xi) = 1, \qquad \dot{z}_2(0;\xi) = -i. \tag{C.9}$$

Observe, that for $\xi = 0$ we have

$$(-\partial_\tau^2 + U_\tau + \mathrm{i}V_\tau)z_j(\tau;0) = 0; \tag{C.10}$$

whence it follows that

$$z_2(\tau;0) = -\mathrm{i}z_1(\tau;0). \tag{C.11}$$

We now construct a linear combination (with real coefficients) of $z_1(\tau;\xi)$ and $z_2(\tau;\xi)$ that satisfies the boundary conditions

$$a_1 z_1(\tau_1;\ \xi) + a_2 z_2(\tau_1;\xi) = 0. \tag{C.12}$$

Separating the real and imaginary parts of eq. (C.12) leads to a homogeneous system of equations that has a nontrivial solution only if

$$\mathrm{Im}\,(z_1(\tau_1;\xi)[z_2(\tau_1;\xi)]^*) = x_1(\tau_1;\xi)y_2(\tau_1;\xi) - y_1(\tau_1;\xi)x_2(\tau_1;\xi) = 0. \tag{C.13}$$

Next we compare the eigenvalues above with those derived from

$$W^{(0)} = \frac{1}{m}v''(0) = U^{(0)} + \mathrm{i}V^{(0)},$$

$$U^{(0)} = \omega_0^2, \qquad V^{(0)} = 0, \tag{C.14}$$

and define the ratio (denoting zero boundary conditions by the subscript ZB)

$$R_{\mathrm{ZB}}^2 = \left(\frac{\prod_n^+ \lambda_n^0}{\prod_n^+ \lambda_n}\right)_{\mathrm{ZB}}^2 = \left|\frac{\prod_n \lambda_n^0}{\prod_n \lambda_n}\right|_{\mathrm{ZB}}, \tag{C.15}$$

where the superscript + in the product above denotes restriction to positive eigenvalues λ_n. For further progress, we compare the expressions

$$\left.\frac{\prod_n(\lambda_n^{(0)} - \xi)}{\prod_n(\lambda_n - \xi)}\right|_{\mathrm{ZB}}$$

$$= \left.\frac{\det\left\{(-\partial_\tau^2 + U^{(0)})\begin{pmatrix}1 & 0\\ 0 & -1\end{pmatrix} - V^{(0)}\begin{pmatrix}0 & 1\\ 1 & 0\end{pmatrix} - \xi\begin{pmatrix}1 & 0\\ 0 & 1\end{pmatrix}\right\}}{\det\left\{(-\partial_\tau^2 + U_\tau)\begin{pmatrix}1 & 0\\ 0 & 1\end{pmatrix} - V_\tau\begin{pmatrix}0 & 1\\ 1 & 0\end{pmatrix} - \xi\begin{pmatrix}1 & 0\\ 0 & 1\end{pmatrix}\right\}}\right|_{\mathrm{ZB}} \tag{C.16}$$

and

$$\frac{\operatorname{Im}(z_1^{(0)}(\tau_1;\xi)[z_2^{(0)}(\tau_1;\xi)]^*)}{\operatorname{Im}(z_1(\tau_1;\xi)[z_2(\tau_1;\xi)]^*)}, \tag{C.17}$$

where $z_1^{(0)}$ and $z_2^{(0)}$ are defined analogous to eqs. (C.8) and (C.9), with W_τ replaced by $W^{(0)}$. Evidently, both expressions (C.16) and (C.17) have simple poles for $\xi = \lambda_n$ and zeros at $\xi = \lambda_n^{(0)}$ and approach 1 as $|\xi| \to \infty$.

Therefore, we conclude that both expressions coincide in the entire ξ-plane. Putting $\xi = 0$ and using eq. (C.11) we conclude that

$$R_{\mathrm{ZB}} = \left| \frac{z_1^{(0)}(\tau_1;0)}{z_1(\tau_1;0)} \right|. \tag{C.18}$$

By definition, the time derivative of the classical path $\dot{r}_\tau$ obeys the equation

$$(-\partial_\tau^2 + U + \mathrm{i}V)\dot{r}_\tau = 0; \tag{C.19}$$

whence it follows that

$$z_1(\tau;0) = \dot{r}_0\dot{r}_\tau \int_0^\tau \mathrm{d}\tau' \, \dot{r}_{\tau'}^{-2}. \tag{C.20}$$

Furthermore,

$$z_1^{(0)}(\tau;0) = \frac{1}{\omega_0}\sinh \omega_0\tau, \tag{C.21}$$

and we find that

$$|\dot{r}_0|^2 R_{\mathrm{ZB}} = \frac{\sinh \omega_0\tau_1}{\omega_0 |\int_0^{\tau_1} \mathrm{d}\tau \, \dot{r}_\tau^{-2}|}. \tag{C.22}$$

We observe that for an analytical potential

$$\int_0^{\tau_1} \mathrm{d}\tau \, \dot{r}_\tau^{-2} = \oint \mathrm{d}r \, \dot{r}^{-3} \tag{C.23}$$

does not depend on the particular form of the orbit as long as it encloses the two turning points of classical motion (cf. fig. 4). From the definition of τ_1, we find that

$$\oint \mathrm{d}r \, \dot{r}_\tau^{-3} = -m\frac{\partial \tau_1}{\partial E} = -m\frac{\partial^2 \mathscr{W}(E)}{\partial E^2}. \tag{C.24}$$

Let us now consider fluctuations with periodic boundary conditions. Although the eigenvalues and eigenfunctions are different from those for zero boundary conditions, we will for the sake of simplicity, use the same labeling.

We define

$$R_{\text{PB}}^2 = \left(\frac{\prod_n^+ \lambda_n^{(0)}}{\prod_n^+ \lambda_n} \right)_{\text{PB}}^2 = \left| \frac{\prod_n \lambda_n^{(0)}}{\prod_n' \lambda_n} \right|_{\text{PB}}, \tag{C.25}$$

where $\prod_n^+$ means restriction to positive eigenvalues and $\prod_n'$ means that the eigenvalue zero (corresponding to the eigenfunction $\propto \dot{r}_\tau$) is to be omitted.

In order to control the zero eigenvalue, let us first define a modified ratio

$$\begin{aligned} (R_{\text{PB}}^\varepsilon)^2 &= \left| \prod_n \frac{\lambda_n^{(0)}}{\lambda_n^\varepsilon} \right|_{\text{PB}} \\ &= \left. \frac{\det\left\{ (-\partial_\tau^2 + U^{(0)}) \begin{pmatrix} 1 & 0 \\ 0 & -1 \end{pmatrix} - V^{(0)} \begin{pmatrix} 0 & 1 \\ 1 & 0 \end{pmatrix} \right\}}{\det\left\{ (-\partial_\tau^2 + U_\tau^\varepsilon) \begin{pmatrix} 1 & 0 \\ 0 & -1 \end{pmatrix} - V_\tau \begin{pmatrix} 0 & 1 \\ 1 & 0 \end{pmatrix} \right\}} \right|_{\text{PB}}, \end{aligned} \tag{C.26}$$

where $U_\tau^\varepsilon = U_\tau + \varepsilon$.

We now introduce four Jacobi fields $z_j^\varepsilon(\tau; \xi)$ that satisfy eq. (C.10), with U_τ replaced by U_τ^ε, subject to the initial conditions

$$\begin{aligned} &z_{1,2}^\varepsilon(0; \xi) = 0, \qquad \dot{z}_1^\varepsilon(0; \xi) = 1, \qquad \dot{z}_2^\varepsilon(0; \xi) = -\mathrm{i}, \\ &\dot{z}_{3,4}^\varepsilon(0; \xi) = 0, \qquad z_3^\varepsilon(0; \xi) = 1, \qquad z_4^\varepsilon(0; \xi) = -\mathrm{i}. \end{aligned} \tag{C.27}$$

For $\xi = 0$ we have relations similar to eqs. (C.11) and (C.12). In particular,

$$z_2^\varepsilon(\tau; 0) = -\mathrm{i} z_1^\varepsilon(\tau; 0), \qquad z_4^\varepsilon(\tau; 0) = -\mathrm{i} z_3^\varepsilon(\tau; 0). \tag{C.28}$$

We now form linear combinations of $z_j^\varepsilon(\tau; \xi)$ with real coefficients that satisfy the periodic boundary conditions. This can be done in two steps. First, we form two independent solutions

$$\begin{aligned} &Z_1^\varepsilon(\tau; \xi) = z_1^\varepsilon(\tau; \xi)[1 - z_3^\varepsilon(\tau_1; \xi)] + z_3^\varepsilon(\tau; \xi) z_1^\varepsilon(\tau_1; \xi), \\ &Z_2^\varepsilon(\tau; \xi) = (1, 3) \leftrightarrow (2, 4). \end{aligned} \tag{C.29}$$

For $\xi = 0$ we have, on account of eq. (C.28),

$$Z_2^\varepsilon(\tau; 0) = -\mathrm{i} Z_1^\varepsilon(\tau; 0). \tag{C.30}$$

Secondly, we construct a linear combination of $Z_l^\varepsilon(\tau; \xi)$ that satisfies

$$\sum_{l=1,2} \alpha_l \dot{Z}_l^\varepsilon(0; \xi) = \sum_{l=1,2} \alpha_l \dot{Z}_l^\varepsilon(\tau_1; \xi). \tag{C.31}$$

The condition for this system to have a nontrivial solution is

$$\operatorname{Im}\{[\dot{Z}_1^\varepsilon(\tau_1; \xi) - \dot{Z}_1^\varepsilon(0; \xi)][\dot{Z}_2^\varepsilon(\tau_1; \xi) - \dot{Z}_2^\varepsilon(0; \xi)]^*\} = 0. \tag{C.32}$$

By reasoning similar to that leading to eq. (C.18), we conclude that

$$(R_{\mathrm{PB}}^{\varepsilon})^2 = \left| \frac{\dot{Z}_1^{(0)}(\tau_1; 0) - \dot{Z}_1^{(0)}(0; 0)}{\dot{Z}_1^{\varepsilon}(\tau_1; 0) - \dot{Z}_1^{\varepsilon}(0; 0)} \right|^2, \tag{C.33}$$

where we have made use of eq. (C.30).

Now we eliminate the two eigenvalues close to zero. For $\varepsilon = 0$, let the eigenfunctions to the double-degenerate eigenvalue zero be $(x_0(\tau), y_0(\tau))$ and $(y_0(\tau), -x_0(\tau))$, respectively. To first order in ε, the matrix elements of the perturbation

$$\varepsilon \begin{pmatrix} 1 & 0 \\ 0 & -1 \end{pmatrix}$$

in the two-dimensional subspace are

$$\varepsilon \int_0^{\tau_1} \mathrm{d}\tau \begin{pmatrix} x_0^2 - y_0^2 & 2x_0 y_0 \\ 2x_0 y_0 & y_0^2 - x_0^2 \end{pmatrix}. \tag{C.34}$$

The product of the two eigenvalues is then

$$-\varepsilon^2 \left\{ \left[\int_0^{\tau_1} \mathrm{d}\tau (x_0^2 - y_0^2) \right]^2 + \left[2 \int_0^{\tau_1} \mathrm{d}\tau \, x_0 y_0 \right]^2 \right\} = -\varepsilon^2 \left| \int_0^{\tau_1} \mathrm{d}\tau \, z_0^2 \right|^2, \tag{C.35}$$

which leads to

$$R_{\mathrm{PB}} = \left| \int_0^{\tau_1} \mathrm{d}\tau \, z_0^2 \right| \left| \lim_{\varepsilon \to 0} \varepsilon \frac{\dot{Z}_1^{(0)}(\tau_1; 0) - \dot{Z}_1^{(0)}(0; 0)}{\dot{Z}_1^{\varepsilon}(\tau_1; 0) - \dot{Z}_1^{\varepsilon}(0; 0)} \right|. \tag{C.36}$$

We now make use of the fact that

$$\partial_\tau \{ \dot{Z}_1^{\varepsilon}(\tau; 0) z_0(\tau) - Z_1^{\varepsilon}(\tau; 0) \dot{z}_0(\tau) \} = \varepsilon Z_1^{\varepsilon}(\tau; 0) z_0(\tau). \tag{C.37}$$

Furthermore, since $z_0(\tau)$ solves the equation (C.9) with $\xi = 0$, we may write

$$z_0(\tau) = \dot{z}_0(0) z_1(\tau; 0) + z_0(0) z_3(\tau; 0). \tag{C.38}$$

Since $z_0(\tau)$ is periodic ($z_0(\tau) = \dot{r}_\tau$), we may use the condition $z_0(\tau_1) = z_0(0)$ to deduce from eq. (C.29) that to zeroth order we have

$$Z_1^{\varepsilon}(\tau; 0) = \frac{z_1(\tau_1; 0)}{z_0(0)} z_0(\tau). \tag{C.39}$$

From eq. (C.37) we obtain by integration

$$\dot{Z}_1^{\varepsilon}(\tau_1; 0) - \dot{Z}_1^{\varepsilon}(0; 0) = \frac{\varepsilon}{z_0(0)} \int_0^{\tau_1} \mathrm{d}\tau \, Z_1^{\varepsilon}(\tau; 0) z_0(\tau), \tag{C.40}$$

which, by virtue of eq. (C.39) leads to

$$\dot{Z}_1^{\varepsilon}(\tau_1; 0) - \dot{Z}_1^{\varepsilon}(0; 0) = \varepsilon \frac{z_1(\tau_1; 0)}{z_0^2(0)} \int_0^{\tau_1} d\tau \, z_0^2(\tau). \tag{C.41}$$

Since z_0 is normalized to unity, we have

$$|z_0(0)|^2 = |\dot{r}_0|^2 \Big/ \int_0^{\tau_1} d\tau \, |\dot{r}_\tau^2|. \tag{C.42}$$

Inserting eqs. (C.20), (C.41) and (C.42) into eq. (C.36), we finally arrive at the following expression:

$$R_{PB} \int_0^{\tau_1} d\tau \, |\dot{r}_\tau|^2 = \frac{|\dot{Z}_1^{(0)}(\tau_1; 0) - \dot{Z}_1^{(0)}(0; 0)|}{|\int_0^{\tau_1} d\tau \, \dot{r}_\tau^{-2}|}. \tag{C.43}$$

We observe now that

$$z_1^{(0)}(\tau; 0) = \frac{1}{\omega_0} \sinh \omega_0 \tau, \quad z_3^{(0)}(\tau; 0) = \cosh \omega_0 \tau,$$

$$\dot{Z}_1(\tau_1; 0) - \dot{Z}_1(0; 0) = 4 \sinh^2 \frac{\omega_0 \tau_1}{2}. \tag{C.44}$$

A comparison of eqs. (C.21) and (C.43) together with eq. (C.44) allows us to express R_{ZB} in terms of R_{PB} as follows:

$$R_{ZB} |\dot{r}_0^2| = \frac{1}{2m\omega_0} \coth \frac{\omega_0 \tau_1}{2} \mathscr{S}_0(\tau_1) R_{PB}, \tag{C.45}$$

where we have introduced

$$\mathscr{S}_0(\tau_1) = \mathscr{S}_0([r_\tau]; \tau_1) = m \int_0^{\tau_1} d\tau \, |\dot{r}_\tau|^2. \tag{C.46}$$

At this point, we find it convenient to introduce the notation

$$R_{XB} = \left. \frac{\det^+\{-\partial_\tau^2 + W_\tau^{(0)}\}}{\det^+\{-\partial_\tau^2 + W_\tau\}} \right|_{XB}. \tag{C.47}$$

Evidently, for $W_\tau \to U_\tau$ real, we have

$$\frac{\det^+\{-\partial_\tau^2 + W_\tau^{(0)}\}}{\det^+\{-\partial_\tau^2 + W_\tau\}} = \left| \frac{\det\{-\partial_\tau^2 + U^{(0)}\}}{\det'\{-\partial_\tau^2 + U_\tau\}} \right|. \tag{C.48}$$

Using the above definitions and relations, we may express the statistical matrix ρ in quasiclassical approximation as follows:

$$\rho(R, R; \tau_1) = \left(\frac{m\omega_0}{2\pi\hbar \sinh \omega_0 \tau_1} \right)^{1/2} \left| \frac{\det^+(-\partial_\tau^2 + W^{(0)})}{\det^+(-\partial_\tau^2 + W_\tau)} \right|_{\mathrm{ZB}}^{1/2} \times \exp\left(-\frac{1}{\hbar} \mathscr{S}([r_\tau]; \tau_1) \right). \tag{C.49}$$

We have chosen a representation where a ratio of determinants appears since it is easier to handle. That the overall constant is chosen properly can be seen by inserting the trivial path $r_\tau = 0$, where one expects the result $\rho(0, 0; \tau_1) = (m\omega_0/2\pi\hbar \sinh \omega_0\tau_1)^{1/2}$, which is known to be valid for the harmonic oscillator. Using eq. (C.45) together with $Z_0^{-1} = 2 \sinh \omega_0 \tau_1/2$, we may rewrite eq. (C.49) as follows:

$$\rho(R, R; \tau_1) = Z_0^{-1} \left(\frac{\mathscr{S}_0(\tau_1)}{2\pi\hbar |\dot{r}_0|^2} \right)^{1/2} R_{\mathrm{PB}}^{1/2} \exp\left(-\frac{1}{\hbar} \mathscr{S}([r_\tau]) \right). \tag{C.50}$$

Calculating the decay current $J(R)$ and the decay rate Γ according to eqs. (2.57) and (2.60), we obtain the result given previously by eqs. (2.44), (2.51) and (2.53), namely

$$\bar{\Gamma} = \left(\frac{\mathscr{S}_0(\tau_1)}{2\pi\hbar} \right)^{1/2} \left| \frac{\det^+(-\partial_\tau^2 + W^{(0)})}{\det^+\{-\partial_\tau^2 + W_\tau\}} \right|_{\mathrm{PB}}^{1/2} \exp\left(-\frac{1}{\hbar} \mathscr{S}([r_\tau]; \tau_1) \right). \tag{C.51}$$

Alternatively, we may combine eqs. (C.22), (C.24) and (C.45) to obtain

$$\bar{\Gamma} = Z_0^{-1} \left[2\pi\hbar \frac{\partial^2 \mathscr{W}(E)}{\partial E^2} \right]^{-1/2} \exp\left(-\frac{1}{\hbar} \mathscr{S}([r_\tau]; \tau_1) \right), \tag{C.52}$$

which agrees with eqs. (2.44)–(2.46).

Eventually, we wish to discuss the possibility of deforming the periodic orbit r_τ in the complex r-plane. Evidently, the replacement

$$r_\tau \to r_\tau + \delta \cdot \dot{r}_\tau \tag{C.53}$$

does not change the orbit to $O(\delta)$, and also it leaves the action $\mathscr{S}$ unchanged. In a formal way, we may consider this to be a consequence of the fact that $(\mathrm{Re}\, \dot{r}_\tau, \mathrm{Im}\, \dot{r}_\tau)$ is a solution of the eigenvalue equation (C.4) to the eigenvalue zero.

However, we have also learned that $(\mathrm{Im}\, \dot{r}_\tau, -\mathrm{Re}\, \dot{r}_\tau)$ is also an eigenfunction with eigenvalue zero. On the other hand, $-\mathrm{i}\dot{r}_\tau = \mathrm{Im}\, \dot{r}_\tau - \mathrm{i}\mathrm{Re}\, \dot{r}_\tau$ is, in the complex plane, perpendicular to $\dot{r}_\tau$ and, therefore,

$$r_\tau \to r_\tau - \mathrm{i}\delta \cdot \dot{r}_\tau \tag{C.54}$$

means a real deformation of the orbit which leaves the action unchanged.

Clearly, by successive application of this infinitesimal deformation, we may induce global changes in the orbit. Thus, we may say that it is possible to construct an orbit which passes through any preselected point R.

With some restrictions, the above argument can also be extended to the multidimensional case.

Appendix D. Prefactor for the heavily damped object

We consider the fluctuation operator in the denominator of eq. (4.68). Its eigenvalues Λ and eigenfunctions $v_\tau = v(\tau)$, $v_n = v(\omega_n)$ can be found by solving the equation

$$\{m\omega_n^2 + \eta|\omega_n| + m\omega_0^2\} v_n - 3m\omega_0^2 [\xi_\tau v_\tau]_n = \Lambda v_n, \tag{D.1}$$

which is written down in terms of Fourier transforms [cf. eqs. (4.55) and (4.56)].

In the limit of heavy damping, the extremal path $r_1(\tau) = R_C \xi_\tau$ is given by eq. (4.62). (For convenience, we substitute $\tau - \tau_a$ by τ in the following.) Considering an orbit of multiplicity p, we choose the ansatz

$$v_\tau = \sum_n C_n^k \exp(\mathrm{i}\omega_{pn-k}\tau), \tag{D.2}$$

where the integer k, $0 \leqslant k < p$, plays the role of a "Bloch index". Inserting this expression into eq. (D.1), we obtain*

$$\left[\left(\frac{\omega_1}{\omega_0}\right)^2 (pn+k)^2 + \frac{1}{p}|pn+k|\tanh b + 1\right] C_n^k - 2(\tanh b)\sum_{n'} C_{n'}^k \mathrm{e}^{-b|n-n'|}$$
$$= \frac{\Lambda}{m\omega_0^2} C_n^k, \tag{D.3}$$

where $\omega_1 = 2\pi/\tau_1$.

The case $k = 0$ was discussed by Larkin and Ovchinnikov (1984). We will now briefly discuss their procedure. In the limit $\gamma \gg \omega_0$ we are considering here, and for the lowest eigenvalues, we may neglect the acceleration which is the first term in the brackets of eq. (D.3) and we will call this form the truncated eigenvalue equation.

For $k = 0$, one may also choose the eigenfunctions to have a definite parity. In the even case we start from the ansatz

$$C_n^{\mathrm{e}} = (|n| + C)\mathrm{e}^{-b|n|}. \tag{D.4}$$

* Note that $\tanh b = \tau_B/\tau_p \propto p$ depends on p.

We insert this in the truncated eigenvalue equation (D.3) and arrive at

$$-\left\{C(\tanh b)|n| + C + \tanh b\,\frac{1}{\sinh^2 b}\right\} = \frac{\Lambda^{\mathrm{e}}}{m\omega_0^2}(|n| + C), \tag{D.5}$$

which is satisfied for all n provided that

$$-C\tanh b = \frac{\Lambda^{\mathrm{e}}}{m\omega_0^2}, \qquad -C - \tanh b\,\frac{1}{\sinh^2 b} = \frac{\Lambda^{\mathrm{e}}}{m\omega_0^2}C. \tag{D.6}$$

This leads to a second-order equation for C, which is solved to give the eigenvalues

$$\frac{\Lambda^{\mathrm{e}}_{0,1}}{m\omega_0^2} = -\frac{1}{2}\left[1 \pm \left(1 + \frac{4}{\cosh^2 b}\right)^{1/2}\right]. \tag{D.7}$$

The remaining eigenfunctions of even parity, corresponding to the eigenvalues $\Lambda^{\mathrm{e}}_{2+N}$, $N \geqslant 0$, are of the type

$$C^{\mathrm{e}}_n = (|n| + C)\mathrm{e}^{-b|n|} + d_n, \tag{D.8}$$

where

$$d_n = d_{-n}, \qquad d_N \neq 0, \qquad d_n = 0 \quad \text{for } |n| > N. \tag{D.9}$$

This leads to the following system of equations:

$$-\tanh b\left\{|n|C + (\coth b)C + \frac{1}{\sinh b}\right\} + \{|n|\tanh b + 1\}\,d_n\mathrm{e}^{b|n|}$$
$$-2(\tanh b)\mathrm{e}^{b|n|}\sum_{n'=-N}^{N} d_{n'}\,\mathrm{e}^{-b|n-n'|} = \frac{\Lambda^{\mathrm{e}}_{2+N}}{m\omega_0^2}\{|n| + C + \mathrm{e}^{b|n|}d_n\}. \tag{D.10}$$

Assuming that the equations are valid for $|n| > N$, we conclude that for $n = N$

$$[N\tanh b + 1]\,d_N\,\mathrm{e}^{bN} = \frac{\Lambda^{\mathrm{e}}_{2+N}}{m\omega_0^2}\,\mathrm{d}_N\,\mathrm{e}^{bN}. \tag{D.11}$$

Since $d_N \neq 0$, we must have

$$\frac{\Lambda^{\mathrm{e}}_{2+N}}{m\omega_0^2} = 1 + N\tanh b. \tag{D.12}$$

For the odd eigenfunctions we make the ansatz

$$C^{\mathrm{o}}_n = (n + \bar{C}\,\mathrm{sgn}\,n)\mathrm{e}^{-b|n|} + g_n, \tag{D.13}$$

where

$$g_n = -g_{-n}, \qquad g_M \neq 0, \qquad g_n = 0 \quad \text{for } n > M \tag{D.14}$$

and

$$\operatorname{sgn} n = \begin{cases} 1, & n > 0, \\ 0, & n = 0, \\ -1, & n < 0. \end{cases} \tag{D.15}$$

Thus, we arrive at the system of equations

$$\begin{aligned} &\bar{C} \operatorname{sgn} n - \bar{C}(\tanh b) n + [|n| \tanh b + 1] g_n \mathrm{e}^{b|n|} \\ &\quad - 2(\tanh b) \mathrm{e}^{b|n|} \sum_{n'=-M}^{M} g_{n'} \mathrm{e}^{-b|n-n'|} \\ &= \frac{\Lambda_M^{\circ}}{m\omega_0^2} [n + \bar{C} \operatorname{sgn} n + g_n \mathrm{e}^{b|n|}]. \end{aligned} \tag{D.16}$$

Assuming that the equations are valid for $|n| > M$, we conclude that for $n = M$

$$(M \tanh b + 1) g_M \mathrm{e}^{bM} = \frac{\Lambda_M^{\circ}}{m\omega_0^2} g_M \mathrm{e}^{bM}. \tag{D.17}$$

Since $g_M \neq 0$, we must have

$$\frac{\Lambda_M^{\circ}}{m\omega_0^2} = 1 + M \tanh b \quad (M \neq 0). \tag{D.18}$$

For $M = 0$ we obtain

$$\bar{C}(\operatorname{sgn} n - n \tanh b) = \frac{\Lambda_0^{\circ}}{m\omega_0^2} (n + \bar{C} \operatorname{sgn} n), \tag{D.19}$$

which holds for all n if

$$\bar{C} = \frac{\Lambda_0^{\circ}}{m\omega_0^2} \bar{C}, \qquad -\bar{C} \tanh b = \frac{\Lambda_0^{\circ}}{m\omega_0^2}. \tag{D.20}$$

Either $\bar{C} = 0$, corresponding to $\Lambda_0^{\circ} = 0$ or $\bar{C} \neq 0$ in which case $\Lambda_0^{\circ}/m\omega_0^2 = 1$.

We have now obtained all eigenvalues of the truncated eigenvalue equation. We introduce the following labeling, which includes all even and odd eigenvalues:

$$\begin{aligned} &\tilde{\Lambda}_0 = 0, \\ &\tilde{\Lambda}_{\pm 1} = -\frac{1}{2} m\omega_0^2 \left[1 \pm \left(1 + \frac{4}{\cosh^2 b} \right)^{1/2} \right], \\ &\tilde{\Lambda}_\nu = m\omega_0^2 [(|\nu| - 2) \tanh b + 1], \quad |\nu| \geqslant 2. \end{aligned} \tag{D.21}$$

For states with large quantum numbers, the acceleration term in eq. (D.3), which we have neglected so far (truncation), becomes important. In that case we neglect the last term on the left-hand side of eq. (D.3). Then, the eigenfunction corresponding to the eigenvalue

$$\lambda_\nu^{(0)} = m\omega_0^2\left[p^2\nu^2\left(\frac{\omega_1}{\omega_0}\right)^2 + |\nu|\tanh b + 1\right] \tag{D.22}$$

is simply

$$C_n = \delta_{n\nu}.$$

Including the off-diagonal term by lowest-order perturbation theory, we obtain the eigenvalues

$$\lambda_\nu = m\omega_0^2\left[p^2\nu^2\left(\frac{\omega_1}{\omega_0}\right)^2 + (|\nu| - 2)\tanh b + 1\right]. \tag{D.23}$$

Comparing eqs. (D.21) and (D.23), we recognize that the result (D.21) can be improved if we put

$$\tilde{\Lambda}_\nu = \lambda_\nu, \quad |\nu| \geqslant 2. \tag{D.24}$$

For the prefactor we need also to calculate

$$\begin{aligned}\mathscr{S}_0([r_1(\tau)]; \tau_p) &= m\int_0^{\tau_p} \mathrm{d}\tau(\dot{r}_1(\tau))^2 = \frac{2\pi^2}{\tau_\mathrm{B}}\frac{mR_\mathrm{B}^2}{\cosh^2 b}\\ &= \frac{2\pi^2 mR_\mathrm{B}^2}{\tau_\mathrm{B}\omega_0^4}\left|\frac{\tilde{\Lambda}_1}{m}\frac{\Lambda_{-1}}{m}\right|.\end{aligned} \tag{D.25}$$

As an exercise, we calculate the prefactor $\mathscr{A}_0^{(1)}$ of the decay rate for one primitive orbit completed in time τ_p. There

$$\begin{aligned}\mathscr{A}_0^{(1)} &= \left(\frac{\mathscr{S}_0}{2\pi\hbar}\right)^{1/2}\left\{\frac{\prod_\nu[\omega_{p\nu}^2 + \gamma|\omega_{p\nu}| + \omega_0^2]}{\prod_{\nu\neq 0}\tilde{\Lambda}_\nu/m}\right\}^{1/2}\\ &= \left(\frac{m\omega_0 R_\mathrm{B}^2}{2\hbar}\frac{1}{\gamma\omega_0}\right)^{1/2}\frac{\prod_{\nu=1}^{\infty}[\omega_{p\nu}^2 + \gamma|\omega_{p\nu}| + \omega_0^2]}{\prod_{\nu=2}^{\infty}[\omega_{p\nu}^2 + \gamma|\omega_{p(\nu-2)}| + \omega_0^2]}.\end{aligned} \tag{D.26}$$

We now write the above product in the following form

$$\omega_p^2\frac{\prod_{\nu=1}^{\infty}(\nu - p\nu_1)(\nu - p\nu_2)}{\prod_{\nu=2}^{\infty}(\nu - p\mu_1)(\nu - p\mu_2)}, \tag{D.27}$$

where we have introduced the dimensionless quantities

$$\begin{aligned}\nu_{1,2} &= (\omega_1)^{-1}[-\gamma/2 \pm \sqrt{\gamma^2/4 - \omega_0^2}],\\ \mu_{1,2} &= (\omega_1)^{-1}[-\gamma/2 \pm \sqrt{\gamma^2/4 - \omega_0^2 + 2\gamma\omega_p}].\end{aligned} \tag{D.28}$$

Using the relation

$$\prod_{r=0}^{\infty}\frac{(r + v_1)(r + v_2)}{(r + w_1)(r + w_2)} = \frac{\Gamma(w_1)\Gamma(w_2)}{\Gamma(v_1)\Gamma(v_2)}, \quad v_1 + v_2 = w_1 + w_2, \tag{D.29}$$

the product can be written in terms of Γ-functions:

$$\omega_p^2 \frac{\Gamma(2 - p\mu_1)\Gamma(2 - p\mu_2)}{\Gamma(1 - p\nu_1)\Gamma(1 - p\nu_2)}. \tag{D.30}$$

For very heavy damping, this expression is proportional to γ^4/ω_0^2, which leads to the following limiting form of the prefactor:*

$$\mathscr{A}_0^{(1)} = \left(\frac{m\omega_0 R_{\mathrm{B}}^2}{2\hbar}\right)^{1/2} \left(\frac{\gamma}{\omega_0}\right)^{7/2} \omega_0. \tag{D.31}$$

For $k \neq 0$ parity is not conserved and we need to generalize the ansatz for the eigenfunctions

$$C_n^k = (\alpha|n| + C)\mathrm{e}^{-b|n|} + d_n + (\bar{\alpha}n + \bar{C}\,\mathrm{sgn}\, n)\mathrm{e}^{-b|n|} + g_n, \tag{D.32}$$

where d_n and g_n have the same properties as before. We now insert this ansatz into the truncated form of eq. (D.3). Observing that

$$\left|n + \frac{k}{p}\right| = |n| + \mathrm{sgn}\, n\frac{k}{p} + \frac{k}{p}\delta_{n,0}, \tag{D.33}$$

we obtain

$$\begin{aligned}
&-\{|n|C\tanh b + C + \alpha\tanh b/\sinh^2 b\} - \bar{C}\{n\tanh b - \mathrm{sgn}\, n\}\\
&\quad + \frac{k}{p}\{\alpha n + C\,\mathrm{sgn}\, n + \bar{\alpha}|n| + \bar{C}(1 - \delta_{n,0}) + \delta_{n,0}C\}\tanh b\\
&\quad + \left[\left(|n| + \frac{k}{p}\mathrm{sgn}\, n + \frac{k}{p}\delta_{n,0}\right)\tanh b + 1\right]\{d_n + g_n\}\mathrm{e}^{b|n|}\\
&\quad - 2(\tanh b)\sum_{n'}(d_{n'} + g_{n'})\mathrm{e}^{b|n| - b|n-n'|}\\
&= \frac{\Lambda}{m\omega_0^2}\left\{\alpha|n| + C + \bar{\alpha}n + \bar{C}\,\mathrm{sgn}\, n + (d_n + g_n)\mathrm{e}^{b|n|}\right\}.
\end{aligned} \tag{D.34}$$

For $|n| > N$, the above equation reduces to

$$\begin{aligned}
&-\{|n|C\tanh b + C + \alpha\tanh b/\sinh^2 b\} - \bar{C}\{n\tanh b - \mathrm{sgn}\, n\}\\
&\quad + \frac{k}{p}\{\alpha n + C\,\mathrm{sgn}\, n + \bar{\alpha}|n| + \bar{C}(1 - \delta_{n,0}) + \delta_{n,0}C\}\tanh b\\
&\quad - 2(\tanh b)\sum_{n'}\{d_{n'} + \mathrm{sgn}\, n)g_{n'}\}\,\mathrm{e}^{bn'}\\
&= \frac{\Lambda}{m\omega_0^2}\{\alpha|n| + C + \bar{\alpha}n + \bar{C}\,\mathrm{sgn}\, n\}.
\end{aligned} \tag{D.35}$$

* See also eq. (33) of Larkin and Ovchinnikov (1984).

Comparing terms with the same n-dependence, we must require that

$$-C\tanh b + \frac{k}{p}(\tanh b)\bar{\alpha} = \frac{\Lambda}{m\omega_0^2}\alpha, \tag{D.36a}$$

$$-\bar{C}\tanh b + \frac{k}{p}(\tanh b)\alpha = \frac{\Lambda}{m\omega_0^2}\bar{\alpha}, \tag{D.36b}$$

$$\bar{C} + \frac{k}{p}(\tanh b)C - 2(\tanh b)\sum_{n'} g_{n'}\,\mathrm{e}^{bn'} = \frac{\Lambda}{m\omega_0^2}\bar{C}, \tag{D.36c}$$

$$-C - \alpha(\tanh b)/\sinh^2 b + \frac{k}{p}(\tanh b)\bar{C} - 2(\tanh b)\sum_{n'} d_{n'}\,\mathrm{e}^{bn'} = \frac{\Lambda}{m\omega_0^2}C. \tag{D.36d}$$

For $n = 0$ we have the following condition:

$$-C - \alpha\tanh b/\sinh^2 b + \frac{k}{p}(\tanh b)C + \left(\frac{k}{p}\tanh b + 1\right)d_0$$

$$-2(\tanh b)\sum_{n'}(d_{n'} + g_{n'})\mathrm{e}^{-b|n'|} = \frac{\Lambda}{m\omega_0^2}(C + d_0). \tag{D.36e}$$

We obtain the five eigenvalues $\tilde{\Lambda}_\nu^k$ with $|\nu| \leqslant 2$ from the above equations by putting $d_n = 0$ for $n \neq 0$ and $g_n = 0$ for all n. This leads to the following eigenvalue equation [after subtracting eq. (D.36d) from eq. (D.36e)]:

$$(M^k - \Lambda/m\omega_0^2)\begin{pmatrix}\alpha\\ C\\ d_0\\ \bar{\alpha}\\ \bar{C}\end{pmatrix} = 0, \tag{D.37}$$

where the matrix M^k (with $q^k = (k/m)\tanh b$) is given by

$$M^k = \begin{pmatrix} 0 & -\tanh b & 0 & q^k & 0\\ -\tanh b/\sinh^2 b & -1 & -2\tanh b & 0 & q^k\\ 0 & q^k & 1+q^k & 0 & -q^k\\ q^k & 0 & 0 & 0 & -\tanh b\\ 0 & q^k & 0 & 0 & 1 \end{pmatrix}. \tag{D.38}$$

Obviously, for $k = 0$ the odd subspace decouples from the even one and the product of the three lowest even eigenvalues is $(\tilde{\Lambda}_0\tilde{\Lambda}_1\tilde{\Lambda}_2/m^3\omega_0^6) = -1/\cosh^2 b$, as derived above.

For $k \neq 0$ the eigenvalue $\tilde{\Lambda}_0^k$ is nonzero and we obtain

$$\tilde{\Lambda}_{-2}^k \tilde{\Lambda}_{-1}^k \tilde{\Lambda}_0^k \tilde{\Lambda}_1^k \tilde{\Lambda}_2^k/(m\omega_0^2)^5 = \det M^k$$
$$= (q^k)^2(1 + q^k)(q^k - \tanh b)^2. \tag{D.39}$$

We next consider the case $|\nu| - 2 = N \geqslant 1$. By virtue of eqs. (D.36a–d), the equations for $n = \pm N$ reduce to

$$\left[\left(N + \frac{k}{p}\right)\tanh b + 1\right](d_N + g_N) = \frac{\Lambda}{m\omega_0^2}(d_N + g_N),$$
$$\left[\left(N - \frac{k}{p}\right)\tanh b + 1\right](d_N - g_N) = \frac{\Lambda}{m\omega_0^2}(d_N - g_N). \tag{D.40}$$

We conclude that the eigenvalues are

$$\tilde{\Lambda}_\nu^k = m\omega_0^2\left[\left(\left|\nu + \frac{k}{p}\right| - 2\right)\tanh b + 1\right]. \tag{D.41}$$

As for $k = 0$ we take account of the finite mass by putting

$$\tilde{\Lambda}_\nu^k = m\tilde{L}(|\omega_{p\nu+k}|), \tag{D.42}$$

where

$$\tilde{L}(\omega) = \omega^2 + \gamma(\omega - 2\omega_p) + \omega_0^2. \tag{D.43}$$

We also define

$$L^0(\omega) = \omega^2 + \gamma\omega + \omega_0^2. \tag{D.44}$$

We conclude that the prefactor of an orbit of multiplicity p is similar to eq. (D.26) with an extra product with respect to the Bloch index k and $\mathscr{S}_0$ replaced by $p\mathscr{S}_0$. Thus,

$$\mathscr{A}^{(p)} = \left(\frac{p\mathscr{S}_0([r_1(\tau)];\tau_p)}{2\pi\hbar}\right)^{1/2} \left|\frac{\prod_\nu \prod_{k=0}^{p-1} L^0(|\omega_{p\nu+k}|)}{\prod'_\nu \prod_{k=0}^{p-1} \tilde{L}(|\omega_{p\nu+k}|)}\right|^{1/2}, \tag{D.45}$$

where the prime means that in the product of the denominator the eigenvalue $\tilde{\Lambda}_0^{k=0}$ should be omitted.

Consider now the product

$$\prod_\nu \prod_{k=0}^{p-1} L(|\omega_{p\nu+k}|) = \prod_r L(|\omega_r|) = \frac{1}{L(0)}\left\{\prod_{r=0}^{\infty} L(|\omega_r|)\right\}^2. \tag{D.46}$$

We then write the prefactor in the form

$$\mathscr{A}^{(p)} = \mathscr{A}_1^{(p)}\mathscr{A}_2^{(p)}, \tag{D.47}$$

where

$$\mathscr{A}_1^{(p)} = \left|\frac{\tilde{L}(0)}{L^0(0)}\right|^{1/2} \left|\prod_{r=0}^{\infty} \frac{L^0(|\omega_r|)}{\tilde{L}(|\omega_r|)}\right|, \tag{D.48}$$

and

$$\mathscr{A}_2^{(p)} = \left(\frac{p\mathscr{S}_0(\tau_p)m^4}{2\pi\hbar\Lambda_{-2}^0\Lambda_{-1}^0\Lambda_1^0\Lambda_2^0}\right)^{1/2} \frac{|\prod_{k=0}^{p-1}\prod_{\nu=-2}^{2}\tilde{L}(|\omega_{p\nu+k}|)|^{1/2}}{\prod_{k=1}^{p-1}q^k|q^k - \tanh b|(1+q^k)^{1/2}\omega_0^5}. \tag{D.49}$$

We can, by introducing $\mu_{1,2}$ and $\nu_{1,2}$ as defined in eq. (D.28), write $\mathscr{A}_1^{(p)}$ as follows:

$$\mathscr{A}_1^{(p)} = [1 - 2\tanh b]^{1/2} \left|\frac{\Gamma(-\mu_1)\Gamma(-\mu_2)}{\Gamma(-\nu_1)\Gamma(-\nu_2)}\right|. \tag{D.50}$$

Using the results (D.25) and (D.41) for $\mathscr{S}_0$ and for $\tilde{\Lambda}_\nu^0$, we may write $\mathscr{A}_2^{(p)}$ as

$$\mathscr{A}_2^{(p)} = \left(\frac{mpR_B^2}{2\hbar\gamma}\right)^{1/2} \frac{\omega_1^{5p}}{\omega_0^{5p-2}} \frac{|\prod_{k=0}^{p-1}\prod_{\nu=-2}^{2}[|p\nu+k| - \mu_1][|p\nu+k| - \mu_2]|^{1/2}}{\prod_{k=1}^{p-1}p^{-5/2}(\tanh^2 b)(p-k)k(k\tanh b + p)^{1/2}}. \tag{D.51}$$

Since the five smallest eigenvalues $\tilde{\Lambda}_\nu^k$ were calculated for zero mass, we must, for the sake of consistency, in the expression for $\mathscr{A}_2^{(p)}$ consider $-\mu_2 \to \gamma/\omega_1$ to be very large, leading to

$$\prod_{k=0}^{p-1}\prod_{\nu=-2}^{2}[|p\nu+k| - \mu_2] \Rightarrow (\gamma/\omega_1)^{5p/2}. \tag{D.52}$$

In the same limit we have

$$\mu_1 = 2p - \frac{\omega_0^2}{\gamma\omega_1} = p(2 - \coth b). \tag{D.53}$$

After some algebraic manipulations, we obtain

$$\mathscr{A}_2^{(p)} = \left(\frac{mp}{2\hbar\gamma}\right)^{1/2} R_B \frac{\omega_0^2\tanh b}{p^2\Gamma^2(p)} \left|\frac{\Gamma(2p - p\coth b)\Gamma(p\coth b)}{\Gamma(-p\coth b)\Gamma(-2p + p\coth b)}\right|^{1/2}. \tag{D.54}$$

Let us now introduce the following abbreviations:

$$y = T_B/T = p\coth b, \qquad \rho = \gamma^2/\omega_0^2. \tag{D.55}$$

In the limit of strong damping,

$$\mu_1 = 2p - y, \qquad -\mu_2 = 2p + (\rho - 1)y,$$
$$\nu_1 = -y, \qquad -\nu_2 = (\rho - 1)y; \tag{D.56}$$

after some manipulations, we obtain for the prefactor

$$\mathscr{A}^{(p)} = \left(\frac{m}{2\hbar\gamma}\right)^{1/2} \omega_0^2 R_{\mathrm{B}} \frac{p^{1/2}}{y^2 \Gamma^2(p)} \left| \frac{\Gamma(1 + \rho y + 2p - y)}{\Gamma(1 + \rho y - y)} \right|. \tag{D.57}$$

Eventually, we introduce

$$2\sigma = 1 + \rho y - y = 1 + \frac{T_{\mathrm{B}}}{T}\left(\frac{\gamma^2}{\omega_0^2} - 1\right); \tag{D.58}$$

whence we may write

$$\mathscr{A}^{(p)}(\tau_1) = \left(\frac{9\pi v_{\mathrm{B}}}{\hbar\tau_{\mathrm{B}}}\right)^{1/2} \left(\frac{\tau_{\mathrm{B}}}{\tau_1}\right)^2 \frac{p^{1/2}}{\Gamma^2(p)} \frac{\Gamma(2p + 2\sigma)}{\Gamma(2\sigma)}, \tag{D.59}$$

as quoted in eq. (4.69) of the text.

Appendix E. Calculations of the decay rate of a heavily damped object

For orientation, we first review some characteristic features of the multiple orbit summation problem already considered in section 2.6 and later in section 4.3. We have

$$\bar{\Gamma} = \sum_{p=1}^{\infty} (-1)^{p+1} A^{(p)}(\tau_1) \exp\left[-\frac{1}{\hbar} \mathscr{S}^{(p)}(\tau_1) \right]. \tag{E.1}$$

However, we do not wish to define $\bar{\Gamma}_p(E)$ directly as done in eq. (2.66). Rather, we prefer to consider in addition to eq. (E.1) a representation of the form

$$\bar{\Gamma} = \sum_{p=1}^{\infty} (-1)^{p+1} \int_0^{\infty} \frac{\mathrm{d}E}{2\pi} \tilde{A}^{(p)}(E; \tau_1) \exp\left\{ -\frac{1}{\hbar} [E\tau_1 + \mathscr{W}^{(p)}(E)] \right\}, \tag{E.2}$$

where $\tilde{A}^{(p)}(E; \tau_1)$ has to be so chosen that a steepest descent evaluation of the energy integral leads to a term-by-term agreement of both series (E.1) and (E.2). We will see that there is some arbitrariness in the energy and time dependence of $\tilde{A}^{(p)}(E; \tau_1)$, and we may exploit this ambiguity in order to obtain an expression for $\tilde{A}^{(p)}$ which is only weakly dependent on energy. This seems to be a reasonable choice in a steepest descent approximation, where only E-values near the minimum of $[E\tau_1 + \mathscr{W}^{(p)}(E)]$ are supposed to contribute.

For orientation, we first note that

$$\mathscr{S}^{(p)}(\tau_1) = p\mathscr{S}(\tau_p), \qquad \tau_p = \tau_1/p, \qquad \mathscr{W}^{(p)}(E) = p\mathscr{W}(E) \tag{E.3}$$

are pairs of Legendre transforms

$$\frac{\partial \mathscr{S}(\tau)}{\partial \tau} = E(\tau), \qquad \frac{\partial \mathscr{W}}{\partial E} = -\tau(E) \equiv -\tau_E. \tag{E.4}$$

For illustration, we have for a heavily damped object [see eqs. (4.64) and (4.65)]

$$\mathscr{S}(\tau) = v_{\mathrm{B}}\tau_{\mathrm{B}}[\tfrac{3}{2} - (\tau_{\mathrm{B}}/\tau)^2],$$
$$\mathscr{W}(E) = \tfrac{3}{2}v_{\mathrm{B}}\tau_{\mathrm{B}}[1 - (E/v_{\mathrm{B}})^{2/3}],$$
$$\tau_{\mathrm{E}} = \tau_{\mathrm{B}}(E/v_{\mathrm{B}})^{-1/3}. \tag{E.5}$$

Evaluating the integral in eq. (E.2) by steepest descent, we find that the energy is implicity given by

$$\tau_p = \tau_1/p = \tau_E \tag{E.6}$$

and that

$$\tilde{A}^{(p)}(E;\tau_1) = \left(-2\pi\hbar p\frac{\partial^2 \mathscr{W}}{\partial E^2}\right)^{1/2} A^{(p)}(\tau_1). \tag{E.7}$$

It is possible to eliminate the energy dependence on the right-hand side of the relation above by means of eqs. (E.5) and (E.6). Thus, one arrives at the expression (4.73).

In a heuristic way, let us now eliminate all the p-dependence in eq. (E.7). Thus, we put $p = \tau_1/\tau_E$ and obtain ($\rho = \gamma^2/\omega_0^2$)

$$\tilde{\tilde{A}}^{(p)}(E) = \frac{\tau_E}{\tau_1\Gamma^2(\tau_1/\tau_E)}\,\frac{\Gamma(1 + 2\tau_1/\tau_E + (\rho-1)\tau_1/\tau_{\mathrm{B}})}{\Gamma(1 + (\rho-1)\tau_1/\tau_{\mathrm{B}})}. \tag{E.8}$$

For temperatures close to the crossover temperature $\tau_1 \sim \tau_{\mathrm{B}}$, we expect that energies $E \sim v_{\mathrm{B}}$ are of importance. Thus, we put $\tau_E = \tau_1 = \tau_{\mathrm{B}}$ and obtain

$$\tilde{\tilde{A}}^{(p)}(E \approx v_{\mathrm{B}}) = \frac{\Gamma(2+\rho)}{\Gamma(\rho)}. \tag{E.9}$$

Now, the series (E.2) can easily be summed over. For the sake of consistency, we also replace $\mathscr{W}(E)$ by the quadratic approximation $\mathscr{W}_2(E)$ as shown in eq. (2.77), where for the present case

$$\left(\frac{1}{kT_2}\right)^2 = \frac{2}{3}\frac{\tau_{\mathrm{B}}}{\hbar v_{\mathrm{B}}}. \tag{E.10}$$

Proceeding further according to Affleck's prescription of eq. (2.76), we arrive at eq. (4.76); whence we obtain [cf. eq. (2.79)]

$$\bar{\Gamma}^{\mathrm{A}} = \frac{kT_2}{2\pi\hbar}\frac{\Gamma(2+\rho)}{\Gamma(\rho)}\pi^{1/2}\operatorname{erfc}\left(\frac{T_2}{T_{\mathrm{B}}} - \frac{T_2}{T}\right)\exp\left[-\frac{v_{\mathrm{B}}}{T} + \frac{1}{2}\left(\frac{T_2}{T_{\mathrm{B}}} - \frac{T_2}{T}\right)^2\right], \tag{E.11}$$

which agrees essentially with eq. (43) of Larkin and Ovchinnikov (1984), and eq. (46) of Larkin and Ovchinnikov (1992). Note, however, that the strong energy dependence in eq. (E.8) casts some doubts on the above procedure.

Therefore, we wish to draw attention to the fact that there is a direct way to calculate $\tilde{A}^{(p)}$ for $E \sim v_{\mathrm{B}}$. In order to understand the reasoning, we should recall that the presentation in eq. (E.1) requires different primitive traversal times τ_p for the different periodic orbits of multiplicity p, and that such a flexibility is given only in the case of a nonlinear potential. On the other hand, the presentation in eq. (E.2), requires only data of periodic orbits at a given energy, which can also be obtained in the limit of a harmonic potential. Thus, for $E \sim v_{\mathrm{B}}$ and $\boldsymbol{R} \sim \boldsymbol{R}_{\mathrm{B}}$, we may represent $V(\boldsymbol{R})$ by a quadratic form which is based on the matrix

$$U_{kk'}^{(\mathrm{B})} = \frac{1}{m} \frac{\partial^2 V(\boldsymbol{R})}{\partial R_k \partial R_{k'}}\bigg|_{\boldsymbol{R}=\boldsymbol{R}_{\mathrm{B}}}. \tag{E.12}$$

Then the problem is separable, and putting $\hat{\omega}_{\mathrm{B}}^2 = \hat{U}^{(\mathrm{B})}$, we may single out the unstable mode $\omega_{\mathrm{B}_1}^2 < 0$ from the remaining ones. Then we obtain for the relevant part of the action

$$\frac{1}{\hbar}\mathscr{W}(E) = -\frac{1}{kT_{\mathrm{B}}}(E - v_{\mathrm{B}}), \tag{E.13}$$

and for the prefactor

$$\tilde{A}(\tau_1) = \frac{1}{2\pi\hbar}\sinh(\omega_{\mathrm{B}_1}\tau_1/2)\frac{\det(\sinh\hat{\omega}_0\tau_1/2)}{\det(\sinh\hat{\omega}_{\mathrm{B}}\tau_1/2)}. \tag{E.14}$$

Turning now our attention to the Caldeira–Leggett model (1981, 1983, 1984), we find $\omega_{\mathrm{B}_1} = \mathrm{i}\lambda_1\omega_1$ $(\omega_1 = 2\pi kT/\hbar)$, where

$$\lambda_{1,2} = (\omega_1)^{-1}(-\tfrac{1}{2}\gamma \pm [\tfrac{1}{4}\gamma^2 + \omega_0^2]). \tag{E.15}$$

Concerning the stable modes, we need

$$\Lambda_{1,2} = (\omega_1)^{-1}(-\tfrac{1}{2}\gamma \pm [\tfrac{1}{4}\gamma^2 - \omega_0^2]^{1/2}). \tag{E.16}$$

Thus, we obtain

$$\begin{aligned}\tilde{A}(\tau_1) &= \frac{\sin\lambda_1}{2\pi\hbar}\left|\prod_{n=-\infty}^{\infty}\frac{[1-\Lambda_1/n][1-\Lambda_2/n]}{[1-\lambda_1/n][1-\lambda_2/n]}\right| \\ &= \frac{\sin\lambda_1}{2\pi\hbar}\frac{\Gamma(1-\lambda_1)\Gamma(1-\lambda_2)}{\Gamma(1-\Lambda_1)\Gamma(1-\Lambda_2)}.\end{aligned} \tag{E.17}$$

We observe that (i) $\tilde{A}$ depends only on τ_1, and neither on the multiplicity nor on the energy, as to be expected for a harmonic potential, (ii) in the heavy damping limit, $\tilde{A}(\tau_{\mathrm{B}}) \rightarrow \tilde{\tilde{A}}(v_{\mathrm{B}})$, as given by eq. (E.8) and (iii) that $\tilde{A}(\tau_1)$ together with

$\mathscr{W}_1(E)$ leads to the well-known expression

$$\bar{\Gamma} = \frac{kT_\mathrm{B}}{\hbar} \frac{\Gamma(1-\lambda_1)\Gamma(1-\lambda_2)}{\Gamma(1-\Lambda_1)\Gamma(1-\Lambda_2)} \mathrm{e}^{-v_\mathrm{B}/kT}. \tag{E.18}$$

This expression agrees with eq. (47) of Larkin and Ovchinnikov (1984).

Note also that $\bar{\Gamma}$ of eq. (E.18) can be obtained from the imaginary part of the free energy calculated from the trivial orbit $\boldsymbol{R} = \boldsymbol{R}_\mathrm{B}$ if an additional factor T_B/T is inserted as shown in eq. (2.74).

Since the Γ-functions depend very sensitively on the arguments, the approximations in the transition from eq. (E.8) to eq. (E.9) may not be so good. Thus, we have introduced in section 4.3 the function $\phi(z)$ according to eq. (4.75), which, in terms of the variable

$$t = (1 + 4z)^{-1/2}, \tag{E.19}$$

can be expressed through the Legendre function (Abramowitz and Stegun 1968) as follows:

$$\phi(z) = -\sigma \frac{2\sigma + 1}{4\sigma + 1} t^{2\sigma} [P_{2\sigma+1}(t) - P_{2\sigma-1}(t)]. \tag{E.20}$$

Using standard numerical routines for the Legendre function, Ludviksson (1989) has integrated eq. (4.74) numerically, with some of the results given in Table 1.

References

Abramowitz, M., and I.A. Stegun, 1968, Handbook of Mathematical Functions (U.S. Government Printing Office, Washington, DC).

Affleck, I., 1981, Phys. Rev. Lett. **46**, 388.

Banks, T., and C.M. Bender, 1973, Phys. Rev. D **8**, 3366.

Banks, T., C.M. Bender and T.T. Wu, 1973, Phys. Rev. D **8**, 3346.

Bender, C.M., and S.A. Orszag, 1978, Advanced Mathematical Methods for Scientists and Engineers (McGraw-Hill, New York).

Caldeira, A.O., and A.J. Leggett, 1981, Phys. Rev. Lett. **46**, 211.

Caldeira, A.O., and A.J. Leggett, 1983, Ann. Phys. (New York) **149**, 374.

Caldeira, A.O., and A.J. Leggett, 1984, Ann. Phys. (New York) **153**, 445(E).

Chang, L.-D., and S. Chakravarty, 1984a, Phys. Rev. B **29**, 130.

Chang, L.-D., and S. Chakravarty, 1984b, Phys. Rev. B **30**, 1566.

Coleman, S., 1979, in: The Whys of Subnuclear Physics, ed. A. Zichichi (Plenum Press, New York) p. 805.

Coleman, S., 1985, The Uses of Instantons, in: Aspects of Symmetry (Cambridge University Press, Cambridge, UK) ch. 7.

Courant, R., and D. Hilbert, 1962, Methods of Mathematical Physics, Vol. II (Interscience, New York).

Dashen, R.F., B. Hasslacher and A. Neveu, 1974, Phys. Rev. D **10**, 4114.

de Vega, H.J., J.L. Gervais and B. Sakita, 1978, Nucl. Phys. B **139**, 20.
Feynman, R.P., and A.R Hibbs, 1965, Quantum Mechanics and Path Integrals (McGraw-Hill, New York).
Gervais, J.L., and B. Sakita, 1977, Phys. Rev. D **16**, 3507.
Grabert, H., and U. Weiss, 1984a, Z. Phys. B **56**, 171.
Grabert, H., and U. Weiss, 1984b, Phys. Rev. Lett. **53**, 1787.
Grabert, H., P. Olschowski and U. Weiss, 1987, Phys. Rev. B **36**, 1931.
Hänggi, P., and W. Hontscha, 1988, J. Chem. Phys. **88**, 4094.
Hänggi, P., P. Talkner and M. Borkovec, 1990, Rev. Mod. Phys. **62**, 251.
Kapur, P.L., and R. Peierls, 1937, Proc. R. Soc. (London) A **163**, 606.
Kramers, H.A., 1940, Physica **7**, 284.
Landau, L.D., and E.M. Lifshitz, 1975, Quantum Mechanics, 3rd Ed. (Pergamon Press, Oxford).
Landauer, R., 1950, Phase Integral Approximations in Wave Mechanics (Thesis, Harvard University, USA).
Langer, J.S., 1967, Ann. Phys. (New York) **41**, 108.
Larkin, A.I., and Yu.N. Ovchinnikov, 1983a, Pis'ma Zh. Eksp. & Teor. Fiz. **37**, 322 [JETP Lett. **37**, 382].
Larkin, A.I., and Yu.N. Ovchinnikov, 1983b, Zh. Eksp. & Teor. Fiz. **85**, 1510 [Sov. Phys.-JETP **57**, 876].
Larkin, A.I., and Yu.N. Ovchinnikov, 1984, Zh. Eksp. & Teor. Fiz. **86**, 719 [Sov. Phys.-JETP **59**, 420].
Larkin, A.I., and Yu.N. Ovchinnikov, 1992, ch. 4, this volume.
Leggett, A.J., 1978, J. Phys. C **6**, 1264.
Leggett, A.J., 1980, Prog. Theor. Phys. Suppl. **69**, 80.
Lifshitz, I.M., and Yu. Kagan, 1972, Zh. Eksp. & Teor. Fiz. **62**, 385 [Sov. Phys.-JETP **35**, 206].
Ludviksson, A., 1987, J. Phys. A **20**, 4733.
Ludviksson, A., 1989, Quantum Tunneling at Finite Temperatures and Complex Classical Paths. Thesis, Karlsruhe, Germany.
Mel'nikov, V.I., 1991, Phys. Rep. **209**, 1.
Miller, W.H., 1975, J. Chem. Phys. **62**, 1899.
Schmid, A., 1986, Ann. Phys. (New York) **170**, 333.
Schulman, L.S., 1981, Techniques and Applications of Path Integration (Wiley-Interscience, New York).
Waxman, D., and A.J. Leggett, 1985, Phys. Rev. B **32**, 4450.
Whitham, G.B., 1974, Linear and Nonlinear Waves (Wiley, New York).
Wigner, E., 1932, Z. Phys. Chem. B **19**, 203.
Wolynes, P.G., 1981, Phys. Rev. Lett. **47**, 968.

CHAPTER 4

Dissipative Quantum Mechanics of Josephson Junctions

A.I. LARKIN and Yu.N. OVCHINNIKOV

L.D. Landau Institute for Theoretical Physics
Academy of Sciences
Moscow, Russia

Quantum Tunnelling in Condensed Media
Edited by
Yu. Kagan and A.J. Leggett

Contents

1. Introduction

The Josephson effect, like superconductivity, represents a macroscopic quantum phenomenon. This phenomenon is due to the fact that a macroscopically large number of Cooper pairs exist in a single quantum state. The wave function of this state is the order parameter Δ. The value of the current through the Josephson junction is determined by the phase difference 2φ of the order parameters in the two superconductors. Usually, the order parameter Δ and its phase 2φ are considered as classical quantities. In this case the current I through the junction is given by

$$I = I_{\mathrm{cr}} \sin(2\varphi) + \frac{V}{R} + C\frac{\mathrm{d}V}{\mathrm{d}t}, \tag{1.1}$$

where C is the junction capacitance, R is an ohmic resistance, and V is the voltage imposed on the junction, related to the phase difference 2φ by the Josephson relation

$$V = \frac{\hbar}{e}\frac{\partial\varphi}{\partial t}. \tag{1.2}$$

Equation (1.1) looks like the equation of motion of a particle with mass $m = \hbar^2 C/e^2$ and viscosity $\eta = \hbar^2/Re^2$ in the potential field

$$U(\varphi) = -\left\{\frac{\hbar I}{e}\varphi + \frac{\hbar I_{\mathrm{cr}}}{2e}\cos(2\varphi)\right\}. \tag{1.3}$$

This equation has the form

$$m\frac{\partial^2\varphi}{\partial t^2} + \eta\frac{\partial\varphi}{\partial t} + \frac{\partial U(\varphi)}{\partial\varphi} = 0. \tag{1.4}$$

Recently, there have been obtained junctions with a small capacitance, where quantum properties of the order parameter are observed. Quantum tunnelling through the potential barrier is one manifestation of these properties. At zero viscosity the tunnelling process can be considered by the usual quantum-mechanical method. A study of the influence of the viscosity on the tunnelling probability is of special interest.

The physical reason for the viscosity is the interaction of the relevant degree of freedom φ with a thermal bath. In the case considered the thermal bath consists of single-particle excitations in the superconductors. The tunnelling probability depends on the thermal bath temperature. At low temperatures there occurs underbarrier tunnelling, while at high temperatures activated overbarrier transitions are more probable.

Quantum tunnelling of a particle interacting with a thermal bath is essential in many phenomena, e.g., low-temperature chemical reactions, fission of atomic nuclei, quantum growth of the nucleus of a new phase, motion of a heavy particle in a solid, quantum creep of vortices and dislocations. For an experimental and theoretical investigation of this phenomenon the Josephson effect is convenient, since the form of the potential energy is well known, and such parameters as mass and viscosity can be found from independent experiments.

The first study of the quantum kinetics of a phase transition at zero temperature was made by Lifshitz and Kagan (1972).

Progress in the study of the process of quantum tunnelling of a particle interacting with a thermal bath was made by Caldeira and Leggett (1981). Then, a functional integration method was used to investigate this process. In this method an important quantity is the effective action. In section 2 we will obtain the expression for the decay probability of a metastable state in the one-dimensional case. In section 3 the results of section 2 will be generalized for the multidimensional case and the lifetime of a metastable state against decay into a continuous spectrum will be found via the imaginary part of the partition function. In section 4 the expression for the effective action of the tunnelling junction is obtained. In section 5 a narrow temperature region is investigated, in the vicinity of which the classical high-temperature regime of overbarrier transitions is replaced by the quantum regime of underbarrier tunnelling. In section 6 the case of high viscosities, which has been most thoroughly investigated, is described. In section 7 the limiting case of low viscosities is studied. In section 8 we study the influence of level quantization on the lifetime of a metastable state. In section 9 the resonance reduction of the lifetime of the metastable state under the action of an external high-frequency current is investigated. Finally, concluding remarks are made in section 10.

2. Josephson junction with a large resistance

One can neglect dissipation in the limit of large resistance of the junction. In this case the Josephson junction can be considered as a particle moving in the field of the potential forces given by eq. (1.3) (see fig. 1). Quantization can be performed in the standard way. In practically all cases the junction capacitance is large enough that the motion in the potential (1.3) can be regarded as quasiclassical. In such a potential there exist quasistationary energy levels E_n with width $\frac{1}{2}\gamma_n$,

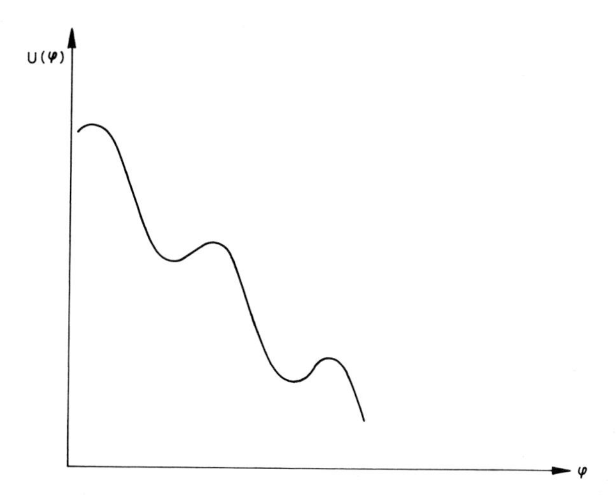

Fig. 1. Dependence of the potential $U(\varphi)$ on the coordinate φ.

where n is the level number. For a sufficiently large current through the junction the potential slope is large, and a particle which escapes from a potential well goes over the next hump. We shall restrict ourselves to the consideration of this case only, since in the majority of the experimental papers the junctions have been investigated at a current I close to the critical one.

With quasiclassical accuracy, the line width can be written as

$$\gamma_n = A_n \exp(-2S(E_n)/\hbar), \tag{2.1}$$

where $S(E_n)$ is the value of the action when the particle moves under the barrier with energy E_n:

$$\begin{aligned} S(E) &= \int [2m(U(\varphi) - E)]^{1/2}\, \mathrm{d}\varphi \\ &= \left(\frac{2mI_{\mathrm{cr}}\hbar}{e}\right)^{1/2} \int \mathrm{d}\varphi \left\{ -\frac{I}{I_{\mathrm{cr}}}\varphi + \frac{I}{2I_{\mathrm{cr}}} \arcsin(I/I_{\mathrm{cr}}) \right. \\ &\quad \left. - \frac{1}{2}\cos(2\varphi) + \frac{1}{2}[1 - (I/I_{\mathrm{cr}})^2]^{1/2} - \frac{eE}{\hbar I_{\mathrm{cr}}} \right\}^{1/2}. \end{aligned} \tag{2.2}$$

The integral in eq. (2.2) is taken between the turning points where $U(\varphi) = E$. For $n \gg 1$ the motion is quasiclassical not only under the barrier, but also in the classically accessible region as well. In this case the pre-exponential factor A_n is equal to the number of particle collisions with the barrier per unit time:

$$A_n = \frac{\omega(E_n)}{2\pi}, \tag{2.3}$$

where $\omega(E)$ is the frequency of classical oscillations of a particle with the energy E (Landau and Lifshitz 1974).

For numbers n of the order of unity the motion in the classically accessible region is not quasiclassical, but the potential is close to a parabolic one. In this case the coefficients A_n can be found by matching the quasiclassical wave function under the barrier to that of a parabolic cylinder (see Erdelyi et al. 1953). As a result, we get

$$A_n = \frac{\omega_0}{(2\pi)^{1/2}} \exp\{(n+\tfrac{1}{2})[\ln(n+\tfrac{1}{2}) - 1]\}/\Gamma(n+1),$$

$$\omega_0 = \omega_j[1-(I/I_{cr})^2]^{1/4}, \tag{2.4}$$

where $\Gamma(n)$ is the Euler gamma function, ω_0 is the frequency of small oscillations in the vicinity of the bottom of the potential well, and $\omega_J = (2I_{cr}\hbar/em)^{1/2}$ is the plasma frequency of the Josephson junction.

At values of the energy E much smaller than the height of the potential barrier U, the dependence of the exponent on energy E is given by

$$S(E) = S(0) - \frac{E}{2(1-(I/I_{cr})^2)^{1/4}}\left(\frac{me}{2\hbar I_{cr}}\right)^{1/2} \times\left[D + \ln\left\{\frac{144 I_{cr}\hbar}{eE}[1-(I/I_{cr})^2]^{3/2}\right\}\right] \tag{2.5}$$

The quantities $S(0)$ and D are functions of I/I_{cr} and for currents close to critical are equal to

$$S(0) = \frac{3}{5}\left(\frac{2mI_{cr}\hbar}{e}\right)^{1/2}[1-(I/I_{cr})^2]^{5/4},$$

$$D(1) = 1.$$

The energy E in eq. (2.5) is measured from the bottom of the potential well. If the viscosity is not too small, thermal equilibrium in the system is established before the metastable state decays. Following Affleck (1981), we shall find the dependence of the lifetime of the metastable state on temperature.

The lifetime of the system Γ^{-1} can be found by averaging the lifetime of the system in the state with energy E with the Boltzmann distribution function:

$$\Gamma = \sum_{n=0} \gamma(E_n)N(E_n) \Big/ \sum_{n=0} N(E_n), \tag{2.6}$$

where $N(E)$ is a distribution function. In thermodynamic equilibrium $N(E) = \exp(-E/kT)$. At low temperatures, the partition function Z is equal to

$$Z = \sum_n \exp(-E_n/kT) = \frac{1}{2\sinh(\omega_0\hbar/2kT)}. \tag{2.7}$$

A wide temperature region ($kT < \hbar\omega_0/2\pi$) exists, where the decay with the largest probability occurs through the states with energy rather far both from the bottom and top of the potential well. This energy is found from the minimization condition on the exponent B of eq. (2.6):

$$B = \frac{E}{kT} + 2S(E)/\hbar. \tag{2.8}$$

The extremum condition yields

$$\frac{\partial S}{\partial E} = -\frac{\hbar}{2kT}. \tag{2.9}$$

Condition (2.9) implies that the period of the particle motion in the inverted potential is equal to $\hbar/kT$.

Replacing the first of the sums in eq. (2.6) by an integral and calculating it by the saddle-point method, we get

$$\Gamma = \frac{\sinh(\omega_0 \hbar/2kT)}{(\pi\hbar(\partial^2 S/\partial E^2))^{1/2}} \exp(-B). \tag{2.10}$$

The value of the energy E in eq. (2.10) is determined from eq. (2.9). At low temperatures ($T \ll \hbar\omega_0/2\pi$), the lifetime is determined by the level width for $n = 0$ (the ground state). In this case the temperature corrections are exponentially small.

Equation (2.10) for the lifetime of a metastable state does not hold at high temperatures either; in this regime the important energy E proves to be in the vicinity of the barrier top U. For such energies

$$S(E) = \frac{\pi(U-E)}{\Omega} + \frac{1}{4}(U-E)^2|T'|,$$

$$U = \frac{\hbar I_{\rm cr}}{e}\left\{\frac{I}{I_{\rm cr}} \arcsin(I/I_{\rm cr}) + [1-(I/I_{\rm cr})^2]^{1/2} - \frac{\pi I}{2I_{\rm cr}}\right\}. \tag{2.11}$$

Here Ω is a frequency of small oscillations in the inverted potential, to be determined by eq. (2.4) for the potential (1.3). For the potential (1.3) the value T' is equal to

$$|T'| = \frac{\pi e}{2\hbar\Omega I_{\rm cr}} \frac{1 + \frac{5}{3}(I/I_{\rm cr})^2/[1-(I/I_{\rm cr})^2]}{[1-(I/I_{\rm cr})^2]^{1/2}}. \tag{2.12}$$

At temperatures T close to $kT_0 = \hbar\Omega/2\pi$, the important energy values E are close to U. For the energy $E > U$ eq. (2.1) is inapplicable. The contribution of this region of energies to Γ is small. The main contribution occurs from a wide region where one can use eq. (2.11) for the value $E < U$. As a result, from eqs.

(2.6), (2.7) and (2.11) we obtain

$$\Gamma = \frac{1}{2Z(2\pi\hbar)^{1/2}} \exp\left[-\frac{U}{kT} + \frac{\hbar}{2|T'|}\left(\frac{2\pi}{\hbar\Omega} - \frac{1}{kT}\right)^2 \right] \times \left\{ 1 + \phi\left[\left(\frac{\hbar}{2|T'|}\right)^{1/2} \left(\frac{1}{kT} - \frac{2\pi}{\hbar\Omega}\right)\right]\right\}, \tag{2.13a}$$

where

$$\phi(x) = \frac{2}{\pi^{1/2}} \int_0^x dt \exp(-t^2) . \tag{2.13b}$$

It is convenient to represent Γ as

$$\Gamma = A \exp(-B). \tag{2.14}$$

In a narrow temperature region of width $\delta T/T_0 \sim kT_0(|T'|/\hbar)^{1/2}$, there occurs a second-order transition. At $T > T_0$ the quantity $B(T)$ is equal to its classical value (the Arrhenius law)

$$B(T) = U/kT, \tag{2.15}$$

while at $T < T_0$ the quantity $B(T)$ is defined by eqs. (2.8) and (2.9) and near T_0 it is equal to

$$B(T) = \frac{U}{kT} - \frac{\hbar}{2|T'|}\left(\frac{1}{kT} - \frac{1}{kT_0}\right)^2 . \tag{2.16}$$

For flat potentials [e.g., for a potential of the form (1.3)], eq. (2.9) has a solution for $T < T_0$, and at $T = T_0$ there occurs a second-order phase transition. For potentials with abrupt walls (e.g., for a rectangular barrier) eq. (2.9) has no solution. In this case the expression for Γ has the form

$$\Gamma \propto \exp(-U/T) + \exp(-2S(0)/\hbar) \tag{2.17}$$

and on raising the temperature there occurs a first-order transition.

The qualitative picture of the transition from a classical overbarrier decay at high temperatures to quantum tunnelling at low temperatures is also preserved when dissipation is present. However, the quantitative results given above hold only in the region of "moderate" viscosity values. For large values of viscosity the probability of quantum tunnelling changes markedly, while for low viscosity thermal equilibrium is not established in the important region of energies near the barrier top.

The pre-exponential factor and the quantitative description of transition from a classical picture of decay to a quantum tunnelling considered in this section are valid only provided that the potential well $U(\varphi)$ contains a great number of levels. At the end of section 8, it will be shown that it is difficult to fulfil this condition experimentally.

3. *Tunnelling probability in the multidimensional case*

Equation (2.1) defines the level width in the one-dimensional case. The interaction of the degree of freedom φ considered with a thermal bath of single-particle excitations makes the problem a multidimensional one. In order to find the lifetime of a metastable state, it is convenient here to use the functional integration method, which holds both in the one-dimensional and multidimensional cases. Equation (2.6) holds in the multidimensional case as well; here E_n are the real levels of the multidimensional problem, and $\gamma(E_n) = -2\mathrm{Im}\, E_n$ are the widths of these levels (Larkin et al. 1984). At sufficiently low temperatures, the important values of the energy E are smaller than the barrier height U. In this case eq. (2.6) can be written as

$$\Gamma = \frac{2kT}{\hbar}\,\mathrm{Im}\sum_n \exp(-E_n/kT)/Z = \frac{2kT}{\hbar}\,\mathrm{Im}\ln Z = -\frac{2k}{\hbar}\,\mathrm{Im}\,F\,, \tag{3.1}$$

where F is the free energy and Z is a partition function,

$$Z = Z_0 + \mathrm{i}Z_1 = \int D\varphi(\tau)\exp\{-B[\varphi]\}. \tag{3.2}$$

The functional integral in eq. (3.2) is taken over all periodic functions $\varphi(\tau)$ with the period $\hbar/kT$. In what follows we give an explicit form for the effective action $B[\varphi]$, taking the interaction of the "particle" with the thermal bath into account.

In order to calculate the quantity Im Z, we use the method given by Langer (1967) and Callan and Coleman (1977). There exists a function $\tilde{\varphi}(\tau)$ where the action takes an extreme value. The function $\tilde{\varphi}(\tau)$ satisfies the equation

$$\frac{\delta B[\varphi]}{\delta\varphi} = 0\,. \tag{3.3}$$

We assume that trajectories close to the extremal one make the main contribution to the expression for the quantity Γ [eq. 3.1]. In the vicinity of the extremal trajectory the function $\varphi(\tau)$ can be represented as

$$\varphi(\tau) = \tilde{\varphi}(\tau) + \sum_n C_n\varphi_n(\tau), \tag{3.4}$$

where $\varphi_n(\tau)$ are the normalized eigenfunctions of the operator $\delta^2 B/\delta\varphi^2$, i.e.,

$$\left(\frac{\delta^2 B}{\delta\varphi^2}\right)\varphi_n = \Lambda_n\varphi_n\,, \tag{3.5}$$

satisfying periodic boundary conditions. One eigenvalue Λ_0 is negative. The contour integration should be shifted to the imaginary axis, which results in the appearance of an imaginary part in the partition function.

As will be shown below, a temperature T_0 exists such that for $T > T_0$ the only solution of eq. (3.3) is $\tilde{\varphi}(\tau) = \text{const}$. For $T < T_0$ there appears a solution $\tilde{\varphi}(\tau)$ differing from const. For an arbitrary value of τ_0 the function $\tilde{\varphi}(\tau - \tau_0)$ is also a solution of eq. (3.3); whence it follows that the function $\partial\tilde{\varphi}(\tau)/\partial\tau$ satisfies eq. (3.5) with zero eigenvalue. For any function $\varphi(\tau)$ we choose τ_0 in such a way that this function should be best approximated by the function $\tilde{\varphi}(\tau - \tau_0)$, i.e., we determine τ_0 from the minimization condition on the functional

$$D(\tau'|\varphi) = \int_{-\hbar/2kT}^{\hbar/2kT} d\tau [\varphi(\tau) - \tilde{\varphi}(\tau - \tau')]^2, \qquad \left.\frac{\partial D}{\partial \tau'}\right|_{\tau' = \tau_0} = 0. \tag{3.6}$$

The method used has been given by Zittartz and Langer (1966) and has been applied to calculate a functional integral of the form (3.2) by Larkin and Ovchinnikov (1984a).

The quantity Z_1 can be written as

$$\begin{aligned} Z_1 &= \operatorname{Im} \int_{-\hbar/2kT}^{\hbar/2kT} d\tau' \int D\varphi(\tau) \exp\{-B[\varphi(\tau)]\}\, \delta(\tau' - \tau_0[\varphi(\tau)]) \\ &= \operatorname{Im} \int_{-\hbar/2kT}^{\hbar/2kT} d\tau' \int D\varphi(\tau) \exp\{-B[\varphi(\tau)]\}\, \delta\left(\frac{\partial D(\tau'|\varphi)}{\partial \tau'}\right) \left|\frac{\partial^2 D(\tau'|\varphi)}{\partial \tau'^2}\right|. \end{aligned} \tag{3.7}$$

It follows from eqs. (3.4) and (3.6) that

$$\begin{aligned} \frac{\partial D(\tau'|\varphi)}{\partial \tau'} &= 2C_1 \left[\int_{-\hbar/2kT}^{\hbar/2kT} d\tau (\partial\tilde{\varphi}/\partial\tau)^2 \right]^{1/2}, \\ \frac{\partial^2 D(\tau'|\varphi)}{\partial \tau'^2} &= 2 \int_{-\hbar/2kT}^{\hbar/2kT} d\tau (\partial\tilde{\varphi}/\partial\tau)^2 . \end{aligned} \tag{3.8}$$

Substituting expressions (3.8) into eq. (3.7), we get

$$\begin{aligned} Z_1 = &\left[\int_{-\hbar/2kT}^{\hbar/2kT} d\tau \left(\frac{\partial\tilde{\varphi}}{\partial\tau}\right)^2 \right]^{1/2} \int_{-\hbar/2kT}^{\hbar/2kT} d\tau' \int_0^\infty \frac{dC_0}{(2\pi)^{1/2}} \exp\left(-\frac{C_0^2}{2}|\Lambda_0|\right) \\ &\times \prod_{n\neq 0} \int_{-\infty}^{\infty} \frac{dC_n}{(2\pi)^{1/2}} \exp\left\{-\frac{\Lambda_n}{2} C_n^2\right\} \delta(C_1) \exp\{-B[\tilde{\varphi}]\}. \end{aligned} \tag{3.9}$$

Similarly, the quantity Z_0 can be written as a Gaussian integral over the region of the values of $\varphi(\tau)$ near the minimum of the effective action. As a result, for the lifetime of the metastable state Γ^{-1} we get expression (2.14), where

$$\begin{aligned} B &= B[\tilde{\varphi}] - B[\varphi_{\min}], \\ A &= \frac{1}{(2\pi)^{1/2}} \left[\int_{-\hbar/2kT}^{\hbar/2kT} d\tau \left(\frac{\partial\tilde{\varphi}}{\partial\tau}\right)^2 \right]^{1/2} \left| \operatorname{Det}'\left(\frac{\delta^2 B}{\delta\varphi^2}\right)_{\varphi=\tilde{\varphi}} \right|^{-1/2} \\ &\quad \times \left| \operatorname{Det}\left(\frac{\delta^2 B}{\delta\varphi^2}\right)_{\varphi=\varphi_{\min}} \right|^{1/2} . \end{aligned} \tag{3.10}$$

In eq. (3.10) the prime on the determinant implies that the zero eigenvalue is omitted in it. Equation (3.10) holds for arbitrary temperatures $T < T_0$ and for any form of the effective action, including dissipative terms.

Equation (3.10) has been used by Affleck (1981) to calculate the tunnelling probability in the one-dimensional nondissipative case. The results obtained in this way coincide with those of a direct quantum-mechanical calculation of the quantity Γ by eq. (2.6). We shall discuss the region of applicability of eqs. (2.6) and (3.10). For zero temperature eq. (2.6) holds in all cases when it is possible to formulate the problem of the decay probability of a metastable state. Equation (3.10) has been obtained by Callan and Coleman (1977). However, they made the following substitution in the pre-exponential factor:

$$\int_{-\hbar/2kT}^{\hbar/2kT} \mathrm{d}\tau \left(\frac{\partial \varphi}{\partial \tau}\right)^2 \to \frac{\hbar}{m} B[\tilde{\varphi}].$$

This substitution is possible only at $T = 0$ in the one-dimensional case. The expression for the decay probability [eq. (3.10)] is applicable in the multidimensional case as well. It can also be used for a particle interacting with a thermal bath. In deriving eq. (3.10) a one-instanton approximation has been used, i.e., we have supposed that a particle escaping from the well never returns into it. In deriving eq. (2.6) at finite temperatures we have assumed that the distribution function is the equilibrium one. In the case of high viscosities this condition is fulfilled for $T < T_0$. At $T > T_0$, for high viscosities the following formula holds:

$$\Gamma = -\frac{2k}{\hbar}\frac{T_0}{T} \operatorname{Im} F.$$

In the case of low viscosities, which is discussed in section 7, eq. (2.6) is inapplicable in the region $T \sim T_0$ as well. In deriving eq. (3.10) we have assumed that the deviations from the extremal trajectory are small. As shown in section 7, this condition is violated in a narrow vicinity of the temperature T_0. Waxman and Leggett (1985) have obtained another expression for the pre-exponential factor. We think that a "natural assumption" made by them, concerning a generalization of their eq. (4.28) for the decay rate in the presence of dissipation is erroneous since, when dissipation is present, an unambiguous relation does not exist between the velocity of the particle in the vicinity of the turning point at the moment of its escape from the barrier and the energy of the particle while approaching the barrier.

4. *Derivation of the effective action*

To derive an expression for the effective action, we shall use a microscopic model of the tunnelling junction, described by the Hamiltonian

$$H = H_\mathrm{L} + H_\mathrm{R} + H_\mathrm{T} + H_\mathrm{Q} + H_\mathrm{M}, \tag{4.1}$$

where H_L and H_R are Hamiltonians describing the left and right superconductors, respectively, H_T is a tunnelling Hamiltonian, H_Q is the Coulomb energy, and H_M is the energy of the magnetic field:

$$H_L = \int d^3r\, \psi^+_{L\sigma}\left(-\frac{1}{2m}\frac{\partial^2}{\partial r^2} - \mu\right)\psi_{L\sigma}$$

$$- \frac{g_L}{2}\int d^3r\, \psi^+_{L\sigma}(r)\, \psi^+_{L-\sigma}(r)\, \psi_{L-\sigma}(r)\, \psi_{L\sigma}(r),$$

$$H_T = \int d^3r_L \int d^3r_R \{\hat{T}(r_L, r_R)\, \psi^+_{L\sigma}(r_L)\, \psi_{R\sigma}(r_R) + \text{c.c.}\},$$

$$H_Q = \frac{1}{2}\sum_{i,k} C^{-1}_{ik}\, \hat{Q}_i \hat{Q}_k, \qquad H_M = \frac{1}{8\pi}\int d^3r(\operatorname{curl} A - H_0)^2, \tag{4.2}$$

where $\hat{Q}_L$ and $\hat{Q}_R$ are charge operators for the left and right superconductors, respectively:

$$\hat{Q}_L = e\int d^3r\, \psi^+_{L\sigma}(r)\, \psi_{L\sigma}(r), \tag{4.3}$$

and C_{ik} is the capacitance matrix.

We shall use the Hubbard–Stratanovich transformation in order to remove the terms ψ^4 in the Hamiltonian (4.1). Here the partition function is written as a functional integral

$$Z = \int D^2\Delta_L\, D^2\Delta_R\, DV_1\, DV_2\, S_p\left\{\hat{T}\exp\left[-\frac{1}{\hbar}\int_0^{\hbar/kT} d\tau\, \mathscr{H}_{\text{eff}}(\tau)\right]\right\}, \tag{4.4}$$

where Δ_L and Δ_R are complex functions of the variables (r, τ), and $V_{1,2}$ are real functions:

$$\mathscr{H}_{\text{eff}}(\tau) = \hat{H}_T + \hat{H}_{L\,\text{eff}} + \hat{H}_{R\,\text{eff}} + i(Q_L V_1 + Q_R V_2) + \tfrac{1}{2}\sum C_{ik} V_i V_k + H_M,$$

$$\hat{H}_{L\,\text{eff}} = \int d^3r\, \psi^+_{L\sigma}\left(-\frac{1}{2m}\frac{\partial^2}{\partial r^2} - \mu\right)\psi_{L\sigma} + \int \Delta^*_L(r, \tau)\, \psi_{L\downarrow}(r)\, \psi_{L\uparrow}(r)\, d^3r$$

$$+ \frac{1}{g_L}\int d^3r |\Delta_L(r, \tau)|^2. \tag{4.5}$$

The quantities V_1, V_2 serve as the potential of the left and right superconductors, respectively.

The trace over the electron operators in eq. (4.4) has been found by Ambegaokar et al. (1982). After averaging over the electron states in the effective

Hamiltonian there appear terms proportional to the quantities

$$\left(eV_{\mathrm{L,R}} - \mathrm{i}\frac{\partial \varphi_{\mathrm{L,R}}}{\partial \tau}\right)^2, \qquad (\nabla \varphi_{\mathrm{L,R}} - eA)^2, \tag{4.6}$$

with coefficients proportional to the volume of superconductor. Thus, the functional integral over the quantities $V_{1,2}$ can be taken by the saddle-point method. In this approximation

$$eV_{\mathrm{L,R}} = \mathrm{i}\frac{\partial \varphi_{\mathrm{L,R}}}{\partial \tau}, \qquad eA = \nabla\varphi\,. \tag{4.7}$$

The functional integral over the modulus of the order parameter is also taken by the saddle-point method. Taking eq. (4.7) into account, to second order in the barrier transparency we get

$$Z = \int D\varphi \exp(-B[\varphi]), \tag{4.8}$$

where

$$\begin{aligned} B[\varphi] = {} & \frac{1}{\hbar}\int_{-\hbar/2kT}^{\hbar/2kT} \mathrm{d}\tau \left[\frac{C}{2e^2}\left(\frac{\partial\varphi}{\partial\tau}\right)^2 - \frac{\hbar I}{e}\varphi\right] \\ & + \frac{\pi\hbar}{2R_{\mathrm{N}}e^2}\int_{-\hbar/2kT}^{\hbar/2kT} \mathrm{d}\tau \int_{-\hbar/2kT}^{\hbar/2kT} \mathrm{d}\tau_1 \\ & \times\left[2\sin^2\left(\frac{\varphi(\tau)-\varphi(\tau_1)}{2}\right) g_{\mathrm{L}}(\tau-\tau_1)g_{\mathrm{R}}(\tau-\tau_1)\right. \\ & \left. - \cos(\varphi(\tau)+\varphi(\tau_1))\, F_{\mathrm{L}}(\tau-\tau_1)F_{\mathrm{R}}(\tau-\tau_1)\right]. \end{aligned} \tag{4.9}$$

The quantity C in eq. (4.9) is equal to the junction capacitance

$$C = \frac{C_{11}C_{22} - C_{12}^2}{C_{11} + C_{22} + 2C_{12}}\,. \tag{4.10}$$

The functions $g(\tau)$, $F(\tau)$ are Matsubara Green functions integrated over the energy variable (Ambeogaokar et al. 1982, Larkin and Ovchinnikov 1983a, c). For superconductors without paramagnetic impurities we get

$$g(\omega_n) = -\frac{\mathrm{i}\omega_n}{(\omega_n^2 + \Delta^2)^{1/2}}, \qquad F(\omega_n) = \frac{\Delta}{(\omega_n^2 + \Delta^2)^{1/2}}, \qquad \omega_n = \pi T(2n+1)\,. \tag{4.11}$$

The “time” of the underbarrier tunnelling is of the order of Ω^{-1} [eq. (2.5)]. Usually, $\Omega \ll \Delta$, which makes it possible to use an adiabatic approximation. As a result we get for the action $B(\varphi)$ the expression

$$B[\varphi] = \frac{1}{\hbar}\left\{ \int_{-\hbar/2kT}^{\hbar/2kT} \mathrm{d}\tau \left[\frac{C^*}{2e^2}\left(\frac{\partial \varphi}{\partial \tau}\right)^2 + U(\varphi) \right.\right.$$
$$\left.\left. + \frac{\pi T^2 \hbar^2 k^2}{Re^2} \int_{-\hbar/2kT}^{\hbar/2kT} \mathrm{d}\tau_1 \frac{\sin^2(\frac{1}{2}(\varphi(\tau) - \varphi(\tau_1)))}{\sin^2((\pi kT/\hbar)(\tau - \tau_1))} \right] \right\}, \tag{4.12}$$

where the potential energy $U(\varphi)$ is defined by eq. (1.3), and C^* is a renormalized value of the junction capacitance (Larkin and Ovchinnikov 1983a):

$$C^* = C + \frac{3 \times 2^{1/2} \pi \hbar (\Delta_L \Delta_R)^2}{8R_N(\Delta_L^2 + \Delta_R^2)^{5/2}} F\left(\tfrac{5}{4}, \tfrac{7}{4}, 2, \left(\frac{\Delta_L^2 - \Delta_R^2}{\Delta_L^2 + \Delta_R^2}\right)^2\right), \tag{4.13}$$

where F is a hypergeometric function. In deriving eq. (4.13) it has been assumed that the current is close to critical and phase φ is close to the value $\frac{1}{4}\pi$. For currents not close to the critical value the effective capacitance C^* also depends on phase φ. However, in this case the adiabatic approximation is valid only at a large value of the capacitance C and renormalization effects are small.

In eq. (4.12) the quantity R is the junction resistance for the normal current. At low temperatures in superconductors with a gap this resistance is exponentially large (Larkin and Ovchinnikov 1966). In many cases the junction is shunted by a normal resistance. If the shunt is a tunnelling junction between normal metals, then in eq. (4.12) one should put $R = R_{sh}$. If the shunt is a direct contact, then its contribution into the action is quadratic in the potential difference between junctions and is equal to

$$\frac{\pi k^2 T^2 \hbar}{4Re^2} \int_{-\hbar/2kT}^{\hbar/2kT} \mathrm{d}\tau \int_{-\hbar/2kT}^{\hbar/2kT} \mathrm{d}\tau_1 \left(\frac{\varphi(\tau) - \varphi(\tau_1)}{\sin((\pi kT/\hbar)(\tau - \tau_1))} \right)^2. \tag{4.14}$$

5. *Transition from classical to quantum decay regime*

At sufficiently high temperatures the only solution of eq. (3.3) is $\varphi(\tau) = \varphi_0 =$ const, where φ_0 is the maximum point of the potential $U(\varphi)$. A nontrivial solution of eq. (3.3) first appears at a temperature $T = T_0$ given by

$$T_0 = \frac{\hbar}{2\pi m^* k}\left[-\frac{\eta}{2} + \left(\frac{\eta^2}{4} - m^* U''(\varphi_0)\right)^{1/2} \right]. \tag{5.1}$$

In the vicinity of T_0 the function $\varphi(\tau)$ differs little from φ_0 and the decay probability Γ is given by (Larkin and Ovchinnikov 1984a, Grabert and Weiss 1984a)

$$\Gamma = \tfrac{1}{2}\mathscr{B}[1 - \phi(x)] \exp\{x^2 - U(\varphi_0)/kT\},$$

where $\phi(x)$ is defined by eq. (2.13b) and

$$x = \frac{\Lambda_1}{2\mathscr{A}^{1/2}}, \quad \mathscr{A} = \frac{kT}{\hbar^2}\left\{\frac{U''''(\varphi_0)}{4} - \frac{(U'''(\varphi_0))^2}{2U''(\varphi_0)}\left(1 + \frac{U''(\varphi_0)}{2\hbar\Lambda_2}\right)\right\},$$

$$\Lambda_n = \frac{m^*}{\hbar}\left(\frac{2\pi kTn}{\hbar}\right)^2 + \frac{U''(\varphi_0)}{\hbar} + 2\pi kT\eta|n|/\hbar^2,$$

$$\mathscr{B} = \frac{4\pi^2 m^*(kT_0)^3}{\hbar^4}\left(-\frac{\pi U''(\varphi_{\min})}{\mathscr{A}U''(\varphi_0)}\right)^{1/2}\frac{\Gamma(3 + \hbar\eta/2\pi kTm^*)}{\Gamma(1 - n_1)\Gamma(1 - n_2)},$$

$$n_{1,2} = \frac{\hbar}{2\pi kTm^*}\left[-\frac{\eta}{2} \pm \left(\frac{\eta^2}{4} - m^* U''(\varphi_{\min})\right)^{1/2}\right]. \tag{5.2}$$

In eq. (5.2) $U(\varphi_{\min})$ is the value of the potential at the minimum. For low viscosities ($\eta \to 0$) eq. (5.2) for Γ reduces to eq. (2.13).

At $T > T_0$ and for $0 < x \gg 1$, we have

$$\Gamma = \frac{\mathscr{B}}{\pi^{1/2}x}\exp(-U(\varphi_0)/kT). \tag{5.3}$$

This expression coincides, in its region of applicability, with the results of Wolynes (1981) and Mel'nikov and Meshkov (1983), where the case $T > T_0$ has been investigated.

6. *High-viscosity limit*

In the limit of high viscosities the transition probability can be found over the whole temperature region. First consider the case of a current I close to the critical value I_{cr}. The potential energy in this case has a form of a cubic parabola:

$$U(\varphi) = 3U_0(q/q_0)^2[1 - 2q/3q_0], \tag{6.1}$$

where $q = \varphi - \frac{1}{4}\pi + \frac{1}{2}q_0$ is the coordinate φ measured from the minimum of the potential $U(\varphi)$, $q_0 = [1 - (I/I_{cr})^2]^{1/2}$, and U_0 is the height of the potential barrier:

$$U_0 = \frac{\hbar I_{cr}}{3e}[1 - (I/I_{cr})^2]^{3/2}, \qquad \Omega = \frac{1}{q_0}\left(\frac{6U_0}{m}\right)^{1/2}. \tag{6.2}$$

In the vicinity of the critical current the "viscosity" term in eq. (4.12) coincides with expression (4.14). For high viscosities ($\eta^2 q_0^2 \gg 6mU_0$) the extremal value of the action $B[\varphi]$ is attained on the function $\tilde{q}(\tau)$ given by

$$q(\tau) = \frac{q_0^3}{3\hbar U_0} \pi k T \eta \sum_{n=-\infty}^{\infty} \exp\{-b|n| + 2i\pi k T n \tau/\hbar\},$$

$$\tanh b = \pi\eta\, q_0^2\, kT/3\hbar U_0\,. \tag{6.3}$$

The quantity $B[\varphi]$ in this case is equal to (Larkin and Ovchinnikov 1983b)

$$B = \frac{U_0}{kT_0}\left[\frac{3}{2} - \frac{1}{2}(T/T_0)^2\right], \qquad kT_0 = \frac{3\hbar U_0}{\pi\eta q_0^2}[1 - 6mU_0/\eta^2 q_0^2]. \tag{6.4}$$

The pre-exponential factor in the expression for Γ has been found by Larkin and Ovchinnikov (1984a)*:

$$\Gamma = \frac{\eta^{7/2} q_0^3}{6(2\hbar)^{1/2} U_0 m^{*2}} \exp(-B). \tag{6.5}$$

The temperature dependence of the exponent [eq. (6.4)] has been experimentally confirmed by Schwartz et al. (1985). We point out that in the limit of high viscosities the pre-exponential factor is large, since fluctuations with large frequencies make a contribution to it. A large pre-exponential factor in eq. (6.5) can be written as a correction to the action B. This correction implies a renormalization of the critical current (Zaikin and Panyukov 1986). Thus, it was impossible to determine the pre-exponential factor in the experiment (Schwartz et al. 1985). The decay probability we have found makes it possible to determine the I–V characteristics of the junction for small values of the voltage (Ovchinnikov et al. 1984, Barone et al. 1985). For $T < T_0$

$$e\langle V\rangle = \pi\hbar\Gamma. \tag{6.6}$$

For temperature $T \gg T_0$ only thermal fluctuations are important, and in this case the I–V characteristic of the tunnel junction has been found by Ambeogaokar and Halperin (1969) and Ivanchenko and Zilberman (1968).

The decay probability into the neighbouring minimum of the potential in the limit of high viscosities can be found also for currents not close to the critical value (Korshunov 1987):

$$\Gamma = \frac{1}{8m^2\hbar^{1/2}}\left[\frac{2\eta^7}{(I\hbar/2e)^2 + (\pi kT/2\eta\hbar)^2}\right]^{1/2}$$
$$\times \exp\left[-\frac{\pi\eta}{\hbar}\left\{1 - \ln\left[\left(\frac{I}{I_{\rm cr}}\right)(1+x^2)^{1/2}\right] - \frac{1}{x}\arctan x\right\}\right], \tag{6.7}$$

where $x = \pi e\eta kT/\hbar^2 I$.

* For a detailed definition of this formula see Appendix D of the chapter by Echern and Schmid.

For currents close to critical Riseborough et al. (1985) obtained the expression for the action on the extremal trajectory for the whole temperature region $(0, T_0)$ for a definite value of viscosity η which depends on T. Numerical values of the decay probability (exponent and pre-exponential factor) have been found by Grabert et al. (1985, 1987) for the whole range of values of viscosity and temperature for a potential in the form of a cubic parabola. At zero temperature, the dependence of the exponent and pre-exponential factor in the expression for the decay probability [eq. (2.14)] has been obtained by Chang and Chakravarty (1984) numerically with a lower accuracy.

7. Low-viscosity limit

Analytic results in the limit of low viscosities have been mainly obtained for currents close to critical, where the potential energy has a form of a cubic parabola [eq. (6.1)]. We assume that the viscosity is small enough for its influence on tunnelling process to be negligible, yet large enough for the distribution function to be considered as the equilibrium one. In this case the decay probability can be found both by the quantum-mechanical [eqs. (2.1), (2.4) and (2.6)] and by the instanton method [eqs. (2.14) and (3.10)]. In the potential of the form of a cubic parabola, the exponent B and pre-exponential factor A have been found by Ivlev and Ovchinnikov (1987):

$$B = g\frac{32\times 2^{1/2}}{15}\left\{\frac{E(k)}{(1-k^2+k^4)^{1/4}} + K(k)\left[\frac{(k^2-2)(13-13k^2+10k^4)}{36(1-k^2+k^4)^{5/4}} + \frac{5}{18}(1-k^2+k^4)^{1/4}\right]\right\},$$

$$A = \omega_0\left(\frac{2^{1/2}}{\pi}g\right)^{1/2}\sinh\left(\frac{\pi T_0}{T}\right)\frac{k^2(1-k^2)^{1/2}}{(1-k^2+k^4)^{7/8}} \times\left[\frac{1-k^2+k^4}{1-k^2}E(k) - \left(1-\frac{k^2}{2}\right)K(k)\right]^{-1/2}, \tag{7.1a}$$

where

$$g = (27/8\times 2^{1/2})\,U_0/\hbar\omega_0, \qquad T_0 = \hbar\omega_0/2\pi k,$$

$$T_0/T = \frac{2}{\pi}(1-k^2+k^4)^{1/4}\,K(k), \tag{7.1b}$$

and $K(k)$, $E(k)$ are total elliptical integrals. Note that the quantity K is implicitly defined by the third of eqs. (7.1b). At zero temperature this expression yields

$$\Gamma = \frac{6\omega_0}{\pi^{1/2}}\left(\frac{6U_0}{\hbar\omega_0}\right)^{1/2}\exp\left(-\frac{36U_0}{5\hbar\omega_0}\right). \tag{7.2}$$

Equation (7.2) coincides with the expression for Γ which follows from eqs. (2.4)–(2.6).

The correction to the extremal action B due to viscosity can be found by perturbation theory. For that, one should substitute the extremal trajectory $\varphi(\tau)$ found in the absence of viscosity into eq. (4.12). For zero temperature (Caldeira and Leggett 1981) it has been found

$$B = \frac{36U_0}{\hbar\omega_0}\left(1 + \frac{45\zeta(3)}{2\pi^3}\frac{\eta}{m\omega_0}\right), \tag{7.3}$$

where $\zeta(3)$ is the Riemann zeta function.

At low temperatures the correction to the action δB is exponentially small in the absence of viscosity, as follows from eqs. (2.6) and (7.1) and is proportional to T^2 when viscosity is present (Grabert et al. 1984). For low viscosities

$$\delta B = -\frac{U_0}{\hbar\omega_0}\left[\frac{9\eta}{\pi m\omega_0}\left(\frac{2\pi kT}{\hbar\omega_0}\right)^2\right] \tag{7.4}$$

As follows from eq. (6.4), for high viscosities the quadratic dependence of the extreme action on temperature holds up to temperature T_0.

At zero temperature the correction to the pre-exponential factor due to viscosity has been found analytically by Freidkin et al. (1988) and Ovchinnikov and Barone (1987, 1988):

$$A = 6\omega_0\left(\frac{6U_0}{\pi\hbar\omega_0}\right)^{1/2}(1 + 1.43\eta/m\omega_0). \tag{7.5}$$

The correction due to viscosity in eq. (7.5) coincides very accurately with the result of a numerical calculation by Grabert et al. (1985, 1987) and Chang and Chakravarty (1984).

It follows from eqs. (6.4) and (7.3) that the transition from the case of small viscosity to that of a large one is realized at $\eta \sim \eta_2 = m\omega_0$. For temperatures $T \gtrless T_0$, where T_0 is the transition temperature from the classical to the quantum decay regime, the region of low viscosities is divided into two subregions. The typical value of viscosity η_1 separating these two subregions is

$$\eta_1 = m\omega_0\, kT/U_0\,. \tag{7.6}$$

In the region of viscosity $\eta_1 \ll \eta \ll \eta_2$ the results given in section 2 hold. For viscosity $\eta \lesssim \eta_1$ the depopulation of the distribution function is essential for the energies close to the barrier top. At high temperatures $T \gg T_0$ the case of low viscosities has been considered by Kramers (1940) and the case $\eta \sim \eta_1$ by Mel'nikov (1984) and Mel'nikov and Meshkov (1986).

For low viscosities (in the region $\eta < \eta_1$) the temperature T_0, at which occurs the transition from the classical to the quantum decay regime, proves to be lower than $\hbar\Omega/2\pi k$ (Larkin and Ovchinnikov 1984b). This is connected with the fact that the distribution function is depopulated for the upper levels, for which the tunnelling probability is larger than the transition probability between the levels. The boundary E_c of the region, where the distribution function drops rapidly, can be found from the condition

$$\gamma(E_c) = \sum_j W_{ji}. \tag{7.7}$$

Here the value $\gamma(E_c)$ is defined by eq. (2.1), and W_{ji} is the transition probability from state i into state j per unit time:

$$W_{ji} = |\langle j|\varphi|i\rangle|^2 \frac{\eta(E_j - E_i)}{\hbar^2}\left[\coth\left(\frac{E_j - E_i}{2kT}\right) - 1\right]. \tag{7.8}$$

In the vicinity of the temperature T_0, a wide region of energies $E_c - E \gg kT_0$ is important. Thus, the detailed behaviour of the distribution function in the vicinity of E_c is unimportant. One can suppose that for $E < E_c$ the distribution function is the equilibrium one, and for $E > E_c$ it is equal to zero.

From eq. (2.6) we find for $|T - T_0| \ll T_0$,

$$\begin{aligned} \Gamma &= \frac{2\pi}{\omega(E_c)}\gamma(E_c)\sinh\left(\frac{\hbar\omega_0}{2kT}\right)\exp\left(-\frac{E_c}{kT}\right) \\ &\quad\times\left\{1 + \phi\left(\frac{[(kT)^{-1} - (kT_0)^{-1}]}{(4S''(E_c)/\hbar)^{1/2}}\right)\right\} \\ &\quad\times\frac{\exp[\hbar[(kT)^{-1} - (kT_0)^{-1}]^2/4S''(E_c)]}{(4\pi\hbar S''(E_c))^{1/2}}, \\ (kT_0)^{-1} &= -\frac{2}{\hbar}S'(E_c), \end{aligned} \tag{7.9}$$

with $\phi(x)$ given by eq. (2.13a).

For $(T_0 - T) \gg kT_0^2[(4/\hbar)S''(E_c)]^{1/2}$ the distribution function in the important region of energies differs slightly from the equilibrium one and the decay probability is defined by eqs. (2.14) and (3.10). In order to determine the lifetime Γ^{-1} within the temperature region $(T - T_0) \gg kT_0^2[(4/\hbar)S''(E_c)]^{1/2}$, it is necessary to solve a kinetic equation for the distribution function. This equation looks like

$$N(E_f) = \sum_i \tilde{W}_{fi}(1 - \gamma(E_i))N(E_i), \tag{7.10}$$

where $N(E)$ is the distribution function of particles approaching the barrier, and $\tilde{W}_{fi}$ is the total transition probability from state i into state f per motion period.

In the Born approximation this probability differs from eq. (7.8) by the factor $2\pi/\omega(E)$ (where $\omega(E)$ is the frequency of classical oscillations with energy E). For $\eta \sim \eta_1$ it is necessary to take rescattering processes into account. In this case the probability $\tilde{W}_{fi}$ has been found by Larkin and Ovchinnikov (1985). They have obtained an expression for the decay probability Γ which is valid at arbitrary values of viscosity and temperature:

$$\Gamma = 2\pi^2 mk^3 T^2 T_0 \frac{\omega_0}{\Omega} Y\left(\frac{\pi}{\mathscr{A}}\right)^{1/2} \frac{\Gamma(2-\chi_1)\Gamma(2-\chi_2)}{\hbar^4 \Gamma(1-n_1)\Gamma(1-n_2)}$$
$$\times [1-\phi(x)] \exp(-U_0/kT + x^2), \tag{7.11}$$

where

$$\chi_{1,2} = \frac{\hbar}{2\pi kT}\left[-\frac{\eta}{2m} \pm \left(\frac{\eta^2}{4m^2} + \Omega^2\right)^{1/2}\right];$$
$$n_{1,2} = \frac{\hbar}{2\pi kT}\left[-\frac{\eta}{2m} \pm \left(\frac{\eta^2}{4m^2} - \omega_0^2\right)\right];$$
$$x = 2\pi^2 mk^2 T^2 (1-\chi_1)(1-\chi_2)/\hbar^3 \mathscr{A}^{1/2};$$
$$\mathscr{A} = \frac{kT}{\hbar^2}\left\{\frac{U^{\mathrm{IV}}_{(\varphi_0)}}{4} + \frac{U'''(\varphi_0))^2}{2m\Omega^2}\left(1 - \frac{\Omega^2}{2((4\pi kT/\hbar)^2 - \Omega^2 + 4\pi kT\eta/m\hbar)}\right)\right\};$$
$$\chi_1 = T_0/T. \tag{7.12}$$

In eq. (7.12)

$$m\omega_0^2 = \left(\frac{\partial^2 U}{\partial \varphi^2}\right)_{\varphi_{\min}}, \qquad m\Omega^2 = -\left(\frac{\partial^2 U}{\partial \varphi^2}\right)_{\varphi=\varphi_0}, \tag{7.13}$$

$\varphi_{\min}$ is the minimum point and φ_0 the maximum point of the potential.

The function $Y(\eta, T)$ has been studied and tabulated by Larkin and Ovchinnikov (1985).

In the classical limit $T \gg T_0$ eq. (7.11) coincides with the analytical results of Mel'nikov (1984) and with a numerical solution of the Langevin equation given by Buttiker et al. (1983). In the same paper, as well as in Matkowsky et al. (1984), an expression for the decay probability has been found by using the Fokker–Planck equation. In fact, in such an approximation it is assumed that a particle escapes from the potential well uniformly in time. It seems to us that such an assumption is incorrect. Equation (7.10) used by us implies that a particle can leave the potential well only at the moment of its approach to the barrier.

8. *The influence of level quantization on the lifetime of metastable states*

In section 7 the number of levels in the potential well has been assumed to be large enough, for the motion of a particle with energy close to the height of the potential barrier to be considered as quasiclassical. In the potential described by eqs. (1.3) and (6.1) the well width is of the same order as that of the barrier. One might think that the decay probability of a metastable state would be small in a quasiclassical potential only when the number of levels in it is large. However, for numerical reasons this is not so. The difference of the effective actions on two neighbouring levels is equal to 2π. The tunnelling probability is defined by the value of the action on the underbarrier trajectory. In potentials with the same width of the well and barrier the ratio of decay probabilities from the two neighbouring levels is close to $\exp(-2\pi) = 0.00187$. Thus, even for a small number of levels the lifetime of a metastable state is large. In the experimental papers of Voss and Webb (1981), Jackel et al. (1981), Devoret et al. (1985) and Martinis et al. (1985), where quantum tunnelling was investigated, the number of levels was not large (1–10). Level quantization influences most strongly the lifetime of a metastable state within a temperature region T of the order of the distance between the levels. In this case the pre-exponential factor in the expression for the decay probability of a metastable state is an oscillating function of the depth of the potential well. As the viscosity increases, the oscillation amplitude decreases.

Usually, when investigating particle tunnelling through a potential barrier it is assumed that the motion of a particle in the classically accessible region can be described by wave packets. Such an approximation holds for sufficiently wide potential wells. The basis for this approximation is the fact that at energies close to the barrier height the period of the classical motion is large, the distance between the levels is small, and one can construct wave packets. However, for numerical reasons this approximation is suitable only for very high potential barriers. The distance δE between the levels in the vicinity of the barrier top is equal to

$$\frac{\delta E}{\hbar\Omega} = \left(1 + \frac{1}{2\pi}\ln N\right)^{-1}, \tag{8.1}$$

where N is the number of levels in the potential well and the frequency Ω is defined by eq. (7.13).

Quantum tunnelling can be observed only if the number of levels N is not large. In such potentials no crowding of levels takes place. On the other hand, the tunnelling probability of a particle through the potential barrier depends very strongly on the particle energy E and for energies close to the barrier height U_0 we get

$$\gamma(E) = \frac{\delta E}{2\pi\hbar}\exp(-(U_0 - E)/kT_0), \quad kT_0 = \frac{\hbar\Omega}{2\pi}. \tag{8.2}$$

In eq. (8.2) δE is the difference between neighbouring levels.

Thus, one cannot turn to a continuous distribution over energies and for low viscosities one should write a kinetic equation for the probability ρ_j of finding a particle in the jth level:

$$\frac{\partial \rho_j}{\partial t} = \sum_k (W_{jk}\rho_k - W_{kj}\rho_j) - \gamma_j \rho_j, \tag{8.3}$$

where the probability W_{jk} of transition from state k into state j is defined by eq. (7.8), and the tunnelling probability from state j is defined by eqs. (2.1) and (2.5). For an equilibrium thermal bath the matrix elements W_{jk} satisfy the condition

$$W_{jk} = W_{kj} \exp[(E_k - E_j)/kT], \tag{8.4}$$

where T is the thermal bath temperature. As will be shown below, in the potential (6.1), which has the form of a cubic parabola, only the transitions between the nearest levels are important. For sufficiently deep levels $j < n$, the tunnelling probability γ_j is small compared to the transition matrix elements $W_{j-1,j}$ and it can be neglected when solving eq. (8.3). In such an approximation the stationary solution of eq. (8.3) has the form

$$\rho_j = \exp(-E_j/kT) - C \sum_{k \leqslant j} \exp[-(E_j - E_k)/kT]/W_{k-1,k}, \quad j \leqslant n, \tag{8.5}$$

where C is a constant which is found from the solution of the system of equations (8.3) for $n \leqslant j < n + \nu$, where ν is an integer. For these states both processes, tunnelling and dissipation, are essential. For the states with $j \geqslant \nu$ tunnelling is considerably more probable than the process of a dissipative transition and the values ρ_j for $j \geqslant n + \nu$ can be put equal to zero. Due to the rapid increase of the tunnelling probability γ_j with the number j, we can restrict ourselves to small values of quantity ν. The solutions of the system of equations (8.3) for $\nu = 2$ and $\nu = 3$ practically coincide.

The decay probability of a metastable state Γ in this approximation is equal to

$$\Gamma = \left\{ \sum_{j<n+\nu} \gamma_j \rho_j + W_{n+\nu,n+\nu-1}\, \rho_{n+\nu-1} \right\} \Big/ \sum_{j<n+\nu} \rho_j . \tag{8.6}$$

For a quantitative comparison with the experimental data it is necessary to find the position of the levels and their widths. If a level is not very close to the top of the potential barrier ($U_0 - E > 0.4\hbar\Omega$), then its position can be found by the quasiclassical formula

$$\tilde{S}(E_n) = \pi\hbar(n + \tfrac{1}{2}), \tag{8.7}$$

where $\tilde{S}(E)$ is the action in the classically accessible region, and the double width (decay probability) is defined by eqs. (2.1), (2.2) and (2.4).

For a potential of the form (6.1) we find

$$S(E) = \frac{\pi}{4} q_0 (mU_0)^{1/2} (x_2 - x_3)^2 (x_1 - x_3)^{1/2} F\left(-\tfrac{1}{2}; \tfrac{3}{2}; 3; \frac{x_2 - x_3}{x_1 - x_3}\right),$$

$$\tilde{S}(E) = \frac{\pi}{4} q_0 (mU_0)^{1/2} (x_1 - x_2)^2 (x_1 - x_3)^{1/2} F\left(-\tfrac{1}{2}; \tfrac{3}{2}; 3; \frac{x_1 - x_2}{x_1 - x_3}\right), \tag{8.8}$$

where F is a hypergeometric function and $x_3 < x_2 < x_1$ are roots of the cubic equation

$$x^3 - \tfrac{3}{2} x^2 + E/2U_0 = 0. \tag{8.9}$$

The transition matrix element between the states $j, j-1$ in the potential (6.1) is equal to

$$\langle j|\varphi|j-1\rangle = -\frac{\pi^2 q_0 (x_2 - x_3)}{2k^2 K^2(k) \sinh[\pi K'(k)/K(k)]}, \quad k = \left(\frac{x_2 - x_3}{x_1 - x_3}\right)^{1/2}. \tag{8.10}$$

In eq. (8.10) $K(k)$ and $K'(k) = K((1-k^2)^{1/2})$ are total elliptical integrals. The energy E in eq. (8.9), at which the roots $x_{1,2,3}$ are calculated to find the matrix element in eq. (8.10) is equal to $E = \frac{1}{2}(E_j + E_{j-1})$. For levels close to the barrier top the width is not very small. Thus, the transition probability to this level should be described by the quantum-mechanical formulae for transition into a continuous spectrum. The transition probability density within the interval dE is determined as before by eq. (7.8), where now the wave function of the upper level should be replaced by the wave function of a continuous spectrum, normalized by a δ-function over energy. Using for the wave functions a quasi-classical approximation, we get for the matrix element the following expression:

$$|\langle j|\varphi|E\rangle|^2 = \frac{9q_0^2}{(\hbar\Omega)^4} \frac{\pi(E_j - E_{j-1})}{2m} \frac{(E - E_j)^2}{\sinh^2[\pi(E - E_j)/\hbar\Omega]} |C|^2, \tag{8.11}$$

where

$$|C|^2 = \frac{2m}{\pi} \exp(2\pi y) \left| 1 + \frac{(2\pi)^{1/2}}{\Gamma(\frac{1}{2} + \mathrm{i}y)} \exp\left[\frac{\pi}{2} y + \mathrm{i}\frac{6 \times 6^{1/2}}{5\hbar} q_0 (mU_0)^{1/2} \right.\right.$$
$$\left.\left. + 2\mathrm{i}y \ln(6q_0(2m\Omega/\hbar)^{1/2})\right]\right|^{-2},$$

$$y = \frac{E - U_0}{\hbar\Omega}. \tag{8.12}$$

Expression (8.11) has a sharp maximum. The position of this maximum defines the level energy E_{j+1} and its width defines the tunnelling probability $\gamma(E_{j+1})$. Equation (8.11) can also be applied for transitions into states with energy $E > U_0$. It proves that at these energies there exists one virtual level, whose width is small of the order of $\hbar\Omega/2\pi$). This level should be taken into account in eq. (8.6), besides the levels which are under the barrier. The results given in this section were obtained by Larkin and Ovchinnikov (1985; 1986a, b).

The experimental data for the value of the decay probability are usually represented in the form

$$\Gamma = \frac{\omega_0}{2\pi} \exp(-U_0/kT_{\rm esc}). \tag{8.13}$$

For $T > T_0$ the value $T_{\rm esc}$ differs from the thermal bath temperature T owing to the pre-exponential factor in the decay probability. Thus, the quantity $T_{\rm esc}$, as well as the pre-exponential factor, is an oscillating function of the depth of the potential well. In their experiment Devoret et al. (1985) obtained a variation of the depth of the potential well U_0 by varying the current I through the junction. Figure 2 shows the theoretical and the experimental data for $T_{\rm esc}$. The points in fig. 2 represent the experimental results of Devoret et al. (1985). Curve 1 represents the results of the calculations according to eqs. (8.5)–(8.8), (8.10) and (8.11). In the calculations the following junction parameters have been used:

$$R = 190\,\Omega, \qquad C = 6.35\ \mathrm{pF}, \qquad I_{\rm cr} = 9.489\ \mu\mathrm{A}, \qquad T = 0.151\,.$$

The parameters of the potential are defined by eqs. (6.1) and (6.2). Curve 2 is constructed using eq. (7.11), in the derivation of which crowding of levels close to the barrier top has been assumed. Curve 2 is a smooth one, without oscillations, but it is located very close to Curve 1. We have taken Curve 3 from the paper of Devoret et al. (1985) and it reproduces the theoretical results of Buttiker et al. (1983). Changing the junction parameters within the possible experimental errors ($R = 190 \pm 100\,\Omega$, $C = 6.35 \pm 0.4$ pF) one can get a still better agreement of the theoretical and experimental results. The value of the oscillations (Curve 1) proved to be small. This relates to the fact that the shunt resistance was

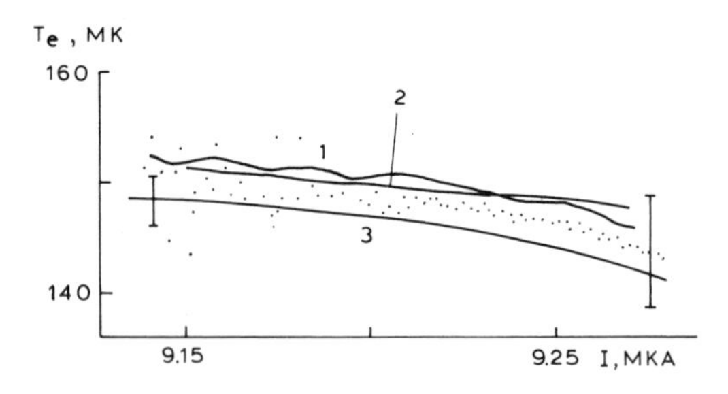

Fig. 2. Influence of the level quantization on the lifetime of metastable state. The points represent the experimental results of Devoret et al. (1985). Curve 1 represents the results of calculations [eqs. (8.5)–(8.10)]. Curve 2 is constructed using eq. (7.11), taking into account a crowding of levels close to the barrier top. Curve 3 reproduces the results of Buttiker et al. (1983).

not large [case of intermediate viscosity: the viscosity η is of order η_1 and the temperature T is high compared to T_0 ($T \sim 3T_0$)]. Under these conditions the decay mainly goes through the resonance level with the energy $E > U_0$. The width of this level depends slightly on energy. As the temperature T decreases within the region $(T - T_0) > T_0/2\pi$ the magnitude of the oscillation increases. The magnitude of the oscillation also grows when the shunt resistance increases, since for low viscosities the decay occurs from deeper levels, whose lifetime depends exponentially on energy. The quantity T_{esc}/T, at currents close to critical, is a universal function of three dimensionless parameters:

$$T_{esc}/T = \psi(z;\, T/T^*;\, R/R^*), \tag{8.14}$$

where

$$Z = (I_{cr} - I)/I^*; \quad I^* = I_{cr}(e^3/\hbar I_{cr} C)^{2/5},$$
$$T^* = k^{-1}(\hbar I_{cr} e/C)^{1/2}\,(e^3/\hbar I_{cr} C)^{1/10},$$
$$R^* = (\hbar I_{cr} C/e^3)^{1/10}\,(\hbar/e I_{cr} C)^{1/2}. \tag{8.15}$$

Curve 1 of fig. 2 gives the quantity T_{esc} for the parameter values: $T/T^* = 0.01725$, $R/R^* = 13.825$.

The quantity Z is connected with the number of levels N in the potential well by the relationship

$$N = \frac{1}{2} + \frac{3 \times 2^{7/4}}{5\pi} Z^{5/4}. \tag{8.16}$$

The period of oscillations δI in the current variable is determined from the condition that the number of levels in the potential well changes by unity:

$$\delta I = \tfrac{1}{3}\pi I^* (2/Z)^{1/4}. \tag{8.17}$$

9. *Resonance reduction of lifetime of a metastable state under the action of an external high-frequency current*

The existence of levels manifests itself most vividly in the resonance reduction of the lifetime of a metastable state under the action of an alternating current with frequency equal to the distance between the levels. Experimentally, these resonances have been observed by Martinis et al. (1985). Theoretically, this phenomenon has been predicted and investigated by Larkin and Ovchinnikov (1986a, b). A detailed analysis of the lifetime of a metastable state of the tunnel junction at external currents close to the critical value I_{cr} and a large value of the shunt resistance has been given by Larkin and Ovchinnikov (1986b), Chow et al. (1988), and Kopietz and Chakravarty (1988).

For a quantitative description of the phenomenon we shall use the system of equations for the density matrix:

$$\frac{\partial \rho_f^j}{\partial t} = \frac{\mathrm{i} I_1}{e} \cos(\omega t) \sum_m \{ \langle j|\varphi|m\rangle \rho_f^m \exp(-\mathrm{i}(E_m - E_j)t/\hbar) - \langle f|\varphi|m\rangle \rho_m^j \exp(\mathrm{i}(E_m - E_f)t/\hbar)\} - \tfrac{1}{2}\sum_m (W_{mj}^{mj} + W_{mf}^{mf})\rho_f^j, \quad (9.1)$$

where ω is the frequency of the external high-frequency field and I_1 is its amplitude. The matrix elements W_{jm}^{jm} have been found by Larkin and Ovchinnikov (1986a) and for a junction shunted by a normal resistance R are equal to

$$W_{fn}^{jm} = \frac{\hbar\tilde{\omega}}{2Re^2}\left(1 + \coth\left(\frac{\hbar\tilde{\omega}}{2kT}\right)\right)\{\langle j|\exp(\mathrm{i}\varphi)|m\rangle \langle f|\exp(-\mathrm{i}\varphi)|n\rangle + \langle j|\exp(-\mathrm{i}\varphi)|m\rangle \langle f|\exp(\mathrm{i}\varphi)|n\rangle\},$$

$$\hbar\tilde{\omega} = (E_m - E_j + E_n - E_f)/2. \quad (9.2)$$

In the experiment of Martinis et al. (1985) the external-field frequency ω was close to the distance between neighbouring levels, and the temperature T was small compared to the distance between the levels. With exponential accuracy in this parameter the nondiagonal elements of the density matrix are equal to

$$\rho_{j+1}^j = -\frac{I_1}{2e}\frac{\langle j|\varphi|j+1\rangle (\rho_j - \rho_{j+1})}{\omega - (E_{j+1} - E_j)/\hbar - (\mathrm{i}/2)\Gamma_j}\exp[-\mathrm{i}(E_{j+1} - E_j - \hbar\omega)t/\hbar], \quad (9.3)$$

where

$$\Gamma_j = \gamma_j + \gamma_{j+1} + W_{j+1,j} + W_{j-1,j} + W_{j,j+1} + W_{j+2,j+1}. \quad (9.4)$$

Taking eq. (9.3) into account, eq. (9.1) for the diagonal elements of the density matrix takes the form

$$\sum_k (W_{jk}\rho_k - W_{kj}\rho_j) - \gamma_j\rho_j - \frac{I_1^2}{4e^2}\langle j|\varphi|j+1\rangle^2 \langle\rho_j - \rho_{j+1}) \times \frac{\Gamma_j}{[\omega - (E_{j+1} - E_j)/\hbar]^2 + \Gamma_j^2/4} + \frac{I_1^2}{4e^2}\langle j|\varphi|j-1\rangle^2 (\rho_{j-1} - \rho_j) \times \frac{\Gamma_{j-1}}{[\omega - (E_j - E_{j-1})/\hbar]^2 + \Gamma_{j-1}^2/4} = 0. \quad (9.5)$$

The system of equations (9.5) is to be solved in the same approximation as the system of equations (8.3). We solve, exactly, two equations of this system: for the virtual level above the barrier and the uppermost underbarrier level. For deeper levels we shall consider the tunnelling probability γ_j as equal to zero. In this approximation the solution of the system of equations (9.5) can be represented as follows:

$$\rho_j = \left(\prod_{\nu<j} G_\nu \right) \left\{ 1 - C \sum_{k<j} \left[(1+b_k) W_{k,k+1} \prod_{\nu<k+1} G_\nu \right]^{-1} \right\}, \tag{9.6}$$

where

$$G_\nu = \frac{b_\nu + \exp(-(E_{\nu+1} - E_\nu)/kT)}{1+b_\nu},$$

$$b_\nu = \frac{I_1^2}{4e^2} \frac{\langle \nu|\varphi|\nu+1\rangle^2}{W_{\nu,\nu+1}} \frac{\Gamma_\nu}{[\omega - (E_{\nu+1} - E_\nu)/\hbar]^2 + \Gamma_\nu^2/4}. \tag{9.7}$$

The constant C is determined from the condition of matching with the exact solution of eqs. (9.5) for the two upper levels. The lifetime of the metastable state is, as before, given by eq. (8.6) and at $T > T_0$ is determined by the distribution function on upper levels.

In the linear approximation in pumping power we get from eq. (9.6)

$$\rho_j = \rho_j^{(0)} + \sum_{\nu \leqslant j} b_{\nu-1} (\rho_{\nu-1}^{(0)} - \rho_\nu^{(0)}) \exp(-(E_j - E_\nu)/kT), \tag{9.8}$$

where the functions $\rho_j^{(0)}$ are defined by eq. (8.5). As mentioned above, for the last two levels considered the system of equations (9.5) should be solved exactly.

In the approximation considered there occur in the quantity Γ weak breaks when a level passes through the barrier top. This break should be smoothened by averaging the value of the quantity Γ by two schemes of calculation (N and $N+1$ level), capturing a small vicinity of the value of the relevant parameter (the current through the junction), at which a level passes through the barrier top. For high quality factors (Q) the function b_ν has a sharp maximum. The change of the distribution function in the high levels due to pumping is equal to the sum of resonance contributions. The linear approximation [eq. (9.8)] is violated for two reasons. For narrow resonances the quantity b_ν near resonance can be larger than or of the order of unity. This will result in equalization of the populations of the ν and $\nu+1$ levels. Further increase of pumping does not result in a growth of the maximum. At low temperatures, the effect is exponentially large:

$$\Delta\Gamma/\Gamma = b_\nu \exp\{(E_{\nu+1} - E_\nu)/kT\}. \tag{9.9}$$

For wide resonances overlapping of neighbouring resonances can be important. If the quantity b_ν for two neighbouring levels becomes either larger or of the order of $\exp(-(E_{\nu+1} - E_\nu)/kT)$, then one should use the general equation (9.6).

Under the experimental conditions of Martinis et al. (1985) the linear approximation [eq. (9.8)] holds. The position of the levels and the transition matrix elements between them can be calculated according to the quasiclassical eqs. (8.7), (8.8), (8.10) and (8.11). The position and widths of the levels can be found very exactly by such a method even when the number of levels in the well is ~ 1. For instance, in the case $N = 4$ the energy of the ground state is found with an accuracy of not worse than 0.5% (Chow et al. 1988).

For comparison with the experimental data of Martinis et al. (1985) there was used only one fitting parameter—the pumping power, which is unknown to us. Good agreement is obtained for pumping power P equal to

$$P = RI_1^2/2 = 8.57 \times 10^{-4}\, \hbar\omega^2, \tag{9.10}$$

where ω is the pumping frequency. In fig. 3 the points represent the experimental data of Martinis et al. (1985), and the solid curve represents the results of a numerical calculation using eqs. (8.6) and (9.8). The junction parameters are: $R = 135.45\,\Omega$, $C = 47$ pF, $T = 2$ mK, $I_{cr} = 30.572$ mA. The pumping frequency is $\omega/2\pi = 2 \times 10^9\,\mathrm{s}^{-1}$.

Now consider the various limiting cases, where we can get rather simple expressions for the resonance change of lifetime of a metastable state.

For very large values of the Q-factor

$$Q = RC\omega_0; \tag{9.11}$$

the resonance condition can be fulfilled, while maintaining the inequality $Q|\hbar(\partial\omega(E)/\partial E| \gg (kTE/(\hbar\omega_0)^2)$, only for the two levels (E_{l+1}, E_l). In this approximation only those nondiagonal matrix elements $\rho_l^{l+1} = (\rho_{l+1}^l)^*$ are large which are defined by eq. (9.3). In this approximation for all $j \neq l$ the diagonal elements of the density matrix are connected by the same relationship as in the absence of pumping, and for the diagonal elements with $j = l$ we have (Larkin and Ovchinnikov 1986a)

$$\rho_{l+1} = \rho_l \exp[-\hbar\omega_l/kT]\, G, \tag{9.12}$$

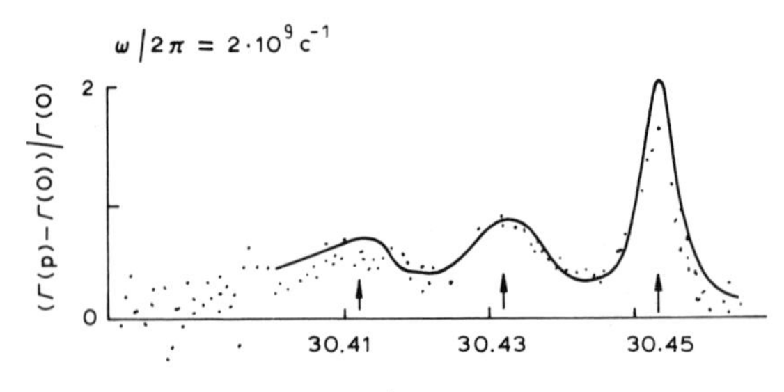

Fig. 3. Resonance reduction of the lifetime of a metastable state: The points represent the experimental data of Martinis et al. (1985); the solid curve represents the results of a numerical calculation [eqs. (8.6) and (9.8)].

where

$$\hbar\omega_l = E_{l+1} - E_l, \qquad b = \frac{I_1^2}{e^2}\frac{\langle l|\varphi|l+1\rangle^2}{(\omega_l - \omega)^2 + \Gamma_l^2/4},$$

$$G = \frac{1 + b\exp(\hbar\omega_l/2kT)\cosh(\hbar\omega_l/2kT)}{1 + b\exp(-\hbar\omega_l/2kT)\cosh(\hbar\omega_l/2kT)}. \tag{9.13}$$

It follows from eqs. (9.12) and (9.13) that pumping increases the population of all levels with $j > l+1$ by a factor of G, so the lifetime of a metastable state decreases by the same factor:

$$\frac{\Gamma(P)}{\Gamma(0)} = G. \tag{9.14}$$

Here we assume that the temperature T is such that decay takes place through the levels with the energy $E > E_{l+1}$.

Thus, in the limit of a large value of the Q-factor the dependence of the lifetime of a metastable state on pumping frequency represents a combination of a large number of abrupt peaks of width $\Gamma_l/2$ at a distance of $\hbar\omega(E)\,(\partial\omega(E)/\partial E)$ from each other.

It is interesting to consider the case when the width of the peaks is comparable with the distance between them. As above, let us denote by l the number of the level for which the resonance condition is best fulfilled. In this case pumping is important only for those levels which are close to a resonant one. In this region the quantity

$$\alpha = -\hbar\omega(E)\frac{\partial\omega(E)}{\partial E} = \omega_l - \omega_{l+1}$$

can be considered as a constant. As has been shown by Larkin and Ovchinnikov (1986a), in the case when in the resonance there are levels far from the barrier top and bottom of the potential well the ratio of the decay probability without pumping $\Gamma(0)$ to the decay probability with pumping $\Gamma(P)$ is equal to

$$\frac{\Gamma(P)}{\Gamma(0)} = 1 + \frac{2\pi I_1^2}{e^2\alpha\Gamma_l}\sinh\left(\frac{\hbar\omega_l}{kT}\right)\langle l|\varphi|l+1\rangle^2 Z, \tag{9.15}$$

where

$$Z = \frac{1 - \exp(D + D^*)}{|1 - \exp D|^2}, \qquad D = -\frac{2\pi}{\alpha}\left[-\mathrm{i}(\omega_l - \omega) + \Gamma_l/2\right]. \tag{9.16}$$

For large values of the Q-factor the value D is small near resonance and expression (9.15) coincides with eq. (9.14) in the linear approximation in pumping.

In another limiting case, that of a not too large value of the Q-factor, i.e., $Q\hbar|(\partial\omega(E))/\partial E| \ll EkT/(\hbar\omega(E))^2$, there are many levels in the resonance. In this case it is convenient to introduce the functions X, Y related to the elements of the density matrix as follows:

$$\rho_j^j = \exp(-E_j/kT)\,X(E);$$
$$\rho_{j+1}^j = \exp(-E_j/kT)\exp(-i(E_{j+1} - E_j - \hbar\omega)t/\hbar)\,Y(E). \tag{9.17}$$

The functions X, Y are smooth functions of the energy, satisfying the system of differential equations (Larkin and Ovchinnikov 1986a):

$$\bar{A}\frac{\partial X}{\partial E} - \frac{\partial}{\partial E}\left(\bar{B}\frac{\partial X}{\partial E}\right) = \frac{\mathrm{i}I_1}{2e}\exp\left(\frac{\hbar\omega(E)}{kT}\right) \times\left\{\langle j|\varphi|j+1\rangle\left(1-\exp\left(-\frac{\hbar\omega(E)}{kT}\right)\right)(Y - Y^*) - \hbar\omega(E)\frac{\partial}{\partial E}[\langle j|\varphi|j+1\rangle(Y - Y^*)]\right\};$$

$$\bar{A}\frac{\partial Y}{\partial E} - \frac{\partial}{\partial E}\left(\bar{B}\frac{\partial Y}{\partial E}\right) - \mathrm{i}(\omega(E) - \omega)\,Y = -\frac{\mathrm{i}I_1}{2e}\langle j|\varphi|j+1\rangle \times[X(1 - \exp(-\hbar\omega(E)/kT)) - \hbar\omega(E)\exp(-\hbar\omega(E)/kT)\partial X/\partial E], \tag{9.18}$$

where

$$\omega(E) = (E_{j+1} - E_j)/\hbar,$$
$$\bar{A} = \frac{\hbar^2\omega^2(E)}{Re^2}\sum_{n=-\infty}^{\infty} n|\langle j|\exp(\mathrm{i}\varphi)|j+n\rangle|^2,$$
$$\bar{B} = \frac{\hbar^3\omega^3(E)}{2Re^2}\sum_{n=-\infty}^{\infty} n^2\coth\left(\frac{\hbar\omega(E)n}{2kT}\right)|\langle j|\exp(\mathrm{i}\varphi)|j+n\rangle|^2. \tag{9.19}$$

For high temperatures $kT \gg \hbar\omega(E)$ the coefficients $\bar{A}, \bar{B}$ are related by the following:

$$\bar{B} = kT\bar{A}. \tag{9.20}$$

At not too high levels of pumping, the effective temperature differs only slightly *from the thermal bath temperature* T ($\partial \ln X/\partial E \ll 1/kT$). Here in the system of equations (9.18) one can omit the terms with the second derivatives over energy. In the region of high temperatures $[kT \gg \hbar\omega(E)]$ the system of equations (9.18) takes the form

$$\bar{A}\frac{\partial X}{\partial E} = \frac{\mathrm{i}\hbar I_1\omega(E)}{2ekT}\langle j|\varphi|j+1\rangle(Y - Y^*),$$

$$\bar{A}\frac{\partial Y}{\partial E} - \mathrm{i}(\omega(E) - \omega)\,Y = -\frac{\mathrm{i}\hbar I_1\omega(E)}{2ekT}\langle j|\varphi|j+1\rangle X. \tag{9.21}$$

For sufficiently large values of the Q-factor the quantities X, Y change markedly only in the narrow region of energy $|E - E_0| \ll U_0$, where E_0 is the value of the energy at which there is an exact resonance ($\omega(E_0) = \omega$). In this case we get for the value of the relation $\Gamma(P)/\Gamma(0)$ (Larkin and Ovchinnikov 1986a):

$$\ln(\Gamma(P)/\Gamma(0)) = \frac{\pi I_1^2 \exp(\hbar\omega(E)/kT)}{4\hbar^2 \omega^2(E)|\partial\omega(E)/\partial E|}[1 - \exp(-\hbar\omega(E)/kT)]^2 \times [1 + \exp(-2\pi(\omega_0 - \omega)RC)]^{-1}, \tag{9.22}$$

where C is the junction capacitance and ω_0 is the frequency of small oscillations near the bottom of the potential well [eq. (2.5)]. At high temperatures ($kT \gg \hbar\omega$), the effect of the resonance reduction of lifetime of a metastable state has been also considered by Ivlev and Mel'nikov (1986) (cf. also the chapter by the same authors in this volume).

The results of the numerical solution of the system of equations (9.21) for $Q = 13$ are represented in fig. 4 and are compared with the experimental results of Devoret et al. (1984). The theoretical curve has been calculated for pumping power $P = 0.122\omega_0(kT)^2/U_0$, at which the minimum on the theoretical curve coincides with that observed in the experiment. The broken curve of fig. 4 corresponds to expression (9.22) at the same pumping power. The calculated position of the minimum is $\omega_{min}/\omega_0 = 0.975$. In the calculation a quasiclassical expression for the frequency $\omega(E)$ and matrix element $\langle j|\varphi|j+1\rangle$ has been used (Larkin and Ovchinnikov 1984b).

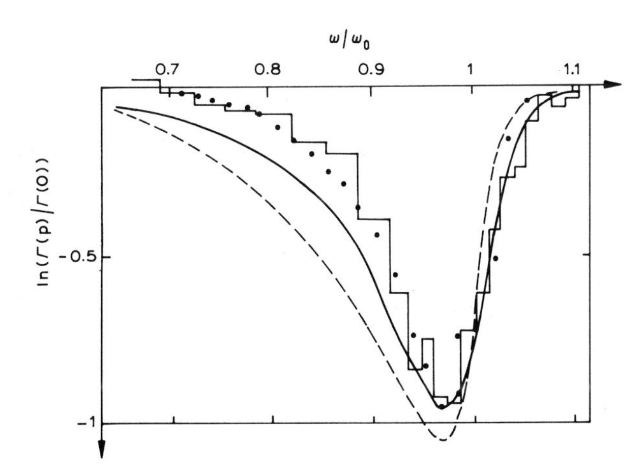

Fig. 4. Resonance in the escape time versus the microwave frequency plot: (•) Experimental values; the solid stepped curve represents the results of the numerical simulation (Devoret et al. 1984); Solid curve represents the numerical solution of eq. (9.21) at $Q = 13$, $p = 0.122\, k^2 T^2 \hbar\omega_0/U_0$; dotted curve represents the large-Q approximation of eq. (9.22).

10. Conclusion

The results presented in this chapter confirm that it is possible to describe a tunnel junction of small size as a quantum particle with finite mass, moving in the potential field $U(\varphi)$ and interacting with a thermal bath. The experiments on resonance stimulation of the decay of a metastable state, as well as the studies of the lifetime of a metastable state at low temperatures, confirm most vividly the possibility of such a description.

Since all the junction parameters can be determined from independent experiments, a tunnel junction is a convenient object for investigating dissipative quantum mechanics. For numerical reasons it is possible to observe quantum effects only in the case when the number of levels in the well is not large. Thus, in the problem of lifetime of a metastable state the quasiclassical approach with the use of wave packets has mainly academic interest. At the same time, the quasiclassical method of finding levels in the system and transition matrix elements between them proves to be unusually exact.

At present, the case when the state of the thermal bath is not an equilibrium one is the least investigated. At low temperatures normal excitations in the superconducting junction electrodes, which serve as a thermal bath, can be easily thrown out of equilibrium ones. An investigation related to this lack of equilibrium would be interesting both from a theoretical and an experimental viewpoint.

References

Affleck, I., 1981, Quantum-statistical metastability, Phys. Rev. Lett. **46**, 388.

Ambegaokar, V.A., and B.I. Halperin, 1969, Voltage due to thermal noise in the DC Josephson effect, Phys. Rev. Lett. **22**, 1364.

Ambegaokar, V.A., U. Eckern and G. Schön, 1982, Quantum dynamics of tunneling between superconductors, Phys. Rev. Lett. **48**, 1745.

Barone, A., C. Camerlingo, R. Cristiano and Yu.N. Ovchinnikov, 1985, Effects of fluctuations on current–voltage characteristics of Josephson tunnel junctions, IEEE Trans. Magn. **MAG-21**(2), 626.

Büttiker, M., E.P. Harris and R. Landauer, 1983, Thermal activation in extremely underdamped Josephson-junction circuits, Phys. Rev. B **28**, 1268.

Caldeira, A.O., and A.J. Leggett, 1981, Influence of dissipation on quantum tunneling in macroscopic systems, Phys. Rev. Lett. **46**, 211.

Callan, C., and S. Coleman, 1977, Fate of the false vacuum II, first quantum corrections, Phys. Rev. D **16**, 1762.

Chang, L.-D., and S. Chakravarty, 1984, Quantum decay in a dissipative system, Phys. Rev. B **29**, 130. Erratum: **30**, 1566.

Chow, K.S., D.A. Browne and V.A. Ambegaokar, 1988, Quantum kinetics of a superconducting tunnel junction: theory and comparison with experiment, Phys. Rev. B **37**, 1624.

Devoret, M.H., J.M. Martinis, D. Esteve and J. Clarke, 1984, Resonant activation from the zero-voltage state of a current-biased Josephson junction, Phys. Rev. Lett. **53**, 1260.

Devoret, M.H., J.M. Martinis and J. Clarke, 1985, Measurements of macroscopic quantum tunneling out of the zero-voltage state of a current-biased Josephson junction, Phys. Rev. Lett. **55**, 1908.

Erdelyi, A., 1953, Higher Transcedental Functions (McGraw-Hill, New York) sect. 8.2.

Freidkin, E., P.S. Riseborongh and P. Hänggi, 1988, The influence of dissipation on the quantal transition state tunneling rate, Solid State Phys. **21**, 1543.

Grabert, H., and U. Weiss, 1984a, Crossover from thermal hopping to quantum tunneling, Phys. Rev. Lett. **53**, 1787.

Grabert, H., and U. Weiss, 1984b, Thermal enhancement of the quantum decay rate in a dissipative system, Z. Phys. B **56**, 171.

Grabert, H., U. Weiss and P. Hänggi, 1984, Quantum tunneling in dissipative systems at finite temperatures, Phys. Rev. Lett. **52**, 2193.

Grabert, H., P. Olschowski and U. Weiss, 1985, Temperature dependence of quantum decay rates in dissipative systems, Phys. Rev. B **32**, 3348.

Grabert, H., P. Olschowski and U. Weiss, 1987, Quantum decay rates for dissipative systems at finite temperatures, Phys. Rev. B **36**, 1931.

Ivanchenko, Yu.M., and I.A. Zil'berman, 1968, The Josephson effect on small size tunnel contacts, Zh. Eksp. & Teor. Fiz. **55**, 2395.

Ivlev, B.I., and V.I. Mel'nikov, 1986, Effect of resonant pumping on activated decay rates, Phys. Lett. A **116**, 427.

Ivlev, B.I., and Yu.N. Ovchinnikov, 1987, Decay of metastable states at presence of close underbarrier trajectories, Zh. Eksp. & Teor. Fiz. **93**, 668.

Jackel, L.D., J.P. Gordon, E.L. Hu, R.E. Howard, L.A. Fetter, D.M. Tennant and R.W. Epworth, 1981, Decay of the zero voltage state in small-area, high-current-density Josephson junctions, Phys. Rev. Lett. **47**, 697.

Kopietz, P., and S. Chakravarty, 1988, Lifetime of metastable voltage states of superconducting tunnel junctions, Phys. Rev. B **38**, 97.

Korshunov, S.E., 1987, Quantum mechanical tunneling with dissipation in tilted sine potential, Zh. Eksp. & Teor. Fiz. **92**, 1828.

Kramers, H.A., 1940, Brownian motion in a field of force and the diffusion model of chemical reactions, Physica **7**, 284.

Landau, L.D., and E.M. Lifshitz, 1974, Quantum Mechanics (Phys. Math. Publ. Corp., Moscow).

Langer, J.S., 1967, Theory of the condensation point, Ann. Phys. (New York) **41**, 108.

Larkin, A.I., and Yu.N. Ovchinnikov, 1966, Tunnel Effect between superconductors in an alternating field, Zh. Eksp. & Teor. Fiz. **51**, 1935.

Larkin, A.I., and Yu.N. Ovchinnikov, 1983a, Decay of supercurrent in tunnel junctions, Phys. Rev. B **28**, 6281.

Larkin, A.I., and Yu.N. Ovchinnikov, 1983b, Quantum tunneling with dissipation, Zh. Eksp. & Teor. Fiz. Pis'ma **37**, 322.

Larkin, A.I., and Yu.N. Ovchinnikov, 1983c, Attenuation of a superconducting current in tunnel junctions, Zh. Eksp. & Teor. Fiz. **85**, 1510.

Larkin, A.I., and Yu.N. Ovchinnikov, 1984a, Quantum mechanical tunneling with dissipation, The pre-exponential factor, Zh. Eksp. & Teor. Fiz. **86**, 719.

Larkin, A.I., and Yu.N. Ovchinnikov, 1984b, Current attenuation in superconducting junctions with a nonequilibrium electron distribution function, Zh. Eksp. & Teor. Fiz. **87**, 1842.

Larkin, A.I., and Yu.N. Ovchinnikov, 1985, The crossover from classical to quantum regime in the problem of the decay of the metastable state, J. Stat. Phys. **41**, 425.

Larkin, A.I., and Yu.N. Ovchinnikov, 1986a, Resonance reduction of the lifetime of the metastable state of tunnel junctions, J. Low Temp. Phys. **63**, 317.

Larkin, A.I., and Yu.N. Ovchinnikov, 1986b, Effect of level quantization on the lifetime of metastable states, Zh. Eksp. & Teor. Fiz. **91**, 318.

Larkin, A.I., K.K. Likharev and Yu.N. Ovchinnikov, 1984, Secondary quantum macroscopic effects in weak superconductivity, Physica B+C **126**, 414.

Lifshitz, I.M., and Yu. Kagan, 1972, Quantum kinetics of phase transition at temperatures close to absolute zero, Zh. Eksp. & Teor. Fiz. **62**, 385.

Martinis, J.M., M.H. Devoret and J. Clarke, 1985, Energy-level quantization in the zero-voltage state of current-biased Josephson junction, Phys. Rev. Lett. **55**, 1543.

Matkowsky, B.J., Z. Shuss and C. Tier, 1984, Uniform expansion of the transition rate in Kramers' problem, J. Stat. Phys. **35**, 443.

Mel'nikov, V.I., 1984, Activated tunneling decay of metastable states, Zh. Eksp. & Teor. Fiz. **87**, 663.

Mel'nikov, V.I., and S.V. Meshkov, 1983, Brownian motion of the quantum partices, Zh. Eksp. & Teor. Fiz. Pis'ma **38**, 111.

Mel'nikov, V.I., and S.V. Meshkov, 1986, Theory of activated rate processes: exact solution of the Kramers problem, J. Chem. Phys. **85**, 1018.

Ovchinnikov, Yu.N., and A. Barone, 1987, Influence of dissipation on the prefactor in the expression of the decay rate of a metastable state, J. Low Temp. Phys. **67**, 323.

Ovchinnikov, Yu.N., and A. Barone, 1988, Erratum, J. Low Temp. Phys. **72**, 195.

Ovchinnikov, Yu.N., R. Cristiano and A. Barone, 1984, Effect of capacitance of I–V characteristics of overdamped Josephson junctions: classical and quantum limits, J. Appl. Phys. **56**, 1473.

Riseborough, P.S., P. Hänggi and E. Freidkin, 1985, Quantum tunneling in dissipative media: intermediate-coupling-strength results, Phys. Rev. A **32**, 489.

Schwartz, D.B., B. Sen, C.N. Archie and J.E. Lukens, 1985, Quantitative study of the effect of the environment on macroscopic quantum tunneling, Phys. Rev. Lett. **55**, 1547.

Voss, R.F., and R.A. Webb, 1981, Macroscopic quantum tunneling in 1-μn Nb Josephson junctions, Phys. Rev. Lett. **47**, 265.

Waxman, D., and A.J. Leggett, 1985, Dissipative quantum tunneling at finite temperatures, Phys. Rev. B **32**, 4450.

Wolynes, P.G., 1981, Quantum theory of activated events in condensed phases, Phys. Rev. Lett. **47**, 968.

Zaikin, A.D., and S.V. Panyukov, 1986, The life time of macroscopic current states, Zh. Eksp. & Teor. Fiz. Pis'ma **43**, 518.

Zittartz, J., and J.S. Langer, 1966, Theory of bound states in a random potential, Phys. Rev. **148**, 741.

CHAPTER 5

Quantum Tunneling in a High-Frequency Field

B.I. IVLEV and V.I. MEL'NIKOV

L.D. Landau Institute for Theoretical Physics
Academy of Sciences
Moscow, Russia

Quantum Tunnelling in Condensed Media
Edited by
Yu. Kagan and A.J. Leggett

Contents

1. Introduction

In the last few years the quantum properties of Josephson junctions have been actively investigated both theoretically and experimentally. The interest in studies of these systems is motivated by progress in the fabrication of small-size junctions (Voss and Webb 1981, Jackel et al. 1981). Their dynamics is described by one variable, the phase difference of the contacting superconductors. Such junctions could be considered as the simplest quantum systems. They are specifically suited to the verification of quantum-mechanical concepts about particle motion in static and alternating fields under interaction with a thermostat. This field of physics is usually covered by the concept of macroscopic quantum tunneling (Caldeira and Leggett 1981, Ambegaokar et al. 1982, Larkin and Ovchinnikov 1983). There are also some other phenomena where the effects of quantum tunneling are of crucial importance, e.g., quantum diffusion (Kagan and Prokof'ev 1987, 1989), kinetics of phase transitions (Lifshitz and Kagan 1972), autoelectronic emission, interband breakdown and charge exchange in semiconductors (see Ziman 1964), tunnel chemical reactions, motion of dislocations in crystals (Petukhov and Pokrovskii 1972) and motion of a charge density wave in Peierls dielectrics (Rice et al. 1976).

In this paper we consider the influence of a high-frequency field on the above-mentioned effects. We suppose that the conditions for the semiclassical approximation for the amplitude and frequency of the alternating field are valid. First a general formulation of the problem of quantum tunneling in an alternating field will be given. Then the limit of a weak alternating field (linear in the amplitude of the field contribution to the tunneling exponent) will be investigated in detail, as well as the decay of a current state of a Josephson junction. We consider also an exact solution of the semiclassical problem of autoelectronic emission and interband tunneling.

The processes of subbarrier transmission and above-barrier reflection, which are forbidden by classical mechanics, acquire finite probabilities when quantum effects are taken into account. As a rule, these probabilities decrease exponentially as the barrier width to particle wavelength ratio increases. The computation of the probability for a classically forbidden process has a certain peculiarity from the mathematical standpoint: there arises necessarily the concept of motion in imaginary time or along a complex trajectory (see Landau and

Lifshitz 1977). This characteristic of semiclassical processes makes their computation quite difficult, and because of this, no detailed investigation, in particular, of the effect of a variable perturbation on classically forbidden processes has thus far been published. The problems that have been solved either assume that the variable perturbation is weak, as in the case of the Franz–Keldysh effect (see Anselm 1978), or are limited by the stipulation that the static potential is a short-range one, a condition which, fortunately, turns out to be adequate for the investigation of the practically important problem of many-photon ionization of an atom (Keldysh 1964, Baz et al. 1969, Delone and Krainov 1984). Thus, a large number of phenomena that arise when semiclassical processes occurring in realistic potential fields are exposed to the nonlinear action of a high-frequency field have as yet not been investigated.

Let us illustrate the subject of the present paper by the process of tunneling. If the amplitude of the alternating field is small, then the passage through the barrier will be determined largely by ordinary tunneling, and the alternating field can be taken into account within the framework of perturbation theory. This means that the probability for tunneling accompanied by absorption of one or several photons is small compared to the probability for tunneling in zero field. But if the alternating-field strength exceeds a certain value, then the tunneling will be insignificant, and the passage will occur as a result of absorption by the particle of as many photons as it needs in order to get to the top of the potential barrier.

These limiting cases are separated by a broad range of alternating-field strength values at which the passage through the barrier is a process of the mixed type, in the sense that it is advantageous for the particle to absorb a certain number of field quanta so as to tunnel in a higher-energy region, where the barrier is more transparent. The optimum number of absorbed quanta is determined in this case by the competition between the growth of the tunneling probability and the decrease of the absorption probability as the number of quanta increases (Ivlev and Mel'nikov 1986a).

This type of problem was originally solved (Ivlev and Mel'nikov 1985a) in a weak-alternating-field approximation. It was shown there that the amplitude of the alternating field $\mathscr{E}\cos(\Omega t)$ enters into the solution in the combination $\mathscr{E}\exp(\Omega\tau_s)$, where τ_s is connected with particle motion in the forbidden region. Thus, in the high-frequency limit $\Omega\tau_s \gg 1$ the barrier penetration probability depends anomalously strongly on the amplitude and frequency of the field (Ivlev and Melnikov 1985a, b).

In spite of the fact that the problem under discussion is a pressing one, in view of the latest experimental advances in the study of the voltage states of Josephson junctions (Dmitrenko et al. 1982, Schwartz et al. 1985, Devoret et al. 1985), the question of the effect of an alternating field on the tunneling processes has almost not been touched upon in the literature. The effect of an alternating field on the motion of a particle in the classically allowed region is considered by

Chakravarty and Kivelson (1983, 1985), Golub (1985), Ivlev and Mel'nikov (1986b) and Larkin and Ovchinnikov (1986). Buettiker and Landauer (1982) have made an attempt to take into account the effect of an alternating field on subbarrier motion, but the general arguments do not lead to specific results.

The purpose of the present paper is to investigate semiclassical processes in a high-frequency field which is not necessarily considered to be weak. In section 2 we qualitatively consider tunneling in an alternating field, and determine the typical orders of magnitude of the quantities involved in the process. In section 3 we formulate the problem in a relatively general situation and indicate a procedure for solving it with the use of the method of complex trajectories. Since Newton's equation in variable and spatially inhomogeneous fields can be solved only in specific cases, the remaining part of the paper is devoted to the analysis of a number of specific problems. In section 4 we investigate the effect of a weak alternating field on tunneling, i.e., the case in which it is sufficient to take account of the linear-in-the-field correction to the argument of the tunneling exponential function. This in no way implies that the effect in question is weak. On the contrary, the condition of applicability of the semiclassical approximation requires that the transmission coefficient should increase by several orders of magnitude. To elucidate more fully the physical picture of the phenomenon in question, we consider the effect of spatially inhomogeneous perturbations in the same linear approximation. In section 5 we study the decay of metastable states in an alternating field, and show that the oscillating dependence of the argument of the tunneling exponential function on the frequency is connected not only with the normal classical but also with a specific quantum resonance, whose frequency is determined by particle motion in the forbidden region. In section 6 we briefly investigate the effect of an alternating field on above-barrier reflection. An allowance in the argument of the exponential function for the terms nonlinear in the field is possible only for relatively simple potentials; therefore, here we shall not generalize the results obtained by Ivlev and Mel'nikov (1985b, c) for a sinusoidal potential to the nonlinear case. We next present the solution to the nonlinear problem in two situations. In section 7 we consider the dependence of the exponential enhancement by an alternating field on the shape of the potential barrier. In section 8 the influence of the alternating component of the current across a Josephson junction on the lifetime of the current state is investigated. In section 9 we obtain and investigate in detail the exact solution to the problem of tunneling through a triangular barrier in an alternating field. These results have a direct bearing on the phenomenon of field emission. With the aid of the simplest band structure model for a semiconductor, we carry out in section 10 a detailed investigation of the interband breakdown in constant and alternating electric fields (the nonlinear Franz–Keldysh effect). In section 11 the tunneling motion of a string under the effect of an alternating field is considered. In section 12 we discuss the results obtained in the paper and the possibility of their experimental observation.

2. A naive classification of tunneling regimes

Let us consider the problem of subbarrier tunneling in a uniform alternating field from a phenomenological point of view, in order to elucidate the physics of the matter, without laying any claims to quantitative results. We shall assume that a particle of energy E is incident from the left on a potential barrier (of height V) of the type shown in fig. 1. In the absence of an alternating field the probability for penetration through the barrier is, with exponential accuracy, equal to $\exp[-A_0(E)]$, where A_0 is the imaginary part of the corresponding action. In the general case, for a barrier of width of the order of a, and for $E \propto V$, the quantity $A_0 \propto V/\omega$, where $\omega = (V/ma^2)^{1/2}$, is the characteristic oscillation frequency in the inverted potential. On the basis of the condition for the semiclassical approximation, we have $A_0 \gg 1$.

When the alternating field is taken into account within the framework of perturbation theory, the transmission probability increases by a term proportional to the square of the field:

$$D \approx \exp[-A_0(E)] + (\mathscr{E}/\tilde{\mathscr{E}})^2 \exp[-A_0(E+\Omega)], \tag{2.1}$$

where $\tilde{\mathscr{E}}$ is some internal field, the magnitude of which will be discussed later. The second term in expression (2.1) corresponds to single-photon absorption and subsequent tunneling with the increased energy $E + \Omega$. We shall assume everywhere below that the alternating-field frequency is small compared to the height of the potential barrier and the initial energy of the particles: $\Omega \ll V, E$. Taking into account the relation

$$\partial A_0/\partial E = -2\tau_0,$$

where $i\tau_0$ is the imaginary time of the motion under the barrier between the

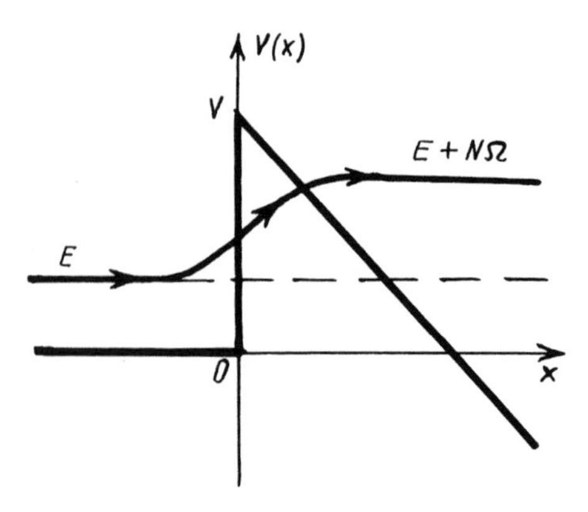

Fig. 1. Subbarrier transmission of a particle with absorption of N quanta of the alternating field.

turning points, we obtain in place of expression (2.1) the expression

$$D(E) \approx D(0)\{1 + [\mathscr{E}\exp(\Omega\tau_0)/\tilde{\mathscr{E}}]^2\}. \tag{2.2}$$

It can be seen from this that perturbation theory ceases to be applicable even in very weak fields, specifically, in fields of the order of $\tilde{\mathscr{E}}\exp(-\Omega\tau_0)$, when we are considering the high-frequency limit $\Omega\tau_0 \gg 1$. Let us note that $\tau_0 \propto 1/\omega$; therefore, for semiclassical potentials (i.e., for $V \gg \omega$), the semiclassicality condition $V \gg \Omega$ can be fulfilled simultaneously with the condition $\Omega\tau_0 \gg 1$.

As the amplitude of the alternating field increases, many-photon processes, i.e., the next terms of the expansion in powers of $(\mathscr{E}/\tilde{\mathscr{E}})^2$ in expression (2.1), become more and more important. The term corresponding to the absorption of N photons has the form

$$D_N \approx (e\mathscr{E}/\tilde{\mathscr{E}}N)^{2N}\exp[-A_0(E + N\Omega)], \tag{2.3}$$

where $N\Omega$ is the energy received by the particle from the field. The change in the perturbation theory parameter is due to the fact that the Nth term in the perturbation theory series contains in the denominator the factor $(N!!)^4$ (Delone and Krainov 1984), which has been included in expression (2.3) after being approximated by the Stirling formula, which is applicable when $N \gg 1$. We find the optimum number of quanta $N_{\rm m}$ by minimizing expression (2.3) with respect to N, as a result of which we obtain

$$D \approx \exp[-A_0(E) + 2\mathscr{E}\exp(\Omega\tau_0)/\tilde{\mathscr{E}}], \tag{2.4}$$

$$N_{\rm m} \approx \mathscr{E}\exp(\Omega\tau)/\tilde{\mathscr{E}}.$$

This expression is applicable when the correction to the argument of the exponential function is much greater than unity, a condition which coincides with the condition for the process to be a multiphoton one. The effective field $\mathscr{E}\exp(\Omega\tau_0)$ should, therefore, be stronger than the characteristic internal field $\tilde{\mathscr{E}}$. Moreover, $N_{\rm m}\Omega$ should be at least significantly smaller than the barrier height $V - E$. When the field $\mathscr{E}$ is increased further, and the energy transfer becomes comparable in order of magnitude to the barrier height $V - E$, passage occurs without the participation of tunneling. Retaining only the pre-exponential factor in expression (2.3) with $N\Omega$ replaced by $V - E$, we obtain in this limit the expression

$$D \approx \exp\left[-\frac{2(V-E)}{\Omega}\ln\frac{\tilde{\mathscr{E}}(V-E)}{\mathscr{E}\Omega}\right]. \tag{2.5}$$

The exact limits of applicability of expressions (2.2) and (2.5) will be indicated in section 9, after the problem has been rigorously solved.

Expressions (2.2) and (2.5) have been obtained on the basis of simple physical arguments. The parameter $\tilde{\mathscr{E}}$ entering into them can easily be estimated by computing, with the aid of perturbation theory, the linear-in-the-field-amplitude correction to the wave function of the transmitted particle. In the case of, e.g., the triangular barrier shown in fig. 1, this calculation yields

$$\tilde{\mathscr{E}} \propto \Omega^2[m/(V-E)]^{1/2}.$$

Naturally, for potentials of general form, the foregoing results are only of a qualitative nature, for the same reason that the total probability cannot in the general case be represented in the form of a product of separate probabilities for absorption and tunneling. Nevertheless, for potential barriers having artificial singularities, e.g., kinks, the qualitative results obtained here are, as shown below, exact in the quantitative sense as well. Only the regions of applicability and the numerical coefficients of $\tilde{\mathscr{E}}$ in the various limiting cases are determined more accurately. Thus, the correctness of expression (2.4) for triangular and rectangular barriers can easily be verified.

Formula (2.4) with the exponential amplification of the field amplitude $\mathscr{E}$, which is based on simplified arguments, is valid qualitatively for potential barriers of a general type. In the exact formula, instead of τ_0, as shown below, some other quantity appears, which depends essentially on the particular type of the potential. For some potentials it has the same order of magnitude as τ_0, but for some special cases it can even be zero.

3. *Starting semiclassical expressions*

Let us proceed to the derivation of the general expression for the probability for tunneling of a particle through an arbitrary time-dependent semiclassical barrier $V(x, t)$. As is well known, in the semiclassical limit the wave functions can be sought with exponential accuracy in the form

$$\psi(x, t) = \exp[iS(x, t)],$$

where $S(x, t)$ is the classical action, and x and t lie on the classical particle trajectory, which can be found from Newton's equation

$$m\mathrm{d}^2x/\mathrm{d}t^2 + \partial V(x, t)/\partial x = 0. \tag{3.1}$$

Let the particle be incident on the barrier from the left. The problem is to find a relation between the values of the wave function at points x_1 and x_2 lying on opposite sides of the barrier. The forbiddenness of the tunneling process in classical mechanics implies, however, that there does not exist an ordinary trajectory connecting such points. For this reason, we shall consider the trajectories in complex time along the contour C_+ in fig. 2. On the symmetrically located contour C_- we have $x(t^*) = x^*(t)$. To the right on $C_\pm$ the quantities x and t are real, x lies to the right of the barrier, and the solution to eq. (3.1) depends on two arbitrary real parameters. We shall assume that the particle emerges from under the barrier at the moment of time t_2 at the point x_2. This

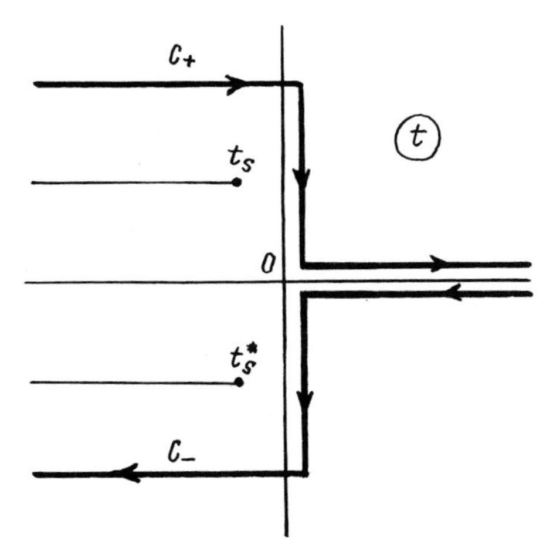

Fig. 2. The integration contour, singular points of the trajectory $x(t)$, and branch cuts for the computation of the subbarrier transmission coefficient.

means that $x(t_2) = x_2$ and $(\mathrm{d}x/\mathrm{d}t)_{t2} = 0$. Then, as the two real parameters, we can take the point x_2 and the instant t_2 at which the particle emerges from under the barrier.

Let us assume that the nonstationary part of the potential is at least adiabatically switched off at $t \to \infty$, i.e., $V(x, t) \to V(x)$. Then at points far to the left on the contour C_+ the function $x(t)$ is a solution to the steady-state equation (3.1) and depends on two parameters, which we can take to be the time shift t_1 and the conserved quantity $E = \frac{1}{2}m(\mathrm{d}x/\mathrm{d}t)^2 + V(x)$, i.e., $x(t) = x(t - t_1, E)$. It is significant that the quantities E and t_1 are, generally speaking, complex, and depend on the point x_2 and the instant t_2 at which the particle emerges from under the barrier.

The physically justified formulation of the problem consists in our prescribing the initial particle energy E and measuring the particle flux emerging from under the barrier at the instant t_2. In the general case there are no grounds for assuming that the trajectory $x(t)$ defined by the real parameters E and t_2 (the latter enters into the problem through the condition $(\mathrm{d}x/\mathrm{d}t)_{t2} = 0$) will itself be real. A more detailed analysis shows, however, that the time-averaged probability for semiclassical processes is determined solely by the real trajectories. The point is that the imaginary part of the action $A(x, t)$ $[=2\,\mathrm{Im}\,S(x, t)]$, obtained from the Hamilton–Jacobi equation is, on the basis of the semiclassicality condition, large, and its variations should also be much greater than unity. Accordingly, the particles should pass through the barrier at instants lying in a narrow neighborhood of that instant at which the function $A(t)$ has its minimum value. With an allowance made for the condition $(\mathrm{d}x/\mathrm{d}t)_{t2} = 0$, the condition for $A(t_2)$ to be the minimum value of $A(t)$ has the form

$$\partial A/\partial t_2 = -2\,\mathrm{Im}\,V(x_2) = 0,$$

from which the reality of x_2 follows. In what follows, we limit ourselves to the computation of the minimum values of the function $A(t)$ and, therefore, we shall consider the trajectory $x(t)$ to be real. The integration contour C_+ in this case consists of a vertical section and two horizontal sections, as shown in fig. 2.

The value of the wave function on the remote left section of the contour C_+ differs from the value of the function on the real axis by the quantity $E\tau_0$ in the index of the exponential function. With an allowance made for this contribution, the effective Lagrangian has the form

$$L = \tfrac{1}{2}m(\mathrm{d}x/\mathrm{d}t)^2 - V(x, t) + E, \tag{3.2}$$

from which we obtain for the transmission coefficient the expression

$$D = \exp(-A), \quad A = -\mathrm{i}\int_C L\,\mathrm{d}t, \tag{3.3}$$

where the contour $C = C_+ + C_-$. For specific calculations it may turn out to be convenient to shift the integration path, allowing for its linkage behind the singular points of the trajectory.

Thus far we have considered the problem of tunneling of a particle with energy E prescribed at $t \to -\infty$, when the potential is stationary. If the tunneling proceeds from a state of thermodynamic equilibrium, then the result should be averaged over the energy E with the Gibbs distribution (Lifshitz and Kagan 1972):

$$\langle D \rangle = \int \exp[-E/T - A(E)]\,\mathrm{d}E, \tag{3.4}$$

where $A(E)$ is the action computed earlier using eqs. (3.2) and (3.3). Here it is important that the energy derivative of the action $A(E)$ be equal to $-2\tau_0$, as was the case in the absence of the alternating perturbation. The point is that because the action is an extremal quantity, only the term with E in eq. (3.3) makes a contribution to $\partial A/\partial E$ and $2\tau_0$ is the distance between the remote ends to the contours $C_\pm$. Of course, the quantity τ_0 may itself depend on the amplitude of the variable field. Thus, in thermodynamic equilibrium the barrier penetration factor is found by substituting into eq. (3.3) the real trajectory satisfying the condition

$$\tau_0 = 1/2T, \tag{3.5}$$

which implicitly selects the energy of the particles tunneling through the barrier.

The scope of the general expressions obtained in the present section is revealed below in a number of specific examples.

4. Tunneling through analytical potentials in a weak alternating field

Let us consider tunneling through potential barriers specified by analytic functions, taking account of only that correction to the action A_0 which is linear in the monochromatic field. We shall consider the field to be homogeneous, so that

$$L = \tfrac{1}{2}m(\mathrm{d}x/\mathrm{d}t)^2 - V(x) + \mathscr{E}x\cos\Omega t + E.$$

We shall also assume that $\Omega \ll V$, a condition which allowed us to use the semiclassical expressions obtained in the previous section. As is well known, an allowance made for a small perturbation in the Lagrangian in the calculations of the action amounts to the substitution into the Lagrangian of the unperturbed trajectory. The action A can, therefore, be represented in the form

$$A = A_0 + A_1,$$

where

$$A_0 = 2(2m)^{1/2}\int_{x_1}^{x_2}[V(x) - E]^{1/2}\,\mathrm{d}x,$$
$$A_1 = -\mathrm{i}\mathscr{E}\int_C x(t)\cos(\Omega\tau + \chi)\,\mathrm{d}t. \tag{4.1}$$

Here x_1 and x_2 are the turning points, $V(x_{1,2}) = E$, and $x(t)$ is the zero-field classical trajectory, fixed, e.g., by the requirement that the particle emerges from under the barrier at the instant $t = 0$, i.e., by the condition that $x(0) = x_2$. The quantity χ then denotes the relative phase of the field.

The action A_1, as a function of the two variables x and t, can be found by computing the linear-in-$\mathscr{E}$ correction $S_1(x, t)$ to the solution of the Hamilton–Jacobi equation:

$$\frac{\partial S}{\partial t} + \frac{1}{2m}\left(\frac{\partial S}{\partial x}\right)^2 + V(x) = \mathscr{E}x\cos\Omega t.$$

We then obtain for A_1 $(=2\,\mathrm{Im}\,S)$ in the region to the right of the barrier the expression

$$A_1(x, t) = -\,\mathrm{i}\mathscr{E}\int_C \mathrm{d}t'\,x(t')\cos\{\Omega(t + t')$$
$$+\,\chi - \Omega\int_{x_2}^{x}(2[E - V(y)]/m)^{-1/2}\,\mathrm{d}y\}, \tag{4.2}$$

which coincides with eq. (4.1) if $t = 0$ and $x = x_2$.

The integration path in eqs. (4.1) and (4.2) is the rectangular contour in fig. 2, where the horizontal straight line on the left is at a distance of $i\tau_0$ from the real axis and corresponds to the motion to the left of x_1, the vertical section from $i\tau_0$ to 0 corresponds to subbarrier motion, and the positive half of the t axis corresponds to the motion to the right of x_2. On such a contour, $x(t)$ is real.

To evaluate the integral (4.1), it is convenient to shift the contour C far to the left (where the field is adiabatically switched off), allowing for its linkage behind the singular points of the trajectory $x(t)$. The importance of studying the singularities of the unperturbed trajectory is apparent from this. As can easily be seen from the following implicit dependence,

$$t(x) = \int_{x_2}^{x} \{2[E - V(y)]/m\}^{-1/2}\,dy, \tag{4.3}$$

the singularities of the function $x(t)$ are connected with those of the function $V(x)$ in the complex x plane.

Let us consider the barriers for which $V(x)$ possesses power-law singularities at some points x_s and x_s^*, becoming infinite there:

$$V(x) \approx \kappa(x - x_s)^{\alpha}, \quad x \to x_s, \tag{4.4}$$

where $\alpha < 0$. Included here are singularities of the type $V \approx \kappa x^{\alpha}$ for $x \to \infty$ and $\alpha > 0$. In the vicinity of x_s the solution to eq. (3.1) has the form

$$x(t) = x_s + [-\kappa(2 - \alpha)^2 (t - t_s)^2/2m]^{1/(2-\alpha)}, \tag{4.5}$$

where t_s is the complex time required for the motion from x_2 to x_s:

$$t_s = \int_{x_2}^{x_s} \{2[E - V(y)]/m\}^{-1/2}\,dy. \tag{4.6}$$

The corresponding singular points and branch cuts are shown in fig. 2. In order of magnitude, $\tau_s \equiv \operatorname{Im} t_s$ is equal to the time τ_0 of subbarrier motion. In the limit of high alternating-field frequency, i.e., for $\Omega\tau_s \gg 1$, the dominant contribution to the integral (4.1) is made by the branch-cut sections close to the singular points t_s and t_s^*. For the transmission coefficient we finally obtain

$$D(\mathscr{E}, t) = D(0)\exp[a_1 \cos(\Omega t + \chi_1)], \tag{4.7}$$

where χ_1 is the phase shift, which is unimportant for the following discussion because the semiclassical approach corresponds to the maximum value of eq. (4.7), and

$$a_1 = \frac{2\pi\mathscr{E}}{\Omega}\left|\Gamma\left(\frac{2}{\alpha - 2}\right)\right|^{-1}\left[\frac{|\kappa|(2-\alpha)^2}{2m\Omega^2}\right]^{1/(2-\alpha)} \exp(\Omega\tau_s). \tag{4.8}$$

This exact result is similar in structure to expression (2.4), but instead of τ_0 it contains the time τ_s, which for analytic potentials is always smaller than τ_0. The maximum value of the field on a real trajectory is, of course, of the order of $\mathscr{E}\exp(\Omega\tau_0)$, but because of the field oscillations its contribution to the action is greatly reduced, so that, e.g., for even potentials $\mathscr{E}_{\rm eff} \propto \mathscr{E}\exp(\Omega\tau_0/2)$.

Averaging eq. (1.7) over time with an allowance made for the inequality $a_1 \gg 1$, we obtain

$$\overline{D(\varepsilon)} = D(0)(2\pi a_1)^{-1/2}\exp(a_1). \tag{4.9a}$$

The use of perturbation theory makes it possible for us to compute the pre-exponential factor.

The results given by (4.7) and (4.9a) are applicable so long as

$$A_0 \gg a_1 \gg 1, \tag{4.9b}$$

where the limitation from above justifies the linear expansion in $\mathscr{E}$, while the limitation from below is due to the use of the semiclassical approximation. The criterion limiting the field amplitude from above becomes much more rigid in a more accurate calculation, as can be seen from the results pertaining to tunneling through a triangular barrier (section 9) and interband tunneling (section 10).

4.1. Particular examples

Let us illustrate the results obtained with potentials of specific form. If $V(x) = V[\cosh(x/a)]^{-2}$, then the trajectory with energy E is given by the relations

$$\sinh(x/a) = [(V-E)/E]^{1/2}\cosh\omega t, \quad \omega^2 \equiv 2E/(ma^2). \tag{4.10}$$

$$\mathrm{d}x/\mathrm{d}t = a\omega\sinh(\omega t)[E/(V-E) + \cosh^2\omega t]^{-1/2}.$$

The appearance of singularities of the solution in the complex time plane can be directly seen from these relations. Substituting into the general expression (4.8) the values

$$x_s = \tfrac{1}{2}\mathrm{i}\pi a, \qquad \kappa = -Va^2, \qquad \alpha = -2, \qquad \tau_s = \pi/(2\omega),$$

we obtain for the transmission coefficient the expression

$$D(\mathscr{E}) = D(0)\exp\left[\frac{a\mathscr{E}}{\Omega}\left(\frac{2\pi\omega}{\Omega}\right)^{1/2}\left(\frac{V}{E}\right)^{1/4}\exp(\Omega\tau_s)\right]. \tag{4.11}$$

For the potential $V(x) = V(1 + x^2/a^2)^{-1}$ we have $x = \mathrm{i}a$, $\kappa = \frac{1}{2}\mathrm{i}Va$ and $\alpha = -1$, and from eqs. (4.8) and (4.9a) we obtain

$$D(\mathscr{E}) = D(0)\exp\left[\frac{4\pi a\mathscr{E}}{\Omega\Gamma(1/3)}\left(\frac{V}{6ma^2\Omega^2}\right)^{1/3}\exp(\Omega\tau_s)\right], \tag{4.12}$$

where τ_s can be expressed in terms of an elliptic integral.

The exponential field enhancement has been noted by V'yurkov and Ryzhii (1980) in the case of electron tunneling between δ-function wells, when τ_s is equal to the time of flight through the forbidden region, and by Sumetskii (1985) in the case of tunneling through a triangular barrier.

4.2. *Distribution of particles after passing the barrier*

Let us compute the energy distribution for the particles that have tunneled through. The time dependence of the wave function can be explicitly found from the solution to the Hamiltonian–Jacobi equation. In the limit $\Omega\tau_s \gg 1$ we have

$$\psi(t) \propto \exp[-\mathrm{i}Et - \tfrac{1}{2}a_1 \exp(-\mathrm{i}\Omega t)]$$

since there remains out of $\cos \Omega t$ only the term that increases in the region $\mathrm{Im}\, t > 0$. Leaving out the structure of the energy spectrum at scales of Ω, we find that the spectral envelope

$$P(\omega) \propto \left| \int \psi(t) \exp(\mathrm{i}\omega t)\, \mathrm{d}t \right|^2$$

$$\propto \exp[-2(\omega - E - \tfrac{1}{2}a_1\Omega)^2/a_1\Omega^2]. \tag{4.13}$$

Thus, the particles that have passed through the barrier gain, on an average, an energy of $\frac{1}{2}a_1\Omega$, and have a Gaussian distribution in an interval of order $a_1^{1/2}\Omega$.

4.3. *Nonhomogeneous alternating field*

Let us now consider the situation in which the intensity of the alternating field is nonuniform in space:

$$V_1(x, t) = V_1(x) \cos \Omega t.$$

Then instead of eq. (4.1), for the correction to the action we write

$$A_1 = \mathrm{i} \int_C V_1(x(t) \cos \Omega t\, \mathrm{d}t, \tag{4.14}$$

where the singularities of both the unperturbed trajectory $x(t)$ and the potential $V_1(x, t)$ as a function of x must be taken into account. Without intending here an investigation of the general case, we consider the class of potentials $V_1(x)$ that possess no singularities. Then in eq. (4.8) we should simply substitute for $\mathscr{E}$ the quantity $|V'_1(x_s)|$, i.e., the amplitude of the additional field at the singular point of the original potential. In the particular case when

$$V'_1(x) = \mathscr{E} \exp[-(x - x_0)^2/\gamma^2], \qquad V(x) = V[\cosh(x/a)]^{-2},$$

we obtain

$$D(\mathscr{E}) = D(0)\exp\left[\frac{a\mathscr{E}}{\Omega}\left(\frac{2\pi\omega}{\Omega}\right)^{1/2}\left(\frac{V}{E}\right)^{1/4} \times \exp\left(\Omega\tau_s + \frac{\pi^2 a^2 - 4x_0^2}{4\gamma^2}\right)\right]. \tag{4.15}$$

The validity of this result is also restricted by the condition (4.9b).

This example shows that the anomalous increase in the contribution of the perturbation to the tunneling exponential function can be the result not only of rapid variations in time, but also of pronounced spatial inhomogeneity of the perturbation. The effect of the perturbation weakens as the intensity of the alternating field in the barrier region is decreased, as follows from eq. (4.15).

Also of interest is the case when the potential is nonstationary as a result of jittering as a whole:

$$V(x, t) = V(x)(1 + \beta\cos\Omega t), \quad \beta \ll 1.$$

Then in eq. (4.14) we have $V_1(x) = \beta V(x)$, and it is convenient to rewrite it in the form

$$A_1 = -\frac{im\beta}{2}\int_C (\mathrm{d}x/\mathrm{d}t)^2 \cos\Omega t\,\mathrm{d}t. \tag{4.16}$$

In place of eq. (4.8) we obtain

$$a_1 = \pi\beta m\Omega\left(\frac{2}{\alpha - 2}\right)^2\left[\frac{|\kappa|(2-\alpha)^2}{2m\Omega^2}\right]^{2/(2-\alpha)}\frac{\exp(\Omega\tau_s)}{|\Gamma[2\alpha/(\alpha-2)]|},$$

from which we can find the answers for potentials of specific form.

Note that the tunneling problem in a high-frequency field ($\Omega \gg \omega$) cannot be solved with the aid of Kapitza's pendulum method (see Landau and Lifshitz 1976), since in complex time the field is not an oscillating one and the amplitude of the alternating field in the region of interest to us is much smaller than the intensity of the constant field.

5. *Tunneling decay of a metastable state in a weak alternating field*

In the preceding section we investigated the effect of an alternating field on the transparency of an isolated potential barrier. If the initial state corresponds to a particle located in a potential well, the particle motion in zero field is finite and periodic: $x(t + t_0) = x(t)$, where t_0 is the period of the particle vibrations in the well. Taking this into account, we can transform the contour $C = C_+ + C_-$ in fig. 2 into a series of closed contours differing from each other by a t_0 shift along the real t axis. The function $x(t)$ has the same form on all the contours; therefore, the summation over the contours corresponds to our going over from eq. (4.1) to the following expression:

$$A_1 = \frac{i\mathscr{E}}{2\sin(\Omega\tau_0/2)} \oint dt\, x(t) \sin \Omega t, \tag{5.1}$$

where the integration is along the contour shown in fig. 3. The solution $x(t)$ is specified by the condition $x(t_0/2) = x_2$.

We would remind the reader that $x(t)$ is a solution of the Newton equation (3.1) in the unperturbed potential. In the limit of high Ω the integral in eq. (5.1) comes from singular points of the function $x(t)$, its positions being determined by the specific shape of the potential energy $V(x)$. That is why no general expression for A_1 exists even in the limit $\Omega \to \infty$. As an example we consider the following potential:

$$\tilde{V}(x) = \frac{m\omega^2}{2} x^2 \left[1 - \left(\frac{x}{x_c}\right)^3\right], \tag{5.2}$$

with a metastable state near $x = 0$. The energy $E = 0$ corresponds to a particle resting at the bottom of the well. For $E > 0$ the location of the turning points is shown in fig. 4. The potential given by eq. (5.2) corresponds to that given by eq. (4.4) with $\kappa = -m\omega^2/2x_c^3$ and $\alpha = 5$. The locations of the singular points can be calculated through eq. (4.6). To this end, it is convenient to investigate the analytic properties of the function $[E - \tilde{V}(x)]^{1/2}$ in the plane of complex x (see

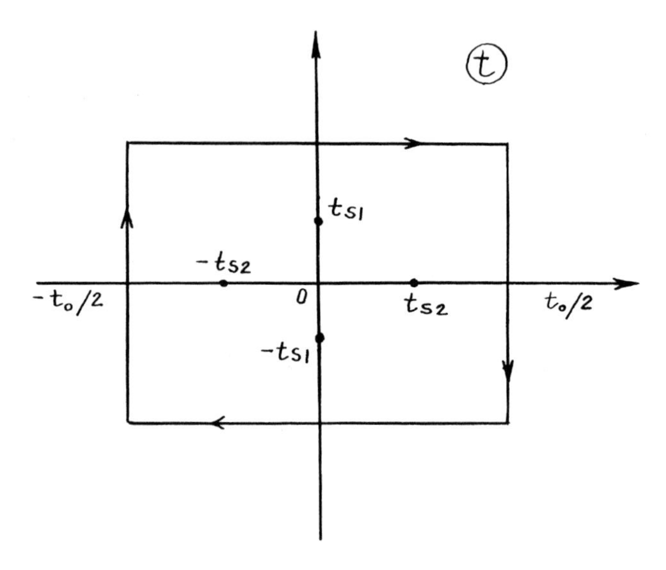

Fig. 3. The integration contour and singular points of the trajectory $x(t)$ in the case of the computation of the probability of decay of a metastable state.

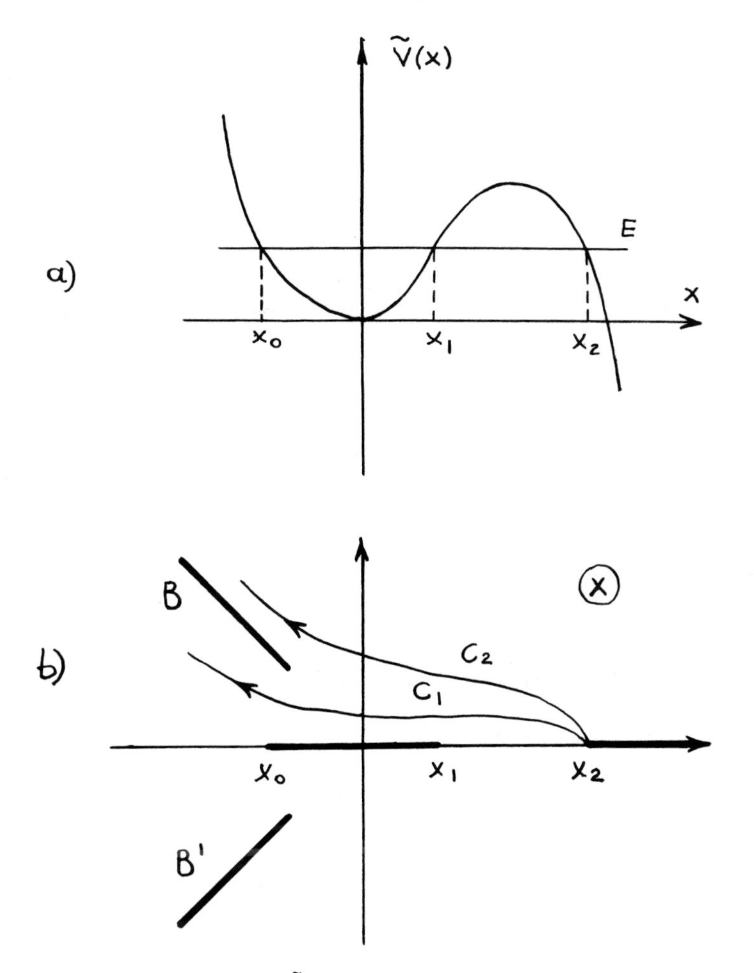

Fig. 4. (a) Tunneling in the potential $\tilde{V}(x)$ with energy E; (b) paths of integration in the complex x plane.

fig. 4). One cut goes from x_0 to x_1 and the other from x_2 to infinity. The contour of integration in eq. (4.6) begins at the point x_2 and goes to infinity.

There are two nontrivial contours C_1 and C_2 going along different sides of the cut B in fig. 4b. The other contours correspond either to a change of sign of t_s or to addition to it of an integer number of periods of oscillation t_0. The period t_0 corresponds to the integral around the cut (x_0, x_1),

$$t_0 = (m/2)^{1/2} \int_{x_0}^{x_1} [E - \tilde{V}(x)]^{-1/2} \, dx. \tag{5.3}$$

Calculating the integrals along the contours C_1 and C_2, we obtain the locations of the singular points t_{s1} and t_{s2}. For energies small compared to the barrier height, $E \ll m\omega^2 x_c^2$, we have found

$$\omega t_{s1} = \mathrm{i}\frac{2E}{m\omega^2 x_c^2}\frac{5\pi}{4\times 3^{1/2}}\left[1 - 2^{4/3}\frac{\Gamma(2/3)}{\Gamma^2(1/3)}\right], \tag{5.4}$$

$$\omega t_{s2} = 2\pi/3.$$

Four singular points, $\pm t_{s1}$, $\pm t_{s2}$, are shown in fig. 3. At high frequencies, when $\Omega \,\mathrm{Im}\, t_{s1} \gg 1$, a contribution to the integral in eq. (5.1) comes only from the points which lie off the real axis. With account taken of the singularities mentioned above, integration in eq. (5.1) gives

$$A_1 = F(\Omega)\frac{2\mathscr{E} x_c}{\omega}\exp(\Omega\tau_s)\frac{\Gamma(1/3)}{(12)^{1/6}}\left(\frac{\omega}{\Omega}\right)^{1/3}, \tag{5.5}$$

which differs from eq. (4.8) by the factor

$$F(\Omega) = \frac{1}{2\sin(\Omega t_0/2)}. \tag{5.6}$$

For particle tunneling from the bottom of the potential given by eq. (5.2) ($E = 0$) formula (5.5) coincides with the result of Fisher (1988). This result demonstrates that dependence of the decay rate on the field frequency has a resonant character. The function $F(\Omega)$ goes to infinity for frequency Ω commensurate with the frequency of internal classical oscillations, $2\pi/t_0$. A finite result could be found in this case by taking into account some nonlinear effects or the finite width of the energy levels, caused either by interaction with the thermostat or by the Breit–Wigner effect. In the last case the resonance tunneling under the effect of an alternating field was considered by Sokolovski (1988).

In the problem considered the singular points $\pm t_{s1}$, which give the main contribution to the integral, lie on the imaginary axis of fig. 3. For a singular point of general location the function $F(\Omega)$ may also contain an oscillatory dependence on Ω in the nominator. As a specific example of this kind we shall consider below the decay of a metastable state in a Josephson junction (see section 8).

6. *Above-barrier reflection in a small alternating field*

In the static situation the coefficient for above-barrier reflection is given by the expression

$$R = \exp(-A_0), \quad A_0 = 4(2m)^{1/2}\,\mathrm{Im}\int_{x_1}^{x_c}[E - V(x)]^{1/2}\,\mathrm{d}x, \tag{6.1}$$

where x_1 is an arbitrary real coordinate and x_c is the complex root of the equation $V(x) = E$ determining the positions of the turning points. The correction due to the alternating field is given, as before, by eq. (4.1), but the integration contour will now be different. The reflection coefficient is given by the ratio of the amplitudes of the incident and reflected waves. Let the particle be incident on the barrier from the left. Then the classical trajectory $x(t)$ describing the reflection of the particle from the barrier is specified in the complex t plane by the contour C_+ in fig. 5. Section 1 corresponds to the incident particle, section 3 to the reflected particle, and section 4 to the transmitted particle. At the remote ends of sections 1 and 3 we have $x < 0$. Section 2 contains the turning points, where $\mathrm{d}x/\mathrm{d}t = 0$ and the coordinate x is complex, in accordance with the fact that the turning point does not exist in classical mechanics. With an allowance made for the foregoing,

$$R(\mathscr{E}) = R(0)\exp\left[-\mathscr{E}\int_{C_+ + C_-} x(t)\cos\Omega t\,\mathrm{d}t\right]. \tag{6.2}$$

As before, in the high-frequency limit the dominant contribution to the integral is made by the singular points of the potential, in the vicinity of which the behavior of $x(t)$ has already been investigated. It is clear that for potentials with power-law singularities the correction to the action will, as before, be given by eq. (4.8), with the only difference that we should now use for the imaginary part of the time of motion to the singular point x_s the expression

$$\tau_s = \mathrm{Im}\int_{x_1}^{x_s} \{2[E - V(x)]/m\}^{-1/2}\,\mathrm{d}x,$$

where x_1 is an arbitrary real coordinate.

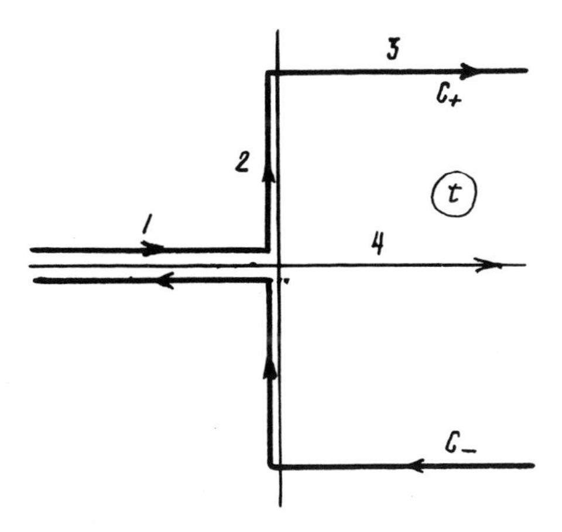

Fig. 5. Integration contour for the computation of the above-barrier reflection coefficient.

For the particular case of a potential of the form $V(x) = V[\cosh(x/a)]^{-2}$, we have

$$\sinh(x/a) = [(E - V)/E]^{1/2} \sinh \omega t, \quad \omega^2 = 2E/ma^2,$$

$$\mathrm{d}x/\mathrm{d}t = (2E/m)^{1/2} [\sinh^2 \omega t + E/(E - V)]^{-1/2} \cosh \omega t.$$

As can be seen, the turning point $t_R = \mathrm{i}\pi/2\omega$ is located in the middle of section 2 in fig. 5, where x is pure imaginary. On sections 1 and 3 the quantity x is real. The correction to the argument of the exponential function in the limit $\Omega\tau_s \gg 1$ has the same form as for the transmission coefficient in the case when $E < V$:

$$R(\mathscr{E}) = R(0) \exp\left[\frac{a\mathscr{E}}{\Omega} \left(\frac{2\pi\omega}{\Omega} \right)^{1/2} \left(\frac{V}{E} \right)^{1/4} \exp(\Omega\tau_s) \right]. \tag{6.3}$$

Let us note that the equality of τ_s and $\operatorname{Im} t_R$ is due to the special choice of potential, and does not hold in the general case.

7. Is exponential enhancement a common feature of all potential barriers?

In the previous sections the principal object of our attention was on the effect of an exponential enhancement of the alternating-field amplitude. This effect is strongly dependent on the specific form of the potential. For the two examples of a symmetric potential barrier, considered in section 4.2, the enhancement coefficient $\exp(\Omega\tau_s)$ contains a parameter τ_s, which is just one-half of the time of subbarrier motion. For the potential given by eq. (5.2), considered in section 5 for energies near the bottom of the well, the time τ_s is proportional to the energy of the particle [see eq. (5.4)]. In other words, τ_s is much less than the time of subbarrier motion. In this way one arrives at the conclusion that no universal expression exists for the parameter τ_s. We recall that we started from a naive picture of tunneling in an alternating field, when the exponential enhancement was connected directly with the fact of subbarrier motion. Instead of τ_s in the factor for exponential enhancement this approach substitutes the total time of the subbarrier motion. Rigorous calculations demonstrate, however, that the real situation is much more complicated, so that simplified considerations may easily give a wrong result.

The extremal example of this kind is a class of potentials, which exhibit no effects of an exponential enhancement of alternating field. The point is that in this case all singular points t_s, calculated from eq. (4.6), lie on the real axis of the complex t plane. A sufficient condition for this is the absence of singular points of the function $p(x) = [E - V(x)]^{1/2}$ off the real axis in the complex x plane. Integration in eq. (4.6) along the contour going from the turning point at the outer side of the barrier to infinity is then reduced to integration along the real cuts in the plane of x. This situation is depicted in fig. 4b, if the cuts B and B' are omitted.

The simplest example of tunneling without an exponential enhancement of an alternating field is the decay of a metastable state in a cubic potential

$$V_c(x) = \tfrac{1}{2} m\omega^2 x^2 (1 - x/x_c). \tag{7.1}$$

According to eq. (4.6), there is only one singular point, lying at the origin of the coordinate system shown in Fig. 3. This can be easily verified by drawing the contour of integration in fig. 4b to the right from the point x_2 along the real axis. As t_s is real, only the real cut (x_0, x_1) will contribute to the integral, and the result equals one-half of the period of oscillation in the classically allowed region, as follows from eq. (5.3). Then $t_s = -t_0$, calculated from eq. (4.6), corresponds to the origin of coordinates in fig. 3.

From eqs. (4.4) and (7.1) it follows that $\kappa = -m\omega^2/2x_c$, $\alpha = 3$. The type of singularity, according to eq. (4.5), corresponds to a pole of second order,

$$x = 4x_c/(\omega^2 t^2). \tag{7.2}$$

Equation (5.1) for the action gives

$$A_1 = \frac{4\mathscr{E} x_c}{\omega} \frac{\pi\Omega/\omega}{\sin(\Omega t_0/2)}. \tag{7.3}$$

The period of oscillation in the classically allowed region depends on the particle energy and is given by eq. (5.3). For energy near the well bottom (see fig. 4a), when $t_0 = 2\pi/\omega$, eq. (7.3) was derived by Fisher (1988) by solution of the Hamilton–Jacobi equation.

Now we consider a model potential in the shape of a truncated parabola,

$$V(x) = \begin{cases} m\omega^2/2, & x < x_c, \\ -\infty, & x > x_c. \end{cases} \tag{7.4}$$

This potential is nonanalytic, which results in a piecewise representation of the function $x(t)$ on the real axis x in the complex t plane,

$$x(t) = \begin{cases} -x_c\left[\exp(\mathrm{i}\omega t) - \dfrac{2\mathrm{i}\sin(\omega t)}{1 + \exp(2\omega\tau_0)}\right], & 0 < \operatorname{Im} t, \\[2ex] -x_c\left[\exp(-\mathrm{i}\omega t) + \dfrac{2\mathrm{i}\sin(\omega t)}{1 + \exp(2\omega\tau_0)}\right], & 0 > \operatorname{Im} t, \end{cases} \tag{7.5}$$

where τ_0 is the time of subbarrier motion at a given energy E,

$$\tanh(\omega\tau_0) = [1 - 2E/(m\omega^2 x_c^2)]^{1/2}. \tag{7.6}$$

Substituting eq. (7.5) into eq. (5.1), we get an expression for the action,

$$A_1 = 2\mathscr{E} x_c \omega |\omega^2 - \Omega^2|^{-1} [1 - 2E/(m\omega^2 x_c^2)]^{1/2}, \tag{7.7}$$

which coincides with the result obtained by Fisher (1988) at $E = 0$.

For the cubic potential given by eq. (7.1) and the truncated parabola given by eq. (7.4) there are no effects of exponential enhancement of an alternating field at any energy of the particle.

For some potentials the effects of exponential enhancement disappear, e.g., if the particle starts tunneling from the well bottom. One of them is the potential considered above [eq. (5.2)]. For zero energy the Newton equation has an explicit solution,

$$x(t) = x_c[\sin(3\omega t/2)]^{-2/3}, \tag{7.8}$$

which corresponds to fusion of the two singular points $\pm t_{s1}$ in fig. 3 into one point $t = 0$. Two other points $\pm t_{s2}$ remain on the real axis. According to eq. (7.8), each of the three singular points in this case represent a branching point of third order. The integral in eq. (5.1) with the function (7.8) is calculated along the cut on the real axis of t in fig. 3 between the points $\pm 3\pi/2\omega$. The result is (see Fisher 1988)

$$A_1 = \frac{4^{1/3}}{3\Gamma(2/3)} \frac{\mathscr{E} x_c}{\omega} \Gamma\left(\frac{1}{3} - \frac{\Omega}{3\omega}\right) \Gamma\left(\frac{1}{3} + \frac{\Omega}{3\omega}\right). \tag{7.9}$$

In many cases a parabolic approximation for the top of the barrier was successfully exploited in order to simplify the solution of tunneling problems. On the other hand, a cubic approximation to the potential became popular when considering the decay of metastable states. Extremely convenient in all other respects, these approximations could well prove to be unsatisfactory when applied to calculations of high-frequency field effects. We have shown already that the probability of tunneling in a high-frequency field is governed rather by singular points of the potential than by its shape near the top of the barrier. For a parabolic barrier the integral in (4.6) for location of the singular point is divergent. Therefore, the approach developed above cannot be applied to this potential. The same conclusion follows from eq. (4.8), which is singular at $\alpha = 2$.

In the case of a cubic potential our formalism gives a finite answer, but without exponential enhancement by an alternating field. Taking account of the more detailed shape of the potential in realistic problems may reestablish the effect of exponential enhancement, lost as a result of the cubic approximation. A specific example of this kind is considered in section 8.

8. *Alternating-current-induced decay of zero-voltage states of a Josephson junction*

Zero-voltage states of Josephson junctions correspond to the minima of the junction energy $V(\phi)$ as a function of the phase difference between the two superconductors (see the chapters by Devoret et al. and by Larkin and

Ovchinnikov). At sufficiently low temperatures the finite lifetime of the zero-voltage states is due to macroscopic quantum tunneling through the potential barrier (Voss and Webb 1981, Jackel et al. 1981).

In this section we consider the stimulation of tunneling decay by an alternating perturbation as applied to a Josephson junction, whose properties are determined by the critical current I_c, the capacitance C and the shunting resistance R.

The height $V(\phi)$ of the potential barrier is of the order of magnitude $I_c/2e$, and the typical frequency of phase oscillations is the plasma Josephson frequency $\omega = (2eI_c/c)^{1/2}$. We shall distinguish below between junctions with weak dissipation ($\eta \ll \omega$) and those with strong dissipation ($\eta \gg \omega$), where $\eta = 1/RC$ is the viscosity coefficient. Estimates show that at $\eta \ll \omega$ we have in the order of magnitude $\tau_s \sim \omega^{-1}$, and in the opposite limiting case $\tau_s \sim \eta\omega^{-2}$. In the case of weak dissipation the argument of the tunneling exponent should be an oscillating function of the direct current through the junction, of the alternating current frequency, and of the temperature, in accordance with the possibility of resonant absorption of AC energy in the Josephson junction.

The effect of an alternating current on the decay rate of zero-voltage states was investigated in an experiment by Devoret et al. (1985), but the zero-voltage states decayed because of activation processes. Such processes were investigated theoretically at zero alternating current by Mel'nikov (1985).

8.1. Decay of zero-voltage states in the limit of weak dissipation

Assume that an alternating current of amplitude I_1 and frequency Ω flows through the junction in addition to the direct current I_0, so that

$$I(t) = I_0 + I_1 \cos \Omega t. \tag{8.1}$$

Let the alternating current be small, $I_1 \ll I_0$, and let the direct current not exceed the critical value, $I_0 < I_c$. Neglecting the tunneling and activated processes, the phase fluctuations are governed by the junction temperature. With tunneling taken into account zero-voltage states decay with lifetimes D^{-1}, where D is the probability of tunneling through the potential barrier. In the static case D can be obtained by quantum mechanics if $\eta = 0$, and if η is arbitrary one can use the methods of the recently developed theory of quantum tunneling in the presence of dissipation (Caldeira and Leggett 1981). Our aim is to generalize the indicated results to include alternating current.

The equation of motion for the phase difference ϕ is known,

$$\frac{d^2\phi}{dt^2} + \eta \frac{d\phi}{dt} + \omega^2(\cos\phi - k_0 - k_1 \cos \Omega t) = 0, \tag{8.2}$$

where $k_0 = I_0/I_c$ and $k_1 = I_1/I_c$. Note that our ϕ corresponds to δ in the notation of Devoret et al. and 2φ in that of Larkin and Ovchinnikov. For this adiabatic description to be valid, it is necessary that the AC frequency be lower than the characteristic relaxation times in the superconductor; we assume, therefore, that $\Omega \ll T$.

With dissipation neglected, eq. (8.2) corresponds to the Lagrangian

$$\mathfrak{L} = \frac{V}{2\omega^2}\left(\frac{\mathrm{d}\phi}{\mathrm{d}t}\right)^2 + V(-\sin\phi + k_0\phi + k_1\phi\cos\Omega t), \tag{8.3}$$

where $V = I_c/2e$. As before, we represent the semiclassical tunneling probability in the exponential approximation as

$$D = \exp(-A), \quad A = -\mathrm{i}\int_C \mathfrak{L}\,\mathrm{d}t, \tag{8.4}$$

where A is the classical action and the integration is along the contour C in the complex time plane (fig. 2). We note that in the quantum-mechanical problem the location of the contour C as $\mathrm{Re}\,t \to -\infty$ is determined by the initial energy of the particle. We shall assume below that the tunneling occurs from a state of thermodynamic equilibrium, so that as $\mathrm{Re}\,t \to -\infty$ the contour C is at a distance $\mathrm{i}/2T$ from the real axis in accordance with eq. (3.5) (see also Larkin and Ovchinnikov 1983).

Regarding the alternating current as a small perturbation, we write the action in the form $A = A_0 + A_1$, where A_0 is the action at $I_1 = 0$ and A_1 is linear in I_1. This expansion is valid so long as $A_1 \ll A_0$. The smallness of A_1 compared to A_0 does not mean at all that the effect of the alternating current on D is small. On the contrary, the semiclassical approximation is applicable only if $A_1 \gg 1$, so that the alternating current amplitude is bounded from below by the condition that D changes, say, by one or two orders.

We write for A_0 and A_1 the expressions

$$A = -\mathrm{i}\int_C \left[\frac{V}{2\omega^2}\left(\frac{\mathrm{d}\phi}{\mathrm{d}t}\right)^2 - V(\phi)\right]\mathrm{d}t, \quad V(\phi) = V(\sin\phi - k_0\phi), \tag{8.5}$$

$$A_1 = -\mathrm{i}Vk_1\int_C \phi(t)\cos\Omega t\,\mathrm{d}t. \tag{8.6}$$

It is important that both A_0 and A_1 are determined by the same function $\phi(t)$, given by the solution of eq. (8.2) at $\eta = 0$ and $k_1 = 0$. The integral in eq. (8.6) constitutes a Fourier transform along the contour C. In the limit of $\Omega \to \infty$ the asymptote of the Fourier integral is determined by singularities of the function $\phi(t)$. In our case the contour C must be shifted in the direction $\mathrm{Re}\,t \to -\infty$, where the alternating current can be regarded as adiabatically turned off. This means that we need consider only the $\phi(t)$ singularities located inside C. Their distance from the real axis will determine the degree of exponential enhancement of the alternating-current amplitude.

We must, therefore, study the singularities of the solution of the unperturbed problem. In the absence of alternating current, the function $\phi(t)$ is implicitly given by

$$\omega t = (2V)^{1/2} \int [E - V(\phi)]^{-1/2} \, \mathrm{d}\phi. \tag{8.7}$$

The nature of the solution $\phi(t)$ becomes clear from fig. 6a, where the turning points are determined by the condition $V(\phi_{1,2,3}) = E$. The energy E must be obtained from the condition that the time of below-barrier motion between points ϕ_2 and ϕ_1 is $\mathrm{i}/2T$. We then get from eq. (8.7)

$$\frac{\omega}{T} = (2V)^{1/2} \int_{\phi_2}^{\phi_3} [V(\phi) - E]^{-1/2} \, \mathrm{d}\phi. \tag{8.8}$$

The last relation determines the temperature dependence of the energy $E(T)$.

The singular points of the trajectory correspond to those instants of the complex time at which $V(\phi)$ becomes infinite. It follows from eq. (8.7) that these singularities are logarithmic, so that near the singularities we have

$$\begin{aligned} \phi_{1,2}(t) &= \pm 2\mathrm{i} \ln[\omega(t - t_{s1,s2})], \\ \phi_{3,4}(t) &= \mp 2\mathrm{i} \ln[\omega(t - t_{s3,s4})]. \end{aligned} \tag{8.9}$$

For location of the quartet of singular points in fig. 3,

$$t_{s1,s2} = \mathrm{i}\tau_s \mp t_1, \qquad t_{s3,s4} = -\mathrm{i}\tau_s \mp t_1, \tag{8.10}$$

in accordance with eq. (8.7). By integrating along a path drawn from the turning point ϕ_3 upward to infinity we obtain

$$\omega(\mathrm{i}\tau_s + t_1 - \tfrac{1}{2}t_0) = (V/2)^{1/2} \int_{\phi_3}^{\mathrm{i}\infty} [E - V(\phi)]^{-1/2} \, \mathrm{d}\phi. \tag{8.11}$$

The oscillation period in the classically allowed region is determined by the relation

$$\omega t = (2V)^{1/2} \int_{\phi_1}^{\phi_2} [E - V(\phi)]^{-1/2} \, \mathrm{d}\phi. \tag{8.12}$$

With this information on the unperturbed trajectory, we calculate the integral given by eq. (8.6) with the aid of eq. (8.9). The result is

$$D(I_1) = D_0 \exp\left[\frac{2\pi I_1}{e\Omega} \left| \frac{\sin \Omega t_1}{\sin(\Omega t_0/2)} \right| \exp(\Omega \tau_s) \right], \tag{8.13}$$

where D_0 is the transmission coefficient in the absence of an alternating current. In the derivation of eq. (8.13) the frequency was assumed high compared with the reciprocal time of below-barrier motion, $\Omega\tau_s \gg 1$. This imposes the frequency limits $V \gg \Omega \gg \tau_s^{-1}$ if the semiclassical approach is valid. Equation

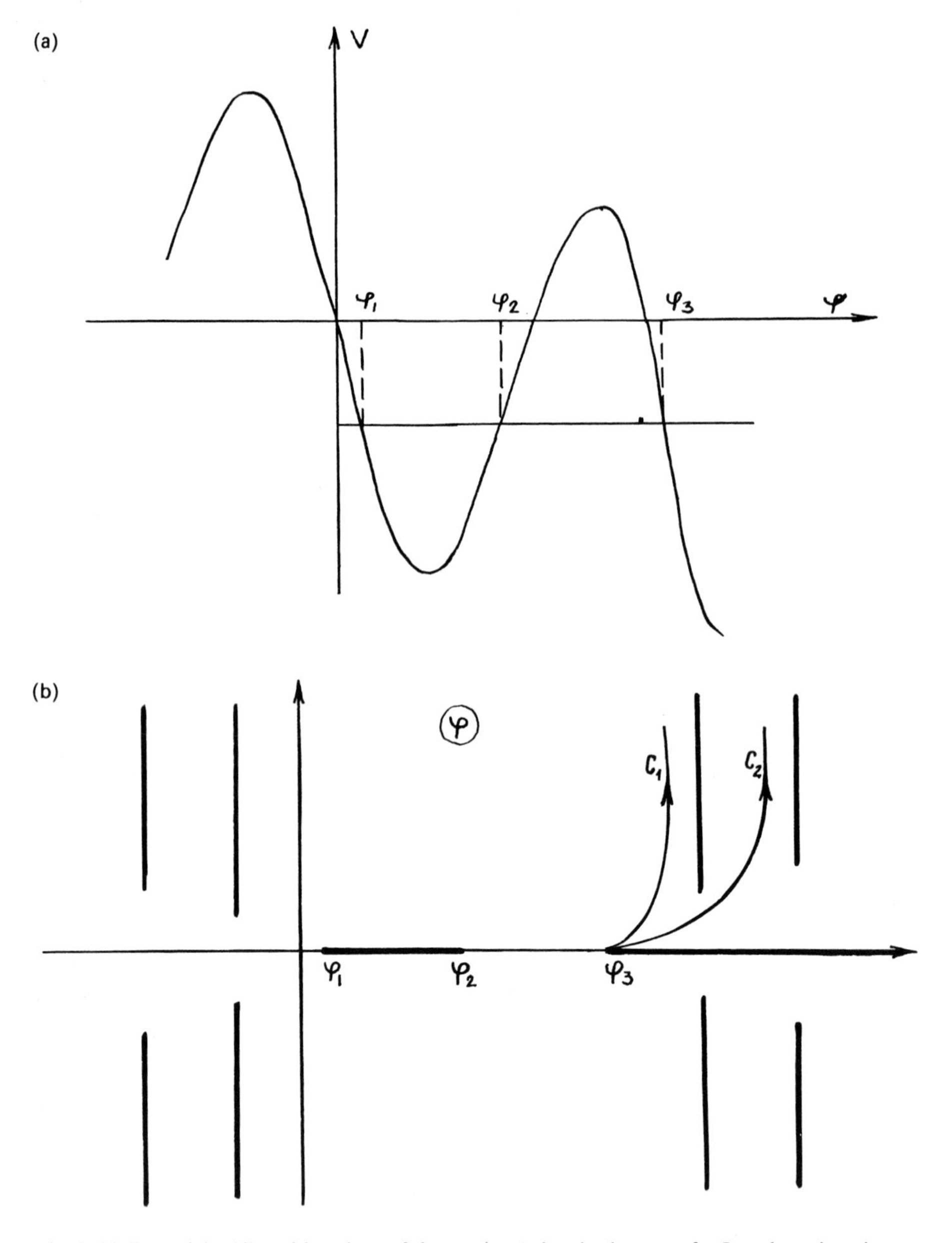

Fig. 6. (a) Potential $V(\phi)$ and locations of the turning points in the case of a Josephson junction; (b) paths of integration in the complex ϕ plane.

(8.13) describes the time-averaged transmission coefficient in the leading exponential approximation. The averaging eliminates the dependence on the alternating-current phase at the instant of passage through the barrier.

As regards the oscillating terms in eq. (8.13), the following remark is in order. The action of even a weak alternating current can alter greatly the state of the junction compared with the unperturbed one, through resonant effects, if the frequency of the alternating current coincides with one of the distances between the quantum levels. In a semiclassical potential, the levels are locally equidistant, and since the alternating-current frequency is low compared with the barrier height V, the resonance condition can be met for states having energies E that satisfy the condition

$$\Omega = 2\pi N / t_0(E), \quad N = 1, 2, 3, \ldots,$$

where $t_0(E)$ is the period of the classical oscillations. This is just the condition under which the linear contribution to the action diverges, as seen from eq. (8.13). To obtain a finite result in this case it is necessary either to take into account the nonlinearity with respect to the alternating current, in the spirit of the theory of resonance in anharmonic systems, or take a small damping into consideration.

A specific feature of eq. (4.8) for the transmission probability through a potential barrier is the exponentially large factor, $\exp(\Omega\tau_s)$. Beyond this factor, eq. (5.5) for the decay probability of a metastable state displays also resonances specific to periodic motion in a potential well. Equation (8.13) describes an even more general result, as in addition to the above-mentioned features it contains also an oscillating factor in the numerator of the exponent. This is a direct consequence of the fact that in the general case the points in fig. 3 do not lie on the coordinate axes.

Expression (8.13) is valid for low-dissipation Josephson junctions at arbitrary values of the direct current through the junction, right up to the critical one, and at any temperature right up to the temperature T_0 at which the macroscopic quantum tunneling goes over into the activation regime (Larkin and Ovchinnikov 1983). The critical temperature T_0 is obtained from relation (8.8) if it is assumed that the energy E tends to the top of the potential barrier,

$$2\pi T_0 = \omega(1 - k_0^2)^{1/4}. \tag{8.14}$$

The calculation of the linear-in-I_1 contribution to the tunneling exponent has, thus, been reduced to a calculation of the current and temperature dependences of the quantities τ_s, t_1 and t_0 that enter into eq. (8.13), using the relations (8.8), (8.11) and (8.12). In the limiting cases when I_0 is close either to the critical value or to zero, simpler relations can be obtained.

We consider first the most important case, when the direct current through the junction is close to the critical value, $I_c - I_0 \ll I_c$, or $1 - k_0 \ll 1$. The potential can be regarded in this case as cubic,

$$V(\phi) = V[(1 - k_0)\phi - \tfrac{1}{6}\phi^3], \tag{8.15}$$

and eq. (8.8) reduces to an elliptic integral. For the oscillation period t_0 we obtain from eq. (8.12) the result

$$t_0 = \frac{1}{\omega}\left(\frac{2}{1-k_0}\right)^{1/4} \begin{cases} 2\pi, & T \ll T_0, \\ \ln[T_0/(T_0 - T)], & T_0 - T \ll T_0. \end{cases} \tag{8.16}$$

Substituting the expansion (8.15) into eq. (8.11) to find the locations of the singularities yields $\tau_s = t_1 = 0$; the four logarithmic singularities coalesce into one second-order pole, in accordance with the results of section 4. To obtain finite values of τ_s and t_1 we must substitute in eq. (8.11) the exact potential $V(\phi)$ from eq. (8.5). Integrating along a contour drawn at $k_0 = 1$ from the turning point $\phi_3 = 0$ to the point $\phi = \phi_0$ and then vertically upward, we get at $\phi_0 = \pi$ the result

$$\omega\tau_s = \omega t_1 = \frac{1}{2}\int_0^\infty \left\{\frac{[\pi^2 + (z + \sinh z)^2]^{1/2} + \pi}{\pi^2 + (z + \sinh z)^2}\right\}^{1/2} dz \cong 1.177. \tag{8.17}$$

In the range of the currents considered, the values of τ_s and t_1 depend little on temperature or current. We note that τ_s and t_1 were found to be substantially smaller than the time $\tau_0 \propto \omega^{-1}(1 - k_0)^{-1/4}$. This is an indirect reflection of the absence of exponential enhancement of the alternating current in the cubic potential approximation.

In the opposite limiting case of low currents, $I_0 \ll I_c$ ($k_0 \ll 1$), we obtain for the oscillation period in the classically allowed region

$$t_0 = \frac{1}{\omega}\begin{cases} 2\pi, & T \ll T_0, \\ 4\ln[T_0/(T_0 - T)], & T_0 - T \ll T_0. \end{cases} \tag{8.18}$$

If the direct current is exactly zero, the potential $V(\phi)$ becomes symmetric. For such potentials, t_1 amounts to half the time of motion, in the classically accessible region between the turning points, while τ_s is half the time of motion in the classically inaccessible region,

$$\tau_s = 1/4T, \qquad t_1 = \tfrac{1}{4}t_0. \tag{8.19}$$

The time τ_s diverges at low temperature, and t_1 at temperatures close to critical. To cut off these divergences we must include a small but finite current I_0. The potential is then no longer strictly symmetric and relations (8.19) no longer hold. It follows in this case, from eq. (8.11), that

$$\tau_s(T = 0) = \frac{1}{2\omega}\ln\frac{I_c}{I_0}, \qquad t_1(T = T_0) = \frac{1}{2\omega}\ln\frac{I_c}{I_0}. \tag{8.20}$$

Numerical results based on relations (8.8), (8.11) and (8.12) are shown in fig. 7 in the form of dependences of the parameters τ_s, t_1 and t_0 on the direct

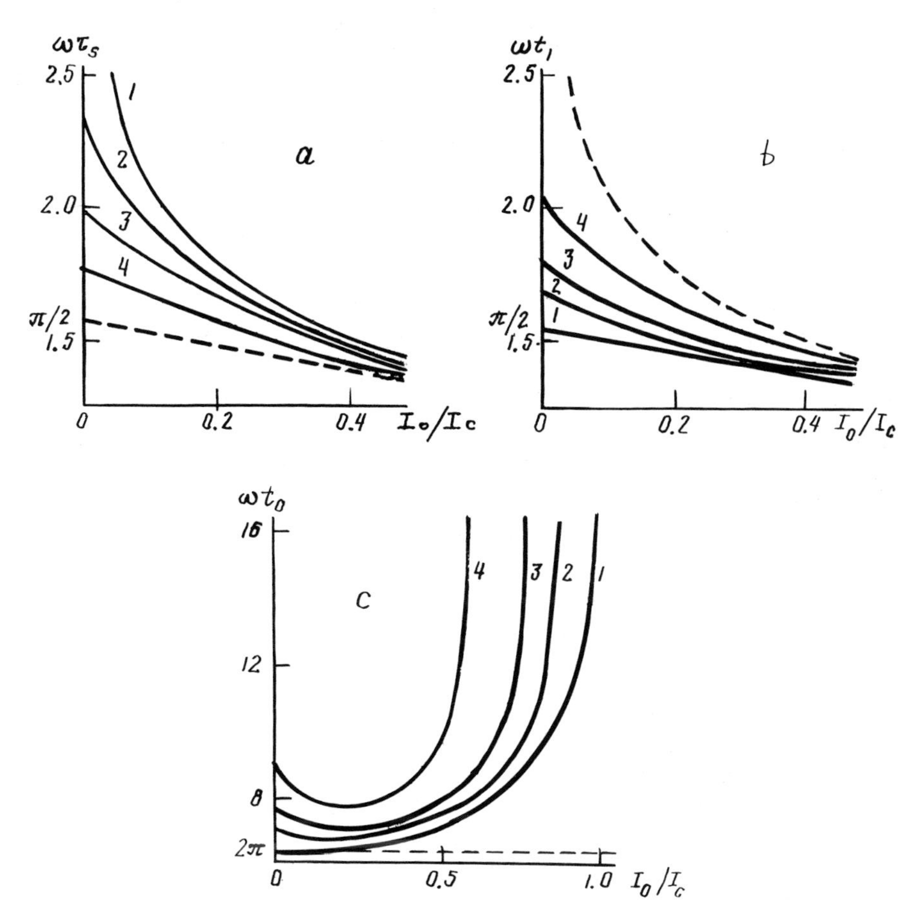

Fig. 7. Plots of (a) $\omega\tau_s$, (b) ωt_1, (c) ωt_0 versus current at different values of the parameter $T^* = 2\pi T/\omega$: 0 (curve 1); 0.7 (curve 2); 0.8 (curve 3); 0.9 (curve 4).

current at different temperatures. Since the critical temperature T_0 is a function of the current [see eq. (8.14)], the curves in fig. 7a, b have termination points corresponding to the change from the tunneling to the activation mechanism. The corresponding boundary curves are shown dashed. It follows from eq. (8.11) that the boundary curve for τ_s coincides with the plot of $t_0(k_0)$ at $T_0 = 0$ and conversely. The solid curves in Fig. 7a, b intersect the dashed ones outside the borders of the figures. We note that for numerical calculations in the entire range of currents it is convenient to use $\phi_0 = \frac{3}{2}\pi$ when calculating τ_s and $\phi_0 = -\frac{3}{2}\pi$ when calculating t_1.

Now some remarks on the computation of the location of the singular points are in order. In the case under consideration the function $[E - V(\phi)]^{-1/2}$ has a large number of branching points in the complex ϕ plane, where the expression under the square-root sign tends to zero. The time t_s corresponds to motion from ϕ_3 to infinity and depends on the position of the path of integration with respect to these points. Some variants of the paths are shown in fig. 6b, the path C giving for t_s eq. (8.10).

Four points, $t_{s1,2,3,4}$, come from the two directions of motion from ϕ_3 (upward and downward) and the two signs of the integrand. Other paths going to infinity, e.g., the path C_2, give smaller values of τ_s. In the limit of critical value of the current, integration along the path C_2 gives $\omega\tau_s = 0.942$, which is smaller than the value [eq. (3.5)] given by the contour C_1. We conclude, therefore, that the potential under consideration generates in fig. 3 an infinite sequence of quartets of singular points t_s. Equation (8.13) corresponds to the quartet with the maximal value of τ_s.

At low temperatures, $T \ll T_0$, the position of the resonances in the denominator of eq. (8.13) is determined by the frequencies of the small oscillations in the potential given by eq. (8.5). Resonances occur at the current values

$$I_{0n} = I_c[1 - (\Omega/n\omega)^4]^{1/2}, \quad n = 1, 2, \ldots$$

We note that to obtain a good resolution of the resonances in eq. (8.13), the junction must have high quality and the ratio Ω/ω must not be too large, otherwise the resonances will be close to one another and the picture becomes smeared out even at relatively weak dissipation. If, on the contrary, we are interested in the exponential enhancement effect, the ratio Ω/ω must be chosen as large as possible at the limit $\Omega < V$ of the validity of the semiclassical approach.

8.2. Decay of zero-voltage states in the limit of strong dissipation

In the case of a Josephson junction with finite dissipation, eq. (8.2) contains a relaxation term $\eta\partial\phi/\partial t$, and the Lagrangian formalism cannot be directly applied. We use, therefore, the theory, developed by Caldeira and Leggett (1981), Ambegaokar et al. (1982) and Larkin and Ovchinnikov (1983), of quantum tunneling in the presence of friction. It follows from these references that the dynamics of a Josephson junction is equivalent, in the semiclassical limit, to a problem with effective action

$$A = -\mathrm{i}V \int_C \left\{ \frac{1}{2\omega^2}\left(\frac{\mathrm{d}\phi}{\mathrm{d}t}\right)^2 - (\sin\phi - k_0\phi - k_1\phi\cos\Omega t) \right.$$
$$\left. + \frac{4\mathrm{i}\pi\eta T^2}{\omega^2} \int_C \frac{\mathrm{d}t_1}{\sinh^2[\pi T(t_1 - t)]} \sin^2\frac{\phi(t) - \phi(t_1)}{4} \right\} \mathrm{d}t_1, \tag{8.21}$$

where the notation is the same as in eq. (8.2), and the contour C is shown in fig. 2. In the absence of alternating current the integrals along the horizontal sections of the contour vanish. This relatively simple form of the effective action corresponds, as does also eq. (8.2), to the adiabatic situation $\Omega \ll T_c$ (Larkin and Ovchinnikov 1983).

Varying the action [eq. (8.21)], we obtain the equation of motion in complex time:

$$\frac{d^2\phi}{dt^2} + \omega^2(\cos\phi - k_0 - k_1\cos\Omega t) = 2\pi i\eta T^2 \int_C \frac{\sin\{\frac{1}{2}[\phi(t) - \phi(t_1)]\}\, dt_1}{\sinh^2[\pi T(t_1 - t)]}, \tag{8.22}$$

where the principal value of integral is understood.

It is possible to transform from the contour C in the t_1 plane to a contour $\tilde{C}$ that differs from C only in that in the vicinity of the point $t_1 = t$ it is a small semicircle located inside the contour C. It is then necessary to add a compensating half-residue to eq. (8.22). The contour $\tilde{C}$ has the advantage that it can be freely displaced when the position of t is fixed. As a result, eq. (8.22) takes the form

$$\frac{d^2\phi}{dt^2} + \eta\frac{d\phi}{dt} + \omega^2(\cos\phi - k_0 - k_1\cos\Omega t) - 2i\pi\eta T^2 \int_{\tilde{C}} \frac{dt_1}{\sinh^2[\pi T(t_1 - t)]} \sin\frac{\phi(t) - \phi(t_1)}{2} = 0. \tag{8.23}$$

At large real $t - t_1$ the integrand is small and the integral can be neglected, so that at real t eq. (8.23) coincides with eq. (8.2).

Just as in the preceding section, we seek a correction, linear in the alternating current, to the action by using eq. (8.6), in which the unperturbed trajectory $\phi(t)$ must be substituted. We consider below the case of strong friction, $\eta \gg \omega$, and of a current close to critical, $I_c - I_0 \ll I_c$ $(1 - k_0 \ll 1)$. Under these conditions the term with the second derivative can be omitted, and we can use for the potential the expansion (8.9), so that eq. (8.23) in terms of imaginary time takes the form

$$\omega^2(\tfrac{1}{2}\phi^2 - 1 + k_0) + \eta T \int_{-1/2T}^{1/2T} d\tau_1 \coth[\pi T(\tau_1 - \tau)] \frac{d\phi}{d\tau_1} = 0. \tag{8.24}$$

The solution of this equation was obtained by Larkin and Ovchinnikov (1983):

$$\phi(\tau) = [2(1 - k_0)]^{1/2} \left[\frac{T^2}{T_0} \frac{2}{T_0 - (T_0^2 - T^2)^{1/2}\cos(2\pi\tau)} - 1 \right], \tag{8.25}$$

where

$$T_0 = \frac{\omega^2}{\pi\eta}[(1 - k_0)/2]^{1/2}.$$

According to eq. (8.6), an exponential enhancement takes place if the singular points of the unperturbed solution do not lie on the real axis of the time t. It follows from eq. (8.25), however, that for a cubic potential the singularity of the trajectory lies exactly on the real axis, as in the previously considered case of weak dissipation. This means that to find τ_s it is necessary, as before, to use the exact form of the potential.

Returning to eq. (8.22), we delete its term with the second derivative, as in the transformation to eq. (8.24), but do not regard ϕ as small. It followed from the preceding section that τ_s at $I_c - I_0 \ll I_c$ is small compared to T_0^{-1} in terms of the parameter $(1-k_0)^{1/4} \ll 1$. Assuming that the same property is preserved also at $\eta \gg \omega$, we suggest that the significant values of t are those for which $|t| \ll T_0^{-1}$. The argument of the hyperbolic sine in eq. (8.22) can then be regarded as small, so that when k_0 is replaced by unity, eq. (8.22) takes the form

$$\eta \frac{d\phi}{dt} + \omega^2(\cos\phi - 1) - \frac{2i\eta}{\pi} \int_{\tilde{C}} \frac{dt_1}{(t-t_1)^2} \sin\frac{\phi(t)-\phi(t_1)}{2} = 0. \tag{8.26}$$

This equation is solved by the function

$$\phi = i \ln\frac{t - i\tau_s}{t + i\tau_s}, \quad \tau_s = \frac{\eta}{\omega^2} = \frac{[2(1-k_0)]^{1/4}}{2\pi T_0}. \tag{8.27}$$

The contour $\tilde{C}$ then contracts into a double vertical section from $-i\tau_s$ to $i\tau_s$.

We note that the reason why we can solve eq. (8.26) exactly is that when the contour $\tilde{C}$ in it is shifted to infinity the integral over the contour tends to zero because of the simplified form of the integral kernel in eq. (8.26) compared to eq. (8.22). By passing the singular point $t_1 = t$ we reverse the sign of $d\phi/dt$, so that eq. (8.26) is equivalent to a relaxation equation with the time reversed,

$$-\eta \frac{d\phi}{dt} + \omega^2(\cos\phi - 1) = 0.$$

At $t \gg \tau_s$ we obtain from eq. (8.27) the pole part of the solution (8.25). It can also be seen that $\tau_s \ll T_0^{-1}$, and the parameter has the meaning of the time that the particle spends under the barrier. These results are valid so long as $T < T_0$.

Substitution of eq. (8.27) into eq. (8.6) yields

$$D(I_1) = D_0 \exp\left(\frac{2\pi I_1}{e\Omega} \sinh\frac{\Omega\eta}{\omega^2}\right). \tag{8.28}$$

It can be seen that for $\eta\Omega \gg \omega^2$ we obtain an exponential enhancement, while for $\eta\Omega \ll \omega^2$ the argument of the exponential in eq. (8.28) coincides with the variation of the static action with respect to the current, since, according to Larkin and Ovchinnikov (1983),

$$A_0 = \frac{4\pi\eta(1-k_0)}{\omega^2}\left(1 - \frac{T^2}{3T_0^2}\right).$$

Evidently, in the case of strong friction the quantity $\tau_s = \eta\omega^{-2}$ greatly exceeds its value, of the order of ω^{-1}, in the nondissipative limit. It is, therefore, better to study the exponential enhancement effect in low-Q junctioins. It must be recognized here that the quantum regime is realized for $T \ll T_0$, while T_0 decreases substantially in the strong-dissipation limit.

We have considered in detail properties of Josephson junctions as they can be looked upon as objects appropriate to the investigation of quantum effects in an alternating field. However, the tunneling decay of the high-Q junctions is practically forbidden by the condition of the validity of the semiclassical approximation, because the corresponding probability is as small as $\exp(-2\pi N)$, where N is the number of energy levels in the allowed region of the potential. Consequently, the field enhancement factor for one-quantum absorption can be estimated as $\exp(2\pi)$. Note that in the limit of strong dissipation the range of validity of the semiclassical approximation is extended due to the absence of discreteness of the energy levels.

Above we considered tunneling in the resistive model of the Josephson junction, when the role of the coordinate of the analogous particle is played by the difference of phase of the order parameter. Quantum effects in the case of electron motion in real coordinate space across the tunneling junction were considered by Lempitskii (1988). It was shown that the change of the critical current of the junction is governed by the exponentially enhanced amplitude of the alternating field, in close analogy with the above-considered effects. The important frequency range is closely related to the inverse time of flight and lies in the infrared region of the spectrum.

9. *Field emission of electrons in an alternating field*

9.1. Exact solution of the semiclassical problem

In the preceding section we computed the corrections to the argument of the tunneling exponential function that are linear in the alternating field. The exact solution in the nonlinear case is possible only for potentials of a specific form. Below we consider in detail the practically important case of the triangular barrier, a case which has a direct bearing on field emission (Neuman 1969). We shall assume that, as shown in fig. 1a, a particle of energy E strikes a barrier of height V from the left, and that a constant, $\mathscr{E}_0$, and an alternating, $\mathscr{E}\cos\Omega t$, field exist beyond the barrier. The assumption that there is no electric field in the region to the left of the barrier not only simplifies the problem but also corresponds to realizable conditions in which the field is screened off in the vicinity of a metal surface. These assumptions reduce the problem to one of computing the imaginary part of the action

$$S = \int_{t_1}^{t_2} \left[\frac{m}{2} \left(\frac{\mathrm{d}x}{\mathrm{d}t} \right)^2 - V + \mathscr{E}_0 x + \mathscr{E} x \cos \Omega t + E \right] \mathrm{d}t, \tag{9.1}$$

where t_1 and t_2 are, respectively, the instants at which the particle enters and emerges from the barrier. The trajectory $x(t)$ can be found from the equation

$$m\mathrm{d}^2 x/\mathrm{d}t^2 = \mathscr{E}_0 + \mathscr{E} \cos \Omega t, \tag{9.2}$$

with the boundary conditions

$$x(t_1) = 0, \qquad (\mathrm{d}x/\mathrm{d}t)_{t1} = \mathrm{i}[2(V - E)/m]^{1/2}. \tag{9.3}$$

As stated in section 3, to compute the minimum value of the imaginary part of the action, it is sufficient to limit ourselves to real trajectories. It is easy to verify that the real solution to eq. (9.2) corresponds to the situation in which the particle emerges from the barrier at the instant when the alternating field has its maximum amplitude. For definiteness, we shall, therefore, assume that $t_2 = 0$. The solution of eq. (9.2) with the boundary condition $(\mathrm{d}x/\mathrm{d}t)_{t=0} = 0$ allows us to find t_1 from the condition (9.3), the quantity $t_1 = \mathrm{i}\tau$ turning out to be purely imaginary. The real trajectory satisfying the indicated conditions has the form

$$x(t) = \mathscr{E}_0(t^2 + \tau^2)/2m + \mathscr{E}(\cosh \Omega\tau - \cos \Omega\tau)/m\Omega^2, \tag{9.4}$$

where t varies along the imaginary axis and the quantity τ can be found from the equation

$$\mathscr{E}_0 \tau + (\mathscr{E}/\Omega) \sinh \Omega\tau = \kappa,$$

where $\kappa = [2m(V - E)]^{1/2}$ is the initial momentum for the subbarrier motion.

Substituting eq. (9.4) into eq. (9.1) and going over to imaginary time, we obtain for the quantity A (=2 Im S), after a number of transformations, the expression

$$A = A_0 \nu_s^{-3} [\tfrac{3}{2}\nu_s^2 \nu - \tfrac{1}{2}\nu^3 - (3\mathscr{E}/\mathscr{E}_0)(\nu \cosh \nu - \sinh \nu) \\ - (\tfrac{3}{8}\mathscr{E}^2/\mathscr{E}_0^2)(\sinh 2\nu - 2\nu)], \tag{9.5}$$

where

$$\nu + (\mathscr{E}/\mathscr{E}_0) \sinh \nu = \nu_s,$$

and the following notations have been introduced:

$$\nu \equiv \Omega\tau, \quad \nu_s \equiv \Omega\tau_s = \Omega\kappa/\mathscr{E}_0.$$

The action A_0 takes, in the absence of an alternating field, the form

$$A_0 = \tfrac{2}{3}\kappa^3/m\mathscr{E}_0. \tag{9.6}$$

The dependence, as given by eqs. (9.5) and (9.6), of A/A_0 on the parameters $\mathscr{E}/\mathscr{E}_0$ and $\Omega\tau_s$ is depicted in fig. 8.

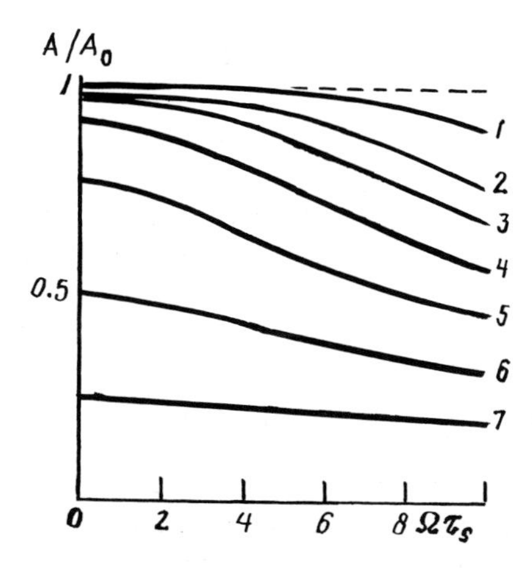

Fig. 8. Dependence of A/A_0 on $\Omega\tau_s$ for the following values of $\mathscr{E}/\mathscr{E}_0$: 0.001 (curve 1); 0.01 (curve 2); 0.03 (curve 3); 0.1 (curve 4); 0.3 (curve 5); 1 (curve 6); 3 (curve 7).

9.2. Successive tunneling regimes

Simpler analytic expressions can be obtained for several limiting cases.

In the limit of high frequencies, when $\nu, \nu_s \gg 1$, the equation for ν takes the form

$$\nu + (\mathscr{E}/2\mathscr{E}_0)e^{\nu} = \nu_s.$$

Ignoring the terms small in the parameters ν_s^{-1} and $\exp(-\nu)$, we obtain for A the expression

$$A = A_0 \nu_s^{-3}[\tfrac{3}{2}\nu_s^2 \nu - \tfrac{1}{2}\nu^3 - (\tfrac{3}{2}\mathscr{E}/\mathscr{E}_0)\nu e^{\nu}], \tag{9.7}$$

where the last term in the brackets is always small compared to the other two, but must be retained since it is precisely the term that gives the linear-in-$\mathscr{E}$ correction in the case of small $\mathscr{E}$. In the latter case we have

$$A = A_0 \left[1 - \frac{1}{(\Omega\tau_s)^2} \frac{3\mathscr{E}}{2\mathscr{E}_0} \exp(\Omega\tau_s) \right], \tag{9.8}$$

where

$$\frac{(\Omega\tau_s)^2}{A_0} \ll \frac{\mathscr{E}\exp(\Omega\tau_s)}{\mathscr{E}_0} \ll 1.$$

The limitation on $\mathscr{E}$ from below is due to the fact that, according to the condition of applicability of semiclassical approximation, the absolute value of the correction to the action should be much greater than unity. The limitation from above indicates that the linear-in-$\mathscr{E}$ correction to the action a_1 is valid so long as $a_1 \ll A_0/(\Omega\tau_s)^2$, which is a much more rigid criterion than the one indicated in section 4.

In the region

$$(\mathscr{E}/\mathscr{E}_0)\exp(\Omega\tau_s) \sim 1,$$

we obtain for A the expression

$$A = A_0\{1 - (\Omega\tau_s)^{-2} f[\mathscr{E}\exp(\Omega\tau_s)/2\mathscr{E}_0]\}, \tag{9.9}$$

where the function $f(x)$ is given by the relations

$$f = 3y + \tfrac{3}{2}y^2, \tag{9.10}$$

$$ye^y = x, \tag{9.11}$$

and its plot is shown in fig. 9. In the limit of small $\mathscr{E}$ eq. (9.9) transforms into eq. (9.8), while at large $\mathscr{E}$ it tends to the expression

$$A = A_0\left\{1 - \frac{3\ln^2[\mathscr{E}\exp(\Omega\tau_s)/2\mathscr{E}_0]}{2(\Omega\tau_s)^2}\right\}, \tag{9.12}$$

which is applicable if

$$\Omega\tau_s \gg \ln[\mathscr{E}\exp(\Omega\tau_s)/2\mathscr{E}_0] \gg 1. \tag{9.13}$$

At still higher $\mathscr{E}$ values we can neglect in eq. (9.7) the last term in the brackets, but the first two should be retained. Solving eq. (9.11) for $y \gg 1$ by an iterative method, and introducing the variable $z = 1 - y/\Omega\tau_s$, we obtain

$$A = A_0[3z/2 - z^3/2], \quad z = (\Omega\tau_s)^{-1}\ln[2\mathscr{E}_0\Omega\tau_s/\mathscr{E}]. \tag{9.14}$$

This expression is applicable if

$$\ln[\mathscr{E}\exp(\Omega\tau_s)/2\mathscr{E}_0] \gg 1, \quad \mathscr{E}/\mathscr{E}_0 \ll \Omega\tau_s,$$

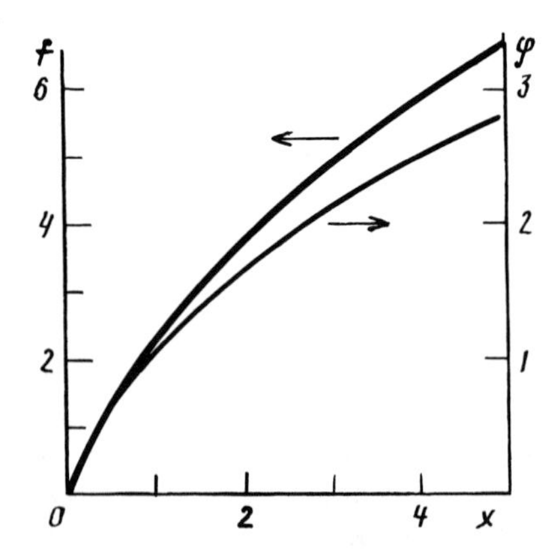

Fig. 9. Plots of the functions f and ϕ defined by eqs. (9.10), (9.11) and (10.8), (10.9).

where the latter inequality follows from the condition $\nu \gg 1$. Equation (9.14) is matched with eq. (9.12) in the region defined by expression (9.13). But if the condition

$$\ln(2\mathscr{E}_0 \Omega\tau_s/\mathscr{E}) \ll \Omega\tau_s$$

is fulfilled, then only the first term in the brackets in eq. (9.14) should be retained. The field $\mathscr{E}_0$ then drops out of the solution, and

$$A = 2[(V - E)/\Omega]\ln(2\Omega\kappa/\mathscr{E}), \tag{9.15}$$

which corresponds to a multiquantum nontunneling penetration of the barrier, when we have for the transmission coefficient,

$$D \approx \exp(-A) = (\mathscr{E}/2\kappa\Omega)^{2(V-E)/\Omega}.$$

An increase in the field amplitude $\mathscr{E}$ leads to a further reduction in the quantity $\nu = \Omega\tau$. When $\nu \ll \nu_s$, eq. (9.6) should be written in the form

$$(\mathscr{E}/\mathscr{E}_0)\sinh\nu = \nu_s,$$

which is valid in the $\nu \sim 1$ case as well. In this situation it is necessary to return to eq. (9.5) for A, in which only the first and the last terms in the square brackets should be retained. It is easy to see that the approximation made corresponds to a field $\mathscr{E}_0$ negligible compared to the strong alternating field, so that the problem reduces to the one solved earlier by Keldysh (1964) and has the solution

$$A = (\kappa^2/m\Omega)[(1 + \mathscr{E}^2/2\Omega^2\kappa^2)\sinh^{-1}(\Omega\kappa/\mathscr{E}) - \tfrac{1}{2}(1 + \mathscr{E}^2/\Omega^2\kappa^2)^{1/2}]. \tag{9.16}$$

In the region of still higher field intensities, where $\mathscr{E} \gg \kappa\Omega$, we find from eq. (9.16) that

$$A = \tfrac{2}{3}\kappa^3/m\mathscr{E}, \tag{9.17}$$

which corresponds to tunneling at the maximum field intensity in a time short compared to the period of the field, so that tunneling occurs quasistatically.

Thus, as the amplitude $\mathscr{E}$ of the high-frequency ($\Omega\tau_s \gg 1$) field increases, the following tunneling regimes occur one after another:

(1) the static regime (9.6), in which the field amplitude $\mathscr{E}$ does not exceed the lower limit set by the inequality (9.8), and the field has no effect on tunneling;

(2) the regime in which the $\mathscr{E}$ field has an anomalously strong effect on the tunneling, and the correction of the order of $(\Omega\tau_s)^{-2}$ to A_0 depends on the combination $\mathscr{E}\exp(\Omega\tau_s)/2\mathscr{E}_0$ (see eqs. (9.9)–(9.11) and fig. 8);

(3) the regime (9.14), in which A depends logarithmically on $\mathscr{E}$, i.e., varies much more slowly with $\mathscr{E}$ than before;

(4) the regime (9.15) of multiquantum nontunneling penetration of the barrier;

(5) the strong-alternating-field regime (9.16) investigated by Keldysh (1968); and

(6) the quasistatic tunneling regime (9.17), in which, because of the high intensity of the alternating field, the tunneling time is short compared to the period of the field, and tunneling occurs at the instant when the barrier is narrowest.

10. The Franz–Keldysh effect in a strong alternating field

10.1. Exact solution of the semiclassical problem

Let us consider the interband breakdown in a semiconductor in the presence of a constant and an alternating electric field. In the case of a very weak alternating field the probability of a one-photon transition can be computed with the aid of perturbation theory (see Anselm 1978). We shall be interested in the case of relatively strong fields, when multiquantum transitions are important. In this case we must use the semiclassical approximation instead of perturbation theory. This method is used to solve the breakdown problem by Bychkov and Dykhne (1970) and Popov (1971), where the constant and alternating fields are considered separately.

For a one-dimensional two-band semiconductor with a spectrum given by

$$\varepsilon(p) = \pm(\varepsilon_g/2)[1 + (2pc/\varepsilon_g)^2]^{1/2},$$

the Lagrangian that takes account of the alternating, $\mathscr{E}\cos\Omega t$, and constant, $\mathscr{E}_0$, fields has the form

$$L = -(\varepsilon_g/2)[1 - (\mathrm{d}x/\mathrm{d}t)^2/c^2]^{1/2} + x(\mathscr{E}_0 + \mathscr{E}\cos\Omega t). \tag{10.1}$$

The classical trajectory can be found in its explicit form

$$\mathrm{d}x/\mathrm{d}t = c[\mathscr{E}_0 t + (\mathscr{E}/\Omega)\sin\Omega t] \times\{(\varepsilon_g/2c)^2 + [\mathscr{E}_0 t + (\mathscr{E}/\Omega)\sin\Omega t]^2\}^{-1/2}. \tag{10.2}$$

The tunneling transition probability in the semiclassical limit can, according to eq. (3.3), be computed with the Lagrangian (10.1), using C_+ and C_- as the contours in fig. 10. The square root in eq. (10.2) has opposite signs on opposite banks of the branch cut, and the interband transition occurs at the instant $i\tau$ corresponding to the zero of the radical in eq. (10.2). As in the preceding section, the instant at which the particle emerges from the forbidden region should correspond to the maximum value of the field. The argument of the tunneling exponential function then has the form

$$A = 4c\int_0^\tau \{(\varepsilon_g/2c)^2 - [\mathscr{E}_0\tau_1 + (\mathscr{E}/\Omega)\sinh\Omega\tau_1]^2\}^{1/2}\,\mathrm{d}\tau_1, \tag{10.3}$$

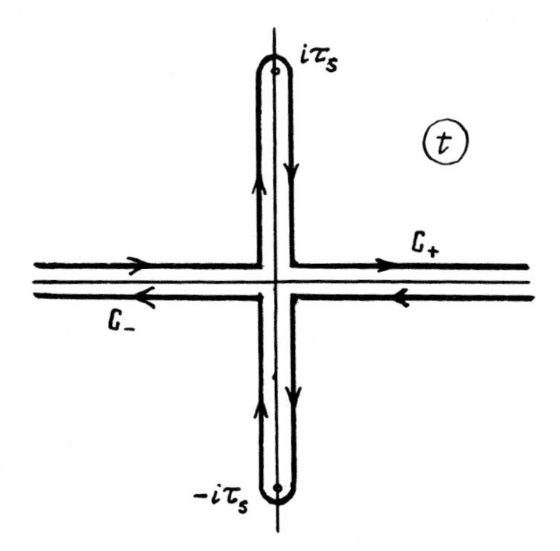

Fig. 10. Integration contour for the computation of the probability for interband tunneling in a semiconductor.

where the upper integration limit corresponds to the root of the integrand. Let us introduce the time

$$\tau_s = \varepsilon_g / 2c\mathscr{E}_0,$$

which is the time required by the particle to gain in the static field $\mathscr{E}_0$ an energy of the order of the forbidden-band width ε_g, and rewrite the action (10.3) in the form

$$A = A_0 (4/\pi \nu_s)^2 \int_0^{\nu} \{\nu_s^2 - [\mu + (\mathscr{E}/\mathscr{E}_0)\sinh \mu]^2\}^{1/2}\, d\mu, \tag{10.4}$$

where

$$A_0 = \pi \varepsilon_g \tau_s / 2$$

is the action in the absence of an alternating field and ν is the root of the equation

$$\nu + (\mathscr{E}/\mathscr{E}_0)\sinh \nu = \nu_s, \quad \nu_s \equiv \Omega \tau_s. \tag{10.5}$$

The family of curves given by eqs. (10.4) and (10.5) is virtually indistinguishable from the family depicted in fig. 8.

10.2. Successive tunneling regimes

The analysis of particular cases with simplified expressions for A is entirely similar to the corresponding analysis carried out in the preceding section. As before, we shall assume that $\Omega \tau_s \gg 1$.

In the limit of weak alternating fields, i.e., for

$$\mathscr{E}\exp(\Omega\tau_s) \ll \mathscr{E}_0,$$

we obtain from eqs. (10.4) and (10.5) the expression

$$A = A_0\left[1 - \left(\frac{2}{\pi}\right)^{1/2} \frac{\mathscr{E}\exp(\Omega\tau_s)}{(\Omega\tau_s)^{1/2}\mathscr{E}_0}\right]. \tag{10.6}$$

As in the preceding section, for the small but nonlinear-in-$\mathscr{E}$ correction to A_0 we obtain

$$A = A_0\{(1 - \Omega\tau_s)^{-1/2}\phi[\mathscr{E}\exp(\Omega\tau_s)/2\mathscr{E}_0]\}, \tag{10.7}$$

where the function ϕ is given by the parametric relations

$$\phi = (2^{5/2}/\pi)\left\{2y^{3/2}/3 + \int_y^\infty [z^{1/2} - (z - y\exp(y - z))^{1/2}]\,\mathrm{d}z\right\}, \tag{10.8}$$

$$y\exp y = x,$$

and its plot is shown in fig. 9. In the limit of weak fields eq. (10.7) reduces to eq. (10.6), and in the limit of strong fields to

$$A = A_0\left\{1 - \frac{2^{7/2}}{3\pi}\frac{\ln^{3/2}[\mathscr{E}\exp(\Omega\tau_s)/2\mathscr{E}_0]}{(\Omega\tau_s)^{3/2}}\right\}, \tag{10.10}$$

which is applicable in the region

$$\Omega\tau_s \gg \ln[\mathscr{E}\exp(\Omega\tau_s)/2\mathscr{E}_0] \gg 1. \tag{10.11}$$

For still higher values of $\mathscr{E}$ we can neglect the hyperbolic sine in the integrand in eq. (10.4) and solve eq. (10.9) for $y \gg 1$ by an iterative method, as a result of which we obtain

$$A = (2A_0/\pi)[\sin^{-1}z + z(1 - z^2)^{1/2}], \tag{10.12}$$

$$z \equiv (\Omega\tau_s)^{-1}\ln(2\mathscr{E}_0\Omega\tau_s/\mathscr{E}).$$

This expression is applicable if

$$\ln[\mathscr{E}\exp(\Omega\tau_s)/2\mathscr{E}_0] \gg 1, \quad \mathscr{E}/\mathscr{E}_0 \ll \Omega\tau_s,$$

where the latter inequality is a consequence of the condition $\Omega\tau \gg 1$. Expression (10.12) is matched with eq. (10.10) under condition (10.11). In the region of still stronger fields $\mathscr{E}$, where

$$\ln(2\mathscr{E}_0\Omega\tau_s/\mathscr{E}) \ll \Omega\tau_s,$$

we have from eq. (10.12) that

$$A = (2\varepsilon_g/\Omega)\ln(\varepsilon_g\Omega/c\mathscr{E}), \tag{10.13}$$

which corresponds to a multiquantum nontunneling surmounting of the forbidden band under conditions when the effect of the constant field can be ignored. Further increase of the field amplitude $\mathscr{E}$ leads to a situation in which the quantity $\nu \ll \nu_s$, and the action is given by the expression

$$A = (2\varepsilon_g/\Omega)\int_0^{\nu} [1 - (2c\mathscr{E}/\varepsilon_g\Omega)^2 \sinh^2\mu]^{1/2}\, d\mu, \tag{10.14}$$

$$(\mathscr{E}/\mathscr{E}_0)\sinh \nu = \nu_s,$$

where the effect of the constant field can be ignored if

$$\ln(\Omega\varepsilon_g/c\mathscr{E}) \ll \Omega\varepsilon_g/c\mathscr{E}_0.$$

In the region of relatively small $\mathscr{E}$ eq. (10.14) reduces to eq. (10.13), while for $\mathscr{E} \gg \varepsilon_g\Omega/c$ the tunneling process is quasistatic, and

$$A = \pi\varepsilon_g/4c\mathscr{E},$$

which is entirely similar to eq. (9.17) in the preceding section.

We shall not enumerate the tunneling regimes, since the corresponding field intensity and frequency regions coincide exactly with the regions given in the preceding section, and the entire difference amounts to some change in the dependence of the action A on the indicated parameters.

11. Tunneling motion of a string under the effect of an alternating field

An elastic string in a potential barrier is of direct relevance to dislocations in a crystal; in this case the velocity of tunneling-activated motion was investigated by Petukhov and Pokrovskii (1972) and to the motion of charged density waves in a Peierls insulator (Rice et al. 1976). The tunneling of a string at zero temperature is analogous to the quantum formation of nuclei in a phase transition (Lifshitz and Kagan 1972, Iordanskii and Finkelstein 1972). The same mechanism results in decay of the metastable vacuum.

The penetration coefficient is known to be determined at zero temperature with exponential accuracy by the following expression:

$$D = \exp\left[-\int_{-\infty}^{\infty} L(\partial y/\partial\tau; y)\, d\tau\right]. \tag{11.1}$$

The Lagrangian of a string is

$$L = \int_{-\infty}^{\infty}\left[\frac{\rho}{2}\left(\frac{\partial y}{\partial\tau}\right)^2 + \frac{\kappa}{2}\left(\frac{\partial y}{\partial x}\right)^2 + U_0 U\left(\frac{y}{a}\right) - \frac{FU_0 y}{a} - U_0 E\right] dx. \tag{11.2}$$

Here $U(z)$ is a symmetric function of the order of unity with minima at the points

$z = \pm 1$ and shown by the dashed curve in fig. 11. The continuous curve represents the potential $V(z) = U(z) - Fz - E$. The constant energy shift E is introduced in order to reduce to zero the potential at the bottom of the initial potential well. Usually, potentials of the form $U(z) = \frac{1}{2}(1 - z^2)^2$ and $U(z) = \cos^2(\frac{1}{2}\pi z)$ are used. The quantity U_0 sets the height of the potential barrier; since F is finite, the degeneracy of the energy minima is lifted. Here $y(x, \tau)$ should be the solution of the classical equation of motion for an imaginary time τ, governed by the condition that the moments in time $\pm\infty$ correspond to the turning points of the string trajectory. Retaining the notations x and τ, we shall measure these quantities in units of

$$x_0 = a(\kappa/U_0)^{1/2}, \qquad \tau_0 = a(\rho/U_0)^{1/2}. \tag{11.3}$$

Then, the penetration coefficient becomes

$$D = \exp[-2a^2(\kappa\rho)^{1/2} A],$$

where

$$A = \int_0^{1/2T} \mathrm{d}\tau \int_{-\infty}^{\infty} \mathrm{d}x[\tfrac{1}{2}(\partial z/\partial\tau)^2 + \tfrac{1}{2}(\partial z/\partial x)^2 + U(z) - Fz - E]. \tag{11.4}$$

Here we have introduced $z = y/a$.

We can see that the semiclassical action is proportional to $V/\omega \gg 1$, where – as deduced from the relationship in eq. (11.3) – we have $\omega \propto \tau_0^{-1}$ and V is

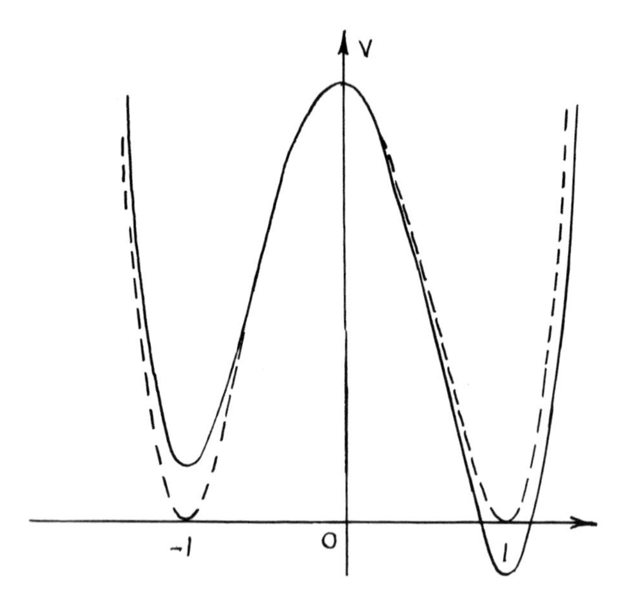

Fig. 11. Cross section of the potential $V(z)$ in the case of string tunneling. The dashed curve is the unperturbed potential $U(z)$.

proportional to the energy of a kink: $V \propto a(\kappa U_0)^{1/2}$. The equation for the classical path $z(x, \tau)$ is obtained by variation of the action (11.4) and in polar coordinates has the form

$$\partial^2 z/\partial r^2 - U'(z) = -(1/r)\partial z/\partial r - F, \tag{11.5}$$

$$r^2 = x^2 + \tau^2,$$

subject to the boundary condition

$$\left.\frac{\mathrm{d}z}{\mathrm{d}r}\right|_{r=0} = 0.$$

The situation resembles the problem of finding a nucleus for a first-order phase transition. The condition $F \ll 1$ corresponds to a small difference between the specific energies of the two phases. Consequently, the function $z(r)$ should describe a thin-walled large-radius nucleus. The problem can, therefore, be tackled in two stages: finding the structure of the boundary and determination of the radius R of a nucleus. Ignoring the right-hand side of eq. (11.5), we obtain the equation

$$\mathrm{d}^2 z_0/\mathrm{d}n^2 - U'(z_0) = 0, \quad n = r - R, \tag{11.6}$$

which describes the structure of the boundary. Allowance for the right-hand side of eq. (11.5) then makes it possible to find the radius R of the nucleus. We shall initially consider the specific case when the line $z(x, \tau) = 0$ can be regarded as the boundary of a nucleus $x(\tau)$. The structure of the boundary is given by the first integral of eq. (11.6),

$$(\mathrm{d}z_0/\mathrm{d}n)^2 = 2U(z_0). \tag{11.7}$$

The radius of the nucleus can be found if we multiply eq. (11.5) by $\mathrm{d}z_0/\mathrm{d}r$ and integrate with respect to r from zero to infinity. Then the integral on the left-hand side vanishes and on the right-hand side we can substitute $r = R$ and take it outside the integral. The result is then

$$R = \alpha/F, \tag{11.8}$$

where α is a number of the order of unity and is governed by the shape of the potential:

$$\alpha = \int_0^1 [2U(z)]^{1/2}\,\mathrm{d}z.$$

In the same approximation we find that the action of eq. (11.4) is given by (Voloshin et al. 1974, Maki 1978)

$$A = \pi\alpha^2/F. \tag{11.9}$$

The dependence of z_0 on r is shown in fig. 12.

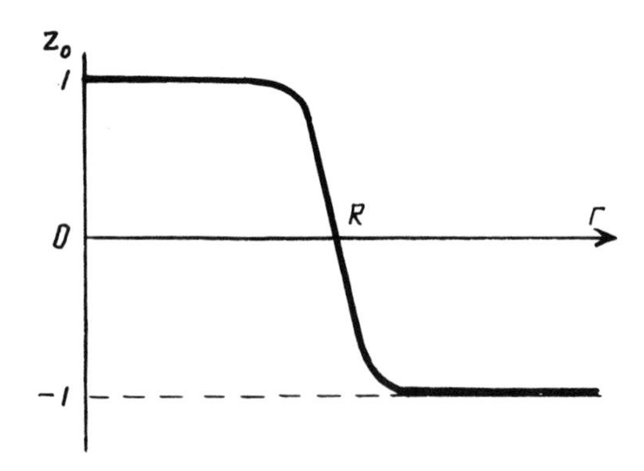

Fig. 12. Radial dependence of the function z_0 at zero temperature.

In contrast to the tunneling of a particle, the penetration of a string across a barrier is entirely due to the lifting of the degeneracy between the potential minima (see fig. 11) because F is finite. The point is this: the increase in the energy of a string of the order of unity caused by the formation of two kinks crossing the barrier should be compensated by an energy reduction $RF \propto 1$, caused by the fact that a part of the string of length $2R$ drops to the lower minimum.

We shall calculate the action using a macroscopic approach based on the fact that the thickness of the boundary of a nucleus is considerably less than its size. The variation of z along the normal n to the boundary is described implicitly by an expression which follows from eq. (11.7):

$$n = \int_0^z [2U(y)]^{-1/2}\,\mathrm{d}y.$$

Using this expression to integrate with respect to n in the equation for the action [eq. (11.4)], we obtain a functional which depends only on the shape of the boundary,

$$A = 2(\alpha l - FS), \tag{11.10}$$

where l and S are the length of the boundary and the area of the nucleus, respectively (see fig. 13). If the boundary is described by the curve $x(\tau)$, then the action of eq. (11.10) can be written in the form

$$A = 4\int_0^{1/2T} \mathrm{d}\tau\{\alpha[1 + (\mathrm{d}x/\mathrm{d}\tau)^2]^{1/2} - Fx\}. \tag{11.11}$$

The condition for an extremum of the action [eq. (11.11)] shows that the boundary of the nucleus is still an arc of a circle of radius α/F, in accordance with eq. (11.8).

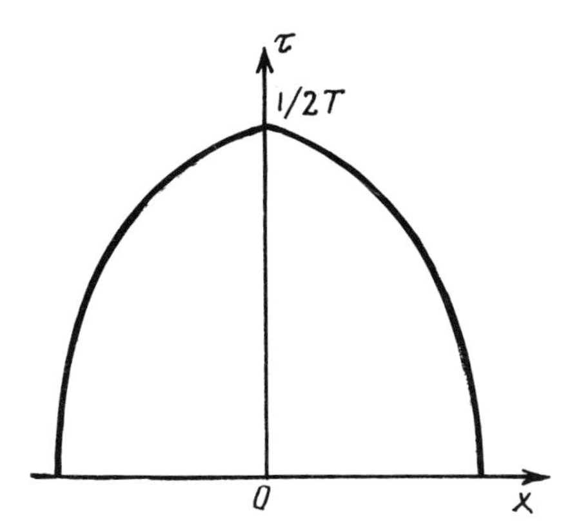

Fig. 13. Shape of the boundary of the nucleus (curve $z(x, \tau) = 0$) in the tunnel-activated regime.

We shall assume now that in addition to a static field F there is also a small alternating component $F_1 \cos \Omega t$. As before, we shall discuss the semiclassical situation when the penetration coefficient can be represented in the form

$$D = \exp(-A_0 + A_1 \cos \Omega t), \tag{11.12}$$

where the correction to the action A should be small compared to A_0 but large compared to unity. The value of A_1 is given by

$$A_1 = \frac{iF_1}{2} \int_c \mathrm{d}t \int_{-\infty}^{\infty} \mathrm{d}x \, z_0(x, t) \cos(\Omega t + \phi). \tag{11.13}$$

Here $z_0(x, \tau)$ is a solution of eq. (11.7), where

$$n = (x^2 - t^2)^{1/2} - \alpha/F.$$

The contour C in fig. 2 at zero temperature follows the imaginary axis and is closed in a distant region of the left half-space. The integral over the whole contour is finite although the partial contribution from the imaginary axis diverges. The constant phase ϕ in eq. (11.13) is selected from the condition that A_0 should be maximal.

We shall consider only the potential

$$U(z) = (1 - z^2)^2/2,$$

where $z(x, \tau) = \tanh[(x^2 - t^2)^{1/2} - 2/3F]$. When considered as a function of t, this solution has poles at points t_k, which are described by the following expression for low values of x:

$$t_k = \pm \frac{2i}{3F}\left(1 - \frac{9F^2x^2}{8}\right) - \frac{\pi}{2}(1 + 2k)\left(1 + \frac{9F^2x^2}{8}\right), \tag{11.14}$$

where k is an integer. Integrating eq. (11.13) with respect to time by the method of residues, we can see that in the limit $\Omega \gg F$ the values $x \propto (\Omega F)^{1/2}$ are important and this justifies the expansion (11.14). Summing with respect to k from zero to infinity, we obtain the following expression for A_1 (Ivlev and Mel'nikov 1986c):

$$A_1 = \pi F_1 \left(\frac{\pi}{3\Omega F} \right)^{1/2} \frac{\exp(2\Omega/3F)}{\sin(\pi\Omega/2)}. \tag{11.15}$$

Two features of this result should be noted. The field F_1 is exponentially enhanced because the time of tunnel motion under the barrier is proportional to $1/F$ and is long compared to the field period $1/\Omega$. The resonance denominator is related to the motion of a string in the classical region where in our units the frequency of small oscillations is $\omega = 2$.

It is important to note that A_1 is calculated allowing implicitly for the structure of the boundary of a nucleus, since the integral in eq. (11.13) is governed by characteristic features of the function z_0 in the complex time plane. The use of the effective Lagrangian of the one-dimensional problem [eq. (11.11)] when a nucleus is described only by the shape of its boundary $x(\tau)$, gives the same exponential dependence as the above equation but it cannot reproduce the resonant denominator of this equation. This macroscopic approach to the solution of the time-dependent problem was used by Maki (1978).

12. Conclusion

Our purpose in the present paper was to investigate in detail the effect of an alternating field on semiclassical processes. The most striking effect here consists in the exponential enhancement of the high-frequency field. We have found that in particular cases of field emission and interband tunneling the argument of the tunneling exponential function depends nonlinearly on the ratio $\mathscr{E} \exp(\Omega\tau_s)/\mathscr{E}_0$ when the latter is of the order of unity. The relative correction to the action in this case is of the same order of smallness as the parameter $(\Omega\tau_s)^{-2}$ or $(\Omega\tau_s)^{-3/2}$, but the absolute value of the correction should be large if the semiclassical approximation is to be applicable. It is, apparently, in this regime that the qualitative difference between the alternating and constant fields is most clearly manifested. As the field amplitude is increased further, the dependence of the action on the field amplitude $\mathscr{E}$ becomes much weaker; specifically, it becomes logarithmic.

In the case of charge exchange between deep-lying centers in semiconductors, when the distance between them is large, typical values of $A_0 \propto \ln(1/D)$ are about 30 (Masterov and Kharchenko 1985). We can then fulfill the condition $A_0 \gg \Omega\tau_s$ by taking $\Omega\tau_s \cong 6$, which corresponds to a field enhancement ratio $\exp(\Omega\tau_s) \cong 400$. But if the action does not attain too large values, i.e., if A_0 lies in the range from 10 to 15, as is possible in tunneling chemical reactions (Zamaraev and Khairutdinov 1978) or in field emission (Neuman 1969), then the exponential enhancement of the alternating field can be observed in the case of the absorption of a small number N of quanta when the quantity $(V-E)/\Omega$ is $\geqslant 1$. For the triangular barrier

$$D \propto \sum_N C_N(\mathscr{E}/\tilde{\mathscr{E}})^{2N} \exp[-A_0(E+\Omega N)],$$

which corresponds to perturbation theory in the case of high frequencies, when the semiclassical approximation is inapplicable. Nevertheless, as can be seen from the expression given above, the effective field turns out to be exponentially enhanced in this limit as well. The high-frequency stimulation of the decay of the current states of Josephson junctions is also described by a dependence of this type.

References

Ambegaokar, V., U. Eckern and G. Shoen, 1982, Phys. Rev. Lett. **48**, 1745.
Ansel'm, A.I., 1978, Introduction to the Theory of Semiconductors (Nauka, Moscow). In Russian.
Baz', A.I., Ya.B. Zel'dovich and A.M. Perelomov, 1969, Scattering, Reactions and Decay in Nonrelativistic Quantum Mechanics (Wiley, New York).
Buettiker, M., and R. Landauer, 1982, Phys. Rev. Lett. **49**, 1739.
Bychkov, Yu.A., and A.M. Dykhne, 1970, Zh. Eksp. & Teor. Fiz. **58**, 1734.
Caldeira, A.O., and A.J. Leggett, 1981, Phys. Rev. Lett. **46**, 211.
Chakravarty, S., and S. Kivelson, 1983, Phys. Rev. Lett. **50**, 1811.
Chakravarty, S., and S. Kivelson, 1985, Phys. Rev. B **32**, 76.
Delone, N.B., and V.P. Krainov, 1984, Atom in a Strong Light Field (Energoatomizdat, Moscow). In Russian.
Devoret, M.H., J.M. Martinis and J. Clarke, 1985, Phys. Rev. Lett. **55**, 1543.
Dmitrenko, I.M., G.M. Tsoi and V.I. Shnyrkov, 1982, Fiz. Nizk. Temp. **8**, 660.
Fisher, M.P.A., 1988, Phys. Rev. B **37**, 75.
Golub, A.A., 1985, Fiz. Nizk. Temp. **11**, 584.
Iordanski, S.V., and A.M. Finkelstein, 1972, Zh. Eksp. & Teor. Fiz. **62**, 403.
Ivlev, B.I., and V.I. Mel'nikov, 1985a, Pis'ma Zh. Eksp. & Teor. Fiz. **41**, 116.
Ivlev, B.I., and V.I. Mel'nikov, 1985b, Phys. Rev. Lett. **55**, 1614.
Ivlev, B.I., and V.I. Mel'nikov, 1985c, Zh. Eksp. & Teor. Fiz. **89**, 2248.
Ivlev, B.I., and V.I. Mel'nikov, 1986a, Zh. Eksp. & Teor. Fiz. **90**, 2208.
Ivlev, B.I., and V.I. Mel'nikov, 1986b, Phys. Lett. A **116**, 427.
Ivlev, B.I., and V.I. Mel'nikov, 1986c, Zh. Eksp. & Teor. Fiz. **91**, 1944.
Jackel, L.D., J.P. Gordon, E.L. Hu, R.E. Howard, L.A. Fetter, D.M. Tennant and R.W. Epworth, 1981, Phys. Rev. Lett. **47**, 697.
Kagan, Yu., and N.V. Prokof'ev, 1987, Zh. Eksp. & Teor. Fiz. **93**, 366.
Kagan, Yu., and N.V. Prokof'ev, 1989, Zh. Eksp. & Teor. Fiz. **96**, 2209.
Keldysh, L.V., 1964, Zh. Eksp. & Teor. Fiz. **47**, 1945.
Landau, L.D., and E.M. Lifshitz, 1976, Mechanics (Pergamon Press, Oxford).
Landau, L.D., and E.M. Lifshitz, 1977, Quantum Mechanics (Pergamon Press, Oxford).
Larkin, A.I., and Yu.N. Ovchinnikov, 1983, Phys. Rev. B **28**, 6281.

Larkin, A.I., and Yu.N. Ovchinnikov, 1986, J. Low Temp. Phys. **63**, 317.
Lempitsky, S.V., 1988, Zh. Eksp. & Teor. Fiz. **94**, 331.
Lifshitz, I.M., and Yu. Kagan, 1972, Zh. Eksp. & Teor. Fiz. **62**, 385.
Maki, K., 1978, Phys. Rev. B **18**, 1641.
Masterov, V.F., and V.A. Kharchenko, 1985, Fiz. & Tekh. Poluprovodn. **19**, 460.
Mel'nikov, V.I., 1985, Zh. Eksp. & Teor. Fiz. **88**, 1429.
Neuman, H., 1969, Physica **44**, 587.
Petukhov, B.V., and V.L. Pokrovskii, 1972, Zh. Eksp. & Teor. Fiz. **63**, 634.
Popov, V.S., 1971, Zh. Eksp. & Teor. Fiz. **61**, 1334.
Rice, M.J., A.R. Bishop, J.A. Krumhansl and S.E. Trullinger, 1976, Phys. Rev. Lett. **36**, 432.
Schwartz, D.B., B. Sen, C.N. Archie and J.E. Lukens, 1985, Phys. Rev. Lett. **55**, 1547.
Sokolovski, D., 1988, Phys. Lett. A **132**, 381.
Sumetskii, M.Yu., 1985, Pis'ma Zh. Tekh. Fiz. **11**, 1080.
Voloshin, M.B., I.Yu. Kobzarev and L.B. Okun', 1974, Yad. Fiz. **20**, 1229.
Voss, R.F., and R.A. Webb, 1981, Phys. Rev. Lett. **47**, 265.
V'yurkov, V.N., and V.I. Ryzhii, 1980, Zh. Eksp. & Teor. Fiz. **78**, 1158.
Zamaraev, K.I., and R.F. Khairutdinov, 1978, Usp. Khim. **47**, 992.
Ziman, J.M., 1971, Principles of the Theory of Solids (Cambridge University Press, Cambridge).

CHAPTER 6

Macroscopic Quantum Effects in the Current-Biased Josephson Junction

Michel H. DEVORET, Daniel ESTEVE and C. URBINA

Service de Physique de l'Etat condensé, CEA Saclay
91191 Gif-sur-Yvette Cedex, France

John MARTINIS

NIST, 325 Broadway
Boulder CO 80303, USA

Andrew CLELAND and John CLARKE

Department of Physics
University of California
Berkeley
and
Materials Science Division
Lawrence Berkeley Laboratory
Berkeley, CA 94720, USA

Quantum Tunnelling in Condensed Media
Edited by
Yu. Kagan and A.J. Leggett

Contents

1. Introduction

Do the laws of quantum mechanics apply to macroscopic degrees of freedom? Until very recently, this question had not been answered experimentally, even though the words "macroscopic quantum effects" have been used in the literature for many years to describe phenomena like superfluidity or superconductivity. In fact, as Leggett (1980a, b, 1984) has emphasized, there are two classes of macroscopic quantum effects. In the first class, to which belong superfluidity and superconductivity, are quantum phenomena that affect only microscopic variables and that add coherently on a macroscopic scale. In the second class are quantum effects displayed directly by a macroscopic variable. A comparison involving a single crystal may help to clarify this point: the discrete orientations of the faces of the crystal are manifestations of the underlying quantum atomic structure revealed on a macroscopic scale by the coherent stacking of atoms. This is an example belonging to the first class of macroscopic quantum effects. On the other hand, a hypothetical observation of the rotation of the crystal as a whole due to zero-point motion would be a macroscopic quantum effect of the second class. The actual observation of phenomena belonging to this second class is the subject of this chapter.

In contrast with a microscopic system, a macroscopic system cannot be decoupled from external influences. It always interacts with an environment containing a large number of degrees of freedom and acting as a thermal reservoir. In almost all cases, the interaction is so strong and/or the reservoir temperature so high compared to the energy level separation of the macroscopic system that quantum coherence is lost. The single crystal mentioned above, even if it is as small as a speck of dust, will normally move spontaneously only by Brownian motion, which is a purely classical phenomenon. However, all macroscopic systems are not hopelessly doomed to behave classically. Some electromagnetic systems, in particular, should display genuine macroscopic quantum effects. Let us consider the simple circuit consisting of an inductor L connected across a capacitor C. Observations on this oscillator are made via leads which can be modeled as an infinitely long transmission line of characteristic impedance Z_c. The flux φ in the inductor is the macroscopic degree of freedom of the system, its conjugate momentum being the charge q on the capacitor. The natural angular frequency of oscillation is $\omega_0 = (LC)^{-1/2}$ and the impedance on resonance is $Z_0 = (L/C)^{1/2}$. The quality factor of the oscillator is $Q = Z_c/Z_0$. To observe quantum effects, two criteria need to be satisfied: (i) The lifetimes of the quantum states of the macroscopic degree of freedom must be sufficiently long on its characteristic time scale. This condition is equivalent to the condition $Q \gtrsim 1$. (ii) The thermal energy must be small compared to the separation of the quantized energy levels. This is expressed by the requirement $\hbar\omega_0 \gg k_B T$, where T is the temperature of the source of damping, here the system of leads. If we take $T = 10$ mK and $Z_c = 50\,\Omega$, these constraints are satisfied for $L = 350$ pH and $C = 15$ pF and the corresponding natural frequency of the oscillator is $\omega_0/2\pi = 2$ GHz. These conditions can nowadays be realized and controlled routinely in the laboratory. Thus, for this system one can hope to challenge the smallness of $\hbar$.

Unfortunately, the harmonic oscillator is a system that does not display quantum effects easily because it is always in the correspondence limit. The average values of the position or the momentum follow classical equations of motion. Quantum mechanics is revealed in higher moments like the variances of the basic variables but these higher moments are very hard to measure. For our LC oscillator, we have, e.g., $\langle q^2 \rangle \approx 10^{-35}\,\mathrm{C}^2 \approx (20e)^2$. Also, there is a more fundamental objection to the use of a linear system to look for macroscopic quantum effects since it is harder to draw the distinction between effects resulting from the coherent addition of microscopic variables and genuine macroscopic quantum effects (Leggett 1980a, b, 1984a, b).

One can circumvent this difficulty by using a nonlinear oscillator involving a Josephson tunnel junction. This component is a sandwich of two superconductors separated by a thin oxide layer (see fig. 1). In contrast with most other nonlinear electronic devices, its relevant properties remain constant at the very low temperatures required by the search for quantum effects. The macroscopic variable that describes the state of the junction is the phase difference δ between the condensates of Cooper pairs in the superconductors on either side of the

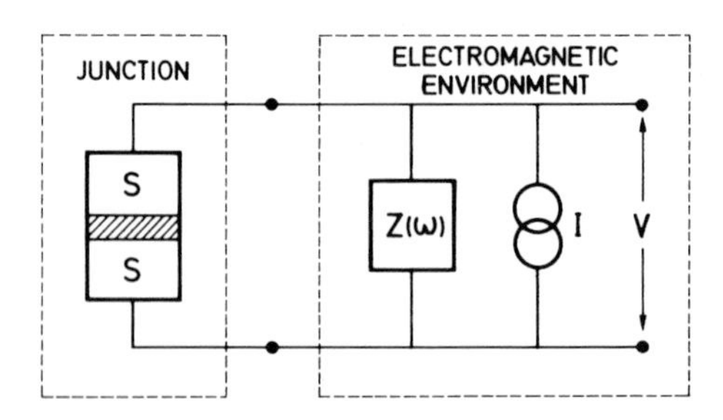

Fig. 1. Schematic representation of a Josephson junction, which consists of two superconducting films separated by a thin oxide barrier. The junction is shunted by its electromagnetic environment which can be represented as a complex frequency-dependent impedance in parallel with a current source.

tunnel barrier. Let us discuss in some detail the nature of the variable δ since it is crucial to the meaning of the macroscopic quantum-mechanical experiments.

At the low temperatures where the experiments are performed, all internal electronic degrees of freedom of the superconductors are, at least in principle, in their ground states and the variables of the junction are the phase differences of the Cooper pairs that tunnel across the barrier. But the rigidity of the superconductive ordering forces the pairs on each side of the junction to have the same relative phase and, hence, all the tunneling pairs have the same phase difference δ (we consider here only the case of a junction much smaller than the Josephson length, which characterizes the rigidity of the phase difference along the junction plane). The variable δ, thus, describes the collective motion of the tunneling pairs and is the single remaining degree of freedom of the junction. Note that this rigidity of the condensates and, thereby, the existence of δ are themselves a macroscopic quantum effect of the first class. Returning to the comparison with a crystal, one sees that the rigidity of the condensates is similar to the interatomic forces that hold the crystal together and whose details are irrelevant to the predictions of the motion of the crystal as a whole. Only macroscopic parameters like moments of inertia need to be known because all internal degrees of freedom are, so to speak, "frozen out".

When a junction is connected via electrical leads to a circuit that performs a measurement, δ couples to the electromagnetic degrees of freedom of the circuit. Apart from the peculiar macroscopic status of δ, the situation is similar to that of a single hydrogen atom which, placed in a cavity, couples to its electromagnetic modes. The role of the position of the electron around the nucleus is played here by the macroscopic variable δ. The role of the cavity is played by the electromagnetic environment of the junction: the measuring circuit and the system of leads linking it to the junction. This environment can be adequately described, from the point of view of the junction, solely in electrical engineering terms as in the previous example of the LC circuit. This is one of the appealing features of Josephson junction systems: In the case of the current-biased Josephson junction, which for simplicity is the only one we consider here,* the environment can, thus, be represented, as shown in fig. 1, by an ideal constant-current source in parallel with a linear impedance $Z(\omega)$ which is, in general, frequency-dependent. Two questions now arise.

Question A. If we engineer the environment such that its influence on the junction is negligible (underdamped case) and if we cool it to sufficiently low temperatures, can we observe the quantum-mechanical behavior of δ as we

* In a variant type of experiment, the Josephson junction is embedded in a superconducting ring and the embraced flux is monitored externally. The macroscopic variable is then the flux through the ring and the role of the current bias is played by the external flux. The environment impedance is in parallel with the ring inductance. See Schwartz et al. (1985) for details.

observe the quantum behavior of the electron in the atom? The situation here is actually more complicated than in the atom, where the properties of each particle (electrons and nucleus) are known independently and where the observations can be directly compared with the predictions obtained from first principles. For a macroscopic system like a Josephson junction, for which a complete description can never be obtained in every detail, the comparison between the observations and the predictions of the quantum theory involves a two-step procedure. In the first step, one measures the time evolution of the system in a regime where thermal fluctuations are large compared with the expected quantum fluctuations. One obtains, thereby, the parameters entering in the classical equations of motion that characterize the macroscopic variables of the system. In the second step, the thermal fluctuations are reduced to well below the level of the expected quantum fluctuations. Of course, because the macroscopic system is prone to all kinds of disturbances, one has to make sure that all other parasitic noises that could mimic quantum fluctuations are eliminated. The new time evolution is then measured. Finally, one compares the results obtained in the second step with the predictions of the quantum theory based on the parameters measured "classically" in the first step. If they agree, one says that the quantum behavior of the macroscopic system has been observed (cf. also Leggett, chapter 1).

Question B. If we now engineer a well-characterized environmental impedance $Z(\omega)$ providing a nonnegligible modification of the junction classical dynamics (damped case), what influence will this have on the quantum behavior? Here, there is a radical difference compared to the atom whose coupling with the electromagnetic modes of the vacuum is "God-given" (Sakurai 1967). The junction could be compared to a hypothetical atomic system for which the fine-structure constant could be adjustable. According to Caldeira and Leggett (1983) and Leggett (1987), the phenomenological knowledge of the environment impedance function $Z(\omega)$ in the classical regime, combined with the phenomenological classical knowledge of the junction itself, suffices to determine theoretically this new quantum behavior. Moreover, they have also proposed a general scheme to compute the effect of environment on quantum properties. Can we test experimentally the predictions of this scheme?

In pioneering experiments (den Boer and de Bruyn Ouboter 1980, Prance et al. 1981, Voss and Webb 1981, Jackel et al. 1981) on macroscopic quantum tunneling (a phenomenon that will be described later) involving Josephson junctions, the temperature of the junctions was lowered so that quantum fluctuations were expected to dominate. The results of these experiments were consistent with a quantum-mechanical interpretation of the behavior of the phase difference of the junctions. Recently, more sophisticated experiments investigated the question of the influence of dissipation on tunneling (Washburn et al. 1985, Schwartz et al. 1985) which had been calculated theoretically

(Caldeira and Leggett 1983, Leggett 1987, Grabert 1985; see also Larkin and Ovchinnikov, chapter 4).

However, a persistent difficulty in all these experiments has been the correct determination of the parameters of the Josephson junctions. Without a precise knowledge of these parameters, no quantitative comparison with quantum theories can be made. In the early experiments, some of the important parameters were treated as adjustable parameters in the predictions or were estimated from measurements of properties not directly connected to the dynamical properties involved in the tunneling experiment. In the experiment of Schwartz et al. (1985) separate measurements of the relevant parameters are made but they are based partially on a quantum analysis of the raw experimental data. This is why we devised new experimental techniques to ensure that all the parameters of the junction would be measured in situ using *classical* phenomena, in particular, under conditions that guarantee that no spurious noise reaches the junction. In a series of experiments that were performed in Berkeley and Saclay we were able to compare theory and experiment with no adjustable parameters. In the first round of experiments, performed in Berkeley, the impedance $Z(\omega)$ was fixed and could be described essentially as a resistor in parallel with a capacitor (Devoret et al. 1984, Martinis et al. 1985, 1987, Devoret et al. 1985, 1987, Cleland et al. 1988). In the second round of experiments, performed in Saclay, we engineered the impedance $Z(\omega)$ so that its frequency dependence was that of a transmission line terminated by a resistor, using the length of the transmission line as a parameter that we could vary in situ (Turlot et al. 1989, Devoret et al. 1989, Esteve et al. 1989).

Since these experiments have been described in detail elsewhere, we will review here only their general principles and their central results. We begin with a short description of the classical dynamics of the Josephson junction biased with a constant current.

2. Classical dynamics of the current-biased Josephson junction

As we have seen, for the purpose of understanding our experiments the underlying physics of the superconductors and their coupling through the oxide barrier can be safely ignored and the junction can be thought of in electrical engineering terms as an ideal *nonlinear* inductor (Josephson element) in parallel with a capacitor (see fig. 2). An ideal linear inductor would be a simple coil made of superconducting wire and would be characterized by an energy quadratic in the flux through the coil. Instead, the Josephson element has an energy given by (Josephson 1962, 1965)

$$E = -I_0(\Phi_0/2\pi)\cos\delta, \tag{2.1}$$

where $\Phi_0 = h/2e$ is the flux quantum, and the variable δ is the phase difference

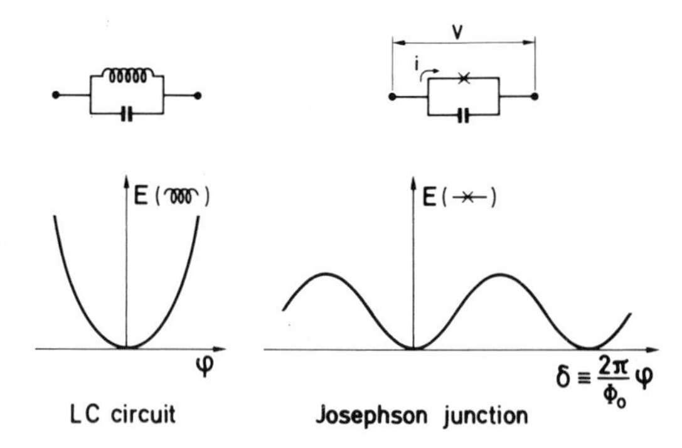

Fig. 2. Comparison between the circuit description of a Josephson junction, which can be represented as a nonlinear inductor in parallel with a capacitor, and a linear LC circuit.

across the junction. Note that δ corresponds to 2φ in the notation of Larkin and Ovchinnikov (chapter 4). The quantity $\varphi = (\Phi_0/2\pi)\delta$ is analogous to the flux through the linear inductor in the sense that the current and voltage for the Josephson element satisfy also $i = \mathrm{d}E/\mathrm{d}\varphi$ and $v = \mathrm{d}\varphi/\mathrm{d}t$. For small δ, the nonlinearity of the junction may be neglected and one recovers the usual relationship between flux and current $\varphi = L_{\mathrm{eff}}i$, where $L_{\mathrm{eff}} = \Phi_0/2\pi I_0$. From the effective inductance L_{eff}, one can define the junction zero-bias impedance $Z_{\mathrm{J0}} = \sqrt{L_{\mathrm{eff}}/C}$, which is analogous to the impedance Z_0 defined above for the LC circuit. This junction impedance has a typical value of a few ohms.

The junction is, thus, characterized by two parameters: the critical current I_0 and the capacitance C. As we have already mentioned, the coupling of the junction to its environment is completely described by the impedance $Z(\omega)$, defined as the voltage response of the environment to a sinusoidal current source of frequency ω which would be connected in place of the junction. We need not consider the response of the environment at optical frequencies and above, where the junction can no longer be thought of simply as a two-terminal device; these higher-frequency modes are never excited during the experiment and their effect is simply to renormalize the parameters I_0 and C. It is important to stress that even if the measuring circuit consists, say, of an ideal current source and voltmeter, each with infinite internal resistance, the impedance due to the electromagnetic coupling between the leads connecting the instruments to the junction will contribute to the environment impedance $Z(\omega)$. Even though these impedances may be negligible at low frequencies and may not affect a quasistatic measurement such as an I–V characteristic, they become of prime importance at the characteristic frequencies of the internal dynamics of the junction which, as we shall see below, are in the microwave range. The junction will, thus, always see in parallel with it a finite effective impedance $Z(\omega)$, which will, in general, be a fraction of the vacuum impedance of 377 Ω.

When it is biased with a constant current I, the junction is equivalent to the model of a particle moving in a tilted washboard potential (see fig. 3). The mass of the particle is the capacitance C and the tilt of the washboard is the ratio I/I_0. When $I < I_0$ the potential has relative minima and the particle can be in one of two states. In the zero-voltage state the particle is confined to within one well (see fig. 3a). The average velocity and, hence, the average voltage across the junction is zero. In the other state, called the voltage state, the particle runs down the washboard at an average constant velocity determined by frictional forces (see fig. 3b) at low frequencies, i.e., by $Z(\omega = 0)$. For the so-called unshunted junctions, which do not have an external resistor connected across their leads, the average velocity is limited solely by the breaking of Cooper pairs in the superconductors and corresponds to the gap voltage, which is of the order of 2 mV. For shunted junctions the velocity depends on the value of the resistor and the bias current. In our experiment with a shunted junction it corresponded to a voltage of about 250 μV which is still easily measurable.

The zero-voltage state is metastable and eventually decays into the voltage state. It is easy to determine the decay rate of the zero-voltage state by monitoring the voltage across the junction. The experiment is performed as follows: Initially, $I = 0$ and the particle representing the phase difference is in a relative minimum of the washboard potential. We ramp the bias current from 0 to $I < I_0$, and wait until a voltage appears across the junction. This voltage is the signature that the particle has escaped from the well and accelerated down the washboard potential. It is worth noting that the acceleration of the particle "amplifies" the very small voltage pulse associated with the escape event. This built-in amplification process is a particularly attractive feature of the current-biased junction.

Because the acceleration process is so fast, the time at which the voltage is measured across the junction can be taken as the time at which the particle escapes. We repeat the experiment a large number of times to obtain the average escape rate.

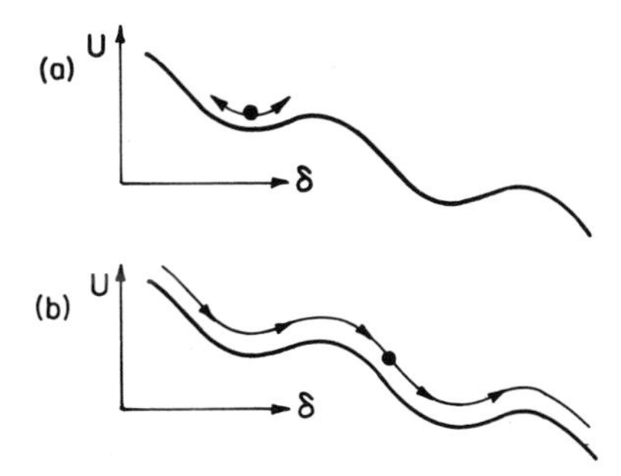

Fig. 3. Tilted washboard analog of a Josephson junction biased with a current $I < I_0$: (a) $\langle\dot{\delta}\rangle = 0$ (zero-voltage state); (b) $\langle\dot{\delta}\rangle \neq 0$ (finite voltage state).

By measuring the escape rate as a function of the bias current, which controls the shape and the size of the potential well, one gains access to some aspects of the behavior of the particle inside the well. However, to obtain a complete picture of the motion of the particle, one needs extra "knobs" on the experiment. This is where our experiments differ fundamentally from all the other experiments: we add to the current bias a weakly perturbing microwave current and measure the *change* in the escape rate induced by this perturbation. The dependence of this change on the frequency and bias current provides additional information that, when combined with the first measurement, enables us to reconstruct the *dynamics* of the particle in the well.

To see what this dynamics is, let us first consider the case of the Berkeley experiments, in which we engineered the environment so that its impedance behaved essentially like a resistor in parallel with a capacitor over a broad frequency range around the characteristic frequency of the particle in the well. The effect of the external shunting capacitance is simply to renormalize the capacitance of the junction. The effect of the resistance is two-fold. One effect is to damp the motion of the particle in the well. The other is to act as a thermal bath and to induce fluctuations in the motion of the particle. The coupling of the junction with the measuring apparatus, thus, involves only two parameters: the resistor R and its temperature T. The temperature of the junction itself is unimportant as long as its intrinsic damping is negligible compared to the damping provided by the external resistor R – this is always the case in our experiment.

The above description of our current-biased Josephson junction corresponds to the circuit of fig. 4 and leads to the classical equation of motion

$$C\left(\frac{\Phi_0}{2\pi}\right)^2 \ddot{\delta} + R^{-1}\left(\frac{\Phi_0}{2\pi}\right)^2 \dot{\delta} + \frac{\partial}{\partial \delta} U(\delta) = \left(\frac{\Phi_0}{2\pi}\right) I_N(t). \tag{2.2}$$

Here $U(\delta)$ is the tilted cosine potential,

$$U(\delta) = (-I_0\Phi_0/2\pi)[\cos\delta + (I/I_0)\delta], \tag{2.3}$$

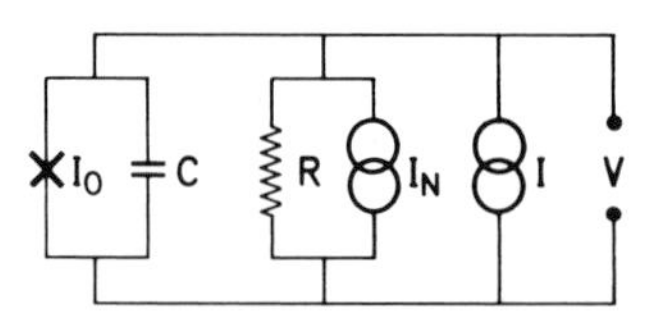

Fig. 4. Equivalent circuit for junctions in the Berkeley experiments, in which the impedance $Z(\omega)$ was a smoothly varying function at microwave frequencies in a broad range around the plasma frequency. The resistor R models the real part of $Z^{-1}(\omega)$. The capacitor C includes a contribution from the imaginary part of the admittance $Z^{-1}(\omega)$. Also represented is the current noise arising from $\mathrm{Re}[Z^{-1}(\omega)]$.

and the current noise $I_N(t)$ satisfies

$$\int_{-\infty}^{+\infty} \langle I_N(t) I_N(0) \rangle_T e^{i2\pi\nu t} dt = 2k_B T/R. \tag{2.4}$$

In practice, the bias current I is very close to the critical current and, as a result, the potential from which the particle escapes is very nearly cubic (see fig. 5). The barrier height ΔU and the oscillation frequency $\omega_p/2\pi$ of the particle at the bottom of the well (referred to in the literature as the plasma frequency) are two useful independent parameters that completely describe the potential. They are given, to a very good approximation, by (Fulton and Dunkleberger 1974):

$$\Delta U = (2\sqrt{2} I_0 \Phi_0 / 3\pi)(1 - I/I_0)^{3/2} \tag{2.5}$$

and

$$\omega_p = (\sqrt{2} 2\pi I_0 / \Phi_0 C)^{1/2} (1 - I/I_0)^{1/4}. \tag{2.6}$$

We can now introduce the current-biased junction impedance $Z_J = 1/\omega_p C = Z_{J0}[2(1 - I/I_0)]^{1/2}$. Finally, the damping due to the resistor is conveniently described by the dimensionless quality factor of the small oscillations at the bottom of the well,

$$Q = R/Z_J. \tag{2.7}$$

An important feature of the particle motion is that, because of the nonharmonicity of the potential, the oscillation frequency decreases when the oscillation amplitude increases. For an oscillation amplitude corresponding to one-half the barrier height, the frequency is approximately 7% lower than the plasma frequency. The oscillation frequency abruptly vanishes when the oscillation reaches the coordinate of the barrier top.

We now turn to the case of a general complex impedance $Z(\omega)$, where $|Z(\omega)| \gg Z_J$, as is the case of the Saclay experiments. The equation of motion (2.2) is modified in two ways. First, the damping term is no longer local in

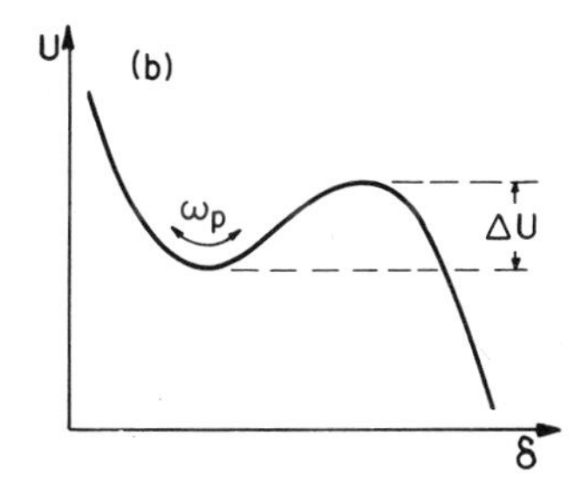

Fig. 5. Cubic potential from which the particle representing the junction phase difference escapes.

time and is replaced by $(\Phi_0/2\pi)^2\int_{-\infty}^{t} F(t-t')\ddot{\delta}(t')\,\mathrm{d}t'$, where $F(t)$ is the current response function of the environment to a voltage step,

$$F(t) = \int_{-\infty}^{+\infty} \frac{\mathrm{d}\omega\,\mathrm{e}^{\mathrm{i}\omega t}}{\mathrm{i}\omega Z(\omega)}. \tag{2.8}$$

Second, the current noise on the right-hand side of eq. (2.2) is no longer white; the right-hand side of eq. (2.4) is replaced by $2k_\mathrm{B}T\,\mathrm{Re}[Z^{-1}(2\pi\nu)]$, where T is the temperature of the source of the impedance $Z(\omega)$, assumed to be in thermal equilibrium.

The motion of the particle at the bottom of the well is not qualitatively different from that in the preceding case. The bare plasma frequency ω_p is renormalized by the factor $1 + \frac{1}{2}Z_\mathrm{J}\,\mathrm{Im}[Z^{-1}(\omega_\mathrm{p})]$, and the quality factor of the small oscillations becomes $\{Z_\mathrm{J}\,\mathrm{Re}[Z^{-1}(\omega)]\}^{-1}$. However, when the particle oscillation energy becomes comparable with the barrier height, the damping of the oscillations becomes amplitude-dependent because of the frequency variations of the real part of $Z^{-1}(\omega)$. This is the main feature that makes the Saclay experiments qualitatively distinct from the Berkeley experiments.

In the discussion of the current-biased Josephson junction, we have treated δ as a classical variable. This is justified at high temperatures, where thermal fluctuations dominate over quantum fluctuations ($k_\mathrm{B}T \gg \hbar\omega_\mathrm{p}$). On the other hand, according to quantum mechanics, at low temperatures δ has to be treated as an operator. When the damping is low enough ($Q \gg 1$), there should be well-defined energy levels in the well (see fig. 6) and the motion of δ is better described in terms of transitions between these levels. Also, in addition to thermal activation, escape from the well can also occur by quantum tunneling from each level through the barrier. Tunneling is a process with no classical analog and has a rate which is temperature-independent. It, thus, becomes predominant at low temperatures. It can be shown that, for low damping,

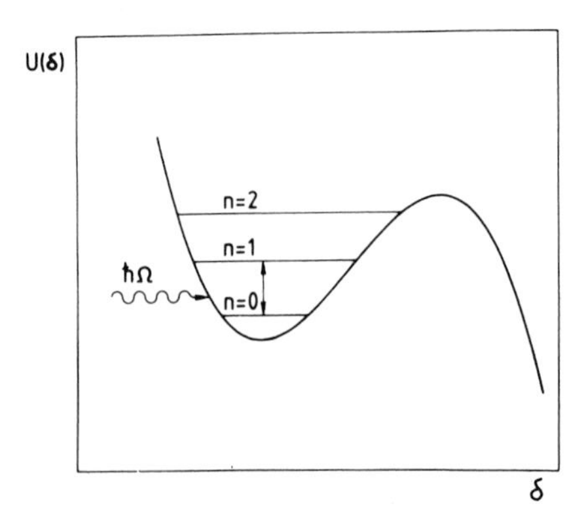

Fig. 6. Quantized energy levels in the well. A microwave photon at frequency Ω can induce transitions between levels.

thermal and tunnel escape coexist with comparable probabilities only in a narrow temperature interval around the crossover temperature given by $T_{co} = \hbar\omega_p/2\pi k_B$ (Grabert 1985).

3. *The junction and its electromagnetic environment*

3.1. Decoupling the junction from ambient noise

It goes without saying that the junction must be completely shielded from parasitic electromagnetic sources like radio stations. But this important precaution is far from being sufficient. Thermal current noise from the measuring apparatus at room temperature – i.e., the current source and the voltage amplifier – must also be prevented from reaching the junction through its leads. This isolation is achieved by interposing a series of filters between the junction and the measuring apparatus. These filters screen out all frequencies except those in a very narrow range near zero so that we are still able to vary the bias current and detect the voltage step as the junction switches to the voltage state. The range is chosen to be sufficiently narrow for the noise in that frequency band to be negligible, yet wide enough to yield a good time resolution of the lifetime of the zero-voltage state.

In practice, such a wideband low-pass filter can be made conveniently only with filter elements having substantial dissipation. Because of this dissipation, the filter itself produces noise and each filter in the chain must be at a lower temperature than the preceding one. In view of the stringent filtering requirements of the experiment, we developed a novel type of filter consisting of a spiral coil of insulated wire inside a copper tube filled with copper powder with a grain size of about 30 μm. Since each grain is insulated from its neighbor by an oxide layer, the effective area of the copper is enormous and, thus, provides substantial skin-effect losses even at the lowest temperatures. The lowest part of the radio-frequency spectrum was excluded with a classic *RC* network. We provided two separate stages of filters in series, one immersed in liquid helium and the other anchored to the mixing chamber of our dilution refrigerator. The chain of filters provided a total attenuation of more than 200 dB at frequencies from 0.1 to 12 GHz. In later experiments at Saclay we developed an improved type of low-pass filter, operating at both microwave and radio frequencies, which consists of a thin-film resistive metallic strip (typically, 30 $\Omega/\square$) deposited on mylar and pressed between two brass blocks. A film of kapton is used as a spacer on top of the metallic thin film.

In both experiments, the last filter was carefully engineered since for unshunted junctions it dominates the junction environment impedance $Z(\omega)$ and, hence, the fluctuations of the phase difference across the junction. For this reason, the last filter is part of the junction mount in all our experiments.

3.2. The junction mount

In the experiments at Berkeley the mount consisted of an attenuating coaxial line made with the same copper powder described above. The copper powder is thermalized by epoxy which is injected between the grains. The junction is mounted as closely as possible to the end of the line to ensure that the impedance discontinuity between the junction and the line occurs over a distance small compared to the wavelength at the plasma frequency. This microwave engineering ensured that the impedance seen by the junction behaved essentially as a parallel RC combination, with no important spurious resonances.

For the experiments at Saclay, we devised a new type of junction mount providing the junction with an adjustable environment. We connect an unshunted junction to a coplanar transmission line partially covered by a microwave-absorbing block, as shown in fig. 7. The distance L between this block and the junction chip can be varied in situ. The portion of the transmission line left uncovered and the horn-shaped junction pads define a delay line, of length l, characteristic impedance Z_L and propagation velocity c_L. The length l is given by $l = L + D$, where the dead length D accounts for the delay due to the junction pads and the indium connections to the coplanar line. In the microwave frequency range the covered portion behaves as a terminating resistor Z_T for the delay line. This circuit presents to the junction an impedance.

$$Z(\omega) = Z_L \frac{(1 - ae^{-2i\omega\Delta t})}{(1 + ae^{-2i\omega\Delta t})}, \tag{3.1}$$

where $\Delta t = l/c_L$ is the propagation delay and $a = (Z_T - Z_c)/Z_T - Z_c$ is the reflection coefficient of the terminating resistor.

The contribution of the other filters to the impedance $Z(\omega)$ is negligible because of the attenuation of the lossy transmission line formed by the section of

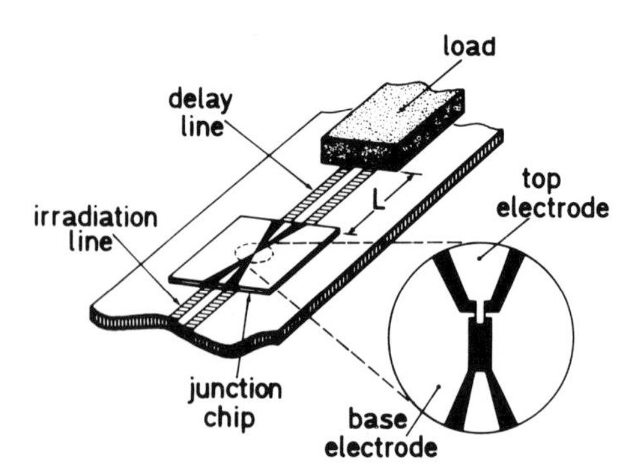

Fig. 7. Junction mount in the Saclay experiments. The junction chip is connected to two coplanar waveguides. One is terminated by a microwave-absorbing block (load) and is used as a delay line. The other is used for exciting the junction with microwaves.

transmission line covered by the block. The block was carefully anchored thermally to the copper box enclosing the junction mount; the temperature of the box was monitored by a calibrated Ge thermometer. The junction was current-biased via the inner and outer conductors of the coplanar line. An insulating coating of the microwave block ensured that the junction, while being loaded at microwave frequencies, remained unshunted for the slowly varying bias current. In addition to the bias current, the junctions in the experiments at both Berkeley and Saclay were submitted to a microwave current. In Saclay this current was made a smooth function of frequency by means of an on-chip irradiation line weakly coupled to the junction by a gap capacitance (see fig. 7).

4. *Experiments with a fixed resistive environment*

4.1. *Determination of parameters in the classical regime*

4.1.1. *Capacitance and resistance*

The parameters ω_p and Q were determined by resonant activation (Devoret et al. 1984). This phenomenon involves the enhancement of the escape by a microwave irradiation of the junction. This enhancement is determined by taking the ratio of the escape rate measured in the presence of microwave irradiation of power P to the escape rate measured under the same conditions without microwaves. Below the plasma frequency, the enhancement is a smoothly increasing function of the microwave frequency. But when a frequency just slightly below the plasma frequency is reached, the enhancement drops steeply to zero. As numerical simulations (Devoret et al. 1984, 1987) and analytical calculations (Fonseca and Grigolini 1986, Ivlev and Mel'nikov 1986, Larkin and Ovchinnikov 1986a, Linkwitz 1990) have shown, the plasma frequency is given, apart from a small correction, by the frequency at which the enhancement drops steeply, while Q is given by the range of frequency over which the drop occurs. Experimentally, it is easier to keep the irradiation frequency fixed and to vary the plasma frequency by varying the bias current [see eq. (2.6)]. An example is shown in fig. 8. At the resonant current I_{res}, the plasma frequency is equal to the

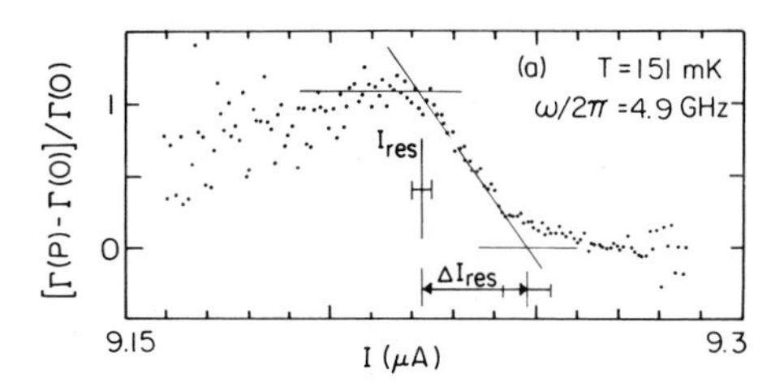

Fig. 8. Microwave-induced enhancement of the escape rate measured as a function of the bias current; $\Omega/2\pi$ is the irradiation frequency. The larger statistical scatter of the data at high and low currents is typical of the results obtained using the technique of Fulton and Dunkleberger (1974).

applied microwave frequency apart from the known correction mentioned earlier. From eq. (2.6) we can deduce the plasma frequency at other values of the bias current. We infer the value of Q from the width in current of the roll-off of the enhancement.

4.1.2. Critical current and temperature

In the thermal regime the escape of the particle from the well occurs by thermal activation. The rate is given by (Hänggi et al. 1990)

$$\Gamma = a_t(\omega_p/2\pi)\exp(-\Delta U/k_B T), \tag{4.1}$$

where the prefactor a_t is of order unity. Although an exact calculation of a_t now exists (Hänggi et al. 1990), one can safely use, given the range of Q and $\Delta U/k_B T$ in the experiment, an approximate expression for a_t (Büttiker et al. 1983) which has the merit of being analytical:

$$a_t = 4/[(1 + Qk_B T/1.8\,\Delta U)^{1/2} + 1]^2. \tag{4.2}$$

We determine I_0 and T from the dependence of the escape rate on the bias current. As is evident from eqs. (2.5) and (4.1), a plot of the experimentally determined quantity $\{\ln[\omega_p(I)/2\pi\Gamma(I)]\}^{2/3}$ versus I should, neglecting departures of a_t from unity, be a straight line with slope scaling as $T^{2/3}$ that intersects the current axis at I_0. Figure 9 shows the three examples of such plots. As expected, the dashed lines drawn through the data intersect the current axis at very nearly the same point. Taking into account the correction due to the departure of a_t from unity, we obtain a temperature-independent value of the critical current (arrow in fig. 9). We find excellent agreement between the temperature measured from the current dependence of the escape rate and the

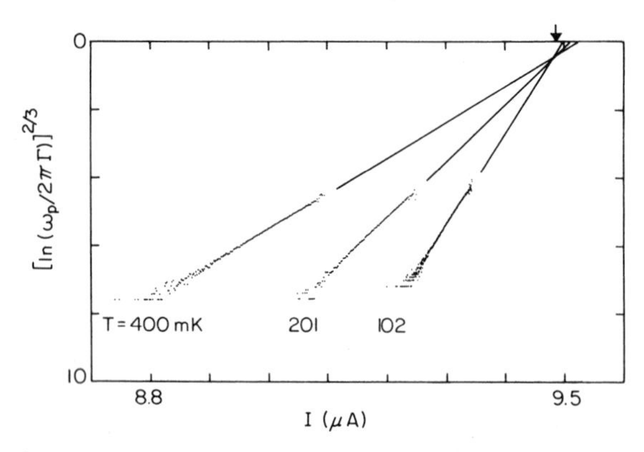

Fig. 9. Plot of the $\frac{2}{3}$ power of the logarithm of the escape rate versus bias current. In this plot, data should fall on a straight line if a_t were equal to unity in eq. (4.1). Dashed lines are the least-square fits to the data. The arrow indicates the value of the critical current.

temperature measured by our thermometers. The fact that the data indeed fall on a straight line is also an additional check that the escape occurs through thermal activation at a well-defined temperature.

We have described how we obtain all the system parameters using the properties of the escape in the classical regime. We now describe the second part of the experiment performed in the quantum regime, obtained by reducing the temperature, where the same properties are remeasured.

4.2. Measurements in the quantum regime

4.2.1. Quantized energy levels

At low temperatures, one has to decrease the barrier height substantially to obtain a lifetime of the zero-voltage state short enough to be measured. Thus, there are only a few levels in the well. As we mentioned earlier, for temperatures T slightly greater than $\hbar\omega_p/2\pi k_B$ the escape in the absence of microwaves occurs essentially via thermal activation through these discrete states to the continuum of states above the potential barrier. In the presence of microwaves, however, the population of the excited states at the top of the well is increased since transitions from one state to another of higher energy are induced by the microwaves. The particle then escapes at a faster rate than in the absence of microwaves. Moreover, one expects a resonant enhancement of the escape rate when the irradiation frequency matches a transition frequency between two energy levels. The quantization of energy levels of the junction zero-voltage state can, thus, be investigated spectroscopically. As in the classical regime, we performed the experiment by keeping the microwave frequency $\Omega/2\pi$ and power P fixed and varying the bias current I, thereby continuously changing all the level spacings. The anharmonic nature of the potential causes the energy level spacings to decrease with increasing energy, so that each resonance corresponding to the transition between a pair of neighboring levels should occur at a distinct value of current.

In fig. 10a we show the change in escape rate versus bias current for a high-Q junction with $C = 47$ pF (junction A in table 1) in the presence of 2.0 GHz microwaves. For the range of current accessible in the experiment, there were five to three energy levels in the well, and the temperature was high enough ($k_B T/\hbar\Omega = 0.29$) for the thermal population of the excited states to be substantial. In fig. 10a we observe three peaks, indicating that the escape rate is resonantly enhanced at certain values of the bias current. These resonances correspond to the transitions shown in the inset. This behavior is in striking contrast to the single, asymmetric resonance observed in the classical regime.

To compare the positions of the resonances with theory, we solved the Schrödinger equation numerically to find the energy levels, using the values of I_0 and C obtained in the classical regime. From these calculations, we obtained the

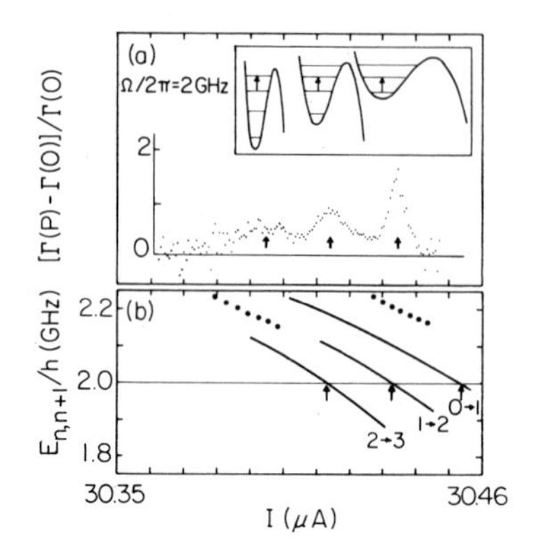

Fig. 10. (a) Microwave-induced enhancement of the escape rate as a function of the bias current for $k_B T/\hbar\omega_p = 0.29$. Arrows indicate the positions of resonances. The inset represents the corresponding transitions between energy levels. (b) Calculated energy level spacings. Dotted lines indicate uncertainties in E_{01} due to errors in the determination of junction parameters. Arrows indicate values of bias current at which resonances are predicted to occur.

Table 1
Junction parameters

Junction	I_0 (nA)	C (pF)	Q
A	30572 ± 17	47 ± 3	75
B	9489 ± 7	6.35 ± 0.4	30
C	24873 ± 4	4.28 ± 0.34	1.8
D	22940 ± 30	44 ± 1	30–80
E	24160 ± 10	11.5 ± 0.5	8–20
F	7040 ± 20	2.7 ± 0.08	2–7

energy spacings corresponding to the $0 \to 1$, $1 \to 2$, and $2 \to 3$ transitions as a function of bias current, as indicated in fig. 10b. The intersection of these curves with the horizontal line corresponding to 2.0 GHz predicts the bias currents at which the peaks should occur. The dotted curves on either side of the $0 \to 1$ curve indicate the uncertainty: the error in current arises from the uncertainty in I_0 [and, hence, in $(I - I_0)/I_0$]. A given error in I_0 shifts the three curves by the same amount. We see that the separations of the measured peaks are in excellent agreement with predictions. The absolute positions of the peaks are shifted along the current axis by about 2 parts in 3000, an error comparable with the indicated uncertainty.

A simple theory (Martinis et al. 1985, 1987, Devoret et al. 1985) for the line shape predicts that the widths of the peaks should be in the ratio 1:3:5 for the $0 \to 1$, $1 \to 2$, and $2 \to 3$ transitions. This prediction is quite well satisfied experimentally. Furthermore, the width of the $0 \to 1$ transition agrees with the predicted value Q^{-1} calculated from the value $Q \approx 75$ estimated from measurements in the classical regime. A more elaborate theory treating the matrix elements between energy levels more exactly arrives at the same conclusion (Larkin and Ovchinnikov 1986b; see also chapter 4 of this book). Finally, a sophisticated theory taking into account the coherence between successive transitions in the well has led to detailed numerical calculations of line shapes that closely fit the data (Chow et al. 1988, Chow and Ambegaokar 1989).

Experiments (Martinis et al. 1985, 1987, Devoret et al. 1985) on other junctions have shown that the position of the peak corresponding to the $0 \rightarrow 1$ transition has the correct dependence on microwave frequency. Resonances corresponding to the $0 \rightarrow 2$ and $1 \rightarrow 3$ transitions have also been observed. These would be strictly forbidden for a simple harmonic oscillator, but are allowed for a cubic potential. Finally, fig. 11 shows the evolution from quantum to classical behavior as the ratio $\hbar\Omega/k_B T$ is increased for a $C = 6$ pF junction (junction B in table 1). At the lowest temperature (curve c), we observe a single, Lorentzian-shaped resonance corresponding to the $0 \rightarrow 1$ transition. At the intermediate temperature (curve b), a shoulder corresponding to the $1 \rightarrow 2$ transition appears as the population of the first excited state becomes significant. At the high temperature (curve a), the resonance is broad and asymmetric: there are several closely spaced levels in the well with substantial thermal population, and the individual transitions overlap to form a continuous response that is reminiscent of classical resonant activation.

4.2.2. *Quantum tunneling*

In the absence of microwaves and for temperatures below $\hbar\omega_p/2\pi k_B$, escape from the well occurs via quantum tunneling through the potential barrier. This process has received the name macroscopic quantum tunneling (MQT) since it affects the state of the junction as a whole and is unrelated to the tunneling of individual Cooper pairs. MQT is analogous to the α decay of a heavy nucleus, the detection of the α particle by a counter being replaced here by the detection

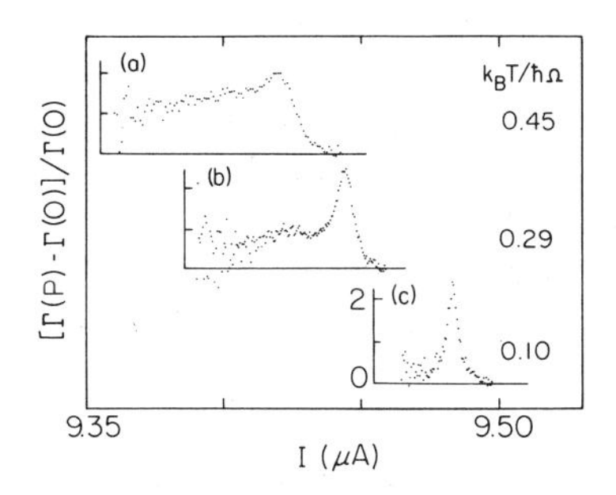

Fig. 11. Microwave-induced enhancement of escape rate versus bias current at three values of $k_B T/\hbar\Omega$. The microwave frequencies are: curve a, 4.5 GHz; curve b, 4.1 GHz; curve c, 3.7 GHz.

of a voltage across the junction. The predicted quantum tunneling rate at $T = 0$ is (Caldeira and Leggett 1983, Leggett 1987)

$$\Gamma = a_q \frac{\omega_p}{2\pi} \exp\left[-7.2 \frac{\Delta U}{\hbar\omega_p} \left(1 + \frac{0.87}{Q} + \cdots \right) \right], \tag{4.3}$$

where

$$a_q = [120\pi(7.2\,\Delta U/\hbar\omega_p)]^{1/2}. \tag{4.4}$$

Equation (4.3) is written so that one can make a direct comparison with eq. (4.1). The usual WKB result is obtained by letting $Q \to \infty$ in eq. (4.3).

The exponential dependence on ΔU common to both eqs. (4.1) and (4.3) makes the absolute value of the escape rate not very meaningful when the temperature is varied. We have, thus, found it useful to transform the escape rate into an "escape temperature" T_{esc} defined through the relation $\Gamma = (\omega_p/2\pi)\exp(-\Delta U/kT_{esc})$. This quantity expresses the magnitude of the fluctuations of the particle in the well, irrespective of their thermal or quantum origin. In contrast with the escape rate, it is nearly independent of the circumstantial parameters ΔU and ω_p in the thermal regime. Neglecting corrections coming from the value of a_t or a_q, it should simply be equal to T in the thermal regime and equal to the temperature-independent expression $\hbar\omega_p/7.2k_B(1 + 0.87/Q)$ in the quantum regime.

Our results for junction B are shown in fig. 12, where T_{esc} is plotted versus temperature (solid circles). The open circles correspond to a measurement with a magnetic field applied to the junction to lower its critical current. With a reduced critical current, the plasma frequency is lower and the junction should behave classically down to all but the lowest temperatures. Indeed, the escape rate followed the classical prediction (solid line), showing that no significant amount of external noise was reaching the junction. When the critical current was restored to its original value, we measured a higher value of the escape rate (solid dots). The escape rate became temperature-independent at low temperatures, as expected when the escape mechanism is dominated by quantum fluctuations. This interpretation is supported by the quantitative agreement between this limiting value of the measured escape rate and its theoretical prediction [the arrow shows the prediction of eq. (4.3); we have also indicated the error uncertainty of this prediction due to the uncertainties in the classical measurements of the parameters]. We note that in our experiment the predicted effect of dissipation on tunneling is too small to be discernible: it is expected to lower T_{esc} by 1.5 mK, a value smaller than the experimental error.

To investigate the effects of dissipation on the tunneling rate, we subsequently performed an experiment in which the junction was shunted by a metallic thin-film resistor (Cleland et al. 1988) to reduce the damping factor Q to 1.8 (see junction C in table 1). We measured a reduction in the tunneling rate by a factor of 300. This result, well outside the uncertainties of the experiment, shows that

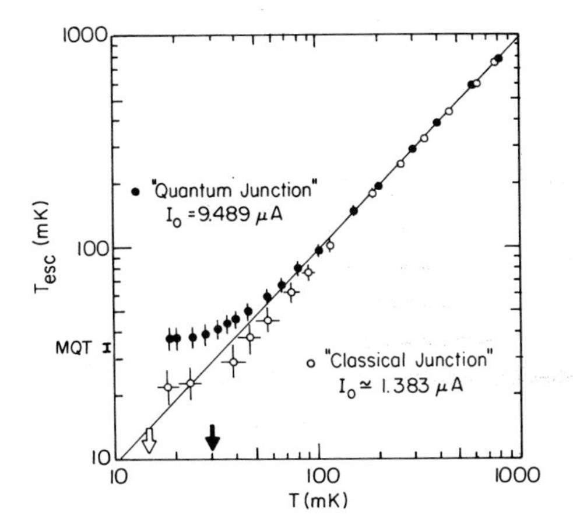

Fig. 12. Escape temperature plotted against temperature for two values of the critical current. The solid line represents the classical prediction $T_{esc} = 0.95T$ [eq. (4.1)]. The error bar to the left of the vertical axis represents the quantum prediction [eq. (4.3)]. Horizontal error bars are a combination of systematic and random errors in the temperature scale; vertical errors indicate primarily systematic uncertainties in the junction parameters. For clarity, error bars for T have been shown for the "classical junction" only; identical errors apply to the "quantum junction". Open and solid arrows indicate the crossover temperatures for the "classical junction" and "quantum junction", respectively.

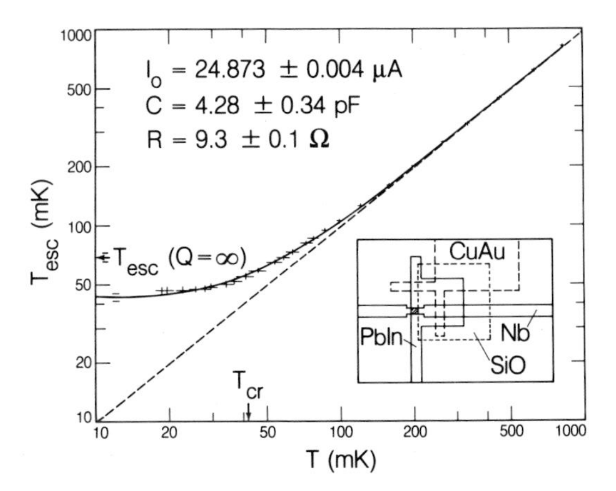

Fig. 13. Escape temperature plotted versus temperature for shunted junction (configuration shown in inset). The solid curve represents theory and the dashed line the classical prediction. The crossover temperature and the predicted $Q = \infty$ escape temperature are indicated by arrows. Error bars are as in fig. 12.

dissipation strongly suppresses tunneling by a factor in very good quantitative agreement with the predictions of the quantum theory (Caldeira and Leggett 1983, Leggett 1987, Grabert 1985).

By replotting the data of fig. 13, we can test two particular aspects of the theory (Grabert and Weiss 1984) in detail. First, for $T < T_{co}$ it is predicted that $\ln \Gamma(T)$ should scale as T^2; to investigate this prediction, in fig. 14 we plot $\ln \Gamma$ at fixed bias current versus T^2. The general behavior of the data and the predictions is rather similar. However, at low temperatures the measured slope is about 30% below the predicted value, for reasons which are not known. We note that if we vary the junction parameters used to obtain the predicted curve within the limits set by the experimental uncertainty, this curve is essentially displaced vertically, with an insignificant change in slope.

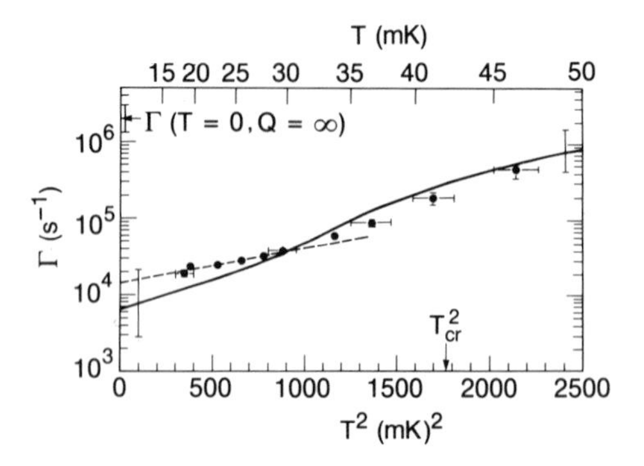

Fig. 14. Same data as in fig. 13 but plotted as escape rate (logarithmic scale) versus the square of temperature. The solid line represents theory which, at low temperature, has a T^2 dependence.

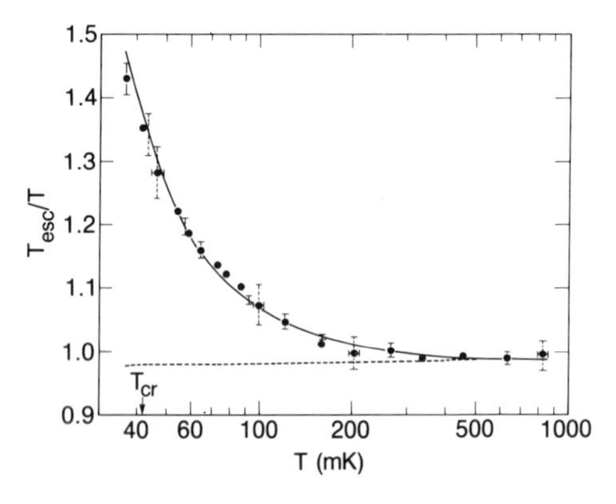

Fig. 15. Same data as in fig. 13 but plotted to show the enhancement of the escape rate above its thermal value (dashed line) at temperatures above the crossover temperature.

A second prediction of the theory is that for $T > T_{co}$, quantum tunneling should enhance the escape rate above the value corresponding to thermal activation. To test this result, in fig. 15 we plot T_{esc}/T versus T. The good agreement between the data (dots) and the theory (solid line) over the entire temperature range provides an unambiguous verification of the Grabert–Weiss theory (Grabert and Weiss 1984).

5. *Experiments on junctions with a variable electromagnetic environment*

5.1. Experiments in the classical regime

We turn now to a series of experiments in which the complex impedance loading the junction was that of a variable-length delay line terminated by a resistor, and whose mathematical description is given by eq. (3.1).

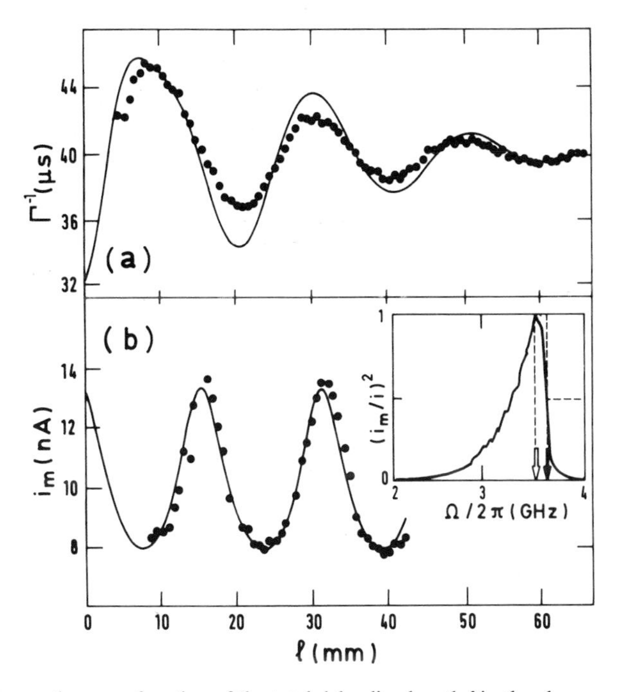

Fig. 16. (a) Escape time as a function of the total delay line length l in the absence of microwaves. The solid line represents the theory from Grabert and Linkwitz (1988). (b) Microwave current needed to decrease the escape time by a factor of 2, plotted as a function of l. The microwave frequency corresponded to the maximum of the resonant activation response curve (open arrow in inset shows curve at minimal length; solid arrow shows plasma frequency). The solid line represents the theory based on Chow et al. (1988), Chow and Ambegaokar (1989) and Esteve et al. (1986).

We measured the parameters Z_L, c_L, and Z_T of the delay line at room temperature with a network analyzer and estimated the dead length parameter D from the junction chip geometry. The parameters of the junction, I_0 and C, were determined as in section 4.1, with the load placed at the minimum distance from the junction so that the delay line behaved almost as a pure resistor (see junction D in table 1). In fig. 16a we show the measured lifetime Γ^{-1} of the zero-voltage state as a function of delay line length (dots) in the absence of microwaves, whereas in fig. 16b we plot the microwave current amplitude for which the lifetime decreases by a factor of two (dots). The microwave frequency was chosen to correspond to the maximum of the resonant activation response curve (see the open arrow in the inset of fig. 16b). This frequency is very close to the plasma frequency (solid arrow). The interpretation of these results is the following: the microwave enhancement of the escape rate by an irradiation at the plasma frequency is directly proportional to the damping at that frequency only, i.e., to $\mathrm{Re}[Z^{-1}(\omega_p)]$, whereas the escape rate itself measures a weighted average of the damping at the various oscillation amplitudes in the well. This explains why the amplitude of the modulation in fig. 16a decreases with increasing L whereas its amplitude stays relatively constant in fig. 16b. The experimental data are in good agreement with theoretical predictions (Grabert and Linkwitz 1988) (solid line). This agreement provides an in situ verification of the parameters of the delay line, in particular of the dead length D which could not be measured at room temperature.

5.2. *Measurements in the quantum regime*

5.2.1. *Quantized energy levels*

We have measured the microwave-induced resonance corresponding to the transition between the ground state and the first excited state with the same techniques as in section 3.2.1. Figure 17 shows an example of the results for a load position $L = 10$ mm and a microwave frequency of 3.6 GHz. The measurements were performed at 20 mK on a junction with parameters close to junction B (see junction E in table 1) and, as expected, the peak in the data is similar to that of curve c in fig. 12. In this section, we define the resonance current I_{RES} as the current corresponding to the maximum of the resonance curve. In fig. 18 we plot I_{RES} as a function of the load position L for three microwave frequencies. We observe the frequency pulling induced by the delay line which acts as an adjustable resonator. A striking feature of the data is that for certain values of the length L, two resonance peaks were observed at the same current for two different frequencies. This is expected if the coupling between the resonator and the junction exceeds a critical value (Devoret et al. 1989). Above this value, the junction first excited-state and the resonator one-photon-state are hybridized. The influence of the line on the junction

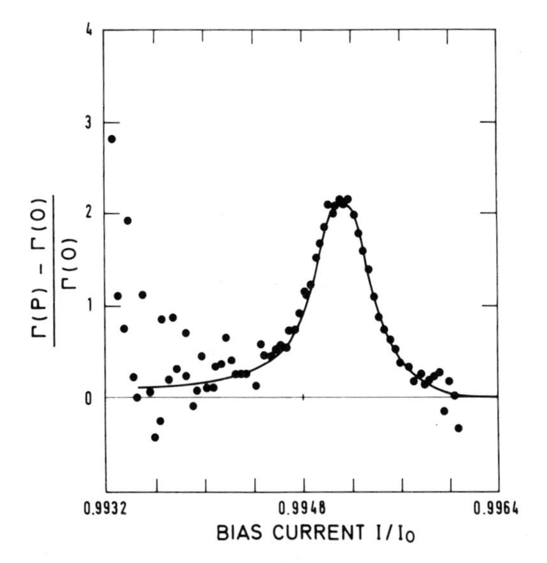

Fig. 17. Microwave-induced enhancement of the escape rate as a function of the bias current at $T = 17$ mK. The irradiation frequency was $\Omega = 3.6$ GHz and the load was positioned at $L = 10$ mm. The solid line is a guide for the eye. The maximum of the resonance defines the current I_{RES}.

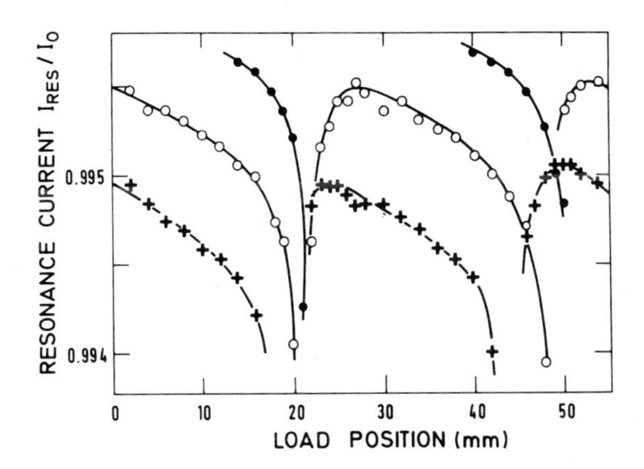

Fig. 18. Resonant current as a function of the position of the load for three irradiation frequencies: Solid dots, 3.4 GHz; open dots, 3.6 GHz; crosses, 3.8 GHz. Solid lines are guides for the eye. The temperature is 20 mK.

cannot then be treated by standard perturbation theory. In fig. 19 we show the predictions of a non-perturbative calculation (Devoret et al. 1989) of I_{RES} as a function of the total electrical length $l = L + D$, the line and junction parameter values being taken from previous experiments. The good agreement between experiment and theory further confirms that the delay line behaves at

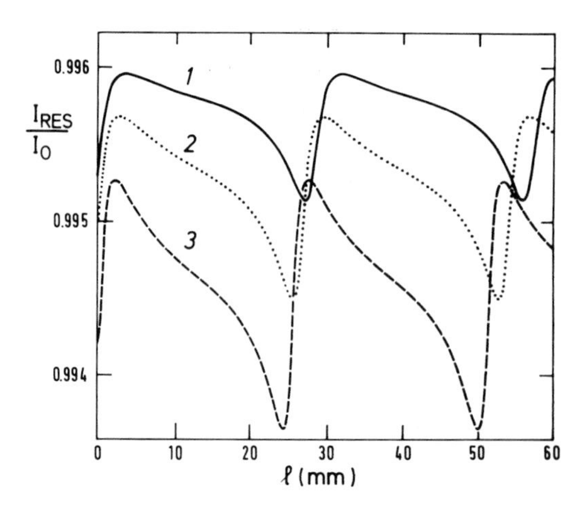

Fig. 19. Predicted resonant current as a function of the electrical length l for the same irradiation frequencies as in fig. 18: curve 1, 3.4 GHz; curve 2, 3.6 GHz; curve 3, 3.8 GHz. The half-wavelength at the three frequencies are, respectively, 30, 32, and 34 mm.

low temperature as expected from room-temperature measurements. Moreover, experiments of this kind provide an additional check on the dead length parameter D, whose value was found to be 6 mm for junctions D and E and 2 mm for junction F (see table 1).

5.2.2. *Quantum tunneling*

The lifetime of the metastable zero-voltage state was then determined for load positions L ranging between 0 and 40 mm and for temperatures between 18 and 150 mK. We focus here first on the behavior of the lifetime when the length L and, hence, the delay time $\Delta t = (L + D)/c_L$ are increased. The experiment was performed on two junctions (junctions E and F in table 1). Figure 20 displays the behavior of junction F at two temperatures for the same bias current. These temperatures, 18 and 65 mK, are, respectively, below and above the crossover temperature $T_{co} = 47$ mK of the junction. At 18 mK (fig. 20b), the lifetime is three times shorter at the longest delay than at the shortest one ($\Delta t = 10$ ps). This decrease in the lifetime is almost complete at a delay of 100 ps. There is, thus, a characteristic delay, beyond which the escape rate becomes independent of the length of the line. This result suggests, independently of any theory, that there is a time associated with tunneling, much shorter than the lifetime, that determines the maximum delay with which the environment must react to affect the tunneling rate. A similar result was observed with junction E, although the lower coupling of the junction to the delay line made the reduction of the lifetime with line length less pronounced.

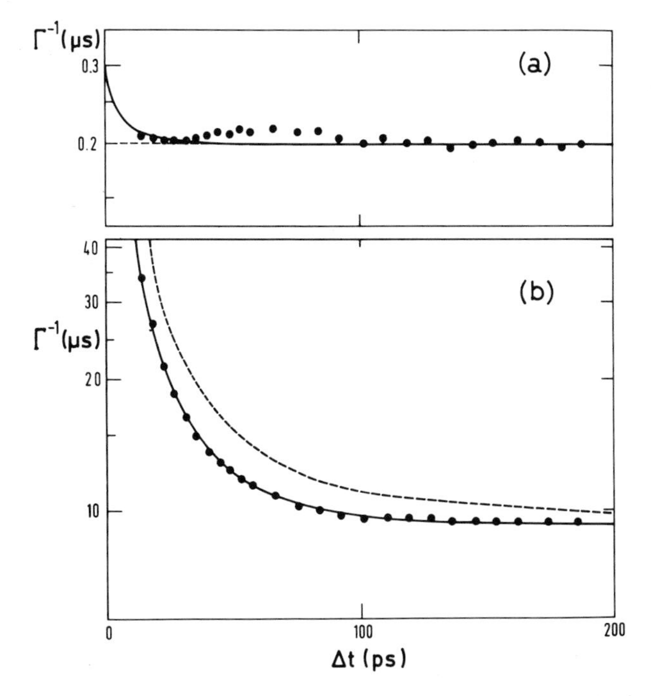

Fig. 20. Escape time versus delay time at two temperatures: (a) $T = 65$ mK and (b) $T = 18$ mK. The dashed line corresponds to the zero-temperature perturbative expression given by eqs. (5.2) and (5.3). The full line represents the prediction of a numerical calculation using the theory developed in Grabert et al. (1987).

At 65 mK, in the thermal regime, we do not observe any important change of the lifetime with the delay (fig. 20a). Similar data taken at 31 and 45 mK show that the influence of the delay on the lifetime gradually decreases when the temperature is raised. Figure 21 shows measurements of the escape temperature of junction F as a function of temperature for the minimum (solid dots) and maximum (open dots) delay line length.

We now compare these results with theoretical predictions. As in our previous experiments, the section of the washboard potential from which the particle tunnels can be very well approximated by a cubic potential with barrier height $\Delta U = (4\sqrt{2}/3)U_0(1 - I/I_0)^{3/2}$. Leggett (1984c) has calculated the effect of a circuit described by a large impedance $Z(\omega)$ with arbitrary frequency dependence on the zero-temperature MQT rate, while Chakravarty and Schmid (1986) have applied this calculation to the case of the transmission line. We have chosen to use a different formulation. Using the step response function $F(t)$ of eq. (2.8), which is the Fourier transform of $[\mathrm{i}\omega Z(\omega)]^{-1}$, and within the cubic potential

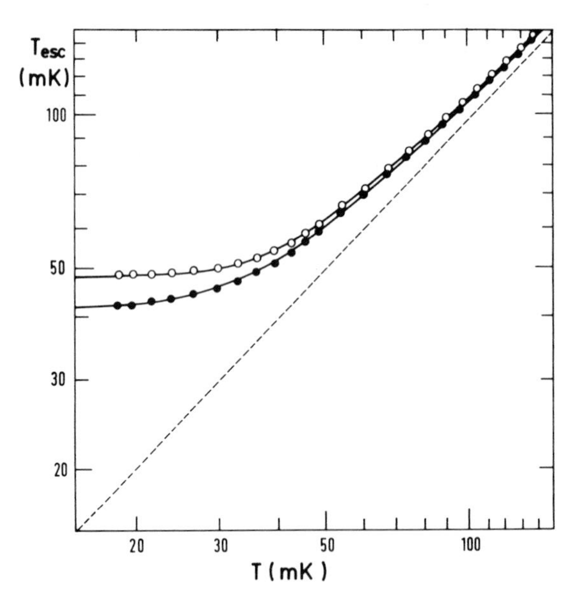

Fig. 21. Escape temperature as a function of the temperatures for minimum delay (solid dots) and maximum delay (open dots). The solid line represents the prediction of a numerical calculation as in fig. 20.

approximation, we can cast Leggett's result into the form (Esteve et al. 1986)

$$\Gamma = \omega_p/2\pi(120\pi B)^{1/2}\exp(-B), \tag{5.1}$$

with

$$B = B_0\left[1 + \int_0^\infty F(t)G(t)\,\mathrm{d}t\right] + \mathrm{O}[F^2(t)]. \tag{5.2}$$

Here $B_0 = 7.2\,\Delta U/\hbar\omega_p$ is the tunneling exponent in the absence of any environment, and the function $G(t)$ is given by

$$G(t) = \frac{45}{\pi^4 C}\sum_{n=1}^{\infty}\frac{n}{[n + (\omega_p t/2\pi)]^5}. \tag{5.3}$$

For the case of the delay line, the function $F(t)$ consists of a series of steps separated by $2\Delta t$, as shown in fig. 22a. In fig. 22b we plot $G(t)$, which we see is a monotonically decaying function. We characterize the time scale of this decay by the latency time t_L defined as

$$t_L = \frac{\int_0^\infty G(t)\,\mathrm{d}t}{G(0)}. \tag{5.4}$$

For the cubic potential [$G(t)$ given by expression (5.3)], the value of the latency

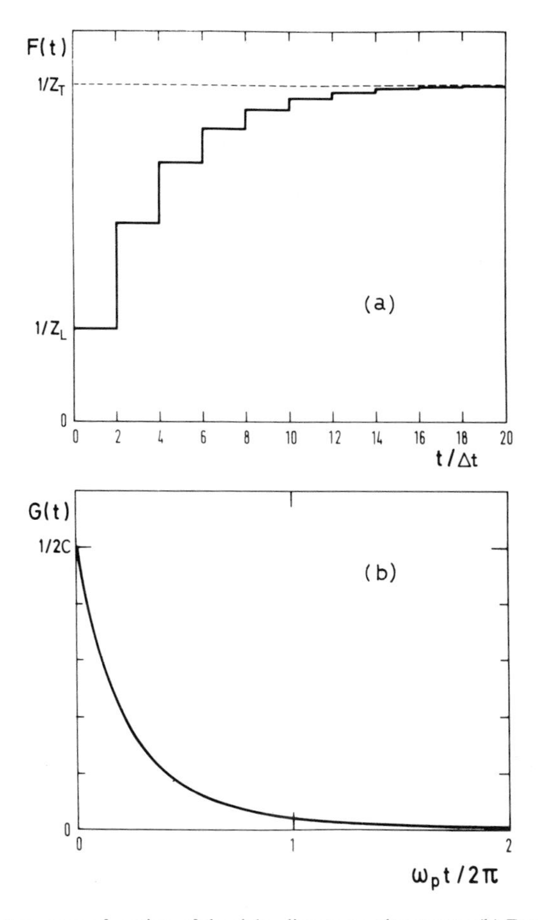

Fig. 22. (a) Current response function of the delay line to a voltage step. (b) Dynamical function of the tunneling out of a cubic potential [eq. (5.3)].

time is $t_L = 1.74\omega_p^{-1}$. The kernel $G(t)$ is also characterized by its total weight $r = \int_0^\infty G(t)\,dt$, which has the dimensions of a resistance and whose value for the cubic potential case is $r = 0.87/C\omega_p$. In the case of frequency-independent damping such that $Z(\omega) = R$, the current response $F(t)$ of the environment has the shape of a perfectly sharp step of height $1/R$ and the integral in eq. (5.2) is simply r/R. One, thus, recovers the result of eq. (4.3) stating that the exponential suppression of tunneling is governed by the magnitude of damping on the scale set by r. In the case of frequency-dependent damping, to affect tunneling not only must the current response $F(t)$ magnitude be of the same order or larger than $1/r$, but also its rise time must be short on the scale of the latency time – hence the name of this latter quantity. This property explains the overall shape of the experimental results: the closer the load, the shorter the response time of the delay line and the longer the lifetime. In fig. 20, the dashed line, which is the prediction of eq. (5.1), is seen to be in qualitative agreement with the data. A difference is expected in view of the fact that the experiment was performed at nonzero temperatures in the presence of nonnegligible friction.

Grabert et al. (1987) have extended Leggett's theory to nonzero temperatures and arbitrary damping strength. They found that the effect of friction on the escape rate decreases strongly as soon as the temperature reaches the crossover temperature defined by $k_B T_{co} = \hbar\omega_p/2\pi$ and that quantitative deviations from eqs. (5.1) and (5.2) arise when one takes into account the response function $F(t)$ at all orders. We have numerically computed the predictions of this theory for the response function $F(t)$ corresponding to our experiment. The continuous line in figs. 20 and 21 are best fits obtained from our full numerical theory using $C = 2.7$ pF and $s = 0.9855$; the measured values were $C = 2.7 \pm 0.08$ pF (see junction F in table 1) and $s = 0.9858 \pm 0.0005$. These fit parameters yield a latency time $t_L = 30$ ps.

5.2.3. Discussion of the latency time

The concept of a tunneling time like the latency time that determines the time scale during which the environment must react to affect the tunneling probability is, of course, not restricted to systems of macroscopic nature. It appears to be relevant to the understanding of the tunneling of microscopic particles such as electrons in the scanning tunneling microscope or in heterostructures. In these cases the problem is not the decay of a bound state as in MQT but the tunneling of a free particle scattered by a potential barrier. Moreover, the environment consists of the electromagnetic plasma modes in the conducting electrodes. These modes, which are analogous to the electromagnetic modes in our delay line, screen the field created by the tunneling electron and can, therefore, react to it. Persson and Baratoff (1988) have shown that this dynamical screening affects the tunneling rate only if it is effective on a time scale shorter than a characteristic time of tunneling, the traversal time introduced by Büttiker and Landauer (1982). These authors interpret the traversal time as the average time spent by the particle while it tunnels through the barrier. Tunneling rate measurements of Gueret et al. (1988) on heterostructures with barriers of various barrier heights have shown that the dynamical screening is indeed effective only when the electronic plasma period in the electrodes is shorter than the traversal time of the electrons through the barrier.

The analogy between the latency time and the traversal time is also supported by their dependence on the potential barrier parameters. The traversal time scales as $[(U - E)/m(\Delta x)^2]^{1/2}$, where U and Δx are the barrier height and width, and E and m are the particle energy and mass. The latency time scales in the same way, ΔU replacing $U - E$.

In conclusion, our measurements of the MQT rate as a function of the retardation of environmental dissipation reveal the latency time of tunneling. This tunneling time determines the effect of the environment response time on the tunneling probability. The latency time plays the same role for the quantum decay of a metastable state as the traversal time does for the tunneling of an unbound particle through a barrier.

6. *Conclusions*

Our measurements of the lifetime of the zero-voltage state of a current-biased Josephson junction agree quantitatively with predictions based on quantum theory, all the relevant parameters being measured in situ. From this we conclude that the phase difference across a Josephson junction, a macroscopic variable, together with the electromagnetic environment seen by the junction do indeed behave quantum-mechanically. The meaning of this result is three-fold.

First, the experimental range of validity of quantum mechanics has been extended. We have shown that the postulate of quantum mechanics that all degrees of freedom, whatever the *complexity* of the system they describe, have to be treated equally remains valid even for a composite, macroscopic system such as a Josephson junction connected to its environment. We are aware that many physicists believe quantum mechanics to be a fundamental theory that cannot "break down". Nevertheless, we hold the view that this should be checked as far as available technology permits. A further challenge in the domain of macroscopic quantum mechanics is, e.g., the observation of macroscopic quantum coherence (Caldeira and Leggett 1983, Leggett 1987, Grabert 1985, Tesche 1990; see also Leggett, chapter 1).

Second, the prediction of Caldeira and Leggett that a phenomenological knowledge of the response of the environment suffices to determine its influence on quantum properties of degrees of freedom that are coupled to it has been experimentally tested. This prediction is of great conceptual significance in the solution of the problems associated with the reduction of the wave packet in quantum measurement theory (Omnes 1990): it can be used to demonstrate quite generally that a measurement, which always involves a macroscopic variable strongly coupled to an environment, will always destroy quantum correlations. In particular, a measurement on a system in a coherent superposition of two states, in an attempt to determine which state the system is in, will always leave the system in an incoherent superposition of the two states, irrespective of whether a conscious observer takes notice or not of the result. In other words, by the time one looks at it, Schrödinger's Cat is no longer dead and alive.

Third, our experiments show that it is indeed possible, given enough filters and shields, to isolate an electronic degree of freedom like a voltage or a current from its environment sufficiently well for its behavior to be entirely dominated by quantum effects. This opens a new area of research on superconducting circuits that would perform quantum signal processing such as squeezing (Slusher et al. 1985) and perhaps implement the so-called back-action evasion amplifiers that are needed for the detection of gravity waves (Caves et al. 1980). The recent experiment of Yurke et al. (1989) on a Josephson parametric amplifier in the quantum regime is a step in this direction. Finally, the quantum electronician can dream of building exotic macroscopic "atoms with wires" that would display new quantum phenomena with no equivalents in the microscopic world.

Acknowledgements

We have benefited from helpful and informative discussions with H. Grabert, A.J. Leggett and R. Omnes. This work was supported by the Director, Office of Energy Research, Office of Basic Energy Sciences, Materials Sciences Division of the US Department of Energy under Contract no. DE-AC03-76SF00098 and by special funding of the Commissariat à l'Energie Atomique.

References

Büttiker, M., and R. Landauer, 1982, Phys. Rev. Lett. **49**, 1739.

Büttiker, M., E.P. Harris and R. Landauer, 1983, Phys. Rev. B **28**, 1268.

Caldeira, A.O., and A.J. Leggett, 1983, Ann. Phys. (New York) **149**, 374.

Caves, C.M., K.S. Thorne, R.V.P. Drever, V.D. Sandberg and M. Zimmermann, 1980, Rev. Mod. Phys. **52**, 341.

Chakravarty, S., and A. Schmid, 1986, Phys. Rev. B **33**, 2000.

Chow, K.S., and V. Ambegaokar, 1989, Phys. Rev. B **38**, 11168.

Chow, K.S., D.A. Browne and V. Ambegaokar, 1988, Phys. Rev. B **37**, 1624.

Cleland, A., J.M. Martinis and J. Clarke, 1988, Phys. Rev. B **37**, 5950.

den Boer, W., and R. de Bruyn Ouboter, 1980, Physica B **98**, 85.

Devoret, M.H., J.M. Martinis, D. Esteve and J. Clarke, 1984, Phys. Rev. Lett. **53**, 1260.

Devoret, M.H., J.M. Martinis and J. Clarke, 1985, Phys. Rev. Lett. **55**, 1908.

Devoret, M.H., D. Esteve, J.M. Martinis, A. Cleland and J. Clarke, 1987, Phys. Rev. B **36**, 58.

Devoret, M.H., D. Esteve, J.M. Martinis and C. Urbina, 1989, Phys. Scr. T **25**, 118.

Esteve, D., M.H. Devoret and J.M. Martinis, 1986, Phys. Rev. B **34**, 158.

Esteve, D., J.M. Martinis, E. Turlot, C. Urbina, M.H. Devoret, S. Linkwitz and H. Grabert, 1989, Phys. Scr. T **29**, 121.

Fonseca, T., and P. Grigolini, 1986, Phys. Rev. A **33**, 122.

Fulton, T.A., and L.N. Dunkleberger, 1974, Phys. Rev. B **9**, 4760.

Grabert, H., 1985, in: Superconducting Devices and Their Applications, eds H.D. Hahlbohm and H. Lübbig (de Gruyter, Berlin) p. 289.

Grabert, H., and S. Linkwitz, 1988, Phys. Rev. A **37**, 963.

Grabert, H., and U. Weiss, 1984, Phys. Rev. Lett. **47**, 1787.

Grabert, H., P. Olschowski and U. Weiss, 1987, Phys. Rev. B **36**, 1931.
Gueret, P., E. Marclay and H. Meier, 1988, Solid State Commun. **68**, 977.
Hänggi, P., P. Talkner and M. Borkovec, 1990, Rev. Mod. Phys. **62**, 251.
Ivlev, B.I., and V.I. Mel'nikov, 1986, Phys. Lett. A **116**, 427.
Jackel, L.D., J.P. Gordon, E.L. Hu, R.E. Howard, L.A. Fetter, D.M. Tennant, R.W. Epworth and J. Kurkijarvi, 1981, Phys. Rev. Lett. **47**, 697.
Josephson, B.D., 1962, Phys. Lett. **1**, 251.
Josephson, B.D., 1965, Adv. Phys. **14**, 419.
Larkin, A.I., and Yu.N. Ovchinnikov, 1986a, J. Low Temp. Phys. **63**, 317.
Larkin, A.I., and Yu.N. Ovchinnikov, 1986b, Sov. Phys.-JETP **64**, 185.
Leggett, A.J., 1980a, Prog. Theor. Phys. (Suppl.) **69**, 80.
Leggett, A.J., 1980b, J. Phys. (Paris) Colloq. **39**, 1264.
Leggett, A.J., 1984a, Contemp. Phys. **25**, 583.
Leggett, A.J., 1984b, Essays in Theoretical Physics in Honour of Dirk Ter Haar, ed. W.E. Parry (Pergamon, Oxford) p. 95.
Leggett, A.J., 1984c, Phys. Rev. B **30**, 1208.
Leggett, A.J., 1987, in: Lecture notes, Les Houches Summer School on Chance and Matter, eds J. Souletie, R. Stora and J. Vannimenus (North-Holland, Amsterdam) and references therein.
Linkwitz, S., 1990, Ph.D. Thesis, Stuttgart.
Martinis, J.M., M.H. Devoret and J. Clarke, 1985, Phys. Rev. Lett. **55**, 1543.
Martinis, J.M., M.H. Devoret and J. Clarke, 1987, Phys. Rev. B **35**, 4682.
Omnes, R., 1990, Ann. Phys. (New York) **201**, 354.
Persson, B.N.J., and A. Baratoff, 1988, Phys. Rev. B **38**, 9616.
Prance, R.J., A.P. Long, T.D. Clark, A. Widom, J.E. Mutton, J. Sacco, M.W. Potts, G. Megaloudis and F. Goodall, 1981, Nature **289**, 543.
Sakuraï, J.J., 1967, Advanced Quantum Mechanics (Addison-Wesley, Reading, CA) ch. 2.
Schwartz, D.B., B. Sen, C.N. Archie and J.E. Lukens, 1985, Phys. Rev. Lett. **55**, 1547.
Slusher, R.E., L.W. Hollberg, B. Yurke, J.C. Metz and J.F. Valley, 1985, Phys. Rev. Lett. **55**, 2409.
Tesche, C.D., 1990, Phys. Rev. Lett. **64**, 2358.
Turlot, E., D. Esteve, C. Urbina, J.M. Martinis, M.H. Devoret, S. Linkwitz and H. Grabert, 1989, Phys. Rev. Lett. **62**, 1788.
Voss, R.F., and R.A. Webb, 1981, Phys. Rev. Lett. **47**, 265.
Washburn, S., R.A. Webb, R.F. Voss and S.M. Faris, 1985, Phys. Rev. Lett. **54**, 2712.
Yurke, B., L.R. Corruccini, P.G. Kaminsky, L.W. Rupp, A.D. Smith, A.H. Silver, R.W. Simon and E.A. Whittaker, 1989, Phys. Rev. A **39**, 2519.

CHAPTER 7

Theory of Nonradiative Trapping in Crystals

Alexei S. IOSELEVICH and Emmanuel I. RASHBA

L.D. Landau Institute for Theoretical Physics
Academy of Sciences
Moscow, Russia
and
Institute for Scientific Interchange
Torino, Italy

Quantum Tunnelling in Condensed Media
Edited by
Yu. Kagan and A.J. Leggett

Contents

List of symbols

a_0	lattice constant
B	pre-exponential factor of the ST rate
br	branching point where a discrete level splits off from the bottom of the band
E_B	half-width of a free particle band
d_I	instanton density
$D_{qs}(\tau)$	phonon Green function
E	initial electron energy
$E_{ST}(Q)$, $E(\tau)$	adiabatic electron energy in a potential produced by a deformed lattice
E_{min}	optimal energy for ST
$\mathcal{E}(\tau)$	dimensionless adiabatic energy
G^R, G^A	retarded and advanced Green functions of an electron, respectively
$\boldsymbol{k}$	electron momentum
L	Lagrangian of a system
L_{lat}	Lagrangian of a free lattice
m	effective mass of a quasiparticle
m_e, m_h	effective masses of an electron and of a hole, respectively (used in section 4.4 only)
N_i	concentration of impurity centers
N_{qs}	phonon occupation number
$P_{tr}(k)$	probability of a particle trapping by an emerging level
$P_e(T)$	temperature factor of a free electron
Q	generalized coordinates of a lattice
$\boldsymbol{q}$	wave factor of a phonon
r_{ST}	spatial scale of the ST state
r_b	spatial scale of a barrier
r_I	spatial scale of an instanton
r_{tr}	trapping radius
S	Hamiltonian action
S_0	truncated action
S_I, S_A	Hamiltonian actions for the instanton and Arrhenius regimes, respectively
$S_{tr}(E)$	exponential in P_{tr} for hot particles
ST	abbreviation for the term "self-trapping"
t	time (complex)
t_0	moment when the tunneling begins
t_c	moment when an electronic wave function collapses
t_*	moment when a discrete level emerges
$t_{tr}(E)$	moment when $E_{ST}(t_{tr}) = E$ for a hot particle
T	temperature
T_c	temperature of the switchover between the thermoactivated and Arrhenius regimes
$u_{ST,\boldsymbol{k}}$	amplitude of an electron transition from a state with the momentum $\boldsymbol{k}$ to the ST state
$U(Q)$	adiabatic potential
$U_{lat}(Q)$	free lattice deformation energy
v	volume of an elementary cell
$v_{tr}(E)$	volume from which a particle with energy E is effectively trapped at a level

V_{ST}	energy gain at ST
V_{LR}	energy of the lattice relaxation
w	ST rate
W	height of a barrier to ST
$\mathscr{W}$	lowest-lying saddle point on $U(Q)$ surface separating the regions of free and ST particles
Z_{lat}	lattice partition function
β	inverse temperature
β_c	$1/T_c$
Γ	contour in a complex t-plane
γ_{qs}	electron–phonon coupling constant
γ_0	the same for the case of coupling to nonpolar dispersionless mode
ε	energy of a tunneling system
$\varepsilon_0, \varepsilon_\infty$	low- and high-frequency dielectric permeabilities (used in section 4.4 only)
λ, Λ	dimensionless constants of electron–phonon coupling in the free and ST states, respectively
ν	dimensionless concentration of trapping centers
τ	imaginary part of time
τ_t	total duration of the under-the-barrier motion
τ_I	duration of an instanton
$\psi(r, \tau)$	adiabatic wave function of an electron
$\bar{\omega}$	characteristic phonon frequency
ω_0	frequency of dispersionless phonons
ω_w	frequency of oscillations (in imaginary time) in the vicinity of the point $\mathscr{W}$
ω_{eff}	effective phonon frequency in the pre-exponential of the ST rate (used in section 5.6)
ω_{soft}	frequency of a soft translational mode
Ω	frequency characterizing the curvature of a discrete level splitting off from the continuum spectrum

1. *Nonradiative trapping – physical fundamentals*

At present a few classes of substances are known where the interaction of quasiparticles (carriers, excitons) with vibrations of a crystalline lattice is strong. This strong interaction results in self-trapping (*ST*) of quasiparticles (i.e., formation of different kinds of the strong coupling polarons). By self-trapped (ST) states we understand self-consistent states with a strong local deformation of the lattice; in these states an electron (exciton) is bound to the field of this deformation. In a perfect lattice such states may occur only if there is a strong coupling of a quasiparticle to lattice phonons. Formation of the ST state of a quasiparticle from the free (band) state is accompanied by a strong lattice relaxation, and the gain in the energy of the system at ST is $V_{ST} \gg \bar{\omega}$, where $\bar{\omega}$ is a characteristic frequency of phonons ($\hbar = 1$ throughout the paper). As an example of ST states may serve holes and excitons in alkali halides and rare gas solids. For surveys on specific substances and general problems, see Williams (1978), Fugol' (1978, 1988), Aluker et al. (1979), Toyozawa (1981), Itoh (1982), Rashba (1982), Lushchik (1982), Schwentner et al. (1985), Ueta et al. (1985) and Zimmerer (1987). ST excitons are usually identified by their luminescence spectra, ST holes – by their transport properties, etc.

In crystals containing impurities and defects one can often observe a related phenomenon, namely, trapping of carriers by centers with a strong lattice relaxation (for reviews, see Langer 1980, Lang 1980). This trapping is observed in a number of semiconductors with a high mobility of charge carriers – e.g., $A_{III}B_V$ compounds and Si (Watkins 1963, Henry 1981) – i.e., in conformity with the conventional classification, in substances with a weak coupling of free carriers to lattice phonons. To the effects caused by strong lattice relaxation centers, one can refer persistent photoconductivity (Lang 1980, Henry 1981), recombination-enhanced motion of defects (Dean and Choyke 1977, Kimerling 1978) and others. The simplest theoretical model of strong lattice relaxation centers is a particle, strongly coupled to a local mode, existing in the vicinity of the center and weakly coupled to crystal phonons.

The existence of ST states of quasiparticles does not at all imply the absence of their free states in the crystal. Under certain conditions, the states of both kinds coexist (Rashba 1957, 1982). The free and ST states are separated by a potential barrier, which is termed a barrier to ST. The conditions of the coexistence are as follows:

(1) A half-width of the band of a free quasiparticle in a rigid lattice is

$$E_B \gg \bar{\omega}. \tag{1.1}$$

Since characteristic electronic frequencies in a localized state are of the order or less than E_B, criterion (1.1) is the necessary condition for adiabacity (i.e., the electronic motion is faster than the lattice motion).

(2) A short-range type of electron–phonon coupling. For excitons, by virtue of their electric neutrality, this condition is always fulfilled, whereas for carriers it is fulfilled only at a nonpolar coupling to phonons.

(3) The system must be three-dimensional (3D).

If (2) or (3) is violated, a barrier to ST is absent and the free state is absolutely unstable. If a barrier to ST does exist, one of the states (usually a free state) is metastable, the other state being stable. A decay of the metastable state occurs when a barrier to ST is being overcome; therefore, if the barrier is high and wide, the rate w of the metastable state decay is exponentially small. The quantity w controls the kinetics of the transformation of free particles into ST.

Applicably to excitons, w fixes the rate of the rise of the luminescence band of ST excitons in the conditions when light generates free excitons.

The only so far known self-consistent way to construct the theory of a barrier to ST is the adiabatic approximation, based on a treatment of the lattice as a slow subsystem and of the particle (below termed electron) as a fast subsystem. The necessary condition of applicability of this approximation (1.1) implies that the frequency of intersite electron transitions considerably exceeds the frequency of lattice vibrations. Everywhere below it is assumed that this condition is fulfilled, therefore, it is natural to introduce adiabatic potential surfaces (APS) $U(Q)$, where Q are generalized coordinates of the lattice. The energy $U(Q) = U_{\mathrm{lat}}(Q) + E_{\mathrm{ST}}(Q)$ contains two contributions: $U_{\mathrm{lat}}(Q)$, the free lattice deformation energy and $E_{\mathrm{ST}}(Q)$, the electron energy in the potential produced by a deformed lattice.

Figure 1a demonstrates a single-mode APS, corresponding to ST in a perfect lattice, i.e., to intrinsic ST. Branch 2 corresponds to the states where the electron is bound on a discrete level, produced by the field of a deformed lattice. This level emerges at a branching point (br) at $Q = Q_{\mathrm{br}}$, where it splits off from the bottom of the electron band. W is the height of a barrier to ST. If the scheme of fig. 1a is generalized to a multimode system, the curves transform into surfaces and the point $\mathscr{W}$ (maximum) – into the lowest-lying saddle point, separating the regions of free and ST states (fig. 1b). The point br also turns into a surface. In three dimensions at a short-range coupling to phonons the quantity $U_{\mathrm{lat}}(Q)$ at small Q always exceeds the gain of the system in energy, emerging when the electron is bound on a local level. Therefore, $W > 0$, i.e., trapping cannot proceed as monotonous lowering of the total energy of the system – unlike for

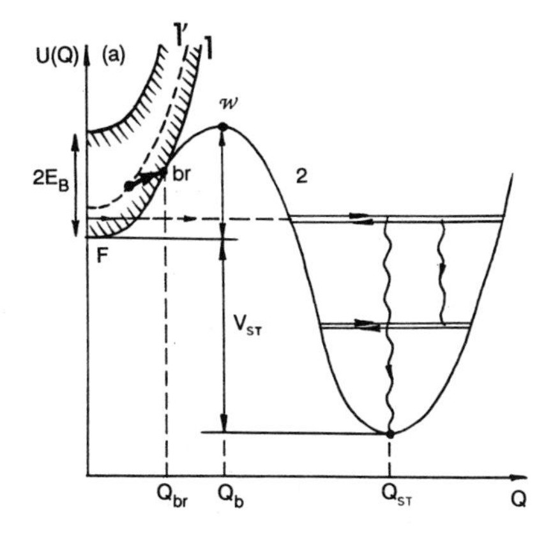

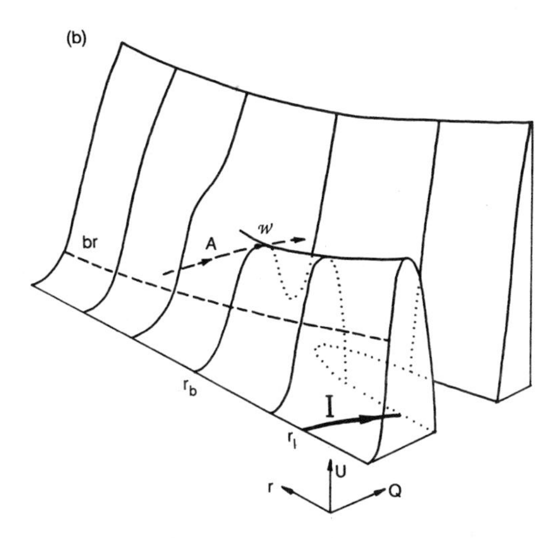

Fig. 1. The schematic dependence of the adiabatic potential U on configurational coordinates for a crystal with an electron. (a) Model with one configurational coordinate Q (deformation): curve 1 represents the potential energy of a crystal with a free electron at the bottom of the electron band (the electron band is dashed) and curve 1′, the same potential energy but shifted by the energy of the free electron; curve 2 represents the adiabatic potential for a crystal with an electron on a local level. W is the height of the barrier to ST, V_{ST} is the lowering of the energy of the system due to ST, E_B is the half-width of the electron band. The following values of the coordinate Q are shown: Q_{ST} corresponding to the ST state, Q_b corresponding to the barrier state (at the point $\mathscr{W}$), Q_{br}, at the branching point br, corresponding to the appearance of a local level for the electron. $Q = 0$ corresponds to the free state F. The heavy arrow illustrates trapping of the electron on the local level, and the wavy arrows, the process of vibrational relaxation; (b) model with two configurational coordinates – the magnitude of the deformation Q and the spatial size of the deformed region r. Activational (A) and tunneling (I) ways of surmounting the barrier are shown. The quantity r_l, which may strongly differ from the radius of the barrier state r_b, corresponds to the tunneling mechanism.

polarons in ionic crystals (Pekar 1947) and for ST states in one dimension (Rashba 1957) – but can proceed only as surmounting of the barrier to ST.

As distinct from the ST in a perfect lattice for a nonradiative trapping by strong lattice relaxation centers, there are two kinds of APS. They are schematically given in fig. 2 for a model with one lattice degree of freedom.

Classification of the centers can be conveniently carried out in accordance with APS types. The trapping by centers with the curves of the type shown in fig. 2a is usually termed as extrinsic ST (Toyozawa 1978) and for the curves, shown in fig. 2b, we, following Lang (1980), shall employ the term "normal defects". In the latter case in the single-mode scheme the points $\mathscr{W}$ and br merge and in the multimode scheme the point br turns into a crest at the bottom of the electron band and the point $\mathscr{W}$ becomes a minimum on this crest. Due to the difference in APS shape, certain properties of the centers belonging to these two categories are different but, in general, the picture of electron trapping by the centers of both kinds is alike. Therefore, the term "ST" will be used in a broad sense, implying intrinsic and extrinsic ST as well as trapping by normal centers. In a perfect lattice the scheme of fig. 2b cannot work since in this scheme a local level is existent already in an undeformed lattice.

To avoid any misunderstanding in what follows we stress here that the term "self-trapping" will be used in any case when a barrier has a shape shown in fig. 2a. The self-trapping will be termed as intrinsic or as extrinsic if it occurs in a perfect lattice or near some defect, respectively. The term "normal center" will be used for systems with the shape of a barrier to ST shown in fig. 2b.

The mechanism of overcoming a barrier to ST is different at low and high temperatures. In the first case this is thermo-activated tunneling, and in the

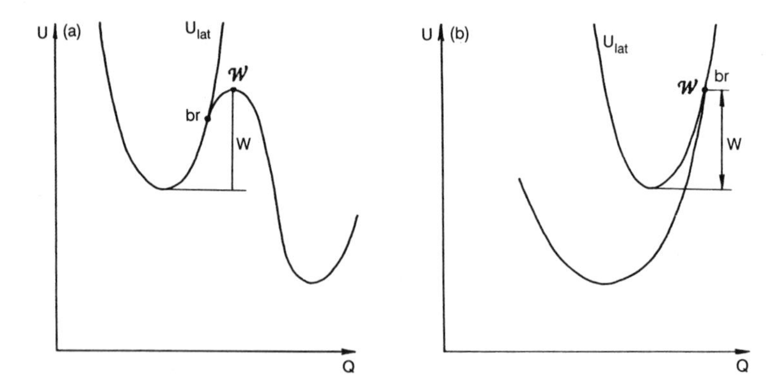

Fig. 2. Two types of adiabatic potential surfaces (APS) for a single-mode system. (a) ST, intrinsic or extrinsic: the barrier to ST is at the point $\mathscr{W}$ on APS, not coinciding with the branching point br, where the bound state of the electron first appears (in the multimode case the point $\mathscr{W}$ is not lying on the branching surface br); (b) Normal (recombinational) centers: the saddle point $\mathscr{W}$ coincides with the point br. In the multimode case the saddle point $\mathscr{W}$ is on the branching surface, which is a sharp crest on APS.

second, activational Arrhenius process; the temperature of the regime switching being $T_c \sim \bar{\omega}$. Both the processes contribute to w an exponentially small factor. It is convenient to single out this factor explicitly and to write down the probability of multiphonon trapping of an electron as

$$w = \bar{\omega} B \nu \exp(-S), \tag{1.2}$$

where the factor ν is a dimensionless concentration of trapping centers. For intrinsic ST, $\nu = 1$, and for trapping by various impurity centers, $\nu = N_i v$, where N_i is a concentration of centers and v is the volume of an elementary cell. Equation (1.2) is applicable both for the activational regime (high T) and for the thermoactivated tunneling regime (low T). The question is what the values of S and B are and how they depend on the lattice temperature T and on the initial electron energy E. For ST in all cases $B \gg 1$. At high temperatures the prefactor B strongly depends on T. This is a very important issue for interpreting experimental data: (i) for a realistic estimate of the prefactor (with the enhancement factor B taken into account), and (ii) for finding a correct value of W from the curve $w(T)$, since it is needed to take into account the T-dependence contained in the prefactor.

The ST mechanism is largely analogous to mechanisms of some other nonradiative processes in solids (Englman 1979, Stoneham 1981, Newmark and Kosai 1983) and also to mechanisms of the metastable state decay, manifesting themselves in some other problems, e.g., in the problem of quantum nucleation (Langer 1967, Lifshitz and Kagan 1972). Here we must point out some peculiarities of the ST process, rendering it specific. These peculiarities originate from the fact that a barrier to ST is to be surmounted not by a particle (electron) which gets self-trapped and not by a free lattice but by a composite system consisting of a local lattice deformation (i.e., virtual phonons) and of a particle trapped by it. This entity is described in terms of a system with a large number of degrees of freedom or in terms of quantum field. It emerges in the course of ST and continuously varies in time. This statement will become more clear if one lists all the main stages of the ST process:

(1) Formation of a lattice fluctuation, thermal or quantum; the deformation caused by it increases with time. The motion of the lattice at this stage will be called free below.

(2) Emergence of a local level for an electron in the field of the fluctuation and trapping of the electron on this level from the continuum spectrum (the heavy arrow in fig. 1a).

(3) Penetration of a barrier to ST by a composite system (a deformation with an electron trapped by it), the barrier being formed by the same interaction which causes the electron dressing (i.e., loaded motion of the lattice as will be termed below).

(4) Relaxation of a strongly nonequilibrium state which emerges after the system passes through the barrier to ST.

This review describes the first three stages of the ST process, which usually control w, from the general point of view, based on a multimode description of the lattice. We single out in w (eq. 1.2) an exponential factor and analyze its temperature dependence and also the temperature dependence of the pre-exponential factor. One can expect that it is the temperature dependence $w(T)$ that permits a most straightforward comparison with experiment. The numerical coefficient in w has not been so far exactly calculated in any ST model.

Stage 4 largely determines not the ST rate but the rate of the energy relaxation of immediate products of the ST process. In fig. 1a it is schematically depicted by the wavy arrows. In principle, this last stage may proceed fast, i.e., in times short in comparison with other characteristic times: the time of trapping, the time of radiative decay of ST excitons, etc. However, if in the course of ST there arise configurations where the frequencies of local vibrations exceed lattice phonon frequencies, a multiphonon decay of local modes controls the process rate. Therefore, the rate of this process is exponentially small and the energy relaxation time is exponentially long. Apparently, this kind of a phenomenon is observed in Ne, where the ST excitons radiatively recombine from nonrelaxed vibronic states (Jortner 1974). For impurity centers the influence of stage 4 on w, including interference effects and reemission, has been studied by Morgan (1983). We shall dwell on this stage only briefly and only in connection with defect-production processes, accompanying ST (section 8).

This review does not touch upon dephasing processes in the initially prepared state (see, e.g., Jortner et al. 1969; Englman 1979), which may take place even prior to stage 1. The effects of phase memory, very important for a number of phenomena in impurity centers, are not quite relevant for nonradiative trapping. Unlike phase relaxation, the rate of energy relaxation may prove to be rather important, as is evident from the strong dependence of the trapping rate on the energy of particles (section 7).

Of great importance is the problem of energy and spatial scales of a barrier to ST. In ST states the energy V_{ST} has an atomic order of magnitude ~ 1 eV ($V_{ST} \sim E_{at}$), whereas the spatial scale of the ST state r_{ST}, like the value of atomic displacements in the core of the ST state, Q_{ST}, is of order of the lattice constant $a_0 (r_{ST} \sim Q_{ST} \sim a_0)$. At such large displacements electron–phonon interaction is nonlinear. General considerations may give rise to a conjecture that for a barrier to ST we have an analogous situation, i.e., $W \sim E_{at}$, and the spatial scale of the barrier is $r_b \sim a_0$. But numerous experimental data testify to the fact that characteristic values of the height of a barrier to ST, $W \gtrsim 0.1$ eV (in alkali halides, rare gas solids, etc.). For excitons at $W \sim 1$ eV ST could not have proceeded for their lifetime. The only so far known argument capable to explain the small value of W is based on the assumption that in investigated crystals

$$\Lambda = V_{ST}/E_B \gg 1 \tag{1.3}$$

(Rashba 1977). Here Λ is a dimensionless constant of electron–phonon coupling describing ST. It is possible to show that

$$W \sim E_B/\Lambda^2, \qquad r_b \sim \Lambda a_0 \tag{1.4}$$

(see section 4.1). Therefore at $\Lambda \gg 1$, the inequality $W \ll E_B$ holds, i.e., W is small and $r_b \gg a_0$, large. At $r_b \gg a_0$, $Q_b \ll a_0$ also holds; therefore, the electron–phonon coupling can be treated as linear. Consequently, at $\Lambda \gg 1$, in describing a barrier to ST, one can employ the continuum approximation and confine oneself to a linear electron–phonon coupling and to a quadratic expansion for $U_{lat}(Q)$. This permits to write out explicit equations, describing the ST dynamics at various types of coupling (Iordanskii and Rashba 1978). Yet, direct experimental evidence of the fact that $r_b \gg a_0$ is absent, and typical values of Λ (estimated as $\sqrt{E_B/W}$) usually amount to $\Lambda \gtrsim 10$, i.e., are not very large. It is not excluded that for some numerical reasons virtually $r_b \sim a_0$. One needs to take into account also the fact that the minimal spatial scale r_I on a tunneling path can be smaller than r_b (see fig. 1b). The ratio r_I/r_b includes a small numerical coefficient (Iordanskii and Rashba 1978) and, under some conditions (see section 4.3), also a small parameter $\Lambda^{-4/5} \ll 1$. That is why it is quite likely that W may be small due to relatively large values of Λ; nevertheless, tunneling paths may pass through configurations with a spatial scale $\sim a_0$, for which nonlinearity is fairly strong. Therefore, it is important to go beyond the scope of the continuum and linear approximations. Of course, then the results become less specific. But at the same time then it is possible to single out general model-independent statements.

The assumption that electron–phonon coupling in free states can be neglected is nontrivial when we deal with crystals where there are ST states. This assumption is justified by the fact that the coupling constants λ and Λ, determining electron–phonon coupling in free and ST states, respectively, strongly differ, i.e., $\lambda \ll \Lambda$. The first principle estimate yields (Rashba 1982)

$$\lambda/\Lambda \sim (m/M)^{1/2} \ll 1. \tag{1.5}$$

Here $(m/M)^{1/2}$ is an adiabatic parameter, m and M are electron and atomic masses, respectively. These arguments are in good agreement with criterion (1.1) if the conventional estimate $\bar{\omega}/E_B \sim (m/M)^{1/2}$, also based on the first principles, is employed.

An efficient method of the theory of nonradiative transitions is the semiclassical approximation, first applied to this problem by Landau (1932), Zener (1932) and by Stückelberg (1932). Further progress in the technique of the semiclassical approximation has made it possible to apply this approach to molecule collisions and simple chemical reactions (Miller 1974, Child 1974). This approach has also been applied to nonradiative transitions in solids, e.g., to the polaron transport problem (Holstein 1978). These attempts, nevertheless, have not been numerous because the semiclassical approximation is rather sophisticated when applied to multimode systems. Major effort has been concentrated on the multiphonon transition (MPT) method, first developed by Lamb (1939), Pekar (1950) and by Huang and Rhys (1950) for inelastic neutron scattering and for optical spectra of impurity centers. The relationship of this method to the semiclassical one has been discussed by Lax (1952), and various versions of the MPT method have been surveyed by Perlin (1963). In their work Huang and Rhys (1950) have first used the MPT method to calculate the rate of nonradiative transitions. However, consecutive works have revealed that for nonradiative transitions the MPT method meets with considerable difficulties (cf. section 9).

In the meantime semiclassical methods have been substantially developed, in particular, in terms of classical solutions of field theory equations [quantum nucleation (Langer 1967), decay of false vacuum (Coleman 1977), quantum statistical metastability (Affleck 1981), etc.]. Reviews have been given by Vainshtein et al. (1982) and by Rajaraman (1982). These methods have opened up real possibilities to construct the theory of nonradiative trapping, in particular, of intrinsic ST, i.e., of the effect existent only in multimode systems. In the adiabatic region, corresponding to a free motion or to a loaded motion at a depth of the discrete level (where the electron resides) exceeding $\bar{\omega}$ (stages 1 and 3), the ST theory is a nonlinear field theory. But in the region of deformations, where the level is shallow (stage 2), it is necessary to study the capture of a free electron by this level under the conditions when it is getting deeper with time (Demkov 1964). The advantages of the method, based on a self-consistent analysis of stages 1–3 (Ioselevich and Rashba 1985a, b, 1986a, b, 1987) are that the method permits to single out the exponential and pre-exponential factors in w. The thermoactivated tunneling regime and Arrhenius regime (low-temperature and high-temperature, instanton and activation solutions) are distinguished by the T-dependence of the exponential factor. In each of the regimes the pre-factor with an accuracy up to a numerical coefficient is found from the first principles. It involves a certain power of the parameter $E/\bar{\omega}$ in the instanton regime and of the parameter E/T in the Arrhenius regime at $T \gg \bar{\omega}$ ($E \sim E_B$ or $E \sim W$, see sections 5 and 6). If the continuum approximation is applicable, the T-dependence of the prefactor is different for intrinsic ST and for trapping by centers. Below we shall concentrate our attention on the theory of the phenomenon, on its physical pattern, shall discuss the results and, in conclusion, perform a comparison with some results quoted in the literature. As for the comparison of the theory with experiment, in our opinion, it is not yet possible to carry it out systematically. In this direction only first steps have been taken, and the reliability of some attempts is rather dubious. Such a comparison is a separate subject lying beyond the scope of this review.

2. Semiclassical approach to the theory of trapping

The adiabatic potential surface is a semiclassical concept. It works at the description of electron processes only if the inequality (1.1) is fulfilled. Naturally, this restriction holds for any structure on APS, including a barrier to ST. Applicably to a barrier to ST, the standard semiclassical condition that the lattice deformation energy (of order W) is large, compared to the phonon frequency $\bar{\omega}$, reads

$$W \gg \bar{\omega}. \tag{2.1}$$

Since the electron binding energy in the barrier state is of order W, the condition (2.1) is also a condition of adiabacity.

The inequality (2.1) ensures the applicability of the semiclassical description not only to the barrier state but also to the ST process on the whole. Under these conditions, to calculate the ST rate, one can make use of the imaginary time technique (Baz' et al. 1975, section V; Miller 1974). Below we shall clarify the essence of this trick for thermoactivated tunneling in a single-mode system.

2.1. The imaginary time method for a single-mode system

The conventional semiclassical expression for the tunneling transparency of the barrier is

$$D(\varepsilon) \propto \exp[-S_0(\varepsilon)], \tag{2.2}$$

where ε is the energy of a tunneling system and

$$S_0(\varepsilon) = 2\int_{Q_1}^{Q_2} \sqrt{2M(U(Q)-\varepsilon)}\,\mathrm{d}Q = 2\int_{Q_1}^{Q_2} (-p^2)^{1/2}\,\mathrm{d}Q, \tag{2.3}$$

Q_1 and Q_2 being the turning points. Under the barrier $\varepsilon < U$, the squared momentum $p^2 = (\partial_t Q)^2 < 0$ (see fig. 3a). For this reason it is convenient to introduce an imaginary time $\tau = -it$. Then the equation of motion in the under-the-barrier region has the form

$$M\partial_\tau^2 Q = \mathrm{d}U/\mathrm{d}Q, \tag{2.4}$$

differing from the standard equation only by the sign of the potential energy $U(Q)$ (see fig. 3b). Then the potential barrier transforms into a potential well, and eq. (2.3) into an expression for a truncated Hamiltonian action:

$$S_0(\varepsilon) = \oint p\,\mathrm{d}Q, \quad p = M\partial_\tau Q. \tag{2.5}$$

The integration (2.5) is performed over the total period of oscillations in the well (fig. 3b). If the distribution over the initial states is quasi-equilibrium, the

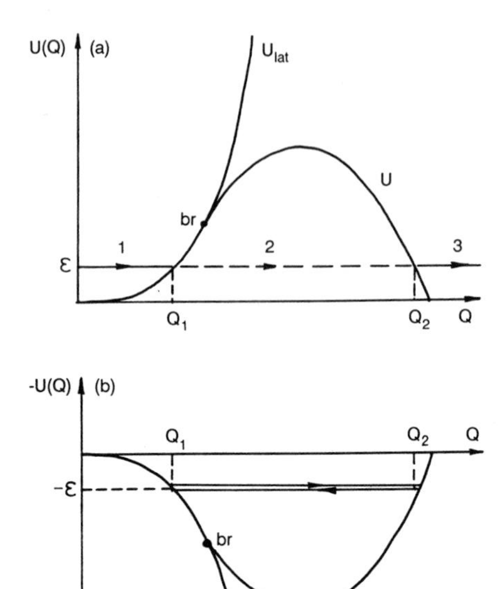

Fig. 3. The imaginary time method. (a) Tunneling of the system with an energy ε through a semi-classical barrier. Q_1 and Q_2 are turning points, lines 1 and 3 represent the motion in the classically accessible region, line 2 represents the motion under the barrier. (b) Classical motion of the system with an energy $-\varepsilon$ in the imaginary time in the potential well $-U(Q)$. The total oscillation period – motion both forward and backward – is given.

transparency (2.2) should be multiplied by the Gibbs factor and integrated over the energy. The tunneling probability is

$$w \propto \exp(-\beta\varepsilon)\exp[-S_0(\varepsilon)] \equiv \exp[-S(\varepsilon)], \tag{2.6}$$

where

$$S(\varepsilon) = S_0(\varepsilon) + \beta\varepsilon, \quad \beta = T^{-1}. \tag{2.7}$$

The largest contribution to tunneling flow comes from the vicinity of the optimal value of the energy, corresponding to the stationary value of $S(\varepsilon)$, i.e., $\mathrm{d}S/\mathrm{d}\varepsilon = 0$, from which it follows that

$$\tau_t = \beta/2, \quad \text{where } \tau_t = -\tfrac{1}{2}\mathrm{d}S_0/\mathrm{d}\varepsilon \tag{2.8}$$

is the duration of the under-the-barrier motion (in one direction). The latter expression is a well-known equation of analytical dynamics, rederived for the imaginary time. So, the quantity $S(\varepsilon)$ has the meaning of the Hamiltonian action for the case when the system covers the under-the-barrier path twice, forward and backward and β is the duration of this motion.

Figure 4 illustrates the dependence of the tunneling duration τ_t on the energy ε. When ε approaches W (tunneling proceeds in the vicinity of the barrier peak), $\tau_t = \pi/\omega_W$, i.e., equals half a period of the oscillation of the system at the bottom of the potential well shown in fig. 3b. Tunneling near the point $\mathscr{W}$ is optimal at $\beta/2 = \pi/\omega_W$. In this case the action $S \approx \beta W$, as it ensues from eq. (2.7), since near $\mathscr{W}$ the energy $\varepsilon \approx W$ and $S_0(\varepsilon) \to 0$. Then the probability of optimal tunneling (in the exponential approximation) coincides with the probability of Arrhenius activation. The specific shape of the function $\tau_t(\varepsilon)$ depends on the shape of the potential $U(Q)$. There are two ways of switching over from the tunneling regime to the activational regime, depending upon the fact whether the function $\tau_t(\varepsilon)$ is monotonous or not (Meshkov 1985, Ioselevich and Rashba 1986a). If the function $\tau_t(\varepsilon)$ is monotonous (fig. 4a), the switching over occurs at the point $\beta_c = 2\pi/\omega_W$ smoothly, without any kink, as has been shown by Affleck (1981) (fig. 5a). If $\tau_t(\varepsilon)$ is not monotonous (fig. 4b), there is a temperature range $\beta_* < \beta < 2\pi/\omega_W$, $\beta_* = 2\tau_{\min}$, where there are two solutions I and I′ of eq. (2.8). In fig. 5b the two branches of $S(\beta)$, I and I′, correspond to them. The solution I′ touches the A-line smoothly (fig. 5b) just as the only solution I in fig. 5a does. But in other aspects the behaviors of these two solutions differ due to the different shapes of $\tau_t(\varepsilon)$-dependences in fig. 4a and fig. 4b, respectively. The solutions I and I′ merge at the point β_* (fig. 5b), which corresponds to $\tau_{\min}$ in fig. 4b. It is important that $S_{I'} > S_I$ for all β; thus, the switchover occurs at the point β_c ($\beta_* < \beta_c < 2\pi/\omega_W$), where the solution I intersects the A-line. Hence, in this case $S(\beta)$, the exponent in formula (1.2), reveals a kink in the point β_c (fig. 5b).

At low temperatures, tunneling occurs near the base of the barrier ($\varepsilon \ll W$). In this case the system spends most part of the time τ_t near the left-hand turning

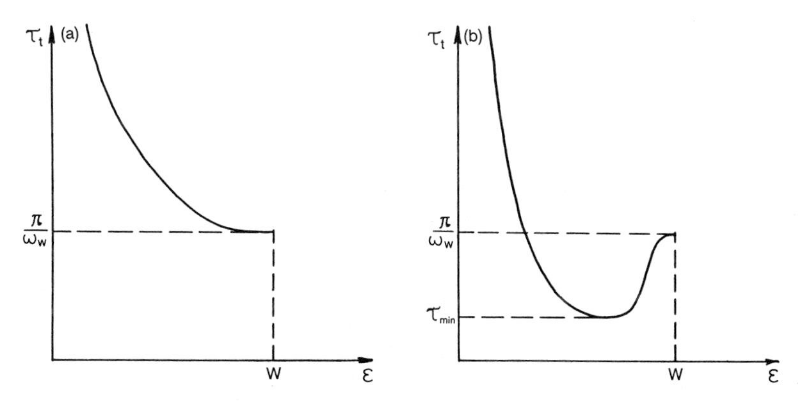

Fig. 4. Possible types of the dependence of the under-the-barrier motion time τ_t on the energy ε of the tunneling system for the barrier illustrated by fig. 2a (ST); ω_W is the particle oscillation frequency near the bottom of the potential well, depicted in fig. 3b.

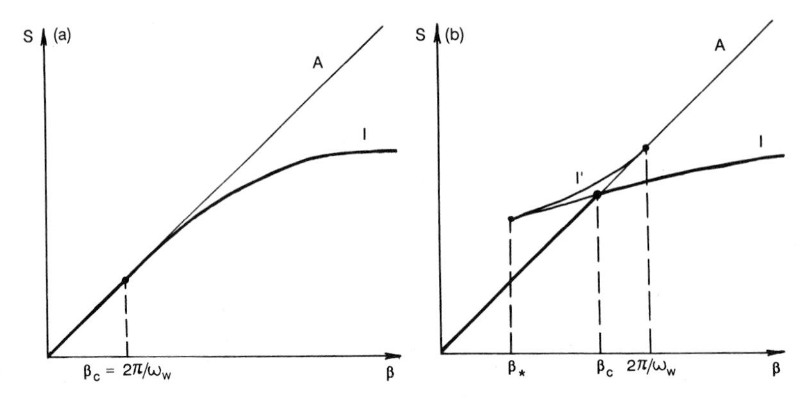

Fig. 5. The qualitative shape of the dependence of the exponent for the ST rate [see eq. (1.2)] on the inverse temperature β: A represents activational mechanism, and I represents instanton mechanism. The heavy line indicates the dominating contribution coming from the smallest of the competing values of S; β_c is the regime switchover point; $\beta_* = 2\tau_{min}$ is the point where the instanton solution disappears. Curve a corresponds to the dependence $\tau_t(\varepsilon)$ given in fig. 4a, and curve b to the dependence $\tau_t(\varepsilon)$ given in fig. 4b.

point Q_1 ($Q_1 \ll Q_2$), where the potential can be treated as quadratic:

$$U_{lat} \approx \tfrac{1}{2} M\omega_0^2 Q^2, \tag{2.9}$$

where ω_0 is the frequency of oscillations of the system in the vicinity of the minimum $Q = 0$ (fig. 3a). Then τ_t can be easily calculated as

$$\tau_t \approx \int_{Q_1}^{Q_2} dQ\,(\omega_0^2 Q^2 - 2\varepsilon/M)^{-1/2} \approx (1/2\omega_0)\ln(W/\varepsilon) \tag{2.10}$$

and the choice of the upper integration limit is irrelevant, since it affects only the numerical factor under the sign of the ln function.

Substitution of eq. (2.10) into eq. (2.8) at $\beta\omega_0 \gg 1$ yields

$$\varepsilon(\beta) \sim W\exp(-\beta\omega_0). \tag{2.11}$$

The temperature-dependent contribution to the action can be easily found from the formula

$$dS/d\beta = \varepsilon(\beta). \tag{2.12}$$

Substituting eq. (2.11) into eq. (2.12), we get for the instanton action

$$S_I(\beta) = S_I(\beta = \infty) - \int_{\beta}^{\infty} d\beta\,\varepsilon(\beta) \approx S_I(\beta = \infty)[1 - b\exp(-\beta\omega_0)], \tag{2.13}$$

where the numerical factor $b \sim 1$ depends on the shape of the barrier. The second term in eq. (2.13) is relevant only when it is large in comparison to unity:

$$S_1(\beta=\infty)\exp(-\beta\omega_0) \gg 1.$$

At the inverse inequality, this calculation is not correct, because the fact that the energy levels in the harmonic potential [eq. (2.9)] are discrete becomes important. Moreover, in this case it is the prefactor that contains the principal temperature dependence of w.

Thus, at low temperatures, in a single-mode system, the temperature-dependent contribution to S is exponentially small in ω_0/T. In section 4.5 it will be shown that in multimode systems in the presence of soft (acoustic phonons) modes there arises a contribution to S, obeying the power law in T.

These calculations can be applied only to the case of APS, given in fig. 2a, when the dependence $U(Q)$ is single-valued in the under-the-barrier region. For a normal center (fig. 2b), in contrast to eq. (2.3), an expression for the truncated action reads

$$S_0(\varepsilon)=2\left\{\int_{Q_1}^{Q_{\mathrm{br}}}\mathrm{d}Q\,\sqrt{2M(U_{\mathrm{lat}}(Q)-\varepsilon)}-\int_{Q_2}^{Q_{\mathrm{br}}}\mathrm{d}Q\sqrt{2M(U(Q)-\varepsilon)}\right\} \tag{2.14}$$

(Markvart 1981b, Abakumov et al. 1985), where Q_1 and Q_2 are turning points (see fig. 6). In terms of the imaginary time, this expression can be interpreted in the following way (Meshkov 1985, Ioselevich and Rashba 1986b). It is clear from fig. 6. that on the tunneling path the point Q_{br} is a turning point. At the same time at this point the imaginary velocity $\partial_\tau Q \neq 0$. Therefore, to ensure that $\partial_\tau Q$ has no discontinuities, one should, at the point Q_{br}, simultaneously alter the direction of the motion in space (fig. 6) and the direction of the imaginary time

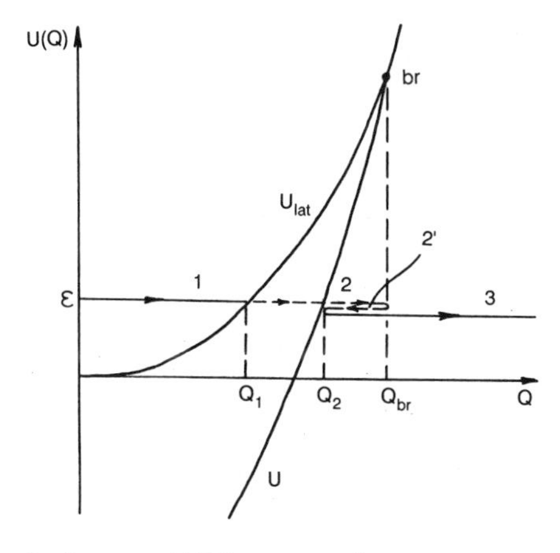

Fig. 6. Tunneling in the case of APS corresponding to a normal center (see fig. 2b).

flow. The latter arguments are consistent with the fact that the tunneling time, found from eq. (2.14) with the use of eq. (2.8), is a difference of times corresponding to the two terms in S_0.

Since the peak of the barrier for normal centers is "spike-like" and the velocity of the system on the peak is nonzero, the tunneling time $\tau_t \to 0$ at $\varepsilon \to W$ (fig. 7a), unlike the case of ST where $\tau_t(\varepsilon = W)$ does not tend to zero (fig. 4). As a result, thermoactivated tunneling becomes more preferable than Arrhenius activation at all T (fig. 7b). However at high temperatures, when $T \gg \bar{\omega}$, the system tunnels near the peak of the barrier, and $S_l(\beta)$ is only slightly distinct from $S_A = W\beta$. Let us analyze this case in more detail. Near the point br an expression for APS can be written as

$$U_{lat} = W - Fq + \tfrac{1}{2}M\omega_1^2 q^2, \tag{2.15}$$

$$U = W - Fq + \tfrac{1}{2}M\omega_2^2 q^2, \tag{2.16}$$

where $q = Q_{br} - Q$, $\omega_1 \gtrsim \omega_2 \sim \bar{\omega}$, $F \sim (WM)^{1/2}\bar{\omega}$. The q-linear terms in eqs. (2.15) and (2.16) coincide, since the parameters of the potential well for an electron depend on q continuously, and the depth of the discrete level, emerging at the point Q_{br}, is q^2-proportional (Baz' et al. 1975). Substituting eqs. (2.15) and (2.16) into eq. (2.14) and regarding the q-quadratic terms as small, we get

$$\begin{aligned} S_0(\varepsilon) &= \int_0^{(W-\varepsilon)/F} [(W-\varepsilon) - Fq]^{-1/2}(2M)^{1/2}\frac{M(\omega_1^2 - \omega_2^2)}{2} q^2\, dq \\ &= (c/\bar{\omega} W^{3/2})(W-\varepsilon)^{5/2}, \end{aligned} \tag{2.17}$$

where

$$c = \frac{8\sqrt{2}}{15}(MW)^{3/2}(\omega_1^2 - \omega_2^2)\bar{\omega}/F^3 \sim 1.$$

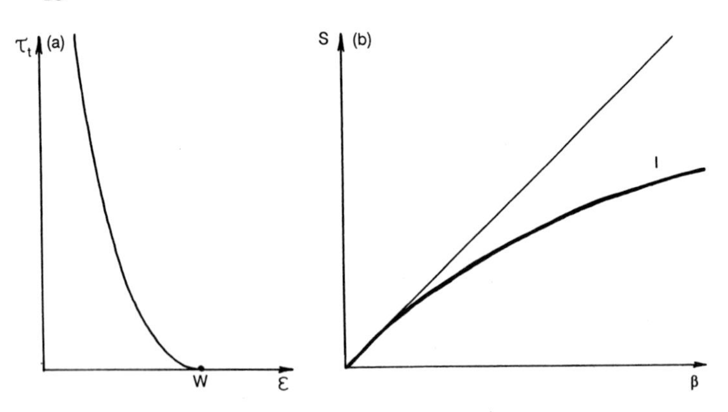

Fig. 7. The dependence of the tunneling time τ_t on the energy (a) and the dependence of the action S [the exponent in eq. (1.2)] on the inverse temperature (b) for trapping by normal centers (APS of the type shown in fig. 2b).

Substituting eq. (2.17) into eq. (2.8), we find $\varepsilon(\beta)$:

$$\varepsilon(\beta) = W(1 - (2\beta\bar{\omega}/5c)^{2/3}), \tag{2.18}$$

and, by means of eq. (2.7), the total action

$$S_{\mathrm{I}} = W\beta[1 - c'(\beta\bar{\omega})^{2/3}], \tag{2.19}$$

where $c' = \frac{3}{5}(2/5c)^{2/3} \sim 1$. So, when the temperature increases, the difference between S_{I} and S_{A} decreases. In some systems, particularly at $\omega_1/\omega_2 \approx 1$, the limit $S_{\mathrm{I}} \approx S_{\mathrm{A}}$ is attained only at very high temperatures (Abakumov et al. 1985).

At low temperatures, all considerations quoted for the ST case are valid for normal centers, and the temperature dependence of ST in a single-mode system is described by eq. (2.13). This behavior is well known also in the multiphonon transition theory (see, e.g., Englman and Jortner 1970).

It is convenient to describe the motion (both under the barrier and beyond it) if to choose an appropriate contour Γ_+ in a complex t-plane (fig. 8). Its sections

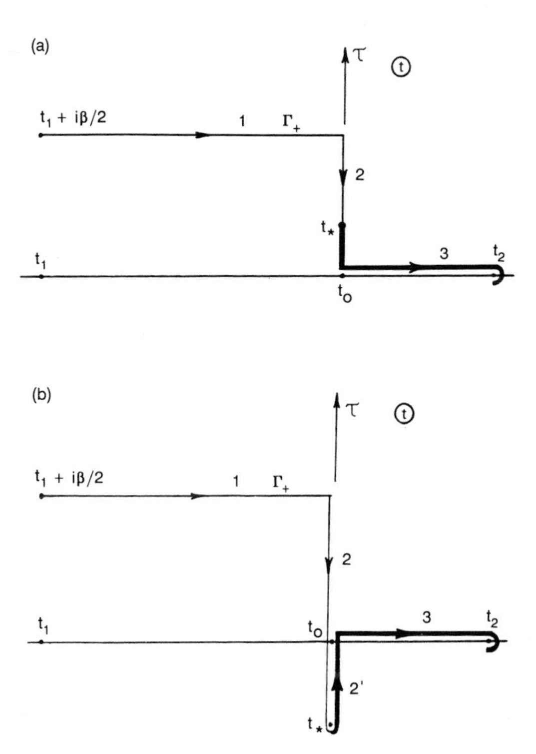

Fig. 8. Contours Γ_+ in the complex t-plane for ST (a) and for normal centers (b). The moments t_1 and t_2 are the beginning and the end of the motion; t_* is the moment of trapping. The shift of the left-hand side of the contour by $\frac{1}{2}i\beta$ is due to averaging over the Gibbs distribution in the initial state. The light line represents the free motion of the lattice, and the heavy line, the loaded motion.

1 and 3, directed along the real time axis, correspond to the motion in classically accessible regions prior to barrier penetration (from t_1 up to the moment $t_0 = \mathrm{Re}\{t_*\}$, when tunneling begins, and after tunneling is completed: from t_0 up to the moment t_2). Section 2, parallel to the imaginary time axis, corresponds to tunneling. Figure 8a gives a contour corresponding to APS of the type of fig. 2a; vertical section 2 has a length $\beta/2$ [eq. (2.8)]. Figure 8b gives a contour corresponding to APS of the type of fig. 2b (or fig. 6); on vertical sections 2 and 2′ the imaginary time flows in opposite directions (cf. the corresponding parts of the path in fig. 6). In both the cases the spacing between sections 1 and 3 equals $\beta/2$. Complex contours with the reverse motion in time have been discussed by Meshkov (1985).

The need of still further complications of the contours may arise if some additional processes with exponentially small probabilities are involved. The example of such a process (a trapping of a hot electron by the emerging level) is described in section 2.3. When applied to the problem of ST, taking into account this process leads to the appearance of additional appendages in Γ (fig. 19). These appendages remove the discontinuity in electron energy which appears if the electron loses it abruptly. The integration of the system Lagrangian over the appendages gives the exponent of a small trapping probability.

2.2. *Multimode tunneling at $T = 0$*

Now we show how the imaginary time technique works for a realistic model of intrinsic ST involving an infinite number of lattice degrees of freedom (Iordanskii and Rashba 1978). To simplify the problem maximally and to clarify the essence of the emerging phenomenon, let us assume that $\Lambda \gg 1$, and, consequently: (i) U_{lat} is quadratic in phonon variable; (ii) electron–phonon coupling is linear in phonon variables; and (iii) the continuum approximation holds. Let us also suppose that $T = 0$.

Under these assumptions, the phonon Lagrangian for the imaginary time has the form

$$L = L_{\mathrm{lat}} + E_{\mathrm{ST}}(Q), \tag{2.20}$$

$$L_{\mathrm{lat}} = \frac{1}{2}\sum_s \int \frac{\mathrm{d}\boldsymbol{q}}{(2\pi)^3}\left(\partial_\tau Q_{qs}\partial_\tau Q_{-qs} + \omega_{qs}^2 Q_{qs}Q_{-qs}\right), \tag{2.21}$$

$$E_{\mathrm{ST}}(Q) = \min_{\psi} \int \mathrm{d}\boldsymbol{r}\left\{\frac{1}{2m}|\nabla\psi|^2 + \sum_s \int \frac{\mathrm{d}\boldsymbol{q}}{(2\pi)^3}\gamma_{qs}Q_{qs}\mathrm{e}^{\mathrm{i}\boldsymbol{q}\cdot\boldsymbol{r}}|\psi(\boldsymbol{r})|^2\right\}. \tag{2.22}$$

Here Q_{qs} are normal complex coordinates, $\boldsymbol{q}$ is the wave vector, s is the index of the phonon branch, L_{lat} is the free phonon Lagrangian, $E_{\mathrm{ST}}(Q)$ is the energy of the ground state of an electron in the field of lattice displacements (i.e., the electronic contribution to the adiabatic potential). E_{ST} consists of the kinetic energy of the electron and of the electron–phonon interaction. It is assumed in eq. (2.22) that the normalization condition, $\|\psi\| = 1$, is fulfilled. The $\boldsymbol{q}$-dependence of the frequencies ω_{qs} and of the coefficients γ_{qs} is determined by the type of phonons and of electron–phonon coupling. Here we shall consider only the case of nonpolar coupling to one branch of dispersionless optical phonons:

$$\omega_{qs} = \omega_0, \qquad \gamma_{qs} = \gamma_0. \tag{2.23}$$

We introduce dimensionless variables

$$r \equiv r_b r', \qquad q \equiv r_b^{-1} q', \qquad \tau \equiv \omega_0^{-1}\tau', \qquad Q_q \equiv Q'_q \gamma_0/\omega_0^2,$$
$$r_b = m\gamma_0^2/\omega_0^2, \tag{2.24}$$

and hereafter omit the primes. The dimensionless Lagrangian $\mathscr{L}$, defined by

$$L = (\omega_0^4/m^3\gamma_0^4)\mathscr{L}, \tag{2.25}$$

may be rewritten in these variables as

$$\mathscr{L} = \tfrac{1}{2}\int \mathrm{d}\boldsymbol{r}\,(\partial_\tau Q)^2 + U(Q), \tag{2.26}$$

where

$$U(Q) = \frac{1}{2}\int \mathrm{d}\boldsymbol{r}\, Q^2 + \min_{\psi} \int \mathrm{d}\boldsymbol{r}\,((1/2)|\nabla\psi|^2 + Q|\psi|^2), \tag{2.27}$$

and $Q(\boldsymbol{r}, \tau)$ is the lattice deformation. The coefficient $\omega_0^4/m^3\gamma_0^4$ determines the scale of all quantities, having the dimensionality of the energy, in particular, of the height of the barrier to ST, and r_b determines the spatial scale of the barrier.

The Hamiltonian action equals

$$S = \sigma_{op}\mathscr{S}, \quad \sigma_{op} = \omega_0^3/m^3\gamma_0^4, \tag{2.28}$$

where

$$\mathscr{S} = \int \mathrm{d}\tau\, \mathscr{L}. \tag{2.29}$$

It is seen from eqs. (2.25) and (2.28) that $\sigma_{op} \sim W/\omega_0$. Consequently, the condition $W \gg \bar{\omega}$ actually coincides with the applicability condition of the semiclassical approximation $S \sim \sigma_{op} \gg 1$. It is clear from eqs. (2.26), (2.27) and (2.29) that the Lagrangian $\mathscr{L}$ and the Hamiltonian action $\mathscr{S}$ have a universal form and do not contain any parameters.

To clarify the dynamics of the lattice and of the electron cloud in the course of tunneling, it is necessary to derive and solve the equations of motion, ensuing from the conditions under which the action is stationary

$$\delta_Q \mathscr{S} = 0 \tag{2.30}$$

and then to calculate the action. At $T = 0$, the total energy of the system

$$\varepsilon = -\left\{\frac{1}{2}\int \mathrm{d}\boldsymbol{r}(\partial_\tau Q)^2 - U(Q)\right\} \tag{2.31}$$

is zero, $\varepsilon = 0$. In this case the Hamiltonian action coincides with the truncated action S_0 entering eq. (2.2). Since calculation of the minimum in eq. (2.27) cannot be performed at arbitrary Q, it is convenient to treat $\mathscr{S}$ as the functional $\mathscr{S}[\psi, Q]$ of independent variables ψ and Q; then eq. (2.30) should be replaced with $\delta_{\psi,Q}\mathscr{S}[\psi, Q] = 0$ at $\|\psi\| = 1$. Then from eqs. (2.26) and (2.27) follows an Euler equation for the lattice

$$\partial_\tau^2 Q - Q - |\psi|^2 = 0, \tag{2.32}$$

which should be solved under the conditions

$$Q|_{\tau=\infty} = 0, \qquad \partial_\tau Q|_{\tau=0} = 0. \tag{2.33}$$

The first of them guarantees that prior to the beginning of ST at $\tau = \infty$, the lattice is not deformed, whereas the second, that in the moment $\tau = 0$ tunneling terminates and the velocity of the lattice is zero at all $\boldsymbol{r}$. The choice of $\tau = \infty$ as the moment when the motion starts is consistent with fig. 8a (at $T = 0$). The solution of eq. (2.32) reads

$$Q(\boldsymbol{r}\tau) = \int_{-\infty}^{\infty} \mathrm{d}\tau' D_0(\tau - \tau')|\psi(\boldsymbol{r}\tau')|^2, \qquad D_0(\tau) = -\tfrac{1}{2}\mathrm{e}^{-|\tau|}. \tag{2.34}$$

It is assumed here that $|\psi|^2$ and Q are symmetrically continued to the region $\tau < 0$. Here $D_0(\tau)$ is the phonon Green function. After Q is eliminated, the action for the entire period of classical motion is represented as a functional

$$\mathscr{S}[\psi] = \min_{\psi} \int \mathrm{d}\boldsymbol{r} \int_{-\infty}^{\infty} \mathrm{d}\tau \left\{\tfrac{1}{2}|\nabla_r\psi(\boldsymbol{r}\tau)|^2 + \tfrac{1}{2}|\psi(\boldsymbol{r}\tau)|^2 \int_{-\infty}^{\infty} \mathrm{d}\tau' D_0(\tau - \tau')|\psi(\boldsymbol{r}\tau')|^2\right\} \tag{2.35}$$

at $\|\psi\| = 1$. Its variation leads to the Euler equation for ψ:

$$-\tfrac{1}{2}\Delta\psi(\boldsymbol{r}\tau) + \psi(\boldsymbol{r}\tau)\int_{-\infty}^{\infty} \mathrm{d}\tau' D_0(\tau - \tau')|\psi(\boldsymbol{r}\tau')|^2 = \mathscr{E}(\tau)\psi(\boldsymbol{r}\tau). \tag{2.36}$$

The integral entering the second term and having the meaning of the potential coincides with $Q(\boldsymbol{r}\tau)$. The Schrödinger eq. (2.36) differs considerably from the static nonlinear Schrödinger equation due to the fact that the eigenvalue $\mathscr{E}(\tau)$ is time-dependent and the nonlinear term involves retardation. The potential $Q(\boldsymbol{r}\tau)$ is large enough to ensure existence of a discrete level $\mathscr{E}(\tau)$ but only at small $|\tau| < \tau_1 \sim 1$. Deformation $Q(\boldsymbol{r}\tau)$ decreases with increasing $|\tau|$, as follows from eq. (2.34). The resulting pattern coincides with the one described in section 1. At

τ decreasing from ∞ to τ_I, the deformation is getting more pronounced but a local level of an electron is still missing; this represents a free motion of the lattice. At $\tau = \tau_I$, a local level arises and the electron gets trapped on it.

In the field theory for localized classical solutions in 4D-space–imaginary-time, the term "instanton" is used (Rajaraman 1982, Vainshtein et al. 1982). We shall also employ this term for solutions of coupled equations (2.36) and (2.32) for the electron and the lattice, respectively. The time τ_I will be termed as the duration of an instanton. It is obvious that $\tau_I < \tau_t$, where $\tau_t = \beta/2$ is the time of tunneling [cf. eq. (2.8)].

To obtain quantitative results for $\mathscr{S}$, one needs to find an extremal $\psi(\boldsymbol{r}\tau)$ by approximate methods. Selecting $\psi(\boldsymbol{r}\tau)$ in the "Π-pulse" approximation,

$$\psi(\boldsymbol{r}\tau) = \psi(\boldsymbol{r})\theta(\tau_I - |\tau|), \quad \theta(\tau) = \begin{cases} 1 & \text{at } \tau > 0, \\ 0 & \text{at } \tau < 0, \end{cases} \tag{2.37}$$

it is possible to find ψ and τ_I from the condition $\delta\mathscr{S} = 0$. In the result, $\tau_I \approx 1.1\,\omega_0^{-1}$, $W \approx 44\omega_0^4/m^3\gamma_0^4$, $S \approx 6.2\,W/\omega_0$. A close estimate, $S = 4W/\omega_0$, has been obtained by Mott and Stoneham (1977) for a single-mode system. The pattern of multimode tunneling, obtained here for $T = 0$, will be generalized to the case $T \neq 0$ in section 4.2.

2.3. Trapping of an electron on the level, emerging in a slowly changing potential well

In series with the tunneling process, studied above, there is another process, which controls the ST rate and can have an exponentially small probability: the process of trapping of a free electron with an energy E on a discrete level in the field of deformation increasing in time (stage 2, section 1). As will be shown below, the process of trapping resembles the above-the-barrier reflection (Landau and Lifshitz 1974). Calculating the ST rate, we see that exponential suppression of the trapping rate may compete, at increasing E, with the increasing rate of barrier penetration caused by the increase of the total energy of the system. This aspect of the problem will be dwelt upon in section 7; here we analyze a model problem on the solution of which section 7 is based.

Usually quantum transitions under the influence of adiabatic perturbations are studied for a discrete spectrum (Dykhne 1961, Landau and Lifshitz 1974). For a two-level problem the Landau–Zener formula is valid. Transitions between discrete and continuous spectra have been studied by Demkov (1964) and by Devdariani (1972) for problems of atomic physics. In what follows the method of deriving the solution of this problem in the exponential approximation is described (Ioselevich and Rashba 1985b).

Let an attracting short-range potential $V(\boldsymbol{r}t)$ slowly vary with t and produce a discrete level at $t = t_*$ (fig. 9). Calculate the probability of an electron to be

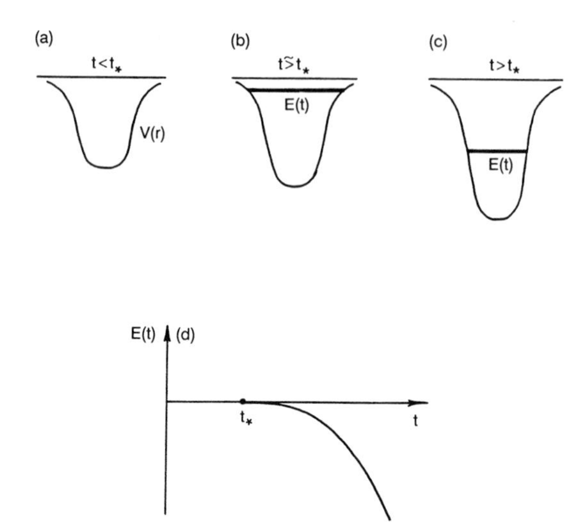

Fig. 9. The model of trapping of a particle on an emerging discrete level. The time evolution of the potential well: (a) for $t < t_*$ there is no level; (b) for $0 < t - t_* \ll \bar{\omega}^{-1}$ there is a shallow level $[E(t) \ll W$, W determines the scale of the well depth$]$; (c) for $0 < t - t_* \sim \bar{\omega}^{-1}$ there is a deep level $[E(t) \sim W]$; (d) the time dependence of the energy of the discrete level.

trapped on the level at $t \to \infty$ under the condition that at $t \to -\infty$ the initial state of the electron belongs to the continuous spectrum and has an energy $E > 0$. Since the energy spectrum changes adiabatically slowly, the probability of electron trapping on the level with $E < 0$ is exponentially small at large initial $E > 0$. It can be found by means of the analytical continuation method.

Figure 10a shows a trajectory $C(0)$ of the discrete level (moving with increasing t) on a two-sheet Riemannian surface. This trajectory lies on a nonphysical sheet (virtual level) at $t < t_*$ and on a physical sheet at $t > t_*$. At $t = t_*$, the trajectory, coming onto the physical sheet, passes through the initial point ($E = 0$) of a cut through the continuous spectrum. The contour $\Gamma(0)$, coinciding with the real axis in the complex t-plane, corresponds to the trajectory $C(0)$ (fig. 10b). To describe trapping of the electron with $E \neq 0$, it is necessary to choose a trajectory $\Gamma(E)$ in the complex t-plane such that the corresponding trajectory $C(E)$ should intersect the cut in the point E. This induces resonant trapping of the electron on a quantum level, which, at the first stage of this process [on the heavy vertical section of the contour $\Gamma(E)$], may be regarded as a pseudolocal level in the complex potential. For real t [on the heavy horizontal section of the contour $\Gamma(E)$], the electron resides on a genuine discrete level. The probability of trapping in the exponential approximation is

$$P_{\mathrm{tr}}(E) \propto \exp(-S_{\mathrm{tr}}(E)], \tag{2.38}$$

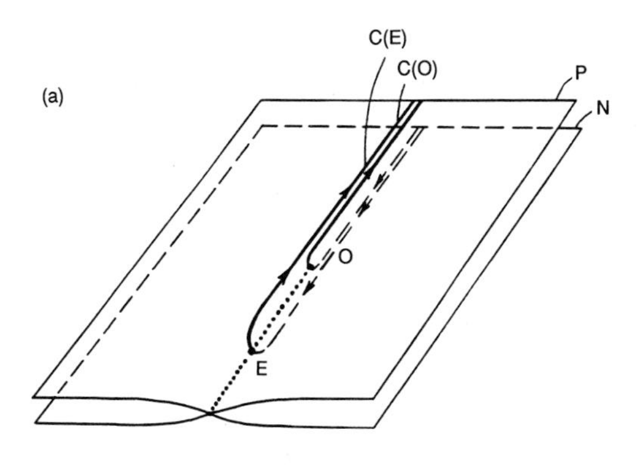

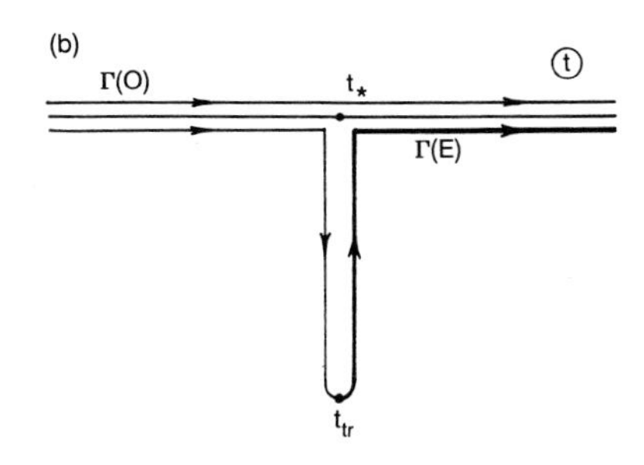

Fig. 10. (a) The trajectories of the energy level on the Riemannian E-surface; E is the electron energy, P is a physical sheet, N is a non-physical sheet, the dotted line indicates the cut outgoing from the point $E = 0$. (b) The corresponding contours in the complex t-plane. $t_{tr}(E)$ is determined by the condition (2.40).

where

$$S_{tr}(E) = 2\operatorname{Im} \int_{t_{tr}}^{t} (E - E(t')]\,dt', \tag{2.39}$$

where t_{tr} is defined from the condition

$$E(t_{tr}) = E. \tag{2.40}$$

Here $S_{tr}(E)$ is the doubled Hamiltonian action on the vertical section of the contour $\Gamma(E)$. Equation (2.39) resembles the Landau–Dykhne formula for adiabatic perturbation in the discrete spectrum. Yet, there is one important distinction. The Landau–Dykhne formula involves a two-valued analytical function and t_{tr} is an isolated branching point. Therefore, the contour can be deformed in such a way that it embraces this point, without passing through it. In the case of the continuous spectrum a large number of levels interact. This becomes particularly clear if the volume of the system is restricted and, as a result, the continuous spectrum transforms into a quasidiscrete spectrum. Then there emerge numerous singularities, lying infinitely closely to t_{tr}, both above and under it. That is why in this case the singular point t_{tr} is not isolated, the contour must necessarily pass through it and cannot be shifted from it. The fact that $P_{tr}(E)$ is exponentially small is a consequence of a strong discordance in oscillatory time dependences of wave functions of the free and bound electron states.

An explicit expression for $S_{tr}(E)$ can be derived at small E. Having in mind application of the result we are arriving at the problem of the ST rate, assume that the characteristic value of the potential is $V(\boldsymbol{r}t) \sim W$ and that the characteristic time of its variation is $\sim \bar{\omega}^{-1}$. The main contribution to the integral in eq. (2.39) comes from the region $|E(t)| < E$. At the moment the discrete level emerges, the potential well is already deep, its depth being $\sim W$. Therefore, if $E \ll W$, the energetic level is shallow (the binding energy is small compared to the depth of the potential well) and the radius of the potential is small compared to the radius of the state. If the size and shape of the well vary smoothly near the point t_* (fig. 9), the t-dependence of the parameters of the well in the vicinity of t_* can be treated as linear; then $E(t) \propto (t - t_*)^2$ [cf. section 2.1, eqs. (2.15) and (2.16)]. It is our assumption here that the symmetry of the level coincides with the symmetry of the band edge. The coefficient of this dependence can be calculated from the condition that by the moment $t \sim \bar{\omega}^{-1}$, the level is no longer shallow, i.e., $E(\bar{\omega}^{-1}) \sim W$. Therefore,

$$E(t) = -\Omega^3 (t - t_*)^2, \tag{2.41}$$

$$\Omega \sim (\bar{\omega}^2 W)^{1/3}. \tag{2.42}$$

This estimate for Ω includes W (but not V_{ST}), since it is W that determines the scale of the energy the lattice possesses near the point Q_{br}. The estimate (2.42) is not applicable in specific situations when the barrier, for some reason, is very steep (see sections 4.3 and 4.4). In these cases the estimate for Ω also includes an extra large parameter characterizing this steepness. We shall not analyze this problem in detail here.

The level becomes adiabatic as soon as $\partial_t \ln|E(t)| < |E(t)|$, i.e., $(t - t_*) > \Omega^{-1}$ and so $|E(t)| > \Omega$. Consequently, trapping lasts a time $\sim \Omega^{-1} \ll \bar{\omega}^{-1}$, and during the entire trapping stage the level remains shallow ($|E| \ll W$) and can be described in terms of the zero-radius potential, which actually has been the assumption made for derivation of eq. (2.41). The adiabaticity condition in the barrier state, $\bar{\omega} \ll W$, is fulfilled if $\bar{\omega} \ll \Omega \ll W$. It follows from eqs. (2.39) and (2.40) that

$$t_{tr} = t_* - \mathrm{i}\Omega^{-3/2} E^{1/2}, \tag{2.43}$$

$$S_{tr} = \tfrac{4}{3}(E/\Omega)^{3/2}. \tag{2.44}$$

Equations (2.38) and (2.44) in the exponential approximation coincides with the exact solution for the zero-radius potential (Demkov 1964), in conformity with which the probability of trapping on the arising level is

$$P_{tr}(\boldsymbol{k}) = v_{tr}(E)/V,$$
$$v_{tr}(E) = 8\pi^2(2m\Omega)^{-3/2}\exp[-S_{tr}(E)], \tag{2.45}$$

where V is the volume of the system. For eq. (2.45) the normalization condition

$$\int d\boldsymbol{r}\frac{d\boldsymbol{k}}{(2\pi)^3}P_{tr}(\boldsymbol{k}) = 1$$

is fulfilled. This condition implies that in the inverse process (when the discrete level merges with the continuous spectrum) the total probability of ionization is unity. Here it is assumed that $E(\boldsymbol{k}) = \boldsymbol{k}^2/2m$ (m is the mass of the electron). The quantity $v_{tr}(E)$ in eq. (2.45) has the meaning of the volume from which a particle is effectively trapped on the level. For slow particles the trapping radius $r_{tr} \equiv [v_{tr}(0)]^{1/3} \sim (m\Omega)^{-1/2}$ and has the meaning of the radius of the wave function on the edge of the adiabaticity region, i.e., at $t \sim \Omega^{-1}$. The applicability criterion of the exponential approximation is $S_{tr} \gg 1$ or $E \gg \Omega$. Equation (2.45) is valid if the electron is trapped when the system is moving in a classically accessible region, i.e., prior to barrier penetration. At trapping on the stage of tunneling, eqs. (2.38) and (2.45) must be modified. In this case the contour given in fig. 10b should be turned by 90°. The system approaches the t_* point along the imaginary time axis and appendage, passing through the point $t_{tr}(E)$, becomes horizontal and gives to the action a purely imaginary contribution, cancelled when the squared modulus of the trapping amplitude is calculated. In this case $P_{tr}(\boldsymbol{k})$ contains no exponential factor.

Equations (2.38) and (2.45) incorporate reemission processes. That is why when they are applied to the problem of nonradiative trapping, reemission on stage 3 (section 1) must not be taken into account again. Reemission on stage 4 is irrelevant if stages 1–3 are the bottleneck of the process, as is assumed in this paper.

3. *General formalism*

In this section we derive a general expression for the trapping rate $w(\boldsymbol{k}T)$, i.e., for the probability of a transition of an electron from the free state into the ST state per unit time.

It will be our assumption that in the initial state an electron has a fixed momentum $\boldsymbol{k}$, and the lattice is in equilibrium and has a temperature T. The probability of transition from the free state with the momentum $\boldsymbol{k}$ to a ST state in a small time $t_2 - t_1$ is equal to the squared amplitude $\mathscr{A}$ of the transition, averaged over phonon variables:

$$w(\boldsymbol{k}T)(t_2 - t_1) = \langle |\mathscr{A}((\mathrm{ST})t_2, \boldsymbol{k}t_1)|^2 \rangle_{\mathrm{ph}}$$

$$= \int \mathrm{d}\boldsymbol{r}_1\, \mathrm{d}\boldsymbol{r}_2\, \mathrm{d}\boldsymbol{r}'_1\, \mathrm{d}\boldsymbol{r}'_2\, \Psi^*_{\mathrm{ST}}(\boldsymbol{r}_2 t_2) \Psi_{\mathrm{ST}}(\boldsymbol{r}'_2 t_2)$$

$$\times F(\boldsymbol{r}_2 t_2, \boldsymbol{r}'_2 t_2; \boldsymbol{r}_1 t_1, \boldsymbol{r}'_1 t_1) \Psi_{\boldsymbol{k}}(\boldsymbol{r}_1 t_1) \Psi^*_{\boldsymbol{k}}(\boldsymbol{r}'_1 t_1). \tag{3.1}$$

Here $\Psi_{\boldsymbol{k}}$ and Ψ_{ST} are Heisenberg wave functions of the electron in the free state with the momentum $\boldsymbol{k}$ and in the ST state, respectively, and

$$F(\boldsymbol{r}_2 t_2, \boldsymbol{r}'_2 t_2; \boldsymbol{r}_1 t_1, \boldsymbol{r}'_1 t_1)$$

$$= Z^{-1}_{\mathrm{lat}} \sum_{i,f} \exp[(-\beta\varepsilon_i)] \langle i | \hat{\psi}(\boldsymbol{r}'_1 t_1) \hat{\psi}^+(\boldsymbol{r}'_2 t_2) | f \rangle \langle f | \hat{\psi}(\boldsymbol{r}_2 t_2) \hat{\psi}^+(\boldsymbol{r}_1 t_1) | i \rangle. \tag{3.2}$$

The states $|i\rangle$ and $|f\rangle$ correspond to purely phonon excitations of a crystal in the absence of a free electron. In this case it is assumed that in the initial state $|i\rangle$ phonons are in equilibrium; Z_{lat} is the normalization factor, $\hat{\psi}$ is the electron annihilation operator. Equation (3.1) can be rewritten as an integral over paths of the lattice coordinates $Q(t)$ (Feynman and Hibbs 1965):

$$(t_2 - t_1) w(\boldsymbol{k}T) = Z^{-1}_{\mathrm{lat}} \int \mathrm{d}\boldsymbol{r}_1\, \mathrm{d}\boldsymbol{r}_2\, \mathrm{d}\boldsymbol{r}'_1\, \mathrm{d}\boldsymbol{r}'_2\, \psi^*_{\mathrm{ST}}(\boldsymbol{r}_2) \psi_{\mathrm{ST}}(\boldsymbol{r}'_2)$$

$$\times \int \mathscr{D}Q\, G^{\mathrm{R}}_{\{Q\}}(\boldsymbol{r}_2 t_2, \boldsymbol{r}_1 t_1) G^{\mathrm{A}}_{\{Q\}}(\boldsymbol{r}'_1 t_1, \boldsymbol{r}'_2 t_2) \exp\left[\mathrm{i} \int_{\Gamma} \mathrm{d}t\, L_{\mathrm{lat}}(Q) \right] \psi_{\boldsymbol{k}}(\boldsymbol{r}_1) \psi^*_{\boldsymbol{k}}(\boldsymbol{r}'_1), \tag{3.3}$$

where

$$Z_{\mathrm{lat}} = \int \mathscr{D}Q \exp\left[\mathrm{i} \int_{\Gamma} L_{\mathrm{lat}}(Q)\, \mathrm{d}t \right]. \tag{3.4}$$

With an appropriate choice of the integration measure $\mathscr{D}Q$, Z_{lat} is equal to the lattice partition function. Here $G^{\mathrm{R}}_{\{Q\}}$ and $G^{\mathrm{A}}_{\{Q\}}$ are the retarded and advanced Green functions of the electron, i.e., functionals of the lattice path $Q(t)$, ψ_{ST} and $\psi_{\boldsymbol{k}}$ are Schrödinger wave functions of the ST and free electron, respectively, L_{lat} is the Lagrangian of a free lattice:

$$L_{\mathrm{lat}}(Q) = \tfrac{1}{2} \sum_{\boldsymbol{n}\rho} (\partial_t Q_{\boldsymbol{n}\rho})^2 - U_{\mathrm{lat}}(Q), \tag{3.5}$$

where $Q_{\boldsymbol{n}\rho}(t)$ are time-dependent atomic displacements, $\boldsymbol{n}$ numbers the unit cells in a crystal and ρ numbers the atoms inside a unit cell; here and elsewhere the masses of atoms are included in the definition of $Q_{\boldsymbol{n}\rho}$. The lattice is not supposed to be harmonic and, therefore, U_{lat} has a more general form than eq. (2.21).

In the case when the initial energy of the electron is not too large, the contours Γ can be obtained from the contour Γ_+, plotted in figs. 8a and 8b for ST (intrinsic and extrinsic) and for normal centers, respectively. The contour Γ should consist of two parts: the contour Γ_+ (upper part) and its mirror reflection Γ_- (lower part) (see fig. 11). On Q are imposed the standard periodic boundary conditions: $Q(t_1 + \frac{1}{2}i\beta) = Q(t_1 - \frac{1}{2}i\beta)$, which, in what follows, are referred to as the periodicity conditions for the imaginary time. In the formalism we are developing, they arise naturally as a consequence of eqs. (3.1) and (3.2). Calculation of the sum over the initial states $|i\rangle$ in eq. (3.2), rewritten in terms of the paths $Q(t)$, is, in fact, integration over the initial coordinate Q_1, common for $\mathscr{A}$ and $\mathscr{A}^*$. The elimination of the Gibbs factor by shifting the edge points of the contour Γ along the imaginary axis by $\pm\frac{1}{2}i\beta$ results in the fact that $Q(t_1 + \frac{1}{2}i\beta) = Q_1$ on the contour Γ_+, and $Q(t_1 - \frac{1}{2}i\beta) = Q_1$ on the contour Γ_-, hence the periodicity condition. The same condition also exists in the Matsubara formalism for bosons (see Abrikosov et al. 1962). Electron Green functions, dependent on lattice paths, have been previously used in other problems (Miller 1974, section V).

Hole bands are usually degenerate; therefore, a hole (and exciton) wave function is multicomponent (Luttinger and Kohn 1965). Equation (3.3) can be easily generalized to this case. For this purpose the factor $\psi_{\mathrm{ST}}^* \psi_{\mathrm{ST}} G^{\mathrm{R}} G^{\mathrm{A}} \psi_k \psi_k^*$ in eq. (3.3) should be regarded as convolution $(\psi_{\mathrm{ST}} \hat{G}^{\mathrm{R}} \bar{\psi}_k)\,(\psi_k \hat{G}^{\mathrm{A}} \bar{\psi}_{\mathrm{ST}})$, where ψ is a wave function (a line) and $\bar{\psi}$ is a wave function (a column) Hermitian conjugate to ψ, $\hat{G}^{\mathrm{R,A}}$ are matrix Green functions. However, it should be borne in mind that the band degeneracy induces spontaneous symmetry breaking of a barrier to ST (Kusmartsev and Rashba 1984) and of extremal paths.

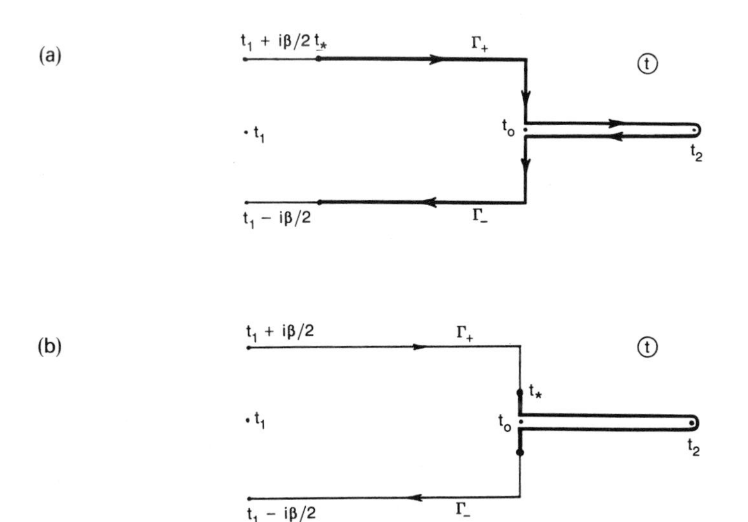

Fig. 11. Contour Γ, consisting of the upper (Γ_+) and lower (Γ_-) parts for ST (cf. fig. 8a): (a) for the static solution; (b) for the instanton solution.

At $t < \mathrm{Re}\{t_*\}$ the electron is free (see fig. 11). Therefore, at the initial moment of time, t_1, one can approximately neglect the influence of the lattice on the electron (cf. section 1). At the final moment of time, t_2, the electron must be in a bound state, following the lattice adiabatically. Therefore, at the moment t_1 it is convenient to introduce a basis of the wave functions $\psi_{k'}(\boldsymbol{r})$ of a free electron, with the momenta $\boldsymbol{k}'$ [the energies $E(\boldsymbol{k}')$ correspond to these functions], and at the moment t_2, a basis of the wave functions $\psi_{m,Q(t)}(\boldsymbol{r})$ adiabatically following on an instantaneous configuration of the lattice [the corresponding energies are $E_m(Q(t))$]. Therefore, the Green functions $G^{\mathrm{R}}_{\{Q\}}$ and $G^{\mathrm{A}}_{\{Q\}}$ can be conveniently written out in the mixed representation as

$$G^{\mathrm{R}}_{\{Q\}}(\boldsymbol{r}_2 t_2, \boldsymbol{r}_1 t_1) = V^{-1/2} \sum_m \int \frac{\mathrm{d}\boldsymbol{k}'}{(2\pi)^3} \psi_{mQ(t_2)}(\boldsymbol{r}_2) u_{mk'}(\{Q\}) \psi^*_{k'}(\boldsymbol{r}_1) \times \exp\left\{ -\mathrm{i} \int_{t_1}^{t_2} E_m[Q(t)]\,\mathrm{d}t \right\}, \tag{3.6}$$

where V is the volume of the crystal. The lowest value of $E_m(Q) \equiv E_{\mathrm{ST}}(Q)$ corresponds to ψ_{ST}. As in eq. (2.22)

$$E_{\mathrm{ST}}(Q) = \min_{\psi} \langle \psi | \hat{H}(Q) | \psi \rangle, \tag{3.7}$$

where ψ is normalized and $\hat{H}(Q)$ is the Hamiltonian of the electron, incorporating the electron–phonon interaction. Yet, in eq. (3.7), in contrast to eq. (2.22), we do not consider displacements Q as small (therefore, $\hat{H}(Q)$ can be nonlinear in Q) nor spatial scales as large (therefore, the electron is not described by the effective mass approximation). In the free motion region one should replace $E_m(Q)$ by $E(\boldsymbol{k})$ in eq. (3.6). The symbol $\{Q\}$ in arguments of various quantities (e.g., $u_{mk'}(\{Q\})$) implies that they depend on the coordinates Q nonlocally in time; in other words, they depend on the entire path of the lattice motion. Local quantities (e.g., $E_m(Q)$) depend only on the value of Q at a given moment of time. The factor $u_{mk'}$ in eq. (3.6) has the meaning of the amplitude of a transition from a $\boldsymbol{k}'$-state into a m-state ($m \equiv ST$ as applied to the ST problem). In this case u is related to the probability P_{tr} of a particle trapping on an emerging level (P_{tr} was calculated in section 2.3.). We shall use this relation below in section 5.2. Substitution of eq. (3.6) into eq. (3.3) yields

$$(t_2 - t_1) w(\boldsymbol{k}T) = (VZ_{\mathrm{lat}})^{-1} \exp\{\beta E(\boldsymbol{k})\} \times \int \mathscr{D}Q\, |u_{\mathrm{ST},k}(\{Q\})|^2 \exp\left\{ \mathrm{i} \int_{\Gamma} L_k(Q)\,\mathrm{d}t \right\}, \tag{3.8}$$

where

$$L_k(Q) = \begin{cases} L_{\mathrm{lat}} - E_{\mathrm{ST}}(Q), & \text{loaded motion,} \\ L_{\mathrm{lat}} - E(\boldsymbol{k}), & \text{free motion.} \end{cases}$$

The factor $\exp[\beta E(\boldsymbol{k})]$ originates from the fact that the integral in the limits (t_1, t_2) entering eqs. (3.3) and (3.6) is replaced by the integral taken along Γ in eq. (3.8).

The functional integral in eq. (3.8) can be calculated by means of the steepest descent method, taking advantage of the presence of a large parameter $W/\bar{\omega} \gg 1$. In the leading approximation it is necessary to retain only exponential factors. For thermalized particles, this is the last factor in eq. (3.8); one can put $E(\boldsymbol{k}) \approx E(0) = 0$. The $\boldsymbol{k}$-dependence of $u_{\mathrm{ST},\boldsymbol{k}}$ is considerable only at $E(\boldsymbol{k}) \gtrsim \Omega$ [see eq. (2.42)]. So, for thermalized particles with $E(\boldsymbol{k}) \sim T$ this dependence can be neglected (i.e., $u_{\mathrm{ST},\boldsymbol{k}} \approx u_{\mathrm{ST},0}$) if $T \gtrsim \bar{\omega} \ll \Omega$. Consequently, for thermalized particles the exponential factor in $w(\boldsymbol{k}T)$ is $\exp(-S)$ [cf. eq. (1.2)]. Here the imaginary action is

$$S = \mathrm{Im} \int_{\Gamma} L(Q)\,\mathrm{d}t \tag{3.9}$$

and the Lagrangian is

$$L(Q) \equiv L_{\boldsymbol{k}=0}(Q) = \tfrac{1}{2} \sum_{\boldsymbol{n}\rho} (\partial_t Q_{\boldsymbol{n}\rho})^2 - U(Q). \tag{3.10}$$

On the free motion section the adiabatic potential $U(Q) = U_{\mathrm{lat}}(Q)$, i.e., equals the potential energy of the lattice (curve 1, fig. 1a), whereas on the loaded-motion section $U(Q) = U_{\mathrm{lat}}(Q) + E_{\mathrm{ST}}(Q)$ (curve 2, fig. 1a). The action acquires the minimum on optimal paths $Q(t)$. The next section 4 is devoted to the analysis of these paths and to the calculation of the extremal action S and to the investigation of its temperature dependence.

4. Exponential factor

Variation of the action (3.9) with respect to Q leads to the standard Newton equation for the lattice:

$$\partial_t^2 Q = -\partial U/\partial Q. \tag{4.1}$$

Equation (4.1) must be solved on the time contour Γ (fig. 11) with the boundary conditions:

$$Q(t_1 + \mathrm{i}\beta/2) = Q(t_1 - \mathrm{i}\beta/2), \qquad \partial_t Q(t_1 + \mathrm{i}\beta/2) = \partial_t Q(t_1 - \mathrm{i}\beta/2). \tag{4.2}$$

The solution of eq. (4.1) is symmetric with respect to the replacement of Γ_+ (the upper part of the contour Γ) by Γ_- (its lower part). Therefore, in the calculation of the action S all contributions of horizontal sections cancel each other, and only the contribution of vertical sections remains. In terms of the imaginary time τ, $t = \mathrm{i}\tau$, we get

$$S = \int_{\Gamma_0} \mathrm{d}\tau \left\{ \tfrac{1}{2} \sum_{np} (\partial_\tau Q_{np}(\tau))^2 + U(Q(\tau)) \right\}, \tag{4.3}$$

where the integration region Γ_0 is given in figs. 12a and 12b for the cases of ST and of trapping by normal centers, respectively. In the former case this is a section $(-\frac{1}{2}\beta, \frac{1}{2}\beta)$, and in the latter, the same section with a fold in its central part. Equation (4.1) on Γ_0 acquires the form

$$\partial_\tau^2 Q = \partial U / \partial Q, \tag{4.4}$$

and the matching condition in the corners of the contour (actually turning points) becomes

$$\partial_\tau Q|_{\tau=0} = \partial_\tau Q|_{\tau=\pm\beta/2} = 0. \tag{4.5}$$

Equation (4.4) with the boundary conditions [eq. (4.5)] provides a rigorous formulation of the problem of calculating the under-the-barrier contribution to the action S and its temperature dependence. However, it proves to be rather difficult to obtain the solution in practice. Actually, to find $U(Q)$ entering eq. (4.4), it is necessary to get a solution of the Schrödinger equation with an arbitrary potential produced by lattice displacements $Q(\boldsymbol{r}t)$:

$$\{E(t) - \hat{H}[Q(\boldsymbol{r}t)]\}\psi(\boldsymbol{r}t) = 0. \tag{4.6}$$

The nonlinear eq. (4.4) may have several solutions.

First, there may exist a static solution (Ioselevich 1982, Ioselevich and Rashba 1984): Q is τ-independent. In this case eq. (4.4) coincides with an equation for stationary points of APS,

$$\partial U / \partial Q = 0. \tag{4.7}$$

The value of U in the lowest-lying saddle point is W, i.e., the height of the barrier to ST. Therefore, it entails from eq. (4.3) and from the equality $\partial_\tau Q \equiv 0$ that for

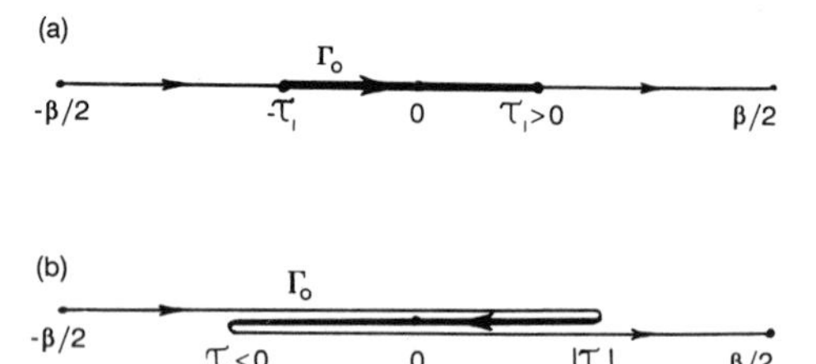

Fig. 12. The time region of the under-the-barrier motion (the vertical part of contour Γ) is represented by the contour Γ_0. The heavy line indicates the loaded-motion section, and the arrow, the direction along Γ. The moment of termination of the loaded under-the-barrier motion determines the instanton duration τ_I: (a) ST; (b) normal centers.

the static solution we have

$$S_A = \beta W, \qquad w_A \sim \exp(-W/T), \tag{4.8}$$

i.e., the ST rate obeys the Arrhenius law. The static solution describes the classical activation mechanism of surmounting a barrier to ST, the point $\mathscr{W}$ on APS (fig. 1b) corresponding to it. In this point the system is in unstable equilibrium. The motion of the system close to this solution describes very slow surmounting of the barrier in the vicinity of the saddle point $\mathscr{W}$ (path A in fig. 1b). In this case the main contribution to the ST rate is given by the paths, passing both slightly higher and slightly lower than $\mathscr{W}$. The particle gets trapped on the horizontal section of the contour Γ before the system passes through the point $\mathscr{W}$ (fig. 11a).

The adiabaticity criterion (2.1) and the condition $S_A \gg 1$, i.e., $W \gg T$, are the validity conditions for eq. (4.8). It is not required that phonons be classical, $T \gg \bar{\omega}$. Nevertheless, it is important that the point $\mathscr{W}$ be stationary, i.e., that it obey eq. (4.7). This holds only for ST (fig. 2a) when the barrier is smooth. For normal centers (fig. 2b) $U(Q)$ has a singularity at the point $\mathscr{W}$. This singularity can be removed if the barrier near $\mathscr{W}$ gets rounded. However, this gives rise to a large frequency ω_W and to the fact that tunneling dominates up to a very high temperature $T_c \approx \omega_W/2\pi$ (section 2.1); therefore, the static solution becomes irrelevant. It is noteworthy that for systems with ST the replacement of a smooth adiabatic potential by a model potential, consisting of two intersecting parabolas (as it is done in the multiphonon transition theory) leads to the loss of the static solution (i.e., of the Arrhenius regime).

Second, at not too high T, in addition to the static solution, there are other solutions of eq. (4.4) for which the action is smaller than S_A, at least at low T. These time-dependent solutions are instantons, already analyzed in section 2.2 at $T = 0$ for one of the models. Instantons cannot be found explicitly at arbitrary $\hat{H}(Q)$ and $U_{\text{lat}}(Q)$. Nevertheless, it is possible to draw some conclusions concerning properties of these solutions. As has been pointed out above, a barrier to ST exists only if the interaction of the particle with phonons is a short-range interaction. In this case the attracting potential for the electron, induced by lattice displacements, rapidly decreases with distance, and a discrete level in a potential well of a depth V and of a width R exists only if $mVR^2 > 1$, where m is the effective mass of the electron. For the time-dependent solution, the discrete level is sure to exist at $\tau = 0$ when the system comes from under the barrier (fig. 8). The depth of the level decreases with $|\tau|$, and at a certain $|\tau| = \tau_I$ the level vanishes. The electron gets trapped exactly at this moment of time, and that is why for an instanton the trapping moment is on the vertical section of the contour. Since the only time scale characterizing the lattice at $T < \bar{\omega}$ is $\bar{\omega}^{-1}$, the duration of the instanton is $\tau_I \sim \bar{\omega}^{-1}$.

The difference in the shape of the contours, shown in fig. 12a, b, can be conveniently interpreted in terms of instantons, as the difference in the sign of τ_I, defined as the moment when tunneling terminates. Then $\tau_I > 0$ for ST, and $\tau_I < 0$ for normal centers. The difference in the shape of the contours Γ_+ in fig. 8 is described analogously and $\tau_I = \text{Im}\{t_*\}$.

The fact that for normal centers $\tau_I < 0$ can also be obtained by means of eqs. (2.8) and (2.14). As is clear from fig. 6, in this case it is the under-the-barrier motion along branch 2 of APS that is loaded. Therefore, the instanton duration can be calculated according to eq. (2.8), if one uses as the action S_0 only the second term in eq. (2.14). Since it enters S_0 with a negative sign, $\tau_I < 0$. Naturally, the total tunneling time, calculated with respect to the total action S_0 [eq. (2.14)], is positive.

The time-dependent (instanton) solutions describe tunneling of the system. In section 2.2 such a tunneling at $T = 0$ was described for one of the models. At finite temperatures tunneling becomes thermoactivated. It means that the energy of the tunneling system ε [see eq. (2.31)] is no longer zero. Mathematically, at finite temperatures the boundary conditions (2.33) must be replaced by eq. (4.5).

In what follows, we study instanton solutions for several models in the continuum approximation.

4.1. Continuum approximation

As mentioned in section 2, in the strong coupling limit ($\Lambda \gg 1$) the lattice in the region of a barrier to ST can be described in the harmonic approximation, and electron–phonon coupling can be regarded as linear in Q. In this case the Lagrangian of the system is described by eqs. (2.20)–(2.22). If, as in section 2, one regards ψ as an independent variable, then from eqs. (4.4), (2.20)–(2.22) it follows that

$$\partial_\tau^2 Q_{qs}(\tau) - \omega_{qs}^2 Q_{qs}(\tau) = \gamma_{qs} \int \mathrm{d}\boldsymbol{r}\, \mathrm{e}^{\mathrm{i}\boldsymbol{q}\cdot\boldsymbol{r}} |\psi(\boldsymbol{r}\tau)|^2. \tag{4.9}$$

The solution of eq. (4.9) subject to the boundary conditions described by eq. (4.5) reads as

$$Q_{qs}(\tau) = \gamma_{qs} \int_{\Gamma_0} \mathrm{d}\tau'\, D_{qs}(\tau - \tau') \int \mathrm{d}\boldsymbol{r}\, \mathrm{e}^{\mathrm{i}\boldsymbol{q}\cdot\boldsymbol{r}} |\psi(\boldsymbol{r}\tau')|^2, \tag{4.10}$$

where

$$D_{qs}(\tau - \tau') = -\frac{1}{2\omega_{qs}} \{(1 + N_{qs}) \exp(-\omega_{qs}[\tau - \tau']_{\Gamma_0}) + N_{qs} \exp(\omega_{qs}[\tau - \tau']_{\Gamma_0})\} \tag{4.11}$$

is the phonon Green function at a finite temperature,

$$N_{qs} = [\exp(\beta\omega_{qs}) - 1]^{-1} \tag{4.12}$$

are phonon occupation numbers, $[\tau - \tau']_{\Gamma_0}$ is the difference of later and earlier times, where the "later–earlier" relationship must be interpreted in terms of ordering along the contour Γ_0 (fig. 12). For ST, $[\tau - \tau']_{\Gamma_0} \equiv |\tau - \tau'|$. The Green function (4.11) can be continued to the entire contour Γ; in this case it is necessary to perform a replacement

$$\int_{\Gamma_0} \mathrm{d}\tau \to \int_{\Gamma} \mathrm{d}t, \qquad D \to \mathrm{i}D, \qquad [\tau - \tau']_{\Gamma_0} \to -\mathrm{i}[t - t']_{\Gamma}.$$

Such a representation appears to be very useful, if not only the tunneling but also the motion of the system in the classically accessible region is studied. For details about the Green functions defined on the contours, see the papers by Ioselevich and Rashba (1985a) and by Kusmartsev and Meshkov (1988).

The total contribution to the action from the under-the-barrier motion (4.3) is

$$S = \min_{\psi}\left\{\int \mathrm{d}\boldsymbol{r} \int_{\Gamma_0} \mathrm{d}\tau \frac{1}{2m} |\nabla\psi(\boldsymbol{r}\tau)|^2 + \frac{1}{2}\iint \mathrm{d}\boldsymbol{r}\,\mathrm{d}\boldsymbol{r}' \int_{\Gamma_0}\int \mathrm{d}\tau\,\mathrm{d}\tau'\, K(\boldsymbol{r} - \boldsymbol{r}', \tau - \tau')|\psi(\boldsymbol{r}\tau)|^2|\psi(\boldsymbol{r}'\tau')|^2\right\}, \tag{4.13}$$

where the kernel is

$$K(\boldsymbol{r}, \tau - \tau') = \sum_s \int \frac{\mathrm{d}\boldsymbol{q}}{(2\pi)^3} \mathrm{e}^{\mathrm{i}\boldsymbol{q}\cdot\boldsymbol{r}} |\gamma_{\boldsymbol{q}s}|^2 D_{\boldsymbol{q}s}(\tau - \tau'). \tag{4.14}$$

The nonlinear Schrödinger equation (4.6) for the wave function ψ is

$$\left\{-\frac{1}{2m}\Delta + V(\boldsymbol{r}t)\right\}\psi(\boldsymbol{r}\tau) = E(\tau)\psi(\boldsymbol{r}\tau), \tag{4.15}$$

where

$$V(\boldsymbol{r}\tau) = \sum_s \int \frac{\mathrm{d}\boldsymbol{q}}{(2\pi)^3} \gamma_{\boldsymbol{q}s} Q_{\boldsymbol{q}s}(\tau)\mathrm{e}^{\mathrm{i}\boldsymbol{q}\cdot\boldsymbol{r}}. \tag{4.16}$$

Substituting eq. (4.10) into eq. (4.16), we get

$$V(\boldsymbol{r}\tau) = \int_{\Gamma_0} \mathrm{d}\tau' \int \mathrm{d}\boldsymbol{r}' K(\tau - \tau', \boldsymbol{r} - \boldsymbol{r}')|\psi(\boldsymbol{r}'\tau')|^2. \tag{4.17}$$

The height of the barrier to ST is a stationary value of the energy functional

$$W = \operatorname*{stat}_{\psi}\left\{\int \mathrm{d}\boldsymbol{r} \frac{1}{2m} |\nabla\psi(\boldsymbol{r})|^2 - \frac{1}{2}\iint \mathrm{d}\boldsymbol{r}\,\mathrm{d}\boldsymbol{r}'|\psi(\boldsymbol{r})|^2|\psi(\boldsymbol{r}')|^2 \sum_s \int \frac{\mathrm{d}\boldsymbol{q}}{(2\pi)^3} \mathrm{e}^{\mathrm{i}\boldsymbol{q}\cdot(\boldsymbol{r} - \boldsymbol{r}')} \frac{|\gamma_{\boldsymbol{q}s}|^2}{\omega_{\boldsymbol{q}s}^2}\right\}. \tag{4.18}$$

After the system goes out from under the barrier (i.e., at $t > t_0$ on section 3 of the contour Γ_+, fig. 8):

$$Q_{qs}(t) = -\frac{\gamma_{qs}}{\omega_{qs}}\int_{t_0}^{t} dt' \sin[\omega_{qs}(t - t')] \int d\boldsymbol{r}\, e^{i\boldsymbol{q}\cdot\boldsymbol{r}} |\psi(\boldsymbol{r}t')|^2$$
$$+ \cos[\omega_{qs}(t - t_0)] Q_{qs}(t_0), \tag{4.19}$$

where $Q_{qs}(t_0) = Q_{qs}(\tau = 0)$. The nonlinear Schrödinger equation for $\psi(\boldsymbol{r}t)$ has the form of eq. (4.15) with the potential (4.16), where τ must be replaced by t, and for Q one should make use of eq. (4.19), which can be derived with the help of the Green function (4.11) extended to the entire contour. Its meaning is: the first term is a particular solution of the equations of loaded motion of the lattice with the initial conditions $Q(t_0) = 0$, $\partial_t Q(t_0) = 0$; the second term is the free motion with the initial conditions $Q(t_0) = Q(\tau = 0)$, $\partial_t Q(t_0) = 0$. The solution (4.19), at the turning point, satisfies the matching condition with the solution describing the motion under the barrier.

In the next subsections the derived expressions will be studied in detail for some concrete types of the phonon spectrum and of the electron–phonon coupling.

In conclusion, it is convenient to obtain the relation between the coupling constant Λ of the continual theory (cf. section 1) and microscopic parameters and to get an estimate for W and r_b. In the barrier state both $\psi(\boldsymbol{r})$ and $Q(\boldsymbol{r})$ have a common scale r_b. Besides, by virtue of the arguments based on the virial theorem (see below section 4.2), it follows that both the terms in eq. (4.18) must have the same order of magnitude. That is why from eq. (4.18) follows an equation for r_b:

$$1/mr_b^2 \sim \gamma_q^2/r_b^3\omega_q^2, \quad q \sim r_b^{-1}.$$

For nonpolar optical phonons, $\omega_q = \omega_0$, $\gamma_q = \gamma_0$, and for acoustic phonons, $\omega_q = cq$, $\gamma_q = \gamma q$. Therefore,

$$r_b \sim m\gamma_0^2/\omega_0^2, \qquad W \sim \omega_0^4/m^3\gamma_0^4$$

for optical phonons, and

$$r_b \sim m\gamma^2/c^2, \qquad W \sim c^4/m^3\gamma^4$$

for acoustic phonons. Since ST occurs on a microscopic scale $\sim a_0$, the continual theory can provide only an estimate for V_{ST}, the lowering of the energy at ST. To get this estimate, one should assume that $\psi(\boldsymbol{r})$ and $Q(\boldsymbol{r})$ are concentrated on a scale $\sim a_0$ and find the value of Q from the condition that the energy of the system should be minimal. Estimating the last term in eq. (2.21) as $\omega_q^2 Q_q^2/a_0^3$ and the last term in eq. (2.22) as $\gamma_q Q_q/a_0^3$, where $q \sim a_0^{-1}$, we find $Q_q \sim \gamma_q/\omega_q^2$ and $V_{ST} \sim \gamma_q^2/\omega_q^2 a_0^3$. Hence,

$$\Lambda = V_{ST}/E_B \sim m\gamma_q^2/a_0\omega_q^2,$$

and

$$\Lambda_{\mathrm{op}} \sim m\gamma_0^2/a_0\omega_0^2, \qquad \Lambda_{\mathrm{ac}} \sim m\gamma^2/a_0c^2.$$

It is easy to verify that in both the cases the relations (1.4) are fulfilled.

4.2. Nonpolar optical phonons

In this section we study ST in the simplest continuum model (Iordanskii and Rashba 1978, Ioselevich and Rashba 1985a). This model, incorporating one branch of dispersionless phonons ($\omega_{\boldsymbol{q}} \equiv \omega_0$) nonpolarly interacting with an electron ($\gamma_{\boldsymbol{q}} \equiv \gamma$), has already been described in section 2. In terms of the dimensionless variables, introduced there, eqs. (4.14) and (4.15) take the form

$$K(\boldsymbol{r}\tau) = \delta(\boldsymbol{r})D_0(\tau), \quad D_0(\tau) = -\tfrac{1}{2}[(1+N)\mathrm{e}^{-|\tau|} + N\mathrm{e}^{|\tau|}],$$
$$N = [\exp(\beta\omega_0) - 1]^{-1} \tag{4.20}$$

and

$$\mathscr{S}(\beta) = \min_{\psi} \int \mathrm{d}\boldsymbol{r} \left\{ \int_{-\omega_0\beta/2}^{\omega_0\beta/2} \mathrm{d}\tau \tfrac{1}{2}|\nabla\psi|^2 + \tfrac{1}{2} \iint_{-\omega_0\beta/2}^{\omega_0\beta/2} \mathrm{d}\tau\,\mathrm{d}\tau' D_0(\tau-\tau')|\psi(\boldsymbol{r}\tau)|^2|\psi(\boldsymbol{r}\tau')|^2 \right\}. \tag{4.21}$$

Thus, for finite T the infinite integration limits in the formulae of section 2 must be replaced by finite and the zero-temperature Green function (2.34) by the Green function (4.20). All the terms in eq. (4.21) are homogeneous with respect to spatial scale transformations. This makes it possible to obtain the virial theorem if one introduces a trial function

$$\chi = \kappa^{3/2}\psi(\kappa\boldsymbol{r}, \tau), \tag{4.22}$$

and to vary $\mathscr{S}$ with respect to κ

$$\left.\frac{\mathrm{d}\mathscr{S}[\chi]}{\mathrm{d}\kappa}\right|_{\kappa=1} = \int \mathrm{d}\boldsymbol{r} \int_{-\omega_0\beta/2}^{\omega_0\beta/2} \mathrm{d}\tau \left\{ \tfrac{1}{2}|\nabla\psi(\boldsymbol{r}\tau)|^2 + \tfrac{3}{4}|\psi(\boldsymbol{r}\tau)|^2 \int_{-\omega_0\beta/2}^{\omega_0\beta/2} D_0(\tau-\tau')|\psi(\boldsymbol{r}\tau')|^2\,\mathrm{d}\tau' \right\} = 0. \tag{4.23}$$

From eqs. (4.23), (4.21) and (4.15) ensues a useful relation,

$$\mathscr{S}(\beta) = -\int_{-\omega_0\beta/2}^{\omega_0\beta/2} \mathscr{E}(\tau)\,\mathrm{d}\tau = \frac{1}{3}\int_{-\omega_0\beta/2}^{\omega_0\beta/2} \mathrm{d}\tau \int \mathrm{d}\boldsymbol{r}\tfrac{1}{2}|\nabla\psi(\boldsymbol{r}\tau)|^2$$
$$= -\frac{1}{4}\iint_{-\omega_0\beta/2}^{\omega_0\beta/2} \mathrm{d}\tau\,\mathrm{d}\tau' D_0(\tau-\tau') \int \mathrm{d}\boldsymbol{r}|\psi(\boldsymbol{r}\tau)|^2|\psi(\boldsymbol{r}\tau')|^2, \tag{4.24}$$

connecting the action with the electron energy and also with its kinetic and potential parts individually. The virial theorem for the action is generalization of the virial theorem for the stationary state of a polaron (Pekar 1951).

The height of the barrier is determined by the lowest-lying saddle point of the function (4.18). In dimensionless units

$$W_0 = \operatorname*{stat}_{\psi} \int \mathrm{d}\boldsymbol{r} \{\tfrac{1}{2}|\nabla\psi(\boldsymbol{r})|^2 - \tfrac{1}{2}|\psi(\boldsymbol{r})|^4\} \approx 44. \tag{4.25}$$

This functional was studied by Mitskevich (1955), Ryder (1967) and Zakharov et al. (1972). A thorough investigation of the functional (4.21) has not so far been carried out. It is possible to calculate $\mathscr{S}_{\mathrm{I}}(\beta)$ approximately, employing the trial functional (2.37). Its insertion into eq. (4.21) leads to the functional

$$\begin{aligned} \mathscr{S}(\beta, \tau_{\mathrm{I}}) &= \operatorname*{stat}_{\psi} \int \mathrm{d}\boldsymbol{r} \{2\tau_{\mathrm{I}}\tfrac{1}{2}|\nabla\psi(\boldsymbol{r})|^2 - f(\tau_{\mathrm{I}})\tfrac{1}{2}|\psi(\boldsymbol{r})|^4\} \\ &= W_0(2\tau_{\mathrm{I}})^3/f^2(\tau_{\mathrm{I}}), \end{aligned} \tag{4.26}$$

where

$$f(\tau_{\mathrm{I}}) = 2\tau_{\mathrm{I}} + N\mathrm{e}^{2\tau_{\mathrm{I}}} + (N+1)\mathrm{e}^{-2\tau_{\mathrm{I}}} - (2N+1). \tag{4.27}$$

The second equality in eq. (4.26) is obtained by means of the scale transformation (4.22) and of eq. (4.25). A verification reveals that at the points $\tau = \frac{1}{2}\beta$ and $\tau = 0$, where the system comes in under the barrier and goes out from under the barrier, the energy of the system takes identical values and its velocity equals zero. Fulfillment of these conditions, evident for the exact solution, requires some control when these solutions are obtained by means of the trial function.

The τ_{I}-dependence of $\mathscr{S}(\beta, \tau_{\mathrm{I}})$ at different values of $\omega_0\beta$ is plotted in fig. 13a. The values of $\mathscr{S}$ must be compared in three extrema: minimum I, maximum I′ (instantons) and edge extremum A (static solution). In fig. 13b the function $S(\beta)$ for all the three extrema is given. Unfortunately, unlike fig. 5b, the solution I′ in fig. 13b does not adjoin the line A and goes into infinity, which happens due to the insufficient flexibility of the trial function (2.37). Improvement of the trial function must bring about alteration of the shape of the instanton curve in fig. 13b in such a way that it becomes tangent to the activation curve. The two ways in which these curves merge are shown in fig. 5. The existence of the two instanton solutions apparently testifies to the nonmonotonous ε-dependence of τ_{t} (figs. 4 and 5). However, we cannot make a final choice between these alternatives before more exact calculations are carried out.

In the low-temperature region the approximation (2.37) is satisfactory. It reproduces eq. (2.13) with the coefficients

$$S_{\mathrm{I}}(\beta = \infty) = 6.2\, W_0\sigma_{\mathrm{op}}, \quad W_0 = 44, \qquad b = 10.3. \tag{4.28}$$

The instanton duration $\tau_{\mathrm{I}}(\beta = \infty) = 1.1\,\omega_0^{-1}$. The point of the regime switch-

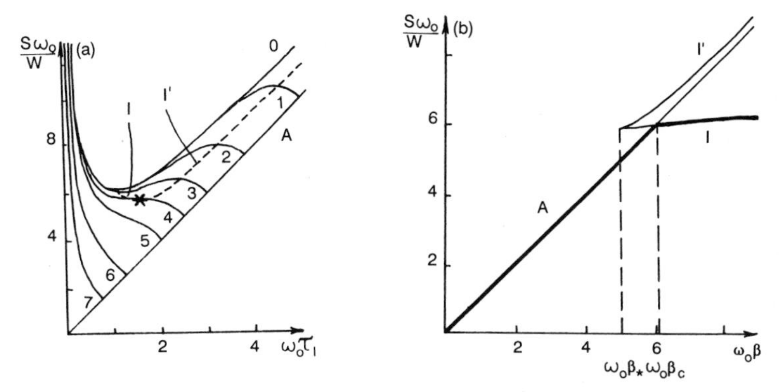

Fig. 13. (a) The dependence of the action (4.21), calculated with the trial function (2.37), on the variational parameter τ_I – the instanton duration (nonpolar coupling to optical phonons). The parameter of the curves is the value of $\omega_0\beta$: ∞ (curve 0); 10 (curve 1); 7.5 (curve 2); 6 (curve 3); 5 (curve 4); 4 (curve 5); 2.5 (curve 6); 1.5 (curve 7). Curve A represents the static solution, and the minima line I and maxima line I′ the instantons. (b) The dependence of the action on the inverse temperature β.

over $\beta_c \approx 6\omega_0^{-1}$ lies in the region $\beta\omega_0 \gg 1$; that is why eq. (2.13) is valid practically in the entire instanton region.

The assertion concerning the exponential behavior of the temperature corrections to the action at $T \neq 0$, discussed in section 2.1 for a single-mode model, holds also for multimode systems if only phonon frequencies are limited from below (i.e., only optical branches are considered). In the model under study one can get an exact bound on the value of the coefficient b. Treating eq. (4.21) as the functional of the variables ψ and of the occupation numbers N, and making use of the fact that $\mathscr{S}$ is extremal with respect to ψ, it is easy to obtain $\mathrm{d}\mathscr{S}/\mathrm{d}N = \partial\mathscr{S}/\partial N$. Hence, in the first order in $N \approx \exp(-\beta\omega_0)$:

$$\mathscr{S}(\beta) \approx \mathscr{S}(\infty) - \tfrac{1}{2}\exp(-\beta\omega_0)\int \mathrm{d}\boldsymbol{r}\left[\int_{-\tau_I}^{\tau_I} |\psi_0(\boldsymbol{r}\tau)|^2 \cosh\tau \,\mathrm{d}\tau\right]^2, \tag{4.29}$$

where $\psi_0(\boldsymbol{r}\tau)$ is the exact extremal of $\mathscr{S}(\infty)$. The last term in eq. (4.29) gives an exact expression for b, but to calculate it, it is necessary to know $\psi_0(\boldsymbol{r}\tau)$. An exact estimate for b can be obtained by means of the virial theorem (4.24)

$$\begin{aligned} 4\mathscr{S}(\infty) &= \frac{1}{2}\iint_{-\infty}^{\infty} \mathrm{d}\tau\,\mathrm{d}\tau' \int \mathrm{d}\boldsymbol{r}\,|\psi_0(\boldsymbol{r}\tau)\psi_0(\boldsymbol{r}\tau')|^2\,\mathrm{e}^{-|\tau-\tau'|} \\ &< \frac{1}{2}\iint_{-\infty}^{\infty} \mathrm{d}\tau\,\mathrm{d}\tau' \int \mathrm{d}\boldsymbol{r}\,|\psi_0(\boldsymbol{r}\tau)\psi_0(\boldsymbol{r}\tau')|^2 \cosh(\tau-\tau') \\ &= \frac{1}{2}\int \mathrm{d}\boldsymbol{r}\left(\int \mathrm{d}\tau\,|\psi_0(\boldsymbol{r}\tau)|^2 \cosh\tau\right)^2. \end{aligned} \tag{4.30}$$

This derivation employs the obvious inequality $e^{-x} < \cosh x$. A comparison of eqs. (4.29), (4.30) and (2.13) yields

$$b > 4. \tag{4.31}$$

After the system goes out from under the barrier (section 3 of contour Γ_+), the ST process enters the stage of fast contraction of the ψ-function (in a time $\sim\omega_0^{-1}$) down to the atomic spatial scale, and of the deformational well deepening. This stage is analyzed in section 8.

The physical pattern of the ST process in accordance with the instanton mechanism studied above is as follows (see fig. 14). First the lattice freely vibrates. The energy of thermal vibrations concentrated in the region of fluctuation and involved in the ST process is $\sim W \exp(-\beta\omega_0)$ [cf. eq. (2.11)]. Then at a certain moment when lattice displacements are large and velocities are zero, the lattice starts tunneling: in the imaginary time displacements increase exponentially. By the moment τ_I, they become so large that in the potential well these deformations produce there emerges a local level on which the electron is trapped (the trapping probability is calculated in section 2.3). Then the lattice tunnels together with the trapped electron, and at the moment $\tau = 0$ goes out from under the barrier, and then in a time of order ω_0^{-1} the electron cloud contracts down to an atomic scale (see section 8).

4.3. Acoustic phonons

In this section we shall study the interaction of longitudinal acoustic phonons with an electron via a deformational potential:

$$\omega_q = cq, \qquad \gamma_q = \gamma q. \tag{4.32}$$

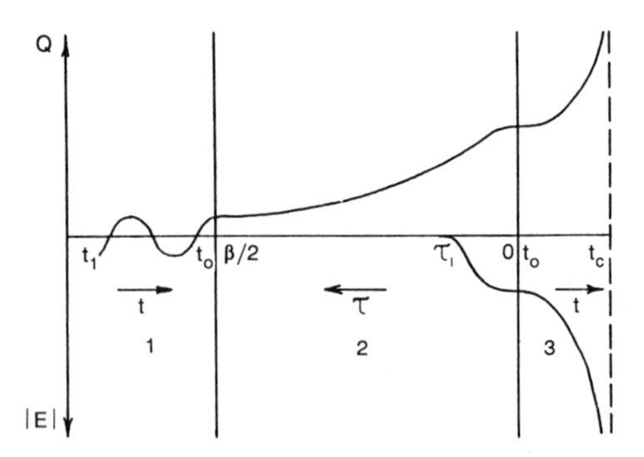

Fig. 14. The time evolution of the deformation $Q(t)$ and of the electron energy $E(t)$ for an instanton at low temperatures (nonpolar coupling to optical phonons). Contour Γ_+ (fig. 8a) is straightened along the horizontal axis. The arrows indicate the direction of time (t and τ) on the corresponding sections.

This model works both for small-radius excitons and for charge carriers. Introducing dimensionless variables

$$\boldsymbol{r} \equiv r_{\mathrm{b}}\boldsymbol{r}', \qquad \tau \equiv \omega_{\mathrm{b}}^{-1}\tau', \qquad \beta \equiv \omega_{\mathrm{b}}^{-1}\beta', \qquad S \equiv \sigma_{\mathrm{ac}}\mathscr{S},$$
$$r_{\mathrm{b}} = m\gamma^2/c^2, \qquad \omega_{\mathrm{b}} = c^3/m\gamma^2, \qquad \sigma_{\mathrm{ac}} = (mr_{\mathrm{b}}^2\omega_{\mathrm{b}})^{-1}, \tag{4.33}$$

and omitting hereafter the primes we get from eq. (4.14)

$$K(\boldsymbol{r}\tau) = \int \frac{\mathrm{d}\boldsymbol{q}}{(2\pi)^3}\, \mathrm{e}^{\mathrm{i}\boldsymbol{q}\cdot\boldsymbol{r}} q[(N_{\boldsymbol{q}} + 1)\mathrm{e}^{-q|\tau|} + N_{\boldsymbol{q}}\mathrm{e}^{q|\tau|}]. \tag{4.34}$$

The height of the barrier, obtained from eq. (4.18), like for optical phonons, reduces to the functional (4.25):

$$W = W_0 c^4/m^3\gamma^4 \approx 44/mr_{\mathrm{b}}^2, \tag{4.35}$$

that is why for the static solution one has

$$S_{\mathrm{A}} \approx 44/mr_{\mathrm{b}}^2 T. \tag{4.36}$$

The spatial scale of the barrier $r_{\mathrm{b}} = m\gamma^2/c^2 \approx \Lambda a_0$ has a macroscopic size, if $\Lambda \gg 1$. Fulfillment of this criterion guarantees the applicability of the continuum approximation for the static solution (the activational regime). However, the spatial (r_{I}) and time (τ_{I}) scales of the instanton in the model (4.32) are equal to zero (Iordanskii and Rashba 1978), this result follows from the virial theorem. In fact, at $T = 0$ all terms in $\mathscr{S}$ are homogeneous with respect to space–time scaling transformations [cf. eq. (4.22), where only spatial transformations were involved]. Introducing a trial function

$$\chi = \kappa^{3/2}\psi(\kappa\boldsymbol{r}, \kappa\tau) \tag{4.37}$$

and performing variation of $\mathscr{S}$ with respect to κ, we get

$$\begin{aligned}\left.\frac{\mathrm{d}\mathscr{S}[\chi]}{\mathrm{d}\kappa}\right|_{\kappa=1} &= \int_{-\infty}^{\infty} \mathrm{d}\tau \int \mathrm{d}\boldsymbol{r}\, \tfrac{1}{2}|\nabla\psi(\boldsymbol{r}\tau)|^2 \\ &\quad + \iint_{-\infty}^{\infty} \mathrm{d}\tau\, \mathrm{d}\tau' \iint \mathrm{d}\boldsymbol{r}\, \mathrm{d}\boldsymbol{r}'\, K(\tau - \tau', \boldsymbol{r} - \boldsymbol{r}')|\psi(\boldsymbol{r}\tau)|^2 |\psi(\boldsymbol{r}'\tau')|^2 \\ &= \int_{-\infty}^{\infty} \mathrm{d}\tau\, \mathscr{E}(\tau) = 0. \end{aligned} \tag{4.38}$$

Since the energy of the bound state of the electron cannot be positive ($\mathscr{E} \leqslant 0$), it follows from eq. (4.38) that $\mathscr{E} \equiv 0$, i.e., that the electron is free during the entire tunneling time, $\tau_{\mathrm{I}} = 0$. The electron is bound during only an infinitely short time near the point $\tau = 0$. In what follows, it will be shown [cf. eq. (4.50)] that in this case simultaneously the instanton radius $r_{\mathrm{I}} \to 0$. It means that in the framework of the standard continuum model (4.32) instanton solutions are singular, i.e., the theory loses its self-consistency. This happens despite the fact that the scale of barrier states at $\Lambda \gg 1$ remains to be macroscopic. The self-consistency is restored only beyond the framework of the standard model.

Incorporation of various corrections (higher-order terms in the expansion of γ_q and ω_q in q and corrections to the effective mass method), violating the virial theorem, results in finite values of $\tau_I \ll 1$ and $r_I \ll 1$. Depending on the sign and value of the corrections, there are two cases:

(1) the corrections reduce the values of r_I and τ_I, then the minimum of the action is attained on a microscopic scale $r_I \sim a_0$ (fig. 15a), and

(2) the corrections restrict the decrease of r_I (fig. 15b), in this case

$$a_0 \ll r_I \ll 1. \tag{4.39}$$

[Recall that according to eq. (4.33) the length is scaled in units of r_b.] In the first case the instanton solution cannot be described in terms of the continuum approximation, which provides only the upper bound for the action. In the second case, which will be studied here (Ioselevich and Rashba 1985a) the continuum approximation holds. It should be stressed, however, that for acoustic phonons, in contrast to optical phonons (section 4.2), at $\Lambda \gg 1$ the size of the instanton r_I is small compared to r_b, the size of the barrier to ST.

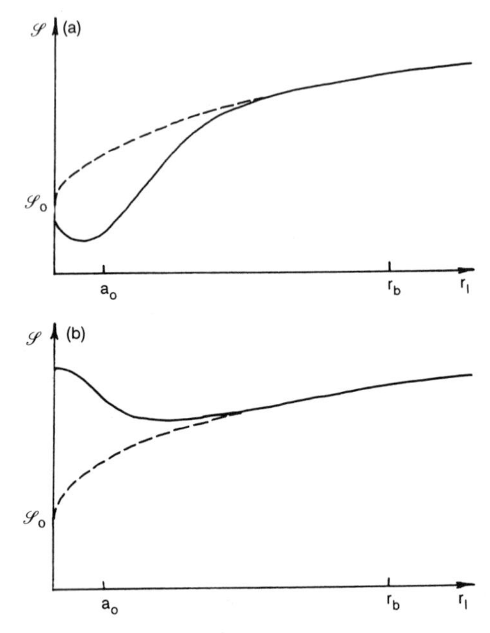

Fig. 15. Possible behavior of the action $\mathscr{S}$ as a function of the instanton size r_I for coupling to acoustic phonons. The dashed line indicates the action calculated in the framework of the standard continuum model. The minimum of the action $\mathscr{S}_0$ is attained at $r_I = 0$. The solid curve is the action including deviations in the region of small spatial scales: (a) the minimum of $\mathscr{S}$ is attained at $r_I \sim a_0$; in this case min $\mathscr{S} < \mathscr{S}_0$; (b) min $\mathscr{S} \gtrsim \mathscr{S}_0$ and is attained at $a_0 \ll r_I \ll r_b$.

It will be our assumption that the spatial scale of the instanton r_I is fixed and obeys the inequality (4.39). Then the form of the instanton and its duration can be expressed in terms of r_I. In particular, it will be shown [cf. eq. (4.50)] that $\tau_\mathrm{I} \sim r_\mathrm{I}^{3/2}$. Since the characteristic momenta q of the phonons, forming the instanton, are of order r_I^{-1}, their frequencies, determining the characteristic time of the change of the well, are $\omega_\mathrm{I} \sim r_\mathrm{I}^{-1}$. Therefore,

$$\omega_\mathrm{I}\tau_\mathrm{I} \sim q\tau_\mathrm{I} \sim r_\mathrm{I}^{1/2} \ll 1 \tag{4.40}$$

[cf. eq. (4.39)]. This inequality shows that for the time τ_I, the discrete level exists, the potential well changes only slightly. On the other hand, close to the moment when the level emerges, this level is very shallow. Therefore, it remains to be shallow during the entire time τ_I. The latter circumstance makes it possible to simplify the problem considerably. The wave function of the shallow level is

$$\begin{aligned} \psi(\boldsymbol{r}\tau) &= r_\mathrm{I}^{-1}\kappa^{1/2}(\tau)\,\mathrm{e}^{-r\kappa(\tau)}\chi(r/r_\mathrm{I}), \\ \chi(\rho) &\approx (2\pi)^{-1/2}\rho^{-1} \quad \text{at } \rho \gg 1. \end{aligned} \tag{4.41}$$

Here

$$\kappa(\tau) = [-2\mathscr{E}(\tau)]^{1/2}, \tag{4.42}$$

and $r_\mathrm{I}\kappa(\tau) \ll 1$ for all relevant values of τ. Employing the inequality (4.40) one can put in eq. (4.34) $\mathrm{e}^{q|\tau|} \approx 1$. Then at $T = 0$, substituting eqs. (4.41) and (4.34) into eq. (4.13), we get

$$\mathscr{S}_\mathrm{I} = Ay - \tfrac{1}{4}B_0 y^2, \quad y = \int_{-\tau_\mathrm{I}}^{\tau_\mathrm{I}} \mathrm{d}\tau\, \kappa(\tau)/r_\mathrm{I}, \tag{4.43}$$

where

$$A = \int \mathrm{d}\boldsymbol{\rho}\, \tfrac{1}{2}|\nabla\chi(\boldsymbol{\rho})|^2, \qquad B_n = \int \frac{\mathrm{d}\boldsymbol{k}}{(2\pi)^3}\, k^{n+1} \left| \int \mathrm{d}\boldsymbol{\rho}\, \mathrm{e}^{\mathrm{i}\boldsymbol{k}\cdot\boldsymbol{\rho}}\chi^2(\boldsymbol{\rho}) \right|^2. \tag{4.44}$$

Deriving eq. (4.44) we have taken into account that the integrals converge at $\rho \sim 1$; that is why in eq. (4.41) one can put $\exp(-r\kappa) \approx 1$.

Requiring that $\mathscr{S}$ be extremal, we get

$$y = 2A/B_0, \qquad \mathscr{S}_0 = A^2/B_0. \tag{4.45}$$

The action $\mathscr{S}$ reaches its stationary value $\mathscr{S}_0$ on the class of functions $\chi(\boldsymbol{\rho})$ with the asymptotics (4.41). For a trial function

$$\chi_0(\boldsymbol{\rho}) = [2\pi(1 + \rho^2)]^{-1/2},$$

it is easy to obtain

$$A = 3\pi/6, \qquad B_n = (n + 1)!/2^{(n+3)}, \qquad \mathscr{S}_0 = 9\pi^2/32 \approx 2.8. \tag{4.46}$$

Thus, in the lowest-order approximation in r_{I} the extremal action is r_{I}-independent. It is entirely determined by the parameters of the standard model

$$S_{\mathrm{I}}(\beta=\infty) \approx 2.8c/m^2\gamma^2 = 2.8\, c\rho/m^2 C^2. \tag{4.47}$$

In the last expression the parameter γ is expressed via the density ρ and the deformational potential C.

To find the instanton duration, it is necessary to perform an expansion of $\mathscr{S}$ in the higher order in r_{I}, retaining in eq. (4.34) the $q|\tau|$-linear term. Taking the Schrödinger equation (4.15), multiplying it by ψ^* and performing integration over the volume with the use of eqs. (4.41) and (4.44), we get

$$\frac{A}{r_{\mathrm{I}}}\kappa(\tau) - \frac{1}{2}\frac{\kappa(\tau)}{r_{\mathrm{I}}^2}\int_{-\tau_{\mathrm{I}}}^{\tau_{\mathrm{I}}} \mathrm{d}\tau'\,\kappa(\tau')\left[B_0 - \frac{B_1}{r_{\mathrm{I}}}|\tau-\tau'|\right] = \mathscr{E}(\tau) \tag{4.48}$$

This, with eq. (4.42) taken into account, leads to an integral equation for $\kappa(\tau)$. Its solution is

$$\kappa(\tau) = \kappa\theta(\tau_{\mathrm{I}} - |\tau|)\cos(\pi\tau/2\tau_{\mathrm{I}}), \tag{4.49}$$

where

$$\tau_{\mathrm{I}} = \pi r_{\mathrm{I}}^{3/2}/2(2B_1)^{1/2}, \qquad \kappa = A(2B_1)^{1/2}/B_0 r_{\mathrm{I}}^{1/2}. \tag{4.50}$$

The correction $\mathscr{S}_1$ to the action $\mathscr{S}_{\mathrm{I}} = \mathscr{S}_0 + \mathscr{S}_1$ is equal to

$$\mathscr{S}_1 = \frac{\pi}{8}(2B_1)^{1/2}(A/B_0)^2 r_{\mathrm{I}}^{1/2}. \tag{4.51}$$

It follows from eqs. (4.50) and (4.51) that in the framework of the model (4.32) the spatial scale of instanton r_{I} goes to zero simultaneously with its time scale τ_{I}; the minimum of the action $\mathscr{S}_{\mathrm{I}}(r_{\mathrm{I}})$ is attained at $r_{\mathrm{I}} = 0$.

The parameter r_{I}, which remains unknown, must be determined from the competition of the contribution $\mathscr{S}_1$ to the action and of the terms arising due to deviations from the standard model (4.32) and from the effective mass method. In the case, illustrated in fig. 15b, the minimum of $\mathscr{S}_{\mathrm{I}}$ corresponds to r_{I}, fulfilling the condition (4.39); in this case an estimate

$$r_{\mathrm{I}}/a_0 \sim (r_{\mathrm{b}}/a_0)^{1/5} \sim \Lambda^{1/5} \tag{4.52}$$

is valid. Here dimensional units have been used. Thus, the condition $\Lambda \gg 1$ makes the macroscopic approximation applicable for determining the size of the instanton. Yet, actually, the criterion is very stringent.

The small instanton duration τ_{I} at $T=0$, which entails from eq. (4.50), means that tunneling occurs in the part of APS (cf. fig. 1b), where the back side of the potential barrier is almost shear. The potential curve, modelling this situation, is depicted in fig. 16. It consists of two parabolas, tangent to each other at the point br. In the region of free motion, the curvature of the parabola is determined by the frequency of phonons, forming the instanton $\omega_0 \sim \omega_{\mathrm{I}} \sim \omega_{\mathrm{b}}(r_{\mathrm{b}}/r_{\mathrm{I}})$, whereas

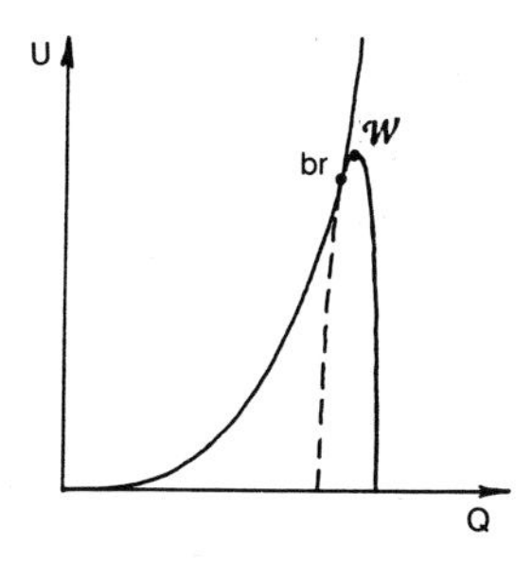

Fig. 16. A model curve of the adiabatic potential with a steep barrier, consisting of two tangent parabolas with different curvatures $\omega_0 \ll \omega_W = \pi/\tau_I$. This model describes the shallow-level situation. It corresponds to the instanton in the following systems: an electron + acoustic phonons (continuum approximation) and Wannier–Mott exciton + polar optical phonons (section 4.4).

in the region of loaded motion the frequency is determined by the instanton duration $\omega_W = \pi/\tau_I \sim \omega_b(r_b/r_I)^{3/2} \gg \omega_0$.

By analogy with section 4.2, one can find the low-temperature correction to the action, expanding $\mathscr{S}$ in N_q:

$$\mathscr{S}(\beta) \approx \mathscr{S}(\infty) - \frac{1}{2}\iint_{-\tau_I}^{\tau_I} \mathrm{d}\tau\, \mathrm{d}\tau' \int \frac{\mathrm{d}\boldsymbol{q}}{(2\pi)^3} qN_q \left| \int \mathrm{d}\boldsymbol{r} |\psi(\boldsymbol{r}\tau)|^2 \,\mathrm{e}^{\mathrm{i}\boldsymbol{q}\cdot\boldsymbol{r}} \right|^2, \tag{4.53}$$

where ψ is determined by eq. (4.41). At $T \to 0$, in contrast to the case of optical phonons, N_q are not small at $q < T$; nevertheless, the temperature-dependent contribution to the action is small because the corresponding phase volume is small. This makes it possible to calculate this contribution in the perturbation theory. In the integral (4.53) the characteristic momenta $q \sim \beta^{-1}$. At $\beta > \kappa^{-1}$ (i.e., at $T < \omega_b(r_b/r_I)^{1/2}$ in dimensional units) the integral over $\boldsymbol{r}$ in eq. (4.53) reduces to the normalization integral, and

$$S(\beta) \approx S(\infty) - (\pi^2/15)(\tau_I^2 C^2/\rho c^5)T^4. \tag{4.54}$$

Thus, the temperature-dependent correction to S due to acoustic phonons is explicitly expressed via the instanton duration τ_I and is T^4-proportional. In section 4.5 it will be shown that the latter statement is of a general character. The inverse inequality $[T > \omega_b(r_b/r_I)^{1/2}]$ in the instanton regime can be satisfied only at extremely high $\Lambda \gtrsim 10^5$.

The temperature dependence of τ_I arises only due to the terms $\sim N_q(q\tau)^2$ in the integrand in eq. (4.34); at $T < \omega_b(\omega_b\tau_I)^{-1/6}$ the dependence $\tau_I(T)$ is weak since τ_I is small. In this temperature range the second term in eq. (4.54) is also small; that is why the curves $S_A(\beta)$ and $S_I(\beta)$ intersect each other (see fig. 5b); in this case $T_c \approx W/S(\infty) \sim \omega_b$. A considerable increase of τ_I becomes noticeable only at $T > T_c$ in the region of "a beak"; in this case $\tau_I \sim \omega_b^{-1}$ on line I′ near the point where it is tangent to line A.

To formulate the quantitative criterion of the validity of the theory, it is important to define the coupling constant in the most adequate way. The definition (1.3) of the coupling constant Λ is based on V_{ST}, i.e., on the energy of the system in the region of strong and, consequently, nonlinear deformation. In the theory of processes proceeding in the region of a barrier to ST, it is natural to base oneself not on V_{ST} but on W, another experimentally measured parameter, and to define the coupling constant by means of a formula

$$W = E_B/\Lambda_b^2. \tag{4.55}$$

Both the definitions of the coupling constant [eqs. (1.3) and (4.55)] are parametrically equivalent but the numerical difference of Λ from Λ_b may prove to be considerable. It follows from eq. (4.35) and from the estimate $E_B \approx (ma_0^2)^{-1}$ that

$$\Lambda_b \approx 0.15 r_b/a_0. \tag{4.56}$$

For rare gas solids typical values of the parameters $W \sim 5\times10^{-3}$ eV and $E_b \sim 0.5$ eV lead to $\Lambda_b \sim 10$. If one defines T_c approximately from the condition $S_A(\beta_c) \approx S_I(\infty)$, then from eqs. (4.35), (4.47) and (4.56) if follows that

$$T_c \approx 16\omega_b \approx 2.4\Lambda_b^{-1}\omega_D, \quad \omega_D \approx c/a_0. \tag{4.57}$$

At $\Lambda_b \approx 10$ we get about the same value of the ratio $T_c/\omega_D \approx 0.2$, which has been found in section 4.2 for optical phonons. One obtains a close value $(2\pi)^{-1}$ for extrinsic ST by a single-mode center, if one puts $\omega_W \approx \omega_D$ (fig. 5a). Apparently a small numerical value of the ratio T_c/ω_D must be observed at realistic values of Λ both for optical and acoustic phonons. It is noteworthy, however, that for acoustic phonons this ratio strongly depends on the value of the coupling constant Λ.

The estimates show that the continual theory of section 4.3 is applicable at $\Lambda_b \gtrsim 10$.

4.4. Wannier–Mott exciton

The Wannier–Mott exciton interacting with polar optical phonons is unique in the sense that this is the only model for which ST states and a barrier to ST can be described in terms of the continual theory.

For this exciton, the scale of the ST state, unlike for an electron (or Frenkel exciton), is larger than atomic. The ST state does exist in the framework of the continual theory and is the ground state of the system if there is a strong inequality of the effective masses of an electron and of a hole:

$$m_e/m_h \ll 1. \tag{4.58}$$

The critical value of the mass ratio depends on the value of ε_0 and ε_∞, low- and high-frequency dielectric permeabilities. Usually, $(m_e/m_h)_{cr} \sim 0.1$ (Dykman and Pekar 1952).

Due to the electric neutrality of the exciton, its interaction with long-wavelength polar phonons with wave vectors $\boldsymbol{q}$ such that $qa_{ex} \ll 1$ (here $a_{ex} = \mu e^2/\hbar^2\varepsilon_\infty$, $\mu^{-1} = m_e^{-1} + m_h^{-1}$, μ being the reduced mass) is strongly suppressed. As a result, ST, if it does take place, is always associated with passage over the barrier to ST, unlike a large polaron, which is self-trapped through monotonous energy decrease. The dependence of the height W of the barrier to ST and of the depth V_{ST} of the ST state on the parameters of the system is shown in fig. 17, which gives the results of the variational calculation performed in the framework of the adiabatic theory (Pekar et al. 1979). It is clear that ST states exist only in the region of small values of m_e/m_h, since at the critical value of the mass ratio $(m_e/m_h)_{cr}$ the saddle point on APS merges with the local minimum on APS, corresponding to a metastable ST state. This occurs at the point where the curves a and b adjoin each other. At $(m_e/m_h) > (m_e/m_h)_{cr}$ ST states do not exist. The ST state is the ground state of the system if its total energy is $J < 0$.

The Wannier–Mott exciton consists of two particles, and that is why has one internal degree of freedom. Since the region of small values of $m_e/m_h \ll 1$ is important for ST, in the problem there are several length scales. One of them, r_{ex}, is the Bohr radius of the exciton. The second is related with the curves b (fig. 17). At $m_e/m_h \to 0$, the ST state is a hole polaron with an energy $J \approx -0.05\, m_h e^4/\hbar^2\bar{\varepsilon}^2$, $\bar{\varepsilon}^{-1} = \varepsilon_0^{-1} + \varepsilon_\infty^{-1}$ and a radius $r_h \approx 3\hbar^2\bar{\varepsilon}/m_h e^2$ plus an electron weakly bound to this polaron. Thus, there arises the smallest scale, $r_h \ll r_{ex}$, and the large ST energy corresponding to it. A manifestation of this energy scale is the steep slope of the curves b. Contrarily, the small value of $W \ll E_{ex}$ and the superlinear behavior of W as a function of m_e/m_h (fig. 17), point

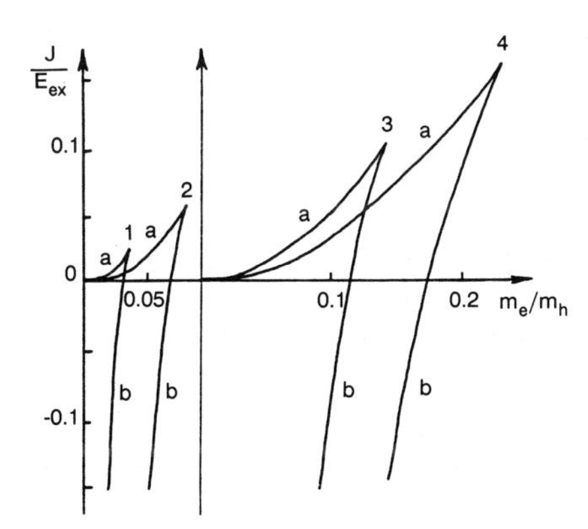

Fig. 17. The total energy J of the system Wannier–Mott exciton + polar lattice as a function of the mass ratio m_e/m_h found by the variational calculation (Pekar et al. 1979). The energy J is referred to from the free exciton level in a rigid lattice and is expressed in units of the exciton binding energy, $E_{ex} = \mu e^4/2\hbar^2\varepsilon_\infty^2$. The figures near the curves are the quantities $4(1 - \varepsilon_\infty/\varepsilon_0)$. Curves a are for the barrier to ST ($J = W$), and the curves b for the ST state ($J = -V_{ST}$).

to (i) the existence of one more, third-largest energy scale for the curves a, and (ii) a qualitative difference between the states corresponding to the curves a and b in the region $m_e/m_h \to 0$. Kusmartsev and Meshkov (1983) have proved that the barrier state corresponds to a shallow level with a potential well radius $\sim r_{ex}$ and with a radius of the quantum state $r_b \sim r_{ex}(m_h/m_e)(\varepsilon_\infty/\bar{\varepsilon}) \gg r_{ex}$. In this state due to $W \ll E_{ex}$ the exciton can be treated as an undeformable particle (the rigid-exciton approximation).

In the theory of tunneling (Kusmartsev and Meshkov 1983) and of thermo-activated tunneling (Ioselevich and Rashba 1985a) it is handy to use dimensionless variables

$$\boldsymbol{r} \equiv r_{ex}\boldsymbol{r}', \qquad \tau \equiv \omega_0^{-1}\tau', \qquad S \equiv \sigma_{ex}\mathcal{S}, \quad \sigma_{ex} = (\varepsilon_0 - \varepsilon_\infty)/m_e\omega_0 r_{ex}^2 \tag{4.59}$$

ω_0 is the phonon frequency; the primes hereafter will be omitted. Everywhere it will be assumed that $m_e \ll m_h$, so only the leading term in the parameter $m_e/m_h \ll 1$ will be retained. If the energy is referred to from $-E_{ex}$, i.e., from the exciton level in a rigid lattice, the Lagrangian is

$$\mathcal{L}[\varphi] = \frac{1}{8\pi}\int \mathrm{d}\boldsymbol{r}[(\partial_\tau \boldsymbol{\nabla}\varphi)^2 + (\boldsymbol{\nabla}\varphi)^2] + \min_{\psi}\int \mathrm{d}\boldsymbol{r}\left[\frac{(\boldsymbol{\nabla}\psi)^2}{2m^*} + \varphi\hat{\rho}|\psi|^2\right], \tag{4.60a}$$

where

$$m^* = (m_h/m_e)(\varepsilon_0 - \varepsilon_\infty)/\varepsilon_0 \gg 1 \tag{4.60b}$$

is a large dimensionless parameter of the theory. The variable $\varphi(\boldsymbol{r}\tau)$ is the electrostatic potential of the lattice, interacting with the electric charge density

$$\rho(\boldsymbol{r}\tau) = \hat{\rho}|\psi|^2 \equiv \int \mathrm{d}\boldsymbol{r}'\{\delta(\boldsymbol{r}-\boldsymbol{r}') - \chi^2(\boldsymbol{r}-\boldsymbol{r}')\}|\psi(\boldsymbol{r}'\tau)|^2, \tag{4.61}$$

where $\psi(\boldsymbol{r}\tau)$ is the wave function of the center of mass of the exciton, which coincides with r_h (i.e., coordinate of the hole) if the conditions (4.58) are fulfilled, and

$$\chi(\rho) = \pi^{-1/2}\mathrm{e}^{-\rho} \tag{4.62}$$

is the wave function of the internal motion in the exciton. In our approximation the Wannier–Mott exciton interacting with polar optical phonons is described by eqs. (4.13)–(4.18), with

$$m = m^*, \qquad \omega_{\boldsymbol{q}} = 1, \qquad \gamma_{\boldsymbol{q}} = \frac{\sqrt{4\pi}}{q}\{1 - [1 + (q/2)^2]^{-2}\}. \tag{4.63}$$

Equations (4.60)–(4.63) give rise to some conclusions concerning the character of barrier states $\psi(\boldsymbol{r})$ and of instanton extremals $\psi(\boldsymbol{r}\tau)$. If to suppose that ψ slowly varies on the scale r_{ex}, then $\hat{\rho}|\psi|^2 \approx -\frac{1}{2}\Delta(\psi^2)$, as follows from eq.

(4.61) or (4.63). But then the last term in eq. (4.60a) is reduced to the form $\frac{1}{2}\int \mathrm{d}\boldsymbol{r}[\nabla(\psi^2)]^2 > 0$, i.e., the nonlinear term corresponds to repulsion, and the bound state is absent. This contradiction testifies to the fact that the state ψ cannot be a single-scale state, and it is natural to assume that the attractive potential is ensured by the presence of a short-scale part in ψ (with the scale r_{ex}). This structure of the wave function corresponds to a shallow level, and this proves to be correct both for the barrier state $\psi(\boldsymbol{r})$ and for the instanton $\psi(\boldsymbol{r}\tau)$. Here there is a serious distinction from an electron coupled to acoustic phonons, where the level is shallow only for the instanton (cf. section 4.3) and the barrier state is a single-scale state.

For a shallow level the bound state exists during a very short period of time $\tau_I \ll 1$, for which the potential well does not change considerably. Following the arguments of section 4.3, we can represent the wave function of the exciton as [cf. eqs. (4.41) and (4.42)]:

$$\psi(\boldsymbol{r}\tau) = (\kappa(\tau))^{1/2}\exp(-r\kappa(\tau))\chi(\boldsymbol{r}),$$

$$\chi(\boldsymbol{r}) \approx (2\pi)^{-1/2} r^{-1} \quad \text{for } r \gg 1, \tag{4.64}$$

where

$$\kappa(\tau) = (-2m^* \mathscr{E}(\tau))^{1/2}. \tag{4.65}$$

If, similar to eqs. (4.43) and (4.44) we introduce the quantities

$$A = \int \mathrm{d}\boldsymbol{r}\tfrac{1}{2}|\nabla\chi(\boldsymbol{r})|^2, \qquad B_0 = \iint \mathrm{d}\boldsymbol{r}\,\mathrm{d}\boldsymbol{r}'\,k(\boldsymbol{r}-\boldsymbol{r}')\chi^2(\boldsymbol{r})\chi^2(\boldsymbol{r}'), \tag{4.66}$$

and put $\exp(-|\tau-\tau'|) \approx 1$ in $D_0(\tau-\tau')$, then

$$\mathscr{S} = \frac{Ay}{m^*} - \frac{1}{4}B_0(1+2N)y^2, \quad y = \int_{-\tau_I}^{\tau_I} \kappa(\tau)\,\mathrm{d}\tau. \tag{4.67}$$

The extremality condition for $\mathscr{S}$ leads to [cf. eqs. (4.45) and (4.46)]:

$$\mathscr{S}(\beta) = \mathscr{S}_0/(1+2N)m^{*2} = (\mathscr{S}_0/m^{*2})\tanh(\omega_0\beta/2), \quad \mathscr{S}_0 = A^2/B \approx 2.14. \tag{4.68}$$

By analogy with section 4.3, the function $\kappa(\tau)$ can be found from the Schrödinger equation (4.15) if in the expansion of $D_0(\tau)$ the $|\tau|$-linear term is retained:

$$E(\tau) = \frac{A\kappa(\tau)}{m^*} - \frac{\kappa(\tau)}{2}B_0\int_{-\tau_I}^{\tau_I} \mathrm{d}\tau'\kappa(\tau')(1+2N-|\tau-\tau'|). \tag{4.69}$$

Solving this equation together with eq. (4.65), we get a solution in the form of eq. (4.49) with

$$\tau_I = \pi/2(2m^*B_0)^{1/2}, \qquad \kappa = [A/(2N+1)](2/m^*B_0)^{1/2}. \tag{4.70}$$

The fact that the quantity $\tau_{\mathrm{I}} \sim (m^*)^{-1/2} \ll 1$ is small confirms the assumption that the level is shallow.

Analogously, one can find the height of the barrier W; in this case in the barrier state the level is also shallow. Substitution of the wave function in the form of eq. (4.64) with $\kappa = \kappa_{\mathrm{b}}$, independent of time, into eq. (4.18) yields

$$W = \kappa_{\mathrm{b}} A/m^* - \tfrac{1}{2}\kappa_{\mathrm{b}}^2 B_0; \tag{4.71}$$

hence, with eq. (4.68) taken into account, we have

$$W = \mathscr{S}_0/2m^{*2}, \qquad E_{\mathrm{b}} = A^2/2B_0^2 m^{*3} \tag{4.72}$$

and the depth of the barrier level $E_{\mathrm{b}} \ll W$. This inequality shows that the shallow level approximation is self-consistent.

The ultimate formulae, written out in terms of dimensional units, read as

$$S_{\mathrm{A}} = \beta W, \quad W = 1.07(\varepsilon_0 - \varepsilon_\infty)/\varepsilon_0 m_{\mathrm{e}} r_{\mathrm{ex}}^2 m^{*2} \tag{4.73}$$

for the activational regime, and as

$$S_{\mathrm{I}} = (2W/\omega_0)\tanh(\omega_0 \beta/2) \tag{4.74}$$

for the instanton regime.

It follows from eqs. (4.73) and (4.74) that $S_{\mathrm{I}} < S_{\mathrm{A}}$ at all temperatures, although at high temperatures both solutions practically coincide ($T_{\mathrm{c}} = \infty$). Incorporation of higher-order corrections in $(m^*)^{-1}$ leads to the pattern of fig. 5 with $\beta_{\mathrm{c}} \approx 2\tau_{\mathrm{I}}$, or in terms of dimensional units, to

$$T_{\mathrm{c}}/\omega_0 \approx 0.5(m^*)^{1/2} \gg 1. \tag{4.75}$$

The dimensionless mass m^* is defined by eq. (4.60b). Thus, the barrier to ST with a shallow level imitates the situation inherent in normal centers. The results of this section can be best understood in terms of APS with a steep barrier (fig. 16).

4.5. *Low-temperature behavior of the action*

As it ensues from eqs. (4.29) and (4.74), the temperature dependence of S_{I} caused by the coupling of an electron to optical phonons is exponentially small at low temperatures. The coupling to acoustic phonons gives a power-like contribution to the action; see eq. (4.54). In real systems all groups of phonons, including short-wavelength phonons, contribute to ST. However, the low-temperature behavior is always determined by long-wavelength acoustic phonons (the only low-frequency modes of the system), irrespective of the fact which groups of phonons prevail in instanton formation. The self-consistency of motions in the system: long-wavelength acoustic phonons + instanton, satisfying the condition that the action is minimal, enables one to find a low-temperature correction,

$\Delta S(\beta)$, to it. Kagan and Klinger (1976) called this consistency of motions of different modes "preparation of the barrier". To find the contribution of acoustic modes to the action (Ioselevich and Rashba 1986b), first write the equations of motion for these modes entailing from eq. (4.4):

$$\partial_\tau^2 Q_{qs} = \omega_{qs}^2 Q_{qs} - q f_{ns}(\{Q^{(0)}(\tau)\}), \quad \boldsymbol{n} = \boldsymbol{q}/q. \tag{4.76}$$

The first term on the right-hand side is linear in Q_{qs}, since deformation is small for long-wavelength modes. The second term is a force exerted by instanton and acting on long-wavelength phonons. This term arises from the electron–phonon interaction as well as from anharmonisms and is determined by all Fourier components $\{Q_{qs}^{(0)}\}$ of the instanton solution at $T=0$. The force is large at $|\tau| < \tau_1$ and decreases rapidly at increasing $|\tau|$. The factor q in this term arises due to the fact that acoustic deformations, and not displacements, interact with the instanton. Solving eq. (4.76) with the boundary conditions described by eq. (4.5) and substituting the solution into eq. (4.3), we get a low-temperature contribution to the action:

$$S_1(\beta) = S_1(\infty) - \Delta S, \tag{4.77}$$

where

$$\Delta S = \frac{\pi^2 T^4}{60} \sum_s \int \frac{\mathrm{d}\Omega_n}{4\pi} c_s^{-5}(\boldsymbol{n}) \left| \int_{-\infty}^{\infty} \mathrm{d}\tau\, f_{ns}(\{Q^{(0)}(\tau)\}) \right|^2 \tag{4.78}$$

and $c_s(\boldsymbol{n})$ is the sound velocity, dependent on the direction $\boldsymbol{n}$. Equation (4.78) agrees with eq. (4.54), derived for the case when f_{ns} is entirely determined by the electron–phonon coupling via the deformational potential.

Equation (4.78) is valid if the coupling of an electron to acoustic phonons is purely deformational. If this interaction involves a piezoelectric component, then

$$\Delta S \sim (e^2\beta^2/\rho c^3 \bar{\omega}^2) T^2, \tag{4.79}$$

where β is the piezoelectric modulus; consequently, the low-temperature asymptotics is determined by this interaction. However, the instanton is predominantly formed by short-range interactions since there is no barrier to ST for a piezopolaron.

5. Pre-exponential factor – low temperatures

In this section the pre-exponential factor $B(T)$ in eq. (1.2) is calculated for the low-temperature case, i.e., for the instanton mechanism of ST. This factor is a product of several multipliers. All of them have been calculated below and their physical meaning has been elucidated.

5.1. Temperature factor of a free electron

If the original distribution of particles is Maxwellian with the temperature coinciding with the lattice temperature, one must perform averaging of eq. (3.8) over this distribution. As noted in section 3, for slow electrons with $E(\boldsymbol{k}) \sim T \ll \Omega$ we can neglect the $\boldsymbol{k}$-dependence of $u_{\mathrm{ST},\boldsymbol{k}}(\{Q\})$ (nonequilibrium particles with large energies $E(\boldsymbol{k})$ will be considered in section 7). Under these conditions the dependence of w on $E(\boldsymbol{k})$ arises in eq. (3.8) only due to the factor $\exp[\beta E(\boldsymbol{k})]$ and to the integral in the exponent taken along the part of the vertical section of Γ, which corresponds to the free motion. Integration over t in eq. (3.8) yields the following result:

$$w(\boldsymbol{k}T) = w(0, T)\exp[2\tau_{\mathrm{I}} E(\boldsymbol{k})]. \tag{5.1}$$

When particles are thermalized, the ST rate, averaged with respect to the Maxwellian distribution, is

$$\begin{aligned} w(T) &= \langle w(\boldsymbol{k}T)\rangle_T \\ &= \int \frac{\mathrm{d}\boldsymbol{k}}{(2\pi)^3} \exp[-\beta E(\boldsymbol{k})] w(\boldsymbol{k}T) \left\{ \int \frac{\mathrm{d}\boldsymbol{k}}{(2\pi)^3} \exp[-\beta E(\boldsymbol{k})] \right\}^{-1} \\ &= P_{\mathrm{e}}(T) w(0, T), \end{aligned} \tag{5.2}$$

where the temperature factor of a free electron is defined as

$$P_{\mathrm{e}}(T) = \langle \exp[2\tau_{\mathrm{I}} E(\boldsymbol{k})]\rangle_T = (1 - 2\tau_{\mathrm{I}} T)^{-3/2}. \tag{5.3}$$

This is the first of the multipliers entering $B(T)$. At very low temperatures, the temperature dependence provided by $P_{\mathrm{e}}(T)$ dominates in $w(T)$. Actually, as will become clear below, $P_{\mathrm{e}}(T)$ is the only multiplier where T enters linearly. Then, due to the opposite sign of τ_{I} for ST ($\tau_{\mathrm{I}} > 0$) and for normal centers ($\tau_{\mathrm{I}} < 0$) (cf. the beginning of section 4), the temperature dependence of $P_{\mathrm{e}}(T)$ must exhibit the opposite behavior, namely, for ST $P_{\mathrm{e}}(T)$ increases with increasing T, and for normal centers it falls.

5.2. Trapping on a local level

Consider the factor $w(0, T)$ entering eq. (5.2). In its calculation by the steepest descent method, the multiplier $|u_{\mathrm{ST},0}(\{Q\})|^2$ can be taken out from under the sign of the functional integral in eq. (3.8) at the value corresponding to an extremal instanton path Q_{I}, which is assumed to be known from section 4. Then

$$w(0, T) = [(t_2 - t_1) V Z_{\mathrm{lat}}]^{-1} |u_{\mathrm{ST},0}(\{Q_{\mathrm{I}}\})|^2 \int \mathscr{D}Q \exp\left[\mathrm{i} \int_\Gamma L_0(Q)\, \mathrm{d}t \right]. \tag{5.4}$$

However, in conformity with the definition (3.6), the factor $|u_{\mathrm{ST},0}(\{Q_{\mathrm{I}}\})|^2 = VP_{\mathrm{tr}}(0)$, where P_{tr} is the trapping probability calculated in section 2.3.* Then, using eqs. (2.42) and (2.45) as well as eqs. (3.8) and (5.2), one gets

$$w(T) = P_{\mathrm{e}}(T)v_{\mathrm{tr}}[(t_2 - t_1)VZ_{\mathrm{lat}}]^{-1} \int \mathscr{D}Q \exp\left[\mathrm{i}\int_{\Gamma} L(Q)\,\mathrm{d}t\right], \tag{5.5}$$

where

$$v_{\mathrm{tr}} \sim (m\Omega)^{-3/2} \sim m^{-3/2}W^{-1/2}\bar{\omega}^{-1} \tag{5.6}$$

is the second multiplier entering $B(T)$.

5.3. Dimensionless concentration of trapping centers

To calculate the functional integral in eq. (5.5) by the steepest descent method, it is necessary to expand the exponent up to the second order in $q(t) = Q(t) - Q_{\mathrm{I}}(t)$:

$$w^{(0)}(T) = P_{\mathrm{e}}(T)v_{\mathrm{tr}}[(t_2 - t_1)VZ_{\mathrm{lat}}]^{-1}\exp(-S_{\mathrm{I}}) \times \int \mathscr{D}q \exp\left[\mathrm{i}\int_{\Gamma} l(q,t)\,\mathrm{d}t\right], \tag{5.7}$$

where

$$l(q,t) = \frac{1}{2}\sum_{n\rho}(\partial_t q_{n\rho})^2 - \frac{1}{2}\sum_{nn'\rho\rho'} q_{n\rho}q_{n'\rho'}\frac{\delta^2 U(Q_{\mathrm{I}}^{(0)}(t))}{\delta Q_{n\rho}\delta Q_{n'\rho'}}. \tag{5.8}$$

The superscript (0) in $w^{(0)}(T)$ means that the result (5.7) concerns some specific instanton solution $Q_{\mathrm{I}}^{(0)}(t)$. Yet, it should be borne in mind that instanton solutions are multiply degenerate. First there is a discrete degeneracy due to spatial lattice translations. It is clear that for intrinsic ST the geometrical center of an instanton can be located at an arbitrary crystal cell. Therefore, to obtain $w(T)$ one must multiply $w^{(0)}(T)$ by V/v, the number of elementary cells in the crystal. Similarly, for trapping by impurity centers, $w^{(0)}(T)$ must be multiplied by $N_{\mathrm{i}}V$, the number of centers. As a result,

$$w(T) = P_{\mathrm{e}}(T)vv_{\mathrm{tr}}[(t_2 - t_1)vZ_{\mathrm{lat}}]^{-1}\exp(-S_{\mathrm{I}}) \int \mathscr{D}Q \exp\left[\mathrm{i}\int_{\Gamma} l(q,t)\,\mathrm{d}t\right], \tag{5.9}$$

* This equality holds also for $\boldsymbol{k} \neq 0$ if $E(\boldsymbol{k}) \ll \Omega$. For fast particles, $E(\boldsymbol{k}) \gtrsim \Omega$, the factor $|u_{\mathrm{ST},\boldsymbol{k}}|^2$ is equal to the pre-exponential factor in eq. (2.45). The exponential factor must be omitted since the trapping occurs at the stage of tunneling (cf. the end of section 2.3).

where ν is the dimensionless concentration of trapping centers:

$$\nu = \begin{cases} 1 & \text{for intrinsic ST,} \\ vN_{\mathrm{i}} & \text{for extrinsic ST.} \end{cases}$$

5.4. *Time zero mode*

Apart from the discrete degeneracy considered above, there is also a continuous degeneracy. It is associated with an arbitrary choice of the moment of time tunneling begins, i.e., the position of the corner t_0 on the contour Γ. Consequences of this degeneracy become clear after peculiarities of the spectrum of eigenvalues of the quadratic form (5.8) (these peculiarities being caused by this degeneracy) are found.

Denote the eigenvectors of the form $l(q, t)$, obeying the periodicity condition in the imaginary time, as $q^{(s)}(t)$, and the respective eigenvalues as $(-\lambda_s)$. By virtue of the periodicity conditions, λ_s are T-dependent. Denote the eigenvalues of the operator l for a free lattice (i.e., with $Q_{\mathrm{I}}(t) \equiv 0$) as $(-\lambda_s^{(0)})$. For a free lattice it is easy to understand the meaning of the subscript "s". It numbers both normal lattice coordinates and all eigenvibrations (periodic in imaginary time), corresponding to each of the coordinates. There is no such simple classification for the loaded motion, but the total number of eigenvalues remains the same. It is possible to show in a standard way (see Langer 1967) that among λ_s there is one negative ($\lambda_0 < 0$) and one zero ($\lambda_1 = 0$) eigenvalue. The respective zero mode in $q^{(1)} \propto \partial_t Q_{\mathrm{I}}(t)$. This zero mode is a time mode. It corresponds to time translations of an instanton and contributes to the integral (5.9) a factor $\sim(t_2 - t_1) S_{\mathrm{I}}^{1/2}$. The latter result is well known in the quantum field theory (see Rajaraman 1982, Vainshtein et al. 1982). Below we briefly give an idea of its derivation.

First, let us verify that $\partial_t Q_{\mathrm{I}}(t)$ is an eigenfunction of the operator $\hat{l}$ [eq. (5.8)] with a zero eigenvalue:

$$\hat{l}\partial_t Q^{\mathrm{I}}_{n\rho}(t) = \left\{ -\frac{1}{2}\partial_t^2 - \frac{1}{2}\sum_{n'\rho'} \frac{\delta^2 U(Q_{\mathrm{I}})}{\delta Q^{\mathrm{I}}_{n\rho}\delta Q^{\mathrm{I}}_{n'\rho'}} \right\} \partial_t Q^{\mathrm{I}}_{n'\rho'}(t)$$

$$= -\frac{1}{2}\partial_t \left\{ \partial_t^2 Q^{\mathrm{I}}_{n\rho}(t) + \frac{\delta U(Q_{\mathrm{I}}(t))}{\delta Q^{\mathrm{I}}_{n\rho}} \right\} \equiv 0. \tag{5.10}$$

The last equality in eq. (5.10) follows from eq. (4.1). A correctly normalized mode has the form

$$q^{(1)}_{n\rho}(t) = a_1 \partial_t Q_{\mathrm{I}}(t),$$

$$a_1 = \left[\sum_{n\rho} \int_\Gamma \mathrm{d}t (\partial_t Q^{\mathrm{I}}_{n\rho}(t))^2 \right]^{-1/2} \sim S_{\mathrm{I}}^{-1/2}. \tag{5.11}$$

The integration over the coefficient c_1 corresponding to the zero mode $q^{(1)}$ in the expansion

$$q_{np}(t) = \sum_i c_i q_{np}^{(i)}(t)$$

is non-Gaussian and formally brings about a diverging result. However, it can be reduced to the integration over the position of the point t_0, i.e., of the moment on the contour Γ tunneling starts. In fact,

$$Q_{\rm I}(t, t_0 + \Delta t_0) \equiv Q_{\rm I}(t - (t_0 + \Delta t_0)) \approx Q_{\rm I}(t, t_0) - (\Delta t_0/a_1)q^{(1)}(t);$$

consequently,

$$c_1 = \Delta t_0/a_1, \qquad \int \mathrm{d}c_1 \to a_1^{-1} \int \mathrm{d}t_0.$$

Since the integration over t_0 is restricted by the interval (t_1, t_2), the contribution of the zero mode is equal to

$$\int \mathrm{d}c_1 = a_1^{-1}(t_2 - t_1) \sim S_{\rm I}^{1/2}(t_2 - t_1). \tag{5.12}$$

Employing eq. (5.12) and inserting $Z_{\rm lat}$ in the form of the product of $[\lambda_{\rm s}^{(0)}]^{-1/2}$, we get

$$w(T) \sim P_{\rm e}(T)\nu(v_{\rm tr}/v)S_{\rm I}^{1/2}\exp(-S_{\rm I})\omega_{\rm eff}, \tag{5.13}$$

$$\omega_{\rm eff} = \left|\prod_{\rm s} \lambda_{\rm s}^{(0)} \bigg/ \prod_{\rm s}{}' \lambda_{\rm s}\right|^{1/2}. \tag{5.14}$$

The prime labeling the sign of the product means that the zero mode is omitted.

So, the presence of the continuous degeneracy with respect to the moment tunneling begins contributes a large factor $\sim S_{\rm I}^{1/2}$ to $B(T)$. This factor is always present in the problems of tunneling from a potential well of the oscillator type. In terms of elementary quantum mechanics it originates from the fact that tunneling starts not from the bottom of the well but from the lowest level with an energy $\frac{1}{2}\omega$.

5.5. *Total pre-exponential factor – discrete symmetry*

Here we estimate the value of $\omega_{\rm eff}$ in eq. (5.13). If among $\lambda_{\rm s}$ there are no other soft modes apart from long-wavelength acoustic phonons, then, since the characteristic $\lambda_{\rm s} \sim \lambda_{\rm s}^{(0)} \sim \bar{\omega}^2$, the estimate

$$\omega_{\rm eff} \sim \bar{\omega} \tag{5.15}$$

is valid. Note that $\omega_{\rm eff}$ is T-independent at $T \to 0$. As a result, bringing together all the multipliers, we get the final estimate for the prefactor:

$$B(T) \sim P_e(T)(v_{tr}/v)S_I^{1/2}, \tag{5.16}$$

or, if the estimate $E_B \sim m^{-1}v^{-2/3}$ holds,

$$B(T) \sim P_e(T)(E_B/\bar{\omega})^{3/2}. \tag{5.17}$$

Since $P_e(T) \lesssim 1$, as is clear from eq. (5.3), and $E_B \gg \bar{\omega}$ by virtue of the condition (1.1), then $B(T) \gg 1$. So, $B(T)$ is always large and, therefore, is the enhancement factor. Having in mind eq. (1.5) and the paragraph following it, it is evident that $B(T) \sim (M/m)^{3/4} \gg 1$. So, this estimate reveals that the prefactor $B(T)$ is large in the adiabatic parameter.

5.6. *Spatial soft modes – continuum limit*

Equations (5.16) and (5.17) hold when the instanton radius is $r_I \sim a_0$. With increasing r_I among λ_s there arise three anomalously small eigenvalues corresponding to three soft translational modes (with frequencies ω_{soft}). Then $\omega_{eff} \sim \bar{\omega}(\bar{\omega}/\omega_{soft})^3$ and $w(T)$ considerably increases. At $\omega_{soft} \to 0$, the expansion (5.8) proves to be insufficient for "dangerous variables", and the increase of ω_{eff} ceases. In the continuum limit ($r_I \gg a_0$) the soft modes transform into three spatial zero modes, since the group of translations transforms from a discrete into continuum one. The procedure, analogous to the one for the zero time mode, yields

$$\omega_{eff} \sim \bar{\omega}(v/r_I^3)S_I^{3/2} \tag{5.18}$$

since the integration over the instanton coordinate is restricted by an elementary cell. For the interaction with nonpolar optical phonons $r_I \sim r_b$, $S_I \sim W/\bar{\omega}$ (see section 4.2) and then

$$B(T) \sim P_e S_I^3 \sim P_e(T)(W/\bar{\omega})^3. \tag{5.19}$$

For acoustic phonons an instanton possesses two characteristic scales in the continuum approximation (see section 4.3). The internal scale is $r_I \ll r_b$, and $r_I \neq 0$ only due to the fact that the lattice is discrete. In the continuum limit when $r_I \to 0$ and under the condition that the electron is coupled only to acoustic phonons, there exists at $T = 0$ additional symmetry with respect to the 4D scale transformation of the instanton (Iordanskii and Rashba 1978). An additional zero mode must arise due to this symmetry. This fact, alongside with the increase of Ω [see the paragraph after eq. (2.42)], can affect the prefactor $B(T)$, probably towards its increase.

The expressions (1.2) and (5.16)–(5.19) for $w(T)$ can be rewritten as

$$w(T) = P_{\mathrm{e}}(T) v_{\mathrm{tr}} d_{\mathrm{I}}, \quad d_{\mathrm{I}} \sim \begin{cases} (\bar{\omega}/v) S_{\mathrm{I}}^{1/2} \exp(-S_{\mathrm{I}}), \\ (\bar{\omega}/r_{\mathrm{b}}^{3}) S_{\mathrm{I}}^{2} \exp(-S_{\mathrm{I}}). \end{cases} \tag{5.20}$$

The quantity d_{I} is called the instanton density (Vainshtein et al. 1982). It has the meaning of the probability of an instanton fluctuation to emerge in a unit volume per unit time. The upper formula for d_{I} describes the discrete, and the lower, the continuum limit.

Naturally, this additional enhancement of B in the continuum limit occurs only for intrinsic ST. For extrinsic ST, there is no translational symmetry and, consequently, no spatial zero modes arise.

In conclusion, we sum up the obtained results. In all cases (in the discrete and continuum limits, for intrinsic and extrinsic ST) $B \gg 1$. The temperature dependence of S_{I} at $T \ll \bar{\omega}$ is weak (the corrections are proportional to T^4 and to T^2 for deformational and piezoelectric coupling to acoustic phonons, respectively, section 4.5). Therefore, the temperature dependence of $P_{\mathrm{e}}(T)$, as has already been mentioned in section 5.1, must dominate in $w(T)$. The ultimate formulae for $B(T)$ can be represented as

$$B(T) \sim P_{\mathrm{e}}(T) \begin{cases} (W/\bar{\omega})^{3}, \\ (E_{\mathrm{B}}/\bar{\omega})^{3/2}, \end{cases} \tag{5.21}$$

where the upper formula is valid for intrinsic ST in the continuum limit, and the lower, for all other cases.

6. *Pre-exponential factor – high temperatures*

Here we calculate the pre-exponential factor $B(T)$ for $T > T_{\mathrm{c}}$, i.e., for the activational mechanism of ST. At $T \gtrsim T_{\mathrm{c}}$ a large part of the flow still tunnels near the peak of the barrier.

It is convenient to rewrite eq. (3.8) via lattice Green functions $D^{\mathrm{R(A)}}$. They must, in principle, carry two indices labeling the APS sheets (fig. 1a) in the initial and final states, respectively. Since only nondiagonal elements (of the type of $D_{21}^{\mathrm{R(A)}}$) correspond to ST, the subscripts can be discarded. The stage of the process corresponding to the passage of the system through the br-point contributes to D the amplitude of the intersheet transition. This is equal to the amplitude of the capture of an electron by the system moving along the classical path $Q_{\mathrm{A}}(t)$. This amplitude is $V^{-1/2} u(\boldsymbol{k})$, where $u(\boldsymbol{k}) \equiv u_{\mathrm{ST},\boldsymbol{k}}(\{Q_{\mathrm{A}}\})$. As for tunneling ST (section 5), for the activational mechanism there is a discrete degeneracy of the solution $Q_{\mathrm{A}}(t)$, caused by spatial translations. Therefore, one needs to calculate a contribution $w^{(0)}(T)$ to $w(T)$, corresponding to a concrete solution $Q_A(t)$, and to multiply the result by the number of cells in the crystal (V/v) for intrinsic ST, and by the number of impurity centers $(N_i V)$ for extrinsic ST (cf. section 5.3). As a result it follows from eq. (3.8) that

$$w(T) = \nu(V/v)w^{(0)}(T) = \nu(V/v)[(t_2 - t_1)Z_{\text{lat}}Z_e]^{-1} \times \int \frac{\mathrm{d}\boldsymbol{k}}{(2\pi)^3} \int \tilde{\mathrm{d}} Q_1 \,\bar{\mathrm{d}} Q_2 \, D^A(Q_1, t_1 - \mathrm{i}\beta/2 | Q_2 t_2) \times D^R(Q_2 t_2 | Q_1, t_1 + \mathrm{i}\beta/2), \tag{6.1}$$

$$Z_e = \int \frac{\mathrm{d}\boldsymbol{k}}{(2\pi)^3} \exp[-\beta E(\boldsymbol{k})].$$

Here the integration over $\tilde{\mathrm{d}}Q_1$ is performed over the region from which the flow passing through the barrier originates and the integration over $\bar{\mathrm{d}}Q_2$ is extended only to the region beyond the barrier from which it is practically impossible for the system to come back to the free state. In the energy representation the integrand reads

$$\int \frac{\mathrm{d}E\, \mathrm{d}\varepsilon}{(2\pi)^2} D^A_{E+\varepsilon/2}(Q_1, Q_2) D^R_{E-\varepsilon/2}(Q_2, Q_1) \exp[-\beta E + \mathrm{i}\varepsilon(t_2 - t_1)]. \tag{6.2}$$

Since in eq. (6.2) the exponential factor rapidly oscillates on the scale $\varepsilon \sim (t_2 - t_1)^{-1}$, the functions $D^{A(R)}_{E\pm\varepsilon/2}$ can be expanded in ε. Since the motion of the system is semiclassical everywhere (except, perhaps, in a narrow vicinity of the point $\mathscr{W}$), the fastest factor in D^R_E is $\exp[\mathrm{i}S_0(E|Q_2 Q_1)]$, where S_0 is the truncated action [see eq. (6.7) below]. Since Q_2 and Q_1 lie on different sides of the barrier, the reflected wave can be neglected. Expanding the action in ε, remembering that $\partial S_0(E|Q_2, Q_1)/\partial E$ is equal to the time $T_E(Q_2, Q_1)$ of motions between Q_1 and Q_2, and performing integration over ε, we reduce eq. (6.2) to the form

$$\int \frac{\mathrm{d}E}{2\pi} |D^R_E(Q_2, Q_1)|^2 \exp(-\beta E)\delta(t_2 - t_1 - T_E(Q_2, Q_1)). \tag{6.3}$$

To perform the integration in eq. (6.1) over Q_1 and Q_2, it is convenient to divide the entire multidimensional space by a hypersurface $\Sigma_{\mathscr{W}}$ such that it passes through the point $\mathscr{W}$ (in whose vicinity the main flow is concentrated) normally to the steepest descent direction q_0, corresponding to the negative eigenvalue $(-\omega^2_{\mathscr{W}0})$ for the motion in the field of the adiabatic potential near the point $\mathscr{W}$. The main flow is directed along $\boldsymbol{q}_0$. Denote as $Q_{\mathscr{W}}(Q_1, Q_2)$ the point where the classical path, going from Q_1 to Q_2, crosses the surface $\Sigma_{\mathscr{W}}$. Then one can write $T_E(Q_2, Q_1) = T_E(Q_2, Q_{\mathscr{W}}) + T_E(Q_{\mathscr{W}}, Q_1)$, and by means of the elementary transformation, one can represent eq. (6.1) as

$$w(T) = v(V/v)[(t_2 - t_1)Z_{\mathrm{lat}}Z_{\mathrm{e}}]^{-1}\int_{t_1}^{t_2}\mathrm{d}t\int\frac{\mathrm{d}\boldsymbol{k}}{(2\pi)^3}\frac{\mathrm{d}E}{2\pi}$$

$$\times\int \bar{\mathrm{d}}Q_1\bar{\mathrm{d}}Q_2|D_E^{\mathrm{R}}(Q_2,Q_1)|^2\exp(-\beta E)\delta(t_2 - t - T_E(Q_2,Q_{\mathscr{W}}))$$

$$\times\,\delta(t - t_1 - T_E(Q_{\mathscr{W}},Q_1)). \tag{6.4}$$

To clarify the meaning of the internal integral, it is useful to consider an auxiliary expression

$$Z_{\mathrm{lat}}^{-1}\frac{\mathrm{e}^{-\beta E}}{2\pi}\tilde{\mathrm{d}}Q_1\,\bar{\mathrm{d}}Q_2\delta(f_1(Q_1))(\boldsymbol{p}_1\cdot\nabla f_1)|D_E^{\mathrm{R}}(Q_2,Q_1)|^2$$

$$\times(\boldsymbol{p}_2\cdot\nabla f_2)\delta(f_2(Q_2)), \tag{6.5}$$

where $f_1(Q_1) = 0$ and $f_2(Q_2) = 0$ are the equations for the two hypersurfaces Σ_1 and Σ_2, and $\boldsymbol{p}_i$ is the multidimensional momentum of the lattice in the point Q_i for the path, going from Q_1 to Q_2. If Σ_1 crosses a volume unit $\tilde{\mathrm{d}}Q_1$, then

$$\mathrm{d}\Pi_1 = Z_{\mathrm{lat}}^{-1}\exp(-\beta E)\,\tilde{\mathrm{d}}Q_1\rho(E,Q_1)\delta(f_1(Q_1))(\boldsymbol{p}_1\cdot\nabla f_1)$$

is equal to the spectral density of the flow near the energy E passing through an element of the surface Σ_1. Here $\rho(E, Q_1) = \pi^{-1}\,\mathrm{Im}\{D_E^{\mathrm{R}}(Q_1, Q_1)\}$ is the density of states. From the equation $(\hat{H} - E + \mathrm{i}0)D_E^{\mathrm{R}}(Q, Q_1) = -\delta(Q - Q_1)$, which is the definition of the function D_E^{R}, it follows that the stationary concentration produced at the point Q by a δ-functional source located at Q_1, equals $|D_E^{\mathrm{R}}(Q, Q_1)|^2$. At the same time, from the nonstationary Schrödinger equation with the identical right-hand part it follows that the source generates the total flow, equal to $2\,\mathrm{Im}\{D_E^{\mathrm{R}}(Q_1, Q_1)\}$ (see, e.g., Baz' et al. 1975). In the semiclassical situation the flow is concentrated near the classical path, and the concentration $\mathrm{d}N_2$ at the point Q_2 produced by the flow $\mathrm{d}\Pi_1$ at the point Q_1 equals

$$\mathrm{d}N_2 = Z_{\mathrm{lat}}^{-1}\exp(-\beta E)\pi^{-1}\mathrm{Im}\{D_E^{\mathrm{R}}(Q_1,Q_1)\}\delta(f_1(Q_1))$$

$$\times(\boldsymbol{p}_1\cdot\nabla f_1)[|D_E^{\mathrm{R}}(Q_2,Q_1)|^2/2\,\mathrm{Im}\{D_E^{\mathrm{R}}(Q_1,Q_1)\}]\,\tilde{\mathrm{d}}Q_1$$

$$= Z_{\mathrm{lat}}^{-1}\exp(-\beta E)(2\pi)^{-1}|D_E^{\mathrm{R}}(Q_2,Q_1)|^2\delta(f_1(Q_1))(\boldsymbol{p}_1\cdot\nabla f_1)\tilde{\mathrm{d}}Q_1. \tag{6.6}$$

Now it is clear that eq. (6.5) is an elementary flow from the surface Σ_1 to the surface Σ_2. One can verify that for the surfaces Σ_1 and Σ_2, defined by the δ-functions entering eq. (6.4), the following equality holds: $|(\boldsymbol{p}_i\cdot\nabla f_i)| = 1$, $i = 1, 2$. In fact, if to single out from the set Q_i the coordinate q_{0i} corresponding to the motion along the classical path, tangent to $\boldsymbol{p}_i$, we have $\boldsymbol{p}_i\cdot\partial f_i/\partial q_{0i} = (\boldsymbol{p}_i\cdot\nabla f_i) = \pm 1$. To obtain this equality, we have employed the known formula of classical dynamics

$$\partial T_E(Q_i, Q_{\mathscr{W}})/\partial q_{0i} = \pm p_i^{-1}.$$

Thus, the integral over Q_1, Q_2 in eq. (6.4) is the spectral density of the flow between the surfaces Σ_1 and Σ_2.

Due to flow conservation, in the semiclassical region the integral is not altered if Σ_1 and Σ_2 are shifted along the flow and, therefore, it is t-independent. Consequently, the integral over t in eq. (6.4) reduces to multiplication by $t_2 - t_1$, and the surfaces Σ_1 and Σ_2 can be shifted closer to $\Sigma_{\mathscr{W}}$ so much so that it could be possible to confine ourselves, near $\mathscr{W}$, to the quadratic expansion of $U(Q)$. Then it is convenient to come over to the oscillator quantum number representation for all the coordinates except q_0: $Q \to \{q_0; \{n_i\}, i \geqslant 1\}$. Similarly one can represent the energy as

$$E = W + E_0 + \sum_{i \geqslant 1} E_i,$$

where E_0 is the energy corresponding to q_0 and referred to from the peak of the barrier. The Green function for the coordinate q_0 is

$$D^{\mathrm{R}}_{E_0}(q_{02}, q_{01}) = [p(q_{01}, E_0)p(q_{02}, E_0)]^{-1} \times \exp[\mathrm{i}S_0(E_0 | q_{02}, q_{01})] V^{-1/2} u(\boldsymbol{k}) d(E_0). \tag{6.7}$$

The first two factors are a semiclassical Green function, the factor $V^{-1/2}u(\boldsymbol{k})$ is the contribution coming from the trapping stage, and $|d(E_0)|^2$ is the transparency of the barrier. The latter factor takes into account the nonclassical behavior of the system near $\mathscr{W}$. Using eqs. (6.7) and (2.45) one can perform integration over $\boldsymbol{k}$ in eq. (6.4). The factor $\exp[-E(\boldsymbol{k})/T]$ should not be taken into account in the integration over $\boldsymbol{k}$ in the numerator of eq. (6.4), since $E(\boldsymbol{k})$ is part of the total energy E, entering the Gibbs factor in eq. (6.4). This energy is an independent variable of the integration, which will be performed below. Since the factor $\exp[-E(\boldsymbol{k})/T]$ is absent, the integral over $\boldsymbol{k}$ in eq. (6.4) converges only due to the $\boldsymbol{k}$-dependence of a factor $|u(\boldsymbol{k})|^2$ (see section 2.3) entering eq. (6.4) through $|D^{\mathrm{R}}_E|^2$. So, the characteristic energies of electrons are of order $E(\boldsymbol{k}) \sim \Omega \sim (\bar{\omega}^2 W)^{1/3}$ but not of order T. For this reason the dimensionless phase volume from which the particles are self-trapped is of order $(\Omega/T)^{3/2}$. The integration over Q_1 and Q_2 in eq. (6.4) reduces to the calculation of the partition function over n_i and to the integrations over q_{0i}, $i = 1, 2$. After the last two integrations the δ-functions including $T_{E_0}(Q_i, Q_{\mathscr{W}})$ in their arguments disappear and, simultaneously, the momenta entering eq. (6.7) are cancelled. As a result

$$w(T) = v\frac{2}{v}\left(\frac{2\pi}{mT}\right)^{3/2} \mathrm{e}^{-W/T} \frac{\prod_i \sinh(\omega_{0i}/2T)}{\prod_{i \neq 0} \sinh(\omega_{Wi}/2T)} \int \frac{\mathrm{d}E_0}{2\pi} |d(E_0)|^2 \mathrm{e}^{-\beta E_0}. \tag{6.8}$$

Here ω_{0i} are frequencies of the free lattice and ω_{Wi} are frequencies of the oscillations near the barrier. Taking for $|d(E_0)|^2$ the well-known expression

$|d(E_0)|^2 = [1 + \exp(-2\pi E_0/\omega_{W0})]^{-1}$ for a parabolic barrier (see Landau and Lifshitz 1974), we ultimately get

$$w(T) = [\nu v^{-1}(2\pi/mT)^{3/2}]\frac{\omega_{\text{eff}}}{2\pi}\exp(-W/T),$$
$$\omega_{\text{eff}} = \{\omega_{W0}/\sin(\omega_{W0}/2T)\}\left\{\prod_i \sinh(\omega_{0i}/2T)\Big/\prod_{i\neq 0}\sinh(\omega_{Wi}/2T)\right\}. \tag{6.9}$$

This expression for $w(T)$ differs from the analogous formula for the activational surmounting of the barrier by a multimode system (Affleck 1981) by the factor in square brackets, responsible for the electron capture in the conditions of the activational regime. The meaning of this factor becomes clear if we rewrite it as $\nu(v_{\text{tr}}/v)(\Omega/T)^{3/2}$. In order of magnitude, it equals $\nu(E_{\text{B}}/T)^{3/2}$. If $T \gtrsim T_{\text{c}}$, then $\omega_{\text{eff}} \sim \bar{\omega}$, and for the pre-exponential factor in eq. (1.2) we get

$$B(T) \sim v^{-1}(mT)^{-3/2} \sim (E_{\text{B}}/T)^{3/2}. \tag{6.10}$$

The estimate (6.10) is valid at a small radius of the barrier state when saddle points are well separated. With increasing r_{b} the potential relief (fig. 18) gets smoother and the height ΔW of the barrier, separating the neighboring saddle points, decreases exponentially rapidly. Then among the frequencies ω_{Wi} $(i \neq 0)$ there emerge three anomalously low frequencies ω_{soft}. The respective soft modes transform into zero modes in the continuum limit (cf. section 5.6). If $\Delta W > T$, then $\omega_{\text{eff}} \sim \bar{\omega}(\bar{\omega}/\omega_{\text{soft}})^3$, and instead of eq. (6.10), we get

$$B(T) \sim (E_{\text{B}}/T)^{3/2}(\bar{\omega}/\omega_{\text{soft}})^3, \tag{6.11}$$

i.e., $B(T)$ increases substantially. If r_{b} is so large that $\Delta W \ll T$ (continuum limit), then three spatial zero modes arise, and ω_{eff} can be obtained from eq. (5.18) by

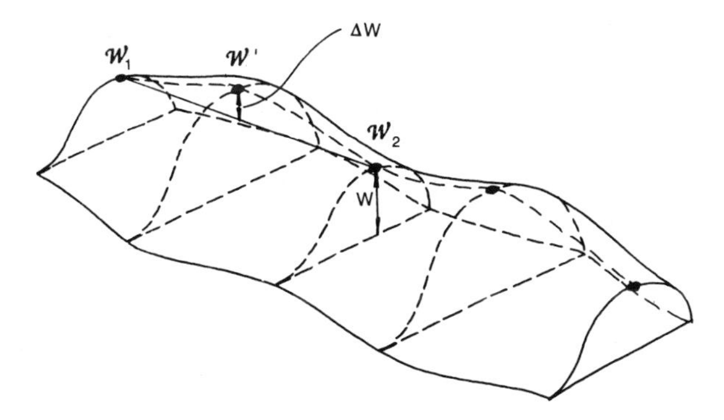

Fig. 18. Schematic shape of APS: $\mathscr{W}_1$ and $\mathscr{W}_2$ are the lowest-lying saddle points, corresponding to the barrier to ST for two neighboring elementary cells; $\mathscr{W}'$ is a higher-lying saddle point, being the saddle point between $\mathscr{W}_1$ and $\mathscr{W}_2$.

means of the replacement $r_\mathrm{I} \to r_\mathrm{b}$, $S_\mathrm{I} \to S_\mathrm{A} = W/T$. As a result

$$\omega_{\mathrm{eff}} \sim \bar{\omega}(v/r_\mathrm{b}^3)S_\mathrm{A}^{3/2}, \qquad B(T) \sim (W/T)^3. \tag{6.12}$$

Thus, in the continuum limit the temperature dependence of $B(T)$ for intrinsic ST enhances.

Ultimately, an estimate for $B(T)$ for high temperatures can be written as

$$B(T) \sim \begin{cases} (W/T)^3, \\ (E_\mathrm{B}/T)^{3/2}, \end{cases} \tag{6.13}$$

where the upper formula holds for intrinsic ST in the continuum limit, and the lower, in all other cases. The estimate (6.13) differs from the analogous estimate (5.21) for low temperatures by the replacement $\bar{\omega} \to T$.

7. *Trapping of hot electrons*

In most experiments nonequilibrium charge carriers and excitons are produced with an initial energy much exceeding T, the lattice temperature. The question is how their thermalization proceeds when free and ST particles coexist and how the process is branched: which part of the particles are thermalized in a metastable free state, and which part are self-trapped in a bypass process, i.e., before thermalization. It is also important what stage of relaxation mostly contributes to the flow of particles into a ST state. To answer all these questions, one must calculate the dependence of the ST rate $w(E, T)$ on the energy of a free particle E.

In this subsection all calculations are performed in the exponential approximation. The calculation of pre-exponential factors is not especially important for the moment, since the methods of exciting fast charge carriers and excitons are hard to be controlled, and their nonequilibrium distribution function is not known.

A barrier to ST is a barrier not for an electron but for lattice degrees of freedom. Nevertheless, the initial energy of an electron can, in principle, be used by the lattice to facilitate the passage over the barrier. However, if a fast electron is loosing its energy gradually, emitting a cascade of incoherent phonons, this cannot in any way facilitate the ST process. The ST rate can be increased only when the energy is coherently transferred to the lattice. But since lattice frequencies are low, the probability of such a transfer is exponentially small. Below we shall describe a version of the technique used above which is needed to calculate this probability (Ioselevich and Rashba 1985b).

For any of the contours of fig. 8a, b the coordinates Q remain real all the time, and the transition from surface 1′ into the point br (belonging to surfaces 1 and 2, fig. 1a) is associated with an abrupt change in the electron energy. It is this change that determines the amount of the energy, which must be coherently transferred to the lattice, and the problem is to calculate the probability of this transfer, i.e., the probability of electron trapping on an emerging discrete level. If $E \ll W$, it is possible to neglect the recoil effects, i.e., in calculating the trapping probability, one can ignore the lattice motion perturbation due to the transfer of the energy from the electron to the lattice. Under these conditions, the problem reduces to the problem of electron trapping in a nonstationary external potential, discussed in section 2.3. Comparing figs. 10b and 11, one can easily understand that for the case $\Omega \ll E \ll W$ the contours in figs. 11a and 11b must be modified in a way shown in figs. 19a and 19b, respectively.

In the activational regime ($T > T_c$, fig. 19a) a discrete level is formed in the course of the motion parallel to the real time axis. The integration over t along the contour Γ results in the action

$$S_A(E, T) = (W - E)/T + \tfrac{4}{3}(E/\Omega)^{3/2}. \tag{7.1}$$

The first term in eq. (7.1) is a modification of the action $S_A = W/T$ [eq. (4.8)]. This modification is a consequence of the absence of the Gibbs averaging over the initial electron energy. The physical meaning of the first term is easy to understand. In the total energy W needed for surmounting the barrier, the energy E is brought in by the electron. Therefore, only the energy $W - E$ must be supplied by the thermal bath.

The second term in eq. (7.1) is S_{tr} [eq. (2.44)], and it determines the probability of the energy E to be coherently transferred from the electron to the lattice. The position of the point t_{tr} on the contour Γ (fig. 19a) is determined by the formula

$$\mathrm{Im}\{t_{tr}\} = \beta/2 - E^{1/2}\Omega^{-3/2}. \tag{7.2}$$

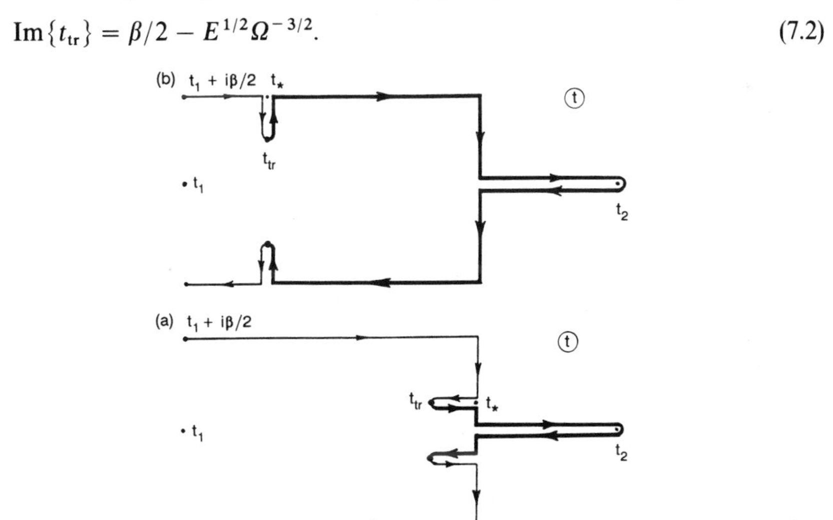

Fig. 19. Modified contours for ST of a hot electron with an energy $\Omega \ll E \ll W$ (cf. figs. 10b, and 11a, b); the length of the appendages is $\Omega^{-3/2}E^{1/2}$: (a) activational surmounting of the barrier; (b) tunneling ST.

The action S_A nonmonotonically depends on E (see fig. 20a). At small E, it always decreases linearly. The minimum is attained at

$$E_{\min} = \Omega^3/4T^2 \sim (\bar{\omega}/T)^2 W; \tag{7.3}$$

in this case

$$S_A(0, T) - S_A(E_{\min}, T) \sim (\bar{\omega}/T)^2 S_A(0, T). \tag{7.4}$$

Equation (7.3) shows that the condition $E \ll W$ is fulfilled at the point $E_{\min}$ only at $T \gg \bar{\omega}$.

In the instanton ST regime ($T < T_c$, fig. 19b) the point t_* is on the vertical section of the contour Γ. Accordingly, the appendages are horizontal, and the integration along them does not contribute to the imaginary action (cf. section 2.3). As a result

$$S_I(E, T) = S_I(T) - 2\tau_I E, \quad \tau_I = \mathrm{Im}\{t_*\} = \mathrm{Im}\{t_{tr}\}. \tag{7.5}$$

Equation (7.5), as well as eq. (7.1), holds at $E \ll W$.

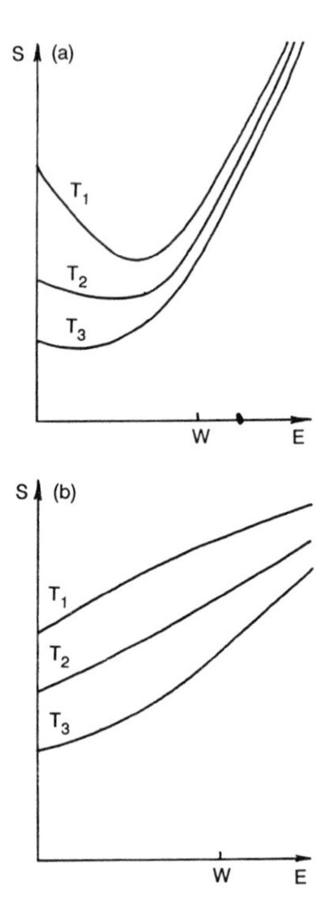

Fig. 20. Schematic dependence of S on the electron energy E at different lattice temperatures ($T_1 < T_2 < T_3$): (a) intrinsic and extrinsic ST; (b) normal centers.

At $E \sim W$ the E-dependence of S in its general form cannot be obtained. Yet, it is possible to establish one general relationship valid at all E and T. The shape of the contour Γ, in particular, the value of t_{tr}, is determined by the variables E and T. However, it is convenient to formally treat the imaginary action S as a function $S(E, T, t_{tr})$. Then t_{tr} is chosen from the condition $\partial S/\partial t_{tr} = 0$ (it is easy to show that eqs. (7.2) and (7.5) obey this condition). Therefore, for the total derivative $dS/dE = \partial S/\partial E$ holds. The explicit form of the E-dependence of S results from the integration of $-E$ over time along the light section of the contour Γ (fig. 19). The corresponding contribution equals $(-2E\,\mathrm{Im}\{t_{tr}\})$. As a result

$$dS/dE = -2\,\mathrm{Im}\{t_{tr}\}. \tag{7.6}$$

Hence, it follows, in particular, that t_{tr} is real at the point $E = E_{min}$. In the region where S increases with E, $\mathrm{Im}\{t_{tr}\} < 0$. This means that at $E \gtrsim W$ the contours get much more complicated compared to the ones depicted in fig. 19a, b.

At $W \ll E \ll E_B$ we can give estimates based on scaling arguments. The results prove to be model-dependent. For the two models (of deformational coupling to acoustic and optical phonons) we shall show that S_A and S_I in this region grow with E according to power laws. Due to the condition $E \ll E_B$, the motion of the system can be described by the equations of the continual theory, namely, eqs. (4.9), (4.15) and (4.16). In this case due to $E \gg W$, the potential energy of the lattice can be neglected in comparison with the kinetic energy [the second term in eq. (4.9) can be omitted]. Introducing $\tau \equiv \mathrm{Im}\{t_{tr}\}$ and estimating all the terms in eqs. (4.9) and (4.15) by the order of magnitude, we get

$$E \sim \gamma_q Q_q q^3, \quad q \sim r^{-1} \sim (mE)^{1/2}, Q_q \sim \tau^2 \gamma_q;$$

hence,

$$\tau^{-1} \sim E^{1/4} m^{3/4} \gamma_q, \qquad S \sim E\tau \sim (E/m)^{3/4}/\gamma_q.$$

Naturally, the phonon frequency spectrum does not enter these expressions. For the interaction via the deformational potential $\gamma_q = \gamma q$ (see section 4.3), and

$$\begin{aligned} &S_{ac}(E) \sim E^{1/4} \rho^{1/2}/m^{5/4} C \sim (E/W)^{1/4}(W/\bar{\omega}), \\ &\bar{\omega}\tau_{ac} \sim (W/E)^{1/4} \ll 1. \end{aligned} \tag{7.7}$$

For nonpolar coupling to optical phonons $\gamma_q = \gamma$ (see section 4.2) and

$$\begin{aligned} &S_{op}(E) \sim (E/W)^{3/4}(W/\bar{\omega}), \\ &\bar{\omega}\tau_{op} \sim (W/E)^{1/4} \ll 1. \end{aligned} \tag{7.8}$$

The fact that τ is small in comparison with the inverse phonon frequency confirms the assumption that the potential energy can be neglected. The derived expressions are temperature-independent, and that is why they are common for the activational and instanton regimes.

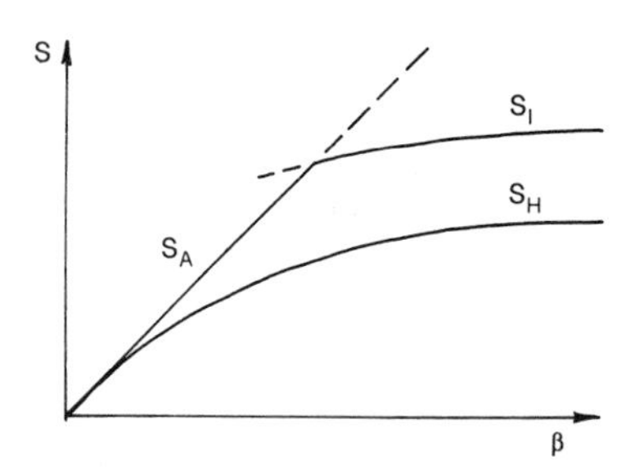

Fig. 21. The dependence of the exponent for the ST rate of hot electrons $S_{H}(T) \equiv S(E_{min}, T)$ in comparison with the analogous exponent (S_I and S_A) for thermalized electrons.

According to eqs. (7.5), (7.7) and (7.8), the dependence of $S(E)$ for the ST case (i.e., at $\tau_I > 0$) is nonmonotonous: $S(E)$ falls at $E \ll W$ and increases at $E \gg W$. Thus, $S(E)$ is nonmonotonous in both the regimes: activational and instanton. For $T < T_c$, the minimum of $S(E)$ is attained at $E_{min} \sim W$, and for $T \gg \bar{\omega}$ at $E_{min} \ll W$. On the other hand, for normal centers (i.e., at $\tau_I < 0$) only the instanton regime exists, section 2.1, and $S(E)$ increases with E at all E and at all temperatures, i.e., $E_{min} = 0$ (fig. 20b). Thus, the dependence $S(E)$ is substantially different for ST and for trapping by normal centers (Ioselevich and Rashba 1985b, Meshkov 1985); this distinction may manifest itself in experiments.

The distribution function of hot electrons usually falls at large energies only in a power-like manner. Therefore, in averaging of the ST rate with respect to this distribution function the main contribution to the trapping flow comes from the energy range near E_{min}. Since in the ST case $E_{min} > 0$, i.e., $S(E_{min}) < S(0)$, the ST rate for hot electrons is much larger than for thermalized electrons (although it remains exponentially small, see fig. 21). This may induce a strong bypass flow in the ST state prior to thermalization. This effect has recently been observed for excitons in pyrene (Furukawa et al. 1989). On the other hand, for trapping by normal centers $E_{min} = 0$ and mean rates of trapping of particles for both nonequilibrium and thermalized distributions do not differ in the exponential approximation.

8. *Relaxation of the system after penetration of a barrier to ST and defect production*

As noted in section 4.3, after the system goes out from under a barrier to ST, the ψ-function rapidly contracts (collapses). The most marked manifestation of this phenomenon occurs when the initial radius of the state $a(t_0)$ in the moment when the system just went out from under the barrier is large; therefore, the continuum approximation is applicable. It will be shown below that in the continuum description of the contraction process the radius of the state $a(t)$ becomes zero in a certain finite moment of time t_c [$(t_c - t_0) \sim \bar{\omega}^{-1}$]. Naturally in the close vicinity of t_c [where $a(t) \sim a_0$], the continuum description does not work. But in the intermediate region of scales, when $a_0 \ll a(t) \ll a(t_0)$ [at $(t_c - t) \ll \bar{\omega}^{-1}$], one can describe the system in terms of the continual theory and, at the same time, can investigate the developed collapse described by the self-similar solution. In what follows, calculations are carried out for nonpolar optical phonons (section 4.2).

At $(t_c - t) \ll \bar{\omega}^{-1}$, the deformation is large; therefore, in eq. (4.19) it is possible to retain only the first (growing) term and to neglect the second (oscillating) term. Then the potential (4.16) in terms of the dimensionless variables [eq. (2.24)] takes the form

$$V(\boldsymbol{r}t) \approx \int_{t_0}^{t} \mathrm{d}t' \sin(t - t')\psi^2(\boldsymbol{r}t'). \tag{8.1}$$

The solution of the Schrödinger equation at $(t_c - t) \ll 1$ will be sought for in the self-similar form:

$$\psi(\boldsymbol{r}t) = a^{-3/2}(t)\psi_c(\boldsymbol{r}/a(t)), \quad a(t) \propto (t_c - t)^{\alpha}. \tag{8.2}$$

Introducing new variables $\boldsymbol{\rho} = \boldsymbol{r}/a(t)$ and $\xi = t - t'$ and expanding the sine function in eq. (8.1), we get

$$\left\{E(t) + \frac{1}{2(t_c - t)^{2\alpha}} \nabla_\rho^2 + \int_0^{t - t_0} \frac{\xi\,\mathrm{d}\xi}{(t_c - t + \xi)^{3\alpha}} \times \psi_c^2\left[\boldsymbol{\rho}\left(\frac{t_c - t}{t_c - t + \xi}\right)^{\alpha}\right]\right\}\psi_c(\boldsymbol{\rho}) = 0. \tag{8.3}$$

The integral converges at $\xi \sim (t_c - t) \ll (t_c - t_0)$; therefore, the integration in eq. (8.3) can be extended to infinite limits. Performing the substitution $x = \xi/(t_c - t)$ and imposing the condition that all terms in eq. (8.3) grow in time according to the same law, we get $\alpha = 2$ and the following equation for $\psi_c(\boldsymbol{\rho})$:

$$\left\{E_c + \frac{1}{2}\nabla_\rho^2 + \int_0^{\infty} \frac{x\,\mathrm{d}x}{(1 + x)^6}\psi_c^2\left[\frac{\boldsymbol{\rho}}{(1 + x)^2}\right]\right\}\psi_c(\boldsymbol{\rho}) = 0. \tag{8.4}$$

The most important quantities increase in conformity with the laws

$$E(t) = E_c(t_c - t)^{-4}, \qquad \int \mathrm{d}\boldsymbol{r}\, Q^2(\boldsymbol{r}t) \sim (t_c - t)^{-2},$$
$$\int \mathrm{d}\boldsymbol{r}(\partial_t Q(\boldsymbol{r}t))^2 \sim (t_c - t)^{-4}. \tag{8.5}$$

For other continuum models the kinetic and potential energy of the lattice as well as the electron energy also diverge at $t \to t_c$ but with different powers.

Important is the fact common for all models that almost all energy of the lattice at $t \to t_c$ is concentrated in the kinetic energy of few atoms. From this fact the possibility of defect production becomes obvious when one takes into account that a finite part of the entire energy released at ST is involved in the collapse process.

Relaxation after the collapse can proceed in different ways. Either a considerable part of the energy will be spent on defect production, or the system will experience numerous oscillations around the ST state, little by little dissipating its energy in phonon emission. The first option is quite probable. Actually, a small number of atoms (in a volume of order of an elementary cell) acquires a large kinetic energy of order of V_{ST} (~ 1 eV). This concentration of the kinetic energy on a small number of the degrees of freedom results in a phenomenon, which resembles, to a great extent, a second-order collision when the energy of electron excitation of colliding atoms transforms into their kinetic energy, rather than the lattice dynamics, formulated in terms of phonons. One can naturally expect that in this case fast atoms will produce lattice defects. Of course, the last stage of this process can be described only on the basis of the microscopic theory.

The mechanism of production of radiation defects belongs to a broad category of processes of electronically stimulated defect production in crystals (see, e.g., Kimerling 1978, Williams 1978, Ito 1982, Klinger et al. 1985, Ueta et al. 1986). In particular, it is well known that ST may be accompanied by defect production. In many cases defect production occurs when the ST state decays; this situation is considered to be typical of alkali halides. The peculiarity of the studied mechanism is that it works at an early stage of the process of formation of ST states. The possibility of defect production at the latter stage of the ST process has been discussed by Kusmartsev and Rashba (1980, 1982). Defect production in rare gas solids at the ST stage has been recently reported (Fugol' 1988).

9. *Comparison of approaches and results*

9.1. *Semiclassical approximation versus multiphonon transition theory*

As stated at the end of section 1, there are two principal methods employed in the theory of nonradiative transitions in solids and molecules. The first of them is the semiclassical approximation, used by Landau (1932) and by Zener (1932), and the second, the multiphonon transition (MPT) method, applied to nonradiative processes by Huang and Rhys (1950), Kubo (1952) and developed in numerous consecutive papers (the MPT method is also known as the generating function method). To start the discussion of certain problems in connection with the application of these methods, it is convenient to consider the simplest two-level system whose configuration-coordinate diagram is given in fig. 22. The noninteracting terms of this system are described by the curves $U^{\mathrm{D}}_{1,2}(Q)$, depicted by the dashed lines and known as diabatic potentials or rough adiabatic potentials. Here we consider the case when the intersection of terms occurs at real Q, since this case is closest to the case of a level splitting off in the branching point (fig. 2), which for us is of primary interest. The part $\hat{V}$ of the electron–phonon interaction Hamiltonian, nondiagonal with respect to electron quantum numbers, gives rise to splitting of the terms in the vicinity of their intersection point Q_{b} and to formation of the adiabatic terms $U^{\mathrm{A}}_{1,2}(Q)$, given in fig. 22 by solid curves. Semiclassical transitions between the terms occur near the point Q_{b}. The probability of these transitions is controlled by the Landau–Zener parameter

$$\gamma = 2\pi V^2/v|F_2 - F_1|, \quad F_{1,2} = \mathrm{d}U_{1,2}/\mathrm{d}Q, \tag{9.1}$$

V is a nondiagonal matrix element of the operator $\hat{V}$ and v is the velocity of the system at the point Q_{b}. The probabilities may be determined by the semiclassical method and cannot be found by perturbation theory. If the terms intersect each other in a classically accessible region, then at $\gamma \ll 1$ the probability of

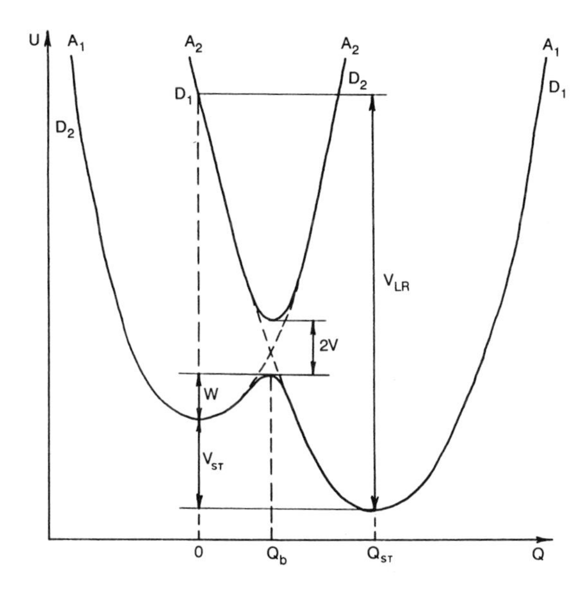

Fig. 22. The configuration-coordinate diagram for a two-level system. The solid curves (A1 and A2) are the adiabatic potentials $U^{\mathrm{A}}_{1,2}(Q)$. The dashed curves (D1 and D2) are the diabatic potentials $U^{\mathrm{D}}_{1,2}(Q)$. The energy $V_{\mathrm{ST}} = p\omega_0$ determines p, the number of phonons emitted at a nonradiative transition; $V_{\mathrm{LR}} = S_{\mathrm{HR}}\omega_0$ is the relaxation energy of the lattice; S_{HR} is the Huang–Rhys parameter characterizing the electron–phonon coupling strength; W is the height of the barrier which is to be surmounted at the transition D2 → D1; V is the matrix element of the operator $\hat{V}$ of the interaction of terms in the Q_{b} configuration. Both the terms have the same symmetry.

a transition between diabatic terms is of order of γ, and at $\gamma \gg 1$ the probability of a transition between adiabatic terms is of order of $e^{-\gamma}$ (Landau and Lifshitz 1974). Equation (9.1) is applicable only in the high-temperature range, when the transition is activational and the system passes through the point Q_b, moving in the direction of the real time. Then, according to eq. (9.1), the quantum mechanical probability is determined by the behavior of the terms exclusively in the vicinity of the point Q_b. Apart from it, the total probability, via the Gibbs factor $\exp(-W/T)$, includes the only additional parameter, namely W, the height of a barrier to ST.

The MPT method aims at: (i) finding theoretical expressions describing $w(T)$ in the entire temperature range, and (ii) using the experimentally found dependence $w(T)$ and also some other experimental data (e.g., Stokes shift) to obtain the basic APS parameters such as V_{ST} and V_{LR} (fig. 22). This method is based on the explicit calculation of some integrals of oscillator wave functions and is applicable only if the variables can be completely separated. In a multimode system the conditions for this are: (i) APS $U^D_{1,2}(Q)$ are parabolic and (ii) for both the surfaces all vibrational frequencies coincide (even a small frequency difference, which can be treated as perturbation, complicates the calculations extremely). Besides, it is assumed that: (iii) $\gamma \ll 1$ and the distortion of APS in the region close to the intersection of the terms can be neglected. The probability of transitions between diabatic terms is calculated by means of the perturbation theory.

Versions of the calculations of the rate of nonradiative transitions for a long time brought controversial results, strongly dependent on details of the calculation procedure (for the analysis of different approximations, see Newmark and Kosai 1983). In earlier works the nonadiabaticity operator is treated as perturbation and calculations are performed in the Condon approximation. This approximation, providing good results for optical transitions, as applied to nonradiative transitions, underestimates the probability by the factor $p^2 \equiv (V_{ST}/\bar{\omega})^2$ (Kovarskii 1962). The large factor $p \gg 1$, equal to the number of emitted phonons, is, in fact, an inverse Bohr–Oppenheimer adiabatic parameter. More recent calculations carried out in the various versions of the non-Condon approximation in the framework of the perturbation theory in the nonadiabaticity parameter (e.g., Ridley 1978, Goto et al. 1980), and in the framework of the perturbation theory in $\hat{V}$ (static approximation, e.g., Pässler 1975) are, on the whole, consistent with the assertion made by Kovarskii (1962). It has also been demonstrated that these two approximations are equivalent (Huang 1981, Gutsche 1982). But it is true only if the calculations are performed in terms of exact wave functions of the adiabatic approximation (note that in these functions the variables cannot be separated). On the contrary, if one uses approximate functions, the difference may appear to be large, as has been shown by Morgan (1983). Thus, although the principal contradictions are now overcome, technical problems, highly complicating the application of the MPT method to nonradiative transitions, still exist.

One of the most important problems is restrictions (i)–(iii). They cannot be eliminated in the framework of the MPT method, not at least in its contemporary form. They are especially marked in the low-barrier ($W \ll V_{LR}$) case, which is of special interest (e.g., for exciton ST), since a low barrier provides a high rate of nonradiative transitions. For parabolic terms

$$W = (V_{ST} - V_{LR})^2/4V_{LR}; \tag{9.2}$$

therefore, from the condition $W \ll V_{ST}$ ensues $V_{LR} \approx V_{ST}$. At small W, to the left of Q_b the potential $U_1^A(Q)$ is drastically different from $U_2^D(Q)$ due to anticrossing of the terms, and to the right of Q_b the potential $U_1^A(Q)$ is different from $U_1^D(Q)$ for the same reason, as well as because Q_b and Q_{ST} are widely apart, so there are no grounds to expect that the parabolic approximation holds near Q_b (fig. 22). This means that the height W and the shape of the barrier and also its transparency could hardly be closely connected with the global behavior of the potential $U_{1,2}^D(Q)$. Nevertheless, it is this behavior that the MPT method is based on; this method employing harmonic diabatic potentials instead of adiabatic potentials in the region of the barrier to ST. This is fraught with major errors even in the exponential factor leading in w.

Another important problem is that it is impossible to perform a consistent quantum-mechanical analysis of the system behavior near the anticrossing point in the framework of the MPT method. The analysis of all those problems brings some authors to pessimistic conclusions, like "the theory is well applicable to systems which are little effective in nonradiative recombination, but has a very limited applicability to systems of primary interest" (Gutsche 1982) and also to the resumption of the Landau–Zener formula (Pässler 1982).

The semiclassical method has the advantage that it permits to single out some general results from calculations of the parameters for concrete models. Its applicability is not restricted by the parabolicity assumption for the chosen ad hoc adiabatic potentials. These potentials are calculated in the framework of the accepted model, which ensures a reasonable shape of the potentials in the barrier to ST region and also a correct analytical behavior of the approximate expressions for the Hamiltonian action entering the exponential factor e^{-S}. In section 5 for the trapping rate problem it has been shown that the semiclassical procedure can incorporate a consistent description of processes, occurring in the nonadiabatic region and can give the temperature dependence of the prefactor in w (at $T \gg \bar{\omega}$) and also the scale of its magnitude, estimated via the adiabatic parameter ($\bar{\omega}/V_{ST}$ or $\bar{\omega}/W$). It seems natural that in the semiclassical theory the adiabatic potentials in the region of a barrier to ST, but not the diabatic potentials $U_{1,2}^D(Q)$ extrapolated from the region near the minima (fig. 22), are involved. Although there are difficulties in the application of the semiclassical procedure to multimode systems, they can be overcome. The reviews by Englman (1979) and by Ranfagni et al. (1984) list some examples. Unfortunately, the pre-exponential factor has not been calculated exactly in the problem of nonradiative trapping in any case. However, the problem of the coefficient has rather an aesthetic than practical significance since (i) the coefficient calculated in the perturbation theory (see, e.g., Abakumov et al. 1985) usually does not differ much from the exact one and (ii) the coupling of an electron to short-wavelength phonons in the initial (free) state, which is always present and cannot be consistently taken into account in any analytical theory, must alter this coefficient.

The efficiency of the semiclassical approach to the nonradiative trapping problem is successfully demonstrated, in particular, in the framework of the single-mode zero-radius potential model, employed by Henry and Lang (1977) and then by Morgan (1983), Abakumov et al. (1985) and by Meshkov (1985). The influence of the Coulomb field has been studied by Abakumov et al. (1988). The use of these models, because they can be solved and then compared with experimental data quantitatively, is rather fruitful, especially when the theory is applied to new phenomena, e.g., to the influence of the electric field on the multiphonon recombination (Karpus and Perel' 1986). The assumptions made when models are formulated are no more restrictive than the ones incorporated in the MPT method. Markvart (1981a) has made an attempt to solve the inverse problem, namely, to reconstruct the adiabatic potential for the B center in GaAs from the temperature dependence of the recombination cross section. It has proved that for this system the frequency difference in the upper and lower states plays a more important role than the shift of equilibrium position of the oscillator.

9.2. *The pre-exponential factor of the multiphonon trapping rate*

In this section we compare the results obtained in sections 5 and 6 for the trapping rate with the results of other studies. The exponent in the high-temperature limit, W/T, is the same in all versions of the theory. At $T \to 0$ the exponent is the Hamiltonian action for tunneling from the bottom of the band. It is different for theories using adiabatic and diabatic potentials, i.e., for the semiclassical theory and MPT method.

There is much more diversity in the results for the pre-exponential factor $B(T)$. One should compare its temperature dependence in the high-temperature limit and also the algebraic factor entering $B(T)$, this algebraic factor being a certain power of the parameter $V_{ST}/\bar{\omega}$ and determining the magnitude of $B(T)$ in the low-temperature limit.

In this section we shall not touch upon specific properties of the systems with a low barrier $W \ll V_{ST}$ (e.g., the continuum case, see section 4.1) and shall assume that all characteristic electron energies (V_{ST}, V_{LR}, E_B) have the same order of magnitude (of the scale of eV). It is assumed that trapping centers are neutral.

1. High temperatures, $T \gg \bar{\omega}$. In conformity with eq. (6.13)

$$B(T) \sim (V_{ST}/T)^{3/2}. \tag{9.3}$$

The estimate of Kovarskii and Sinyavskii (1967), obtained by means of the MPT method, is $B \sim (TV_{ST})^{1/2}/\bar{\omega}$. Pässler (1978), employing the same method, has got $B \sim (V_{ST}/\bar{\omega})(V_{ST}/T)^{1/2}$. The result of Sumi (1980), also obtained by the MPT method, coincides with eq. (9.3), which looks somewhat unexpected, since the method of calculating the electron matrix element is not equivalent to the one described in this survey. Henry and Lang (1977), by means of the two methods (semiclassical and MPT), have obtained the result which in our notation yields $B \sim (V_{ST}/\bar{\omega})(\varepsilon_1/T)^{1/2}$, where ε_1 can apparently be estimated as $\varepsilon_1 \sim (\bar{\omega}W)^{1/2}$.

2. Low temperatures, $T \ll \bar{\omega}$. In conformity with eq. (5.21)

$$B(0) \sim (V_{ST}/\bar{\omega})^{3/2}. \tag{9.4}$$

The results of Pässler (1978) and of Abakumov et al. (1985) are in agreement with eq. (9.4). The formula derived by Henry (1981) yields $B \sim (V_{ST}/\bar{\omega})(\varepsilon_1/\bar{\omega})^{1/2}$.

For intrinsic ST the comparison could be carried out only with the results of Nasu and Toyozawa (1981) and of Kmielik and Schreiber (1987). They have been obtained by means of a modified MPT method, where instead of the potential $U(Q)$ for the ST state (fig. 1a), a certain effective potential, obtained by the variational method, is used. We find it difficult to extract any estimate for $B(0)$ from the cumbersome formulae of their theory, adapted for numerical methods.

In all cases where there is discrepancy in the results, we give preference to eqs. (5.21) and (6.13) because all involved factors have a clear physical meaning. A correct estimate for the algebraic factor in $B(T)$ is important not only as a criterion of the correctness of the theory but for finding a correct scale of this quantity also. Remember that the ratio of trapping rates obtained in the Condon and non-Condon approximations, which have been the subject of numerous studies, involves a factor $(\bar{\omega}/V_{ST})^2$, just of the type considered here.

9.3. Self-trapping of hot particles

The dependence of the ST rate on the energy of particles has recently been discussed by Abe (1990) in his theory of resonant secondary emission, developed within the scope of the dynamical coherent potential method (Sumi 1974, 1975). In his theory there are no explicit formulae for the ST probability, which could have been compared with the results quoted in section 7. Besides, the coherent potential approximation is too rough to give a correct description of exponentially weak processes (tunneling and above-the-barrier reflection). The main qualitative conclusion Abe (1990) has come to is that at $T = 0$ the probability reaches its maximum at the energy of particles $E \gtrsim W$, in conformity with the result obtained by Ioselevich and Rashba (1985b) (see eqs. (7.5) and (7.7) and fig. 20a).

10. Conclusion

In the mid-seventies the coexistence of free and ST excitons in crystals became a reliably well-proven experimental fact. This achievement gave birth to a new branch of research in the physics of ST states, namely, studies of the kinetics of ST from free states, protected by a barrier to ST, controlling the rate of the decay of free states. Development of experimental studies in this field was fairly strongly promoted by the application of up-to-date high spectral and time resolution methods. At present, there have been conducted a large number of experimental studies of the intrinsic ST rate in alkali halides and rare gas solids, and of the extrinsic ST rate and the rate of trapping by normal centers with strong lattice relaxation in semiconductors.

Several attempts have been made to develop the ST rate theory. We believe that the described method, based on a consistent analysis of this problem in the framework of the semiclassical approximation, is the most adequate. The applicability of the semiclassical approximation is an indispensible restriction in the theory of the rate of surmounting of a barrier to ST. This is evident from the fact that the whole concept of APS and, in particular, of a barrier to ST is semiclassical. In the cases when the barrier is so small that it cannot be described semiclassically, its transparency is very large, so large that penetration of the barrier does not restrict the ST rate and the concept of "free" particles makes no sense any longer and the difference between scattering of slow particles and their trapping in the ST state disappears. If the semiclassical approach holds, it permits a general qualitative analysis of the ST rate $w(T)$ behavior and to write out explicit equations for the basic quantities. These equations have the simplest form if ST can be described in terms of the continual theory, the necessary applicability condition of which is a small height of the barrier to ST in comparison with the half-width of the electron band, $W \ll E_B$.

The ST rate of thermalized particles (charge carriers, excitons) can be represented as

$$w(T) = \bar{\omega} B(T) \exp[-S(T)],$$

where $S(T)$ is the temperature-dependent Hamiltonian action. The applicability condition of the semiclassical theory is $S(T) \gg 1$. The advantage of the semiclassical approach is, above all, the fact that it makes it possible to formulate the problem explicitly and to separate the calculation of S and B. In these terms the main conclusions of the theory are:

(1) There are two different regimes of the behavior of the system: instanton (thermoactivated tunneling) and activational (Arrhenius). The action $S_I(T)$ corresponding to the first of them is minimal at $T < T_c$, and at $T > T_c$, the

minimal action is $S_A(T)$, corresponding to the second regime (T_c is the critical temperature). According to general properties of the semiclassical tunneling, the action in the point T_c has no discontinuities $S_I(T_c) = S_A(T_c)$, and its temperature derivative can be either continuous or can experience an abrupt change.

(2) The critical temperature T_c has a scale of the characteristic phonon frequency $\bar{\omega}$. But the analysis of the models reveals that the typical values $T_c < \bar{\omega}$, e.g., $T_c \approx \frac{1}{6}\omega_0$ for a particle interacting with dispersionless nonpolar optical phonons. Anomalously high values of T_c can be expected for Wannier–Mott excitons interacting with polar optical phonons; however, for them the action, already in the instanton region, exhibits a behavior very close to the Arrhenius law.

(3) The low-temperature behavior of $S_I(T)$ is always determined by the interaction with long-wavelength acoustic phonons, irrespective of the fact coupling to which kind of phonons dominates in the ST state. At low temperatures, the temperature-dependent part of the action $S(0) - S(T) \propto T^4$ or T^2, depending upon the fact if the electron-phonon interaction is purely deformational or whether it includes a piezoelectric component. In the Arrhenius regime ($T > T_c$)

$$S_A(T) = W/T.$$

(4) The pre-exponential factor $B(T)$ is the enhancement factor, i.e., always $B(T) \gg 1$. This factor is large in the adiabatic parameter, i.e., in the mass ratio M/m and from the first principles can be estimated as $(M/m)^{3/4}$. The large value of $B(T)$ is largely caused by the high probability of an electron to be trapped on a local level emerging in the potential field of an increasing lattice deformation.

(5) In the instanton regime $B(T) \sim (1 - 2\tau_I T)^{-3/2} (E_B/\bar{\omega})^{3/2}$, where the instanton duration $\tau_I \sim \bar{\omega}^{-1}$. The first multiplier determines the temperature dependence of $w(T)$ in the low-temperature region. In the Arrhenius region

$$B(T) \sim (E_B/T)^{3/2} \quad \text{at } T \gg \bar{\omega}.$$

(6) If $W \ll E_B$ and the continual theory is applicable, the estimate for $B(T)$ changes. For coupling of an electron to nonpolar optical phonons, one must substitute

$$(E_B/\bar{\omega})^{3/2} \to (W/\bar{\omega})^3 \quad \text{and} \quad (E_B/T)^{3/2} \to (W/T)^3.$$

(7) For trapping by normal centers for which the shape of APS is schematically given in fig. 2b, there is only an instanton solution. In the high-temperature limit it behaves as $S_I(T) \approx W/T$, i.e., imitates the Arrhenius behavior. For such centers $\tau_I < 0$; consequently, $w(T)$ at $T \ll \bar{\omega}$ exhibits an unusual behavior – it increases with decreasing T.

The ST rate of fast particles can exceed the ST rate of slow particles (thermal particles) if the energy of a particle is coherently transferred to the lattice; this energy enhances the probability of the barrier to ST to be surmounted. Yet, the probability of this transfer falls exponentially with increasing E at

$E \gg (\bar{\omega}^2 W)^{1/3}$; therefore, these two tendencies may compete. The final conclusions concerning the fast particle trapping are: For intrinsic and extrinsic ST processes there is an optimal value of the kinetic energy, in the Arrhenius region it equals $E \sim (\omega/T)^2 W$, at which the action $S(E, T)$ is minimal and, therefore, the probability $w(E, T)$ is maximal. For nonradiative trapping by normal centers the action $S(E, T)$ reaches its minimal value at $E = 0$, i.e., in the exponential approximation $w(E, T)$ falls with increasing E.

Thus, the semiclassical theory gives a general pattern of the ST process. It reveals the peculiarities associated with coupling to phonons of different kinds as well as with the difference in the shape of APS. The theory solves the problem of the prefactor $B(T)$, the most controversial problem in the nonradiative transition theory. However, note that so far the theory has permitted only to estimate the value of $B(T)$ and to find its temperature dependence. The technically complicated problem of the numerical coefficient in $B(T)$ is still unsolved. Certain specific situations require special attention, e.g., the problem of coupling to acoustic phonons since even at $W \ll E_B$, when the barrier to ST has a macroscopic spatial scale, the tunneling paths, at least partly, have a spatial scale of order of the lattice spacing a_0, and the instanton duration $\tau_I \ll \bar{\omega}^{-1}$. Therefore, one can expect that the results depend on the processes, proceeding on scales $\sim a_0$.

In future, the qualitative analysis of experimental data on the ST rate will require calculations based on models more realistic than those described in section 4. The analysis of these models must be carried out, in our opinion, on the basis of the semiclassical approach described in sections 3–6, since it makes it possible to find $S(T)$, i.e., to determine the exponent playing the most important role in the semiclassical limit. The results of these calculations will permit one to obtain more accurate estimates for $B(T)$ than the one accessible now, particularly, for the intrinsic ST problem. Before this program of studies becomes practical, it is desirable that the analysis of experimental data be performed in terms of the semiclassical theory, e.g., T_c be found from the curves of $\ln w(T)$. The behavior of $w(T)$ in the lowest temperature limit, which may enable one to find the value of τ_I and its sign, is of special interest. Since this method of determining τ_I can be considered reliable only if particles become thermalized before getting self-trapped, it is very important to control their distribution function. Finally, the experiments, making it possible to find the E-dependence of $w(E, T)$ are of particular interest.

References

Abakumov, V.N., I.A. Merkulov, V.I. Perel' and I.N. Yassievich, 1985, Zh. Eksp. & Teor. Fiz. **89**, 1472 [Sov. Phys.-JETP **62**, 853].

Abakumov, V.N., V. Karpus, V.I. Perel' and I.N. Yassievich, 1988, Fiz. & Tekh. Poluprovodn. **22**, 262 [Sov. Phys.-Semicond. **22**, 159].

Abe, S., 1990, J. Lumin. **45**, 272.
Abrikosov, A.A., L.P. Gor'kov and I.E. Dzyaloshinskii, 1962, Methods of Quantum Field Theory in Statistical Physics (Fizmatgiz, Moscow); English Edition: 1963 (Prentice Hall, Englewood Cliffs, NY).
Affleck, I., 1981, Phys. Rev. Lett. **46**, 388.
Aluker, E.D., D.Yu. Lusis and S.A. Chernov, 1979, Electronic Excitations and Radioluminescence (Zinatne, Riga). In Russian.
Baz', A.I., Ya.B. Zel'dovich and A.M. Perelomov, 1975, Scattering, Reactions, and Decays in Non-relativistic Quantum Mechanics (Wiley, New York).
Child, M.S., 1974, Molecular Collision Theory (Academic Press, London).
Coleman, S., 1977, Phys. Rev. D **15**, 2929.
Dean, P.J., and W.J. Choyke, 1977, Adv. Phys. **26**, 1.
Demkov, Yu.N., 1964, Zh. Eksp. & Teor. Fiz. **46**, 1126 [Sov. Phys.-JETP **19**, 762].
Devdariani, A.Z., 1972, Teor. & Mat. Fiz. **11**, 213 [Theor. & Math. Phys. **11**, 460].
Dykhne, A.M., 1961, Zh. Eksp. & Teor. Fiz. **41**, 1324 [Sov. Phys.-JETP **14**, 941].
Dykman, I.M., and S.I. Pekar, 1952, Dokl. Akad. Nauk SSSR **83**, 825.
Englman, R., 1979, Non-Radiative Decay of Atoms and Molecules in Solids (North-Holland, Amsterdam).
Englman, R., and J. Jortner, 1970, Molec. Phys. **18**, 145.
Feynman, R.P., and A.R. Hibbs, 1965, Quantum Mechanics and Path Integrals (McGraw-Hill, New York).
Fugol', I.Ya., 1978, Adv. Phys. **27**, 1.
Fugol', I.Ya., 1988, Adv. Phys. **37**, 1.
Furukawa, M., K. Mizuno, A. Matsui, N. Tamai and I. Yamazaki, 1989, Chem. Phys. **138**, 423.
Goto, H., Y. Adachi and T. Ikoma, 1980, Phys. Rev. B **22**, 782.
Gutsche, E., 1982, Phys. Status Solidi b **109**, 583.
Henry, C.H., 1981, in: Relaxation of Elementary Excitations, Springer Series in Solid State Sciences, Vol. 18 (Springer, Berlin) p. 19.
Henry, C.H., and D.V. Lang, 1977, Phys. Rev. B **15**, 989.
Holstein, T., 1978, Philos. Mag. B **37**, 499.
Huang, K., 1981, Scientia Sinica **24**, 27.
Huang, K., and A. Rhys, 1950, Proc. R. Soc. (London) A **204**, 406.
Iordanskii, S.V., and E.I. Rashba, 1978, Zh. Eksp. & Teor. Fiz. **74**, 1982 [Sov. Phys.-JETP **47**, 975].
Ioselevich, A.S., 1982, Zh. Eksp. & Teor. Fiz. **81**, 1508 [Sov. Phys.-JETP **54**, 800].
Ioselevich, A.S., and E.I. Rashba, 1984, Pis'ma Zh. Eksp. & Teor. Fiz. **40**, 348 [JETP Lett. **40**, 1151].
Ioselevich, A.S., and E.I. Rashba, 1985a, Zh. Eksp. & Teor. Fiz. **88**, 1873 [Sov. Phys.-JETP **61**, 1110].
Ioselevich, A.S., and E.I. Rashba, 1985b, Solid State Commun. **55**, 705.
Ioselevich, A.S., and E.I. Rashba, 1986a, Zh. Eksp. & Teor. Fiz. **91**, 1917 [Sov. Phys.-JETP **64**, 1137].
Ioselevich, A.S., and E.I. Rashba, 1986b, J. Lumin. **34**, 223.
Ioselevich, A.S., and E.I. Rashba, 1987, in: Proc. 18th Int. Conf. on the Physics of Semiconductors, Stockholm, 1986 (World Scientific, Singapore) p. 1385.
Itoh, N., 1982, Adv. Phys. **31**, 491.
Jortner, J., 1974, in: Vacuum Ultraviolet Radiation Physics, eds E.E. Koch, R. Haensel and C. Kunz (Pergamon Press, Braunschweig) p. 263.
Jortner, J., S.A. Rice and R.M. Hochstrasser, 1969, Advances in Photochemistry, Vol. 7, eds J.N. Pitts, G.S. Hammond and W.A. Noyes Jr (Interscience, New York) p. 149.
Kagan, Yu., and M.I. Klinger, 1976, Zh. Eksp. & Teor. Fiz. **70**, 255 [Sov. Phys.-JETP **43**, 132].
Karpus, V., and V.I. Perel', 1986, Zh. Eksp. & Teor. Fiz. **91**, 2319 [Sov. Phys.-JETP **64**, 1376].
Kimerling, L.C., 1978, Solid State Electron. **21**, 1391.

Klinger, M.I., Ch.B. Lushchik, T.V. Mashovetz, G.A. Kholodar', M.K. Sheinkman and M.A. Elango, 1985, Usp. Fiz. Nauk **147**, 523 [Sov. Phys.-Usp. **28**, 994].
Kmiecik, H.J., and M. Schreiber, 1987, J. Lumin. **37**, 191.
Kovarskii, V.A., 1962, Fiz. Tverd. Tela **4**, 1636 [Sov. Phys.-Solid State **4**, 1200].
Kovarskii, V.A., and E.P. Sinyavskii, 1967, Fiz. Tverd. Tela **9**, 1464.
Kubo, R., 1952, Phys. Rev. **86**, 929.
Kusmartsev, F.V., and S.V. Meshkov, 1983, Zh. Eksp. & Teor. Fiz. **85**, 1500 [Sov. Phys.-JETP **58**, 870].
Kusmartsev, F.V., and E.I. Rashba, 1980, in: Proc. Int. Conf. on Radiation Physics of Semiconductors and Related Materials, Tbilisi, 1979 (Tbilisi University Press, Tbilisi) p. 448.
Kusmartsev, F.V., and E.I. Rashba, 1982, Czech. J. Phys. B **32**, 54.
Kusmartsev, F.V., and E.I. Rashba, 1984, Zh. Eksp. & Teor. Fiz. **86**, 1142 [Sov. Phys.-JETP **59**, 668].
Lamb, W., 1939, Phys. Rev. **55**, 190.
Landau, L.D., 1932, Phys. Zs. Sowjet **1**, 88; **2**, 46.
Landau, L.D., and E.M. Lifshitz, 1974, Kvantovaya mekhanika (Nauka, Moscow); English Edition: 1977, Quantum Mechanics (Pergamon Press, Oxford) Sects. 53 and 90.
Lang, D.V., 1980, J. Phys. Soc. Jpn. **49**, Suppl. A, p. 215.
Langer, J.M., 1980, J. Phys. Soc. Jpn. **49**, Suppl. A, p. 207.
Langer, J.S., 1967, Ann. Phys. (New York) **41**, 108.
Lax, M., 1952, J. Chem. Phys. **20**, 1752.
Lifshitz, I.M., and Yu. Kagan, 1972, Zh. Eksp. & Teor. Fiz. **62**, 385 [Sov. Phys.-JETP **35**, 206].
Lushchik, Ch.B., 1982, in: Excitons, eds E.I. Rashba and M.D. Sturge (North-Holland, Amsterdam) p. 505.
Luttinger, J.M., and W. Kohn, 1955, Phys. Rev. **97**, 869.
Markvart, T., 1981a, J. Phys. C **14**, L435.
Markvart, T., 1981b, J. Phys. C **14**, L895.
Meshkov, S.V., 1985, Zh. Eksp. & Teor. Fiz. **89**, 1734 [Sov. Phys.- JETP **62**, 1000].
Miller, W.H., 1974, Adv. Chem. Phys. **24**, 69.
Mitskevich, N.V., 1955, Zh. Eksp. & Teor. Fiz. **29**, 354 [Sov. Phys.-JETP **2**, 197].
Morgan, T.N., 1983, Phys. Rev. B **28**, 7141.
Mott, N.F., and A.M. Stoneham, 1977, J. Phys. C **10**, 3391.
Nasu, K., and Y. Toyozawa, 1981, J. Phys. Soc. Jpn. **50**, 235.
Newmark, G.F., and K. Kosai, 1983, Semiconductors and Semimetals, eds R.K. Willardson and A.C. Beer (Academic Press, New York) p. 1.
Passler, R., 1975, Czech. J. Phys. B **25**, 219.
Passler, R., 1978, Phys. Status Solidi b **85**, 203.
Passler, R., 1982, Czech. J. Phys. B **32**, 846.
Pekar, S.I., 1947, Zh. Eksp. & Teor. Fiz. **17**, 868.
Pekar, S.I., 1950, Zh. Eksp. & Teor. Fiz. **20**, 510.
Pekar, S.I., 1951, Research in the Electron Theory of Crystals (Gostekhizdat, Moscow-Leningrad); German Edition: 1954, Untersuchungen über die Elektronentheorie des Kristallen (Akademie Verlag, Berlin).
Pekar, S.I., E.I. Rashba and V.I. Sheka, 1979, Zh. Eksp. & Teor. Fiz. **76**, 251 [Sov. Phys.-JETP **49**, 129].
Perlin, Yu.E., 1963, Usp. Fiz. Nauk **80**, 553 [Sov. Phys.-Usp. **6**, 542].
Rajaraman, R., 1982, Solitons and Instantons (North-Holland, Amsterdam).
Ranfagni, A., D. Mugnai and R. Englman, 1984, Phys. Rep. **108**, 165.
Rashba, E.I., 1957, Opt. & Spectrosc. **2**, 75, 88.
Rashba, E.I., 1977, Fiz. Nizk. Temp. **3**, 524 [Sov. J. Low Temp. Phys. **3**, 254].
Rashba, E.I., 1982, in: Excitons, eds E.I. Rashba and M.D. Sturge (North-Holland, Amsterdam) p. 543.

Ridley, B.K., 1978, J. Phys. C **11**, 2323.
Ryder, G.H., 1967, Pacif. J. Math. **22**, 477.
Schwentner, N., E.E. Koch and J. Jortner, 1985, Electronic Excitations in Condensed Rare Gases, Springer Tracts in Modern Physics, Vol. 107 (Springer, Berlin).
Stoneham, A.M., 1981, Rep. Prog. Phys. **44**, 1251.
Stuckelberg, E.C.G., 1932, Helv. Phys. Acta **5**, 369.
Sumi, H., 1974, J. Phys. Soc. Jpn. **36**, 770.
Sumi, H., 1975, J. Phys. Soc. Jpn. **38**, 825.
Sumi, H., 1980, Proc. 15th Int. Conf. on Physics of Semiconductors, Kyoto, eds S. Tanaka and Y. Toyozawa (Komiyama Print, Tokyo) p. 226.
Toyozawa, Y., 1978, Solid State Electron. **21**, 1313.
Toyozawa, Y., 1981, in: Relaxation of Elementary Excitations, Springer Series in Solid State Sciences, Vol. 18 (Springer, Berlin) p. 3.
Ueta, M., H. Kanzaki, K. Kobayashi, Y. Toyozawa and E. Hanamura, 1986, Excitonic Processes in Solids, Springer Series in Solid State Sciences, Vol. 60 (Springer, Berlin).
Vainshtein, A.I., V.I. Zakharov, V.A. Novikov and M.A. Shifman, 1982, Usp. Fiz. Nauk **136**, 553 [Sov. Phys.-Usp. **25**, 195].
Watkins, G.D., 1963, Proc. Int. Conf. on Crystal Defects, Conf. J. Phys. Soc. Jpn. **18**. 22.
Williams, R.T., 1978, Semicond. & Insul. **3**, 251.
Zakharov, V.E., V.V. Sobolev and V.S. Synakh, 1972, Prikl. Mat. Tekh. Fiz., No. 1, 92.
Zener, C., 1932, Proc. R. Soc. A **1**, 696.
Zimmerer, G., 1987, in: Excited-State Spectroscopy of Solids, XCVI Corso (Societe Italiana di Fisica, Bologna) p. 37.

CHAPTER 8

Assisted Tunneling in Metallic Systems

A. ZAWADOWSKI*

Institute for Theoretical Physics
Roland Eötvös University
Budapest, Puskin u. 5–7, H-1088, Hungary

and

K. VLADÁR

Central Research Institute for Physics
Budapest, P.O. Box 49, H-1525, Hungary

*Present address: Institute of Physics, Technical University of Budapest, Budafoki út. 8, Budapest, P.O. Box 112, H-1111, Hungary

Quantum Tunnelling in Condensed Media
Edited by
Yu. Kagan and A.J. Leggett

Contents

1. Introduction

The motion of a heavy particle in a fermionic heat bath has attracted considerable interest in the last years, including the case where the heat bath is a degenerate electron gas. Many aspects of this problem are reviewed by Kagan and Prokof'ev in chapter 2 of this volume. The present chapter is devoted to a particular case where the electron-assisted tunneling motion of the heavy particle is also included. This case is briefly reviewed under the title "barrier fluctuation due to the electron interaction" in section 5.6 of the chapter by Kagan and Prokof'ev.

The study of this case has been motivated by different reasons. Cochrane et al. (1975) were the first to suggest that the interaction between a two-level system (TLS) describing a local lattice instability and the electrons in a metal may result in logarithmic corrections to the electrical resistivity due to the electron-assisted tunneling transition in the TLS, showing some analogy with the spin Kondo problem. Kondo (1976b) considered this interaction and concluded that these corrections are likely to be very small. On the other hand, considering the interaction of a heavy particle with different heat baths, a general question has been raised by Caldeira and Leggett (1981, 1983) and Leggett et al. (1987), whether the interaction with an arbitrary heat bath can be described by bosonic variables or not. The answer is yes except if the electron-assisted tunneling also plays an essential role.

The physics of the general problem consists of two phenomena: screening and assisted tunneling.

(i) *Screening.* The interaction between a fixed heavy particle and an electron gas shows up as the Friedel oscillation in the electronic density (fig. 1). The new ground state of the electron gas for a fixed heavy-particle position can be obtained from the unperturbed one by exciting an infinite number of electron–hole pairs with very small energies, thus with very large extension in space. The new ground state of the electron gas is orthogonal to the unperturbed one if the volume is infinite. This orthogonality catastrophe has been first pointed out by Anderson (1967). Considering hopping motion of the heavy particle by tunneling (see fig. 2), the spontaneous hopping rate Δ_0 between two sites is reduced due to the overlap of the electronic screening clouds associated with their positions. In principle, two cases must be distinguished in terms of the

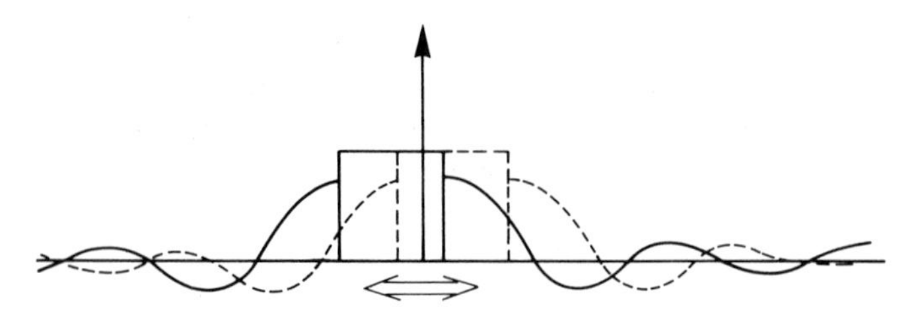

Fig. 1. Changes of the potential due to the tunneling atom is shown. The solid and dotted lines represent the two potentials corresponding to the two degrees of freedom of the TLS, and the associated Friedel oscillations are also represented.

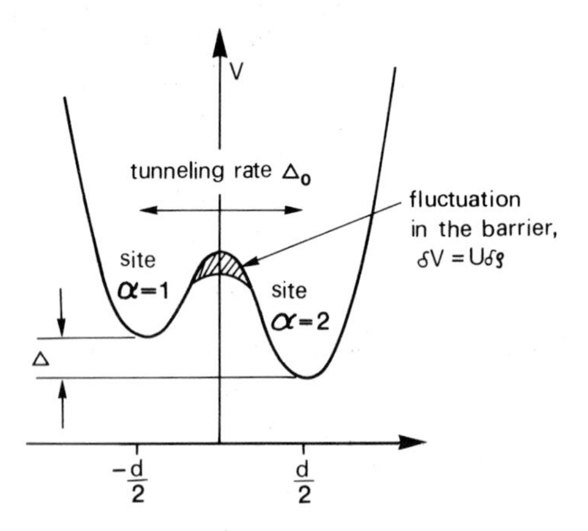

Fig. 2. The potential for the single tunneling atom is shown, which has two minima with energy splitting Δ. The fluctuation in the double-well potential for the TLS is shown. These fluctuations are produced by the density fluctuations $\delta\rho$ in the conduction electron band. The shift of the potential is proportional to the pseudopotential U for the electron–atom scattering.

relevant dimensionless coupling strength K. In the strong-coupling limit, $K > 1$, the renormalized hopping rate $\Delta(T)$ depends on the temperature T, as the effect of the excitations of small energies are smeared out by thermal excitations and

$$\Delta(T) = \Delta_0\left(\frac{T}{D}\right)^K \quad \text{for} \quad K > 1, \tag{1.1}$$

where D is the bandwidth of the electron gas. Thus, at zero temperature the particle is localized in one of the positions. On the other hand, in the weak-coupling limit, $K < 1$, at zero temperature, the excitations forming the screening cloud cannot build up with energies smaller than the renormalized hopping rate Δ_c; thus, for Δ_c we get

$$\Delta_c = \Delta_0\left(\frac{\Delta_c}{D}\right)^K \quad \text{for} \quad K < 1, \tag{1.2}$$

and the particle is not localized. The actual value of coupling K depends on the strength of the dimensionless local interaction v between the heavy particle and the electrons, the spin degeneracy of the electron gas n_s (in reality $n_s = 2$) and $Q = k_F d$, where k_F is the Fermi momentum and d is the separation distance of the two positions of the TLS. The mathematical form of the coupling K depends on the dimensionality of the electron gas, and in the weak-coupling limit (see, e.g., Zimányi et al. 1987) is

$$K = \begin{cases} 4n_s v^2 \sin^2 Q & \text{for the 1D case,} \\ n_s v^2 [1 - J_0^2(Q)] & \text{for the 2D case,} \\ n_s v^2 [1 - j_0^2(Q)] = n_s v^2 (1 - \sin^2 Q/Q^2) & \text{for the 3D case,} \end{cases} \quad (1.3)$$

where J_0 and j_0 are, respectively, the Bessel function and spherical Bessel function of the first kind, and 1D, 2D, etc., stand for one-dimensional, two-dimensional, etc.

It is interesting to note that in the 1D case $K = 0$ for $Q = n\pi$ ($n = 0, 1, 2, \ldots$). This is the case where the distance between the two positions is integer in the units of the wavelength of the Friedel oscillation, thus the long-range part of the Friedel oscillation should not be changed after a hopping. This clearly indicates the role of the Friedel oscillation. In dimensions higher than one the coupling K cannot vanish.

The building up of the screening cloud in the X-ray absorption problem was solved exactly by Nozières and de Dominicis (1969) for arbitrary coupling strength. In their theory the phase shift enters instead of the coupling. The interesting question is whether $K > 1$ can hold in a case of physical interest. In the study of this question s-wave scattering is assumed on the particle. The difference in the scattering amplitudes corresponding to the two positions is expressed by a phase shift $\pm\delta$, depending upon the two positions. Yu and Anderson (1984) studied the problem of the instability of the A-15 compounds by describing the dynamics of each atom by a TLS and they assumed a strong-coupling limit $\delta \approx \frac{1}{2}\pi$. A careful study shows that $|\delta| < \frac{1}{4}\pi$ and $K < 1$ for $n_s = 2$, and $K > 1$ can occur only for unrealistic values of $n_s > 2$ (see, e.g., Tanabe and Ohtaka 1986, Yamada et al. 1986, Kagan and Prokof'ev 1989 and also section 5.6 of the chapter by Kagan and Prokof'ev). This will also play an important role in the interpretation of the results presented in this chapter.

(ii) *Assisted tunneling.* The tunneling of a particle between two states is determined by the size and the shape of the potential barrier. Assuming that the particle interacts with the electrons by a local potential, the fluctuation in the electronic density in the barrier region can be considered as a fluctuation of the barrier. The additional tunneling rate is attributed to the so-called assisted-tunneling processes. This tunneling process is weak as it contains the tunneling matrix element and the coupling between the electrons and the heavy particle, as well.

In comparison with the $S = \frac{1}{2}$ Kondo problem, the TLS is replacing the localized spin and the screening interaction corresponds to the spin–diagonal interaction, while the assisted tunneling shows analogy with the spin flip processes. This is the reason for calling the present problem the orbital Kondo effect (for a review on the spin Kondo problem, see Tsvelik and Wiegmann 1983).

Considering only the screening, logarithmic corrections typical for infrared divergencies appear, which result in the power behavior occurring in eqs. (1.1) and (1.2), which is discussed in section 5 of chapter 2. The coupling constant K is, however, not renormalized in this case. Starting with a weak assisted tunneling and screening, the many-body corrections lead to a very strong renormalization, just like for the spin Kondo problem (Zawadowski 1980). The strength of the renormalization depends very strongly on the initial value of the screening parameter (phase shift). The electrical resistivity due to the electron scattering on the TLS shows a strong temperature dependence, as it increases with lowering the temperature. Whether this increase is observed by experiments or not is still an open question, as other processes can result in similar behavior and it is difficult to separate the different contributions.

The main analogy between the spin Kondo problem and the present problem is the following. In the spin Kondo problem after a spin flip, the spin polarization cloud keeps a long-time memory of the flip; in the TLS problem, however, the shape of the polarization cloud (position of the Friedel oscillations) plays that role. In order to describe the change in the shape of the polarization cloud, at least two angular momentum channels of the electron wave functions must be considered (Zawadowski 1980). After renormalization, actually only two channels remain relevant, e.g., the p- and d-waves, where the centers of the coordinate system and the TLS coincide (Vladár and Zawadowski 1983a). The electrons have two quantum numbers: the angular momentum (p and d) and the real spin with spin degeneracy $n_s = 2$. In the spin Kondo problem there are only the spin variables $\sigma = \pm 1$. The present problem is more closely analogous to the multichannel Kondo problem, where one multiplies the number of channels by an arbitrary number n and thereby associates with the electrons an additional quantum number (see, e.g., Cragg and Lloyd 1978, Nozières and Blandin 1980). Thus, the TLS and the two-channel ($n = 2$) Kondo problems are equivalent after the renormalization of the TLS. The multichannel Kondo problem can be realized also by other realistic models developed to describe the behavior of some of the heavy fermionic systems (Cox 1987) and the high-T_c superconductors (Barnes 1988, Cox 1991), where the crystal split orbits are responsible for the new degrees of freedom. Thus, the study of the present problem may be relevant for other physical systems, too.

The renormalization of the model described can be carried out in four different ways:

(i) Vertex renormalization in the leading logarithmic order by “poor man’s” scaling.

(ii) Vertex and self-energy renormalization in the next to leading order by applying the multiplicative renormalization group technique.

(iii) Mapping to Coulomb gas with interacting charges and dipoles, and applying the Anderson–Yuval–Hamann renormalization technique.

(iv) Bethe ansatz and $1/n_s$ expansion.

Method (i) is very instructive in deciding whether an originally weak-coupling Hamiltonian scales to strong coupling, or not. Method (ii) can treat the renormalization of the dynamics of the heavy object. Method (iii) is applicable even if the screening (polaronic interaction) is strong and only the hopping interactions are weak. The exact methods (iv) enable one to calculate certain quantities like free energy, heat capacity, but not, e.g., the electrical resistivity.

The application of the present theory to well-defined physical systems, as metallic glasses, has difficulties as the parameters used in the theory must be in certain ranges in order to get measurable effects; on the other hand, the TLS are not well defined in glassy systems. According to Kagan and Prokof'ev (see their chapter in the present book), the applicability of the present model to the metallic glasses is further restricted by the facts that the model Hamiltonian to be used is adequate only for energies smaller than the energy spacing ω_0 of the lowest energy levels of atom in the double potential well, which is smaller than the conduction electron bandwidth and, as it has been already mentioned, the phase shift δ calculated for simple model is also limited to the range $|\delta| < \frac{1}{4}\pi$. Our opinion, however, is that the phenomena to be discussed are general enough to have a good chance for its application to realistic systems.

The present chapter is organized as follows. In section 2 the physical model is described. In section 3 the noncommutative nature of the couplings as the origin of the logarithmic correction is discussed. Sections 4 and 5 are devoted to the application of the multiplicative renormalization group method, which shows the crossover from the weak to the strong-coupling region by eliminating the phase space of the electrons far from the Fermi level. The path integral method of Anderson et al., which is adequate for arbitrary value of the phase shift, is reviewed in section 6. The presentation is brief and not self-contained. This serves only as a guide to read the original publications. The following sections can be understood without studying section 6 in detail. In section 7 we review the low-temperature behavior of the multichannel Kondo problem, which is obtained by $1/n_s$ expansion or Bethe ansatz. In section 8 the observable physical quantities of a metallic glass are briefly discussed, without going into details concerning the experimental results for metallic glasses. Section 9 contains the conclusions and other recent related developments.

2. *Hamiltonians*

The simplest realization of a TLS in metals is an atom sitting in a double-well potential and interacting with a free-electron gas (fig. 2). The atom moves between the two positions by tunneling. The atom can be described either as a fermion or a boson. The creation operators at the lowest-lying energy levels at the minima labeled by 1 and 2 are $b_1^\dagger$ and $b_2^\dagger$, respectively. The states $|\uparrow\rangle = b_1^\dagger|0\rangle$ and $|\downarrow\rangle = b_2^\dagger|0\rangle$ can be considered as quasi-spin states, where $|0\rangle$ is the vacuum state. The most general form of the TLS Hamiltonian is

$$H_{\mathrm{TLS}}^0 = \lambda_{\mathrm{ps}} \sum_\alpha b_\alpha^\dagger b_\alpha + \tfrac{1}{2} \sum_{\alpha\beta, i} \Delta^i b_\alpha^\dagger \sigma_{\alpha\beta}^i b_\beta \tag{2.1}$$

in a spin representation, where α, β are the spin indices and $\sigma_{\alpha\beta}^i$ ($i = x, y, z$) are the Pauli matrices. Δ^z describes the energy splitting between the two states, and Δ^x and Δ^y stand for the tunneling transition. The energy $\lambda_{\mathrm{ps}} > 0$ is a fictitious pseudo energy, which is very large to make the simultaneous presence of more than one particle negligible: $\lambda_{\mathrm{ps}} \gg kT$. In this case, however, the population of the double well is very small, and a normalization by $\exp(-\lambda_{\mathrm{ps}}/kT)$ is needed, like in the case of the Kondo problem, where Abrikosov (1965) introduced a similar pseudo energy for the particle representing the spin (for proof of the normalization procedure, see Black et al. 1982). This technique will be used in the application of the multiplicative renormalization group.

In the use of the path integral technique the spin representation is more adequate. The second part of the Hamiltonian (2.1) can be written as a simple spin Hamiltonian

$$H_1^0 = \Delta^z \sigma^z, \tag{2.2}$$

$$H_2^0 = \Delta^+ \sigma^- + \Delta^- \sigma^+, \tag{2.3}$$

where $\sigma^\pm = \frac{1}{2}(\sigma^x \pm \mathrm{i}\sigma^y)$. $\Delta^+ = \Delta^- = \Delta_0$ for real wave functions and $\Delta_0 \sim \omega_0 \mathrm{e}^{-\lambda}$, where ω_0 is of the order of the Debye frequency (which cannot be very different from the spacing to the next level in the double-well potential), and λ is the Gamow factor, which is $\approx w\sqrt{2MV_\mathrm{b}}$, with M the mass of the particle, and w and V_b the width and height of the barrier, respectively.

The Hamiltonian of the free-electron gas is given by eq. (3.1) below. The general form of the electron–TLS Hamiltonian is

$$H_{\mathrm{int}} = \sum_{k_1 k_2, s} \left(a_{k_1 s}^\dagger V_{k_1 k_2}^0 a_{k_2 s} + \sum_i a_{k_1 s}^\dagger V_{k_1 k_2}^i a_{k_2 s} \sigma^i \right), \tag{2.4}$$

where the spin representation is used. The coupling $V_{k_1 k_2}^i$ stands for two processes:

(i) the conduction electrons are scattered by the atom sitting at one of the two positions (V^z),

(ii) the electron scattering induces a transition of the atom between the two positions (V^x, V^y).

The transitions Δ^+ and Δ^- in the Hamiltonian (2.3) are due to spontaneous tunneling, while the process (ii) is known as electron-assisted (induced) tunneling.

An atom at positions 1 and 2 has the wave functions $\phi_1(\boldsymbol{R}) = \phi(\boldsymbol{R} - \frac{1}{2}\boldsymbol{d})$ and $\phi_2(\boldsymbol{R}) = \phi(\boldsymbol{R} + \frac{1}{2}\boldsymbol{d})$, respectively, where a symmetrical well is assumed. The vector $\boldsymbol{d}$ connects the centers of the two positions (fig. 2). Using the conduction electron field operator ψ and the local interaction potential U between the atom and the electrons, the part of the Hamiltonian which is diagonal in the variables of the TLS can be written as

$$\sum_s \int \psi_s^*(\boldsymbol{r})\psi_s(\boldsymbol{r})U(\boldsymbol{r} - \boldsymbol{R})[\tfrac{1}{2}(1 + \sigma^z)\phi_1^2(\boldsymbol{R}) + \tfrac{1}{2}(1 - \sigma^z)\phi_2^2(\boldsymbol{R})]\mathrm{d}^3\boldsymbol{r}\,\mathrm{d}^3\boldsymbol{R}. \tag{2.5}$$

This can be rewritten as $H_{\text{int}}^0 + H_{\text{int}}^z$, the sum of the two terms of eq. (2.4) containing V^z and V^0. The comparison of eqs. (2.4) and (2.5) gives

$$V_{k_1k_2}^0 = U(k_2 - k_1)\int \mathrm{d}^3\boldsymbol{R}\exp[\mathrm{i}(\boldsymbol{k}_2 - \boldsymbol{k}_1)\boldsymbol{R}]\tfrac{1}{2}[\phi_1^2(\boldsymbol{R}) + \phi_2^2(\boldsymbol{R})], \tag{2.6}$$

$$V_{k_1k_2}^z = U(k_2 - k_1)\int \mathrm{d}^3\boldsymbol{R}\exp[\mathrm{i}(\boldsymbol{k}_2 - \boldsymbol{k}_1)\boldsymbol{R}]\tfrac{1}{2}[\phi_1^2(\boldsymbol{R}) - \phi_2^2(\boldsymbol{R})]. \tag{2.7}$$

The first one is a static potential which drops out from the scaling equations of the first order. In this way it can be ignored in the weak-coupling theory (Vladár and Zawadowski 1983a) based on the multiplicative renormalization group (actually, in the previous works of the present authors it has always been neglected, except in Vladár 1991). It has been pointed out by Kagan and Prokof'ev (1989) (see also chapter 2) that this term can be important when the scattering is strong, thus when $|\delta| \ll \frac{1}{2}\pi$ does not hold for the scattering phase shift δ.

The expressions for V^0 and V^z, eqs. (2.6) and (2.7), are even simpler if the wave functions $\phi(\boldsymbol{R})$ are well localized in space and $k_\mathrm{F}d \ll 1$, thus the exponent can be expanded. The result is (see Black et al. 1979)

$$V_{k_1k_2}^0 \approx U(\boldsymbol{k}_2 - \boldsymbol{k}_1), \tag{2.8}$$

$$V_{k_1k_2}^z \approx -\tfrac{1}{2}\mathrm{i}(\boldsymbol{k}_2 - \boldsymbol{k}_1)\boldsymbol{d}U(\boldsymbol{k}_2 - \boldsymbol{k}_1). \tag{2.9}$$

In order to get logarithmic vertex corrections, another vertex with a different structure in momentum space is needed (Kondo 1976b, Zawadowski and Vladár 1980). Such a vertex is due to the assisted tunneling, which can be interpreted in the following way. The local fluctuation in the electron density $\delta\rho(\boldsymbol{r})$ is contributing to the potential of the barrier V_b as $V_\mathrm{b}(\boldsymbol{R}) + \int U(\boldsymbol{R} - \boldsymbol{r})\delta\rho(\boldsymbol{r})\mathrm{d}^3\boldsymbol{r}$ (fig. 2). The fluctuations cause a change in the tunneling rate, if $\delta\rho$ describes fluctuation relative to the averaged electron density at atomic positions $\boldsymbol{R} = \pm\frac{1}{2}\boldsymbol{d}$. In this way the Gamow factor itself fluctuates with $\delta\rho$; it can be expanded in terms of $\delta\rho$ and the linear term contributes to $V_{k_1k_2}^x$, while $V_{k_1k_2}^y = 0$ if the atomic function $\phi(\boldsymbol{R} \pm \frac{1}{2}\boldsymbol{d})$ is real. V^x can be estimated (Vladár and Zawadowski

1983a) as

$$V^x_{k_1k_2} \approx \Delta_0 \frac{\lambda U(\boldsymbol{k}_2 - \boldsymbol{k}_1)}{16 V_b} [(\boldsymbol{k}_2 - \boldsymbol{k}_1)\boldsymbol{d}]^2. \tag{2.10}$$

By comparing eqs. (2.9) and (2.10), the ratio V^x/V^z is obtained as

$$\frac{V^x}{V^z} \sim \Delta_0 \frac{\lambda k_F d}{24 V_b}, \tag{2.11}$$

where $|\boldsymbol{k}_1| \sim |\boldsymbol{k}_2| \sim k_F$ is taken, and k_F is the Fermi wave number. For a TLS of interest $\Delta_0 \sim 1$ K. Typical values for other parameters can be chosen as $\lambda \sim 5$, $k_F d \sim 0.3$ and $M \sim 50\, m_p$, where m_p is the proton mass. With these values we obtain

$$\frac{V^x}{V^z} \sim 10^{-3}\text{–}10^{-4}. \tag{2.12}$$

Thus, the assisted coupling is always very small and is relevant only in the case of enhancement of Kondo type. The dependence of these couplings on the momentum transfer can be expanded in terms of spherical harmonics, and, e.g., the p-wave and s- and d-wave terms are the dominating ones for V^z and V^x, respectively.

3. *Noncommutative Kondo-type vertex corrections*

Since the discovery of the Kondo effect (Kondo 1964) the logarithmic vertex corrections have attracted great interest, as in many cases these are associated with the formation of bound or resonant states. Such corrections are typical for systems where a heavy particle or some degrees of freedom of an object localized in space (impurity spin, TLS) are coupled to an electron gas with large bandwidth. The state of the "heavy" object is characterized by quantum numbers α and the electrons by their momenta and spins (k, s).

The Hamiltonian of the electron gas is

$$H_e = \sum_{ks} \varepsilon_k a^\dagger_{ks} a_{ks}, \tag{3.1}$$

where ε_k is the energy of the electron created and annihilated by the operators $a^\dagger_{ks}$ and a_{ks}, k is the momentum and s stands for the spin. A simplified electron band structure will be used which is spherically symmetric and in which the density of states $\rho(\varepsilon)$ for one spin value is constant; thus,

$$\rho(\varepsilon) = \begin{cases} \rho_0 & \text{for } \; -D < \varepsilon < D, \\ 0 & \text{otherwise,} \end{cases} \tag{3.2}$$

where a cutoff energy D is applied for the band energy ε, which is of the order of the Fermi energy E_F.

As has been suggested by Kagan and Prokof'ev (1986), the application of the model Hamiltonian must be restricted to energies smaller than the energy spacing ω_0 of the levels in the double potential well. In that case $D \approx \omega_0$ must be used.

The Hamiltonian of the "heavy object" is

$$H_0 = \sum_{\alpha} \varepsilon_\alpha b_\alpha^\dagger b_\alpha, \tag{3.3}$$

where b is its annihilation operator of fermionic character and ε_α is the energy. In the case where the heavy object carries momentum, the index α contains the momentum. In the case of a TLS the index α has only two values.

The interaction between the electron gas and the heavy object is described by the Hamiltonian

$$H_{\text{int}} = \sum_{\substack{ks,k's' \\ \alpha\alpha'}} V^{\alpha\alpha'}_{ks,k's'} a^\dagger_{ks} a_{k's'} b^\dagger_\alpha b_{\alpha'}, \tag{3.4}$$

where $V^{\alpha\alpha'}_{ks,k's'}$ is the appropriate coupling. It will be assumed that at each time there is one "heavy object" in the system; thus, they can be treated independently.

In the second order of the perturbation theory there are two vertex corrections. These diagrams are depicted in fig. 3, where the diagrams are time-ordered. The arrows on the heavy line representing the heavy object shows the time flow. The sum of their contribution is

$$\frac{1}{(2\pi)^3} \int \mathrm{d}^3 k'' \sum_{s''\alpha''} \left[V^{\alpha\alpha''}_{ks,k''s''} V^{\alpha''\alpha'}_{k''s'',k's'} \frac{1 - n_F(\varepsilon_{k''})}{\omega - \varepsilon_{k''} - \varepsilon_{\alpha''} + \varepsilon_\alpha} \right.$$
$$\left. - V^{\alpha\alpha''}_{k''s'',k's'} V^{\alpha''\alpha'}_{ks,k''s''} \frac{n_F(\varepsilon_{k''})}{\omega + \varepsilon_{k''} - \varepsilon_{\alpha''} + \varepsilon_\alpha} \right], \tag{3.5}$$

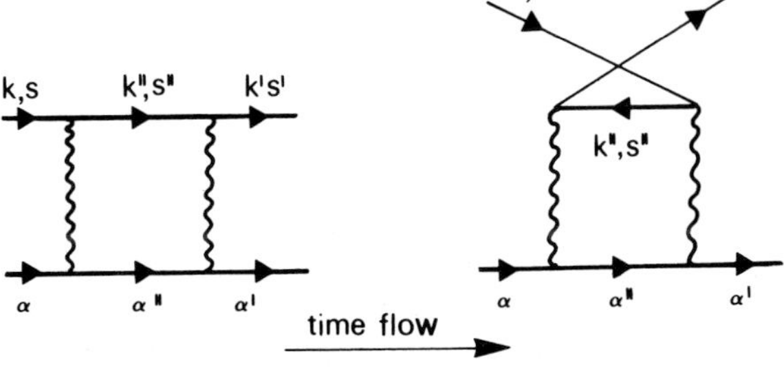

Fig. 3. The scattering of an electron (light line) and a heavy object (heavy line) is depicted by time-ordered diagrams. The direction of the time flow is also indicated. The wavy lines represent the interaction.

where $\omega + \varepsilon_\alpha$ is the initial energy and n_F is the Fermi distribution function at temperature T. The negative sign for the second term is due to the fermionic commutation relation.

The momentum integral can be replaced by

$$\frac{1}{(2\pi)^3}\int \mathrm{d}^3 \boldsymbol{k} \rightarrow \rho_0 \int \frac{\mathrm{d}S_\kappa}{S} \int_{-D}^{D} \mathrm{d}\varepsilon, \tag{3.6}$$

where $\mathrm{d}S_\kappa$ is the surface element of the Fermi surface in the direction $\boldsymbol{\kappa} = \boldsymbol{k}/|\boldsymbol{k}|$, S is the area of the Fermi surface and ε is the energy replacing the absolute value of the momentum. (For the sake of simplicity, the Fermi velocity is assumed constant at the Fermi surface.) It will be assumed that the couplings depend only on $\boldsymbol{\kappa}$; thus, the absolute value of the momentum can be replaced by the Fermi momentum. The energy integrals with respect to ε can be approximated by a simple logarithmic function as

$$\ln\left(\frac{D}{\max\{\omega, T, \Delta\}}\right), \tag{3.7}$$

where Δ stands for $|\varepsilon_{\alpha''} - \varepsilon_\alpha|$. That approximation is justified providing that the logarithmic term is much larger than the unity.

Using these approximations, the scattering amplitude given by eq. (3.5) has a simpler form:

$$-\rho_0 \int \frac{\mathrm{d}S_\kappa}{S} \sum_{\alpha''} \left(V^{\alpha\alpha''}_{\kappa s, \kappa'' s''} V^{\alpha''\alpha'}_{\kappa'' s'', \kappa' s'} - V^{\alpha\alpha''}_{\kappa'' s'', \kappa' s'} V^{\alpha''\alpha'}_{\kappa s, \kappa'' s''}\right) \ln\left(\frac{D}{\max\{\omega, T, \Delta\}}\right). \tag{3.8}$$

The logarithmic term occurs only if the expression in the bracket does not vanish; thus, in that case the coupling must depend on the indices of the electrons. The coupling can depend on the indices in three different ways:

(i) spin-dependent coupling (Kondo effect),

(ii) momentum-dependent coupling (orbital Kondo effect),

(iii) spin- and momentum-dependent coupling (realistic spin Kondo problem or Anderson model with orbital degeneracy).

In the spin Kondo problem the spin of the localized impurity interacts with the spin density of the conduction electrons at the impurity site. The "heavy object" represents the different states of the impurity spin [Abrikosov's (1965) pseudofermion representation]; thus, α is the z-component of that spin S^z. Denoting the matrix elements of the electron and impurity spin operators by $s^i_{s's}$ and $S^i_{\alpha\alpha'}$ ($i = x, y, z$), respectively, the Kondo coupling can be written as

$$V^{\alpha\alpha'}_{ss'} = \sum_i J^i s^i_{ss'} S^i_{\alpha\alpha'}. \tag{3.9}$$

In the realistic spin-rotation invariant case the exchange coupling is $J = J^x = J^y = J^z$.

The simplest case of the orbital Kondo problem is where a TLS is coupled to the conduction electrons. The TLS is described in terms of Pauli matrices, and the two states are characterized by $\sigma^z = \pm 1$. Then the coupling can be decomposed in terms of Pauli matrices as

$$V_{\kappa\kappa'}^{\alpha\alpha'} = \sum_i V_{\kappa\kappa'}^i \sigma_{\alpha\alpha'}^i, \tag{3.10}$$

where there are three new couplings $V_{\kappa\kappa'}^i$ ($i = x, y, z$). Using this notation the scattering amplitude given by eq. (3.8) can be written as

$$-2i\rho_0 \sum_{ij} \int \frac{\mathrm{d}S_{\kappa''}}{S} V_{\kappa\kappa''}^i V_{\kappa''\kappa'}^j \varepsilon^{ijk} \ln\left(\frac{D}{\max\{\omega, T, \Delta\}}\right), \tag{3.11}$$

where ε^{ijk} is the Levi–Civitta symbol. Logarithmic vertex corrections occur if the coupling is noncommutative; thus,

$$\sum_{ij} V_{\kappa\kappa''}^i V_{\kappa''\kappa'}^j \varepsilon^{ijk} \neq 0 \tag{3.12}$$

holds at least for some regions of the momenta. Thus, V^i must be different from zero at least for two values of the index i.

There are more complex situations, where the heavy particle moves by hopping and the coupling depends on the incoming and outgoing momenta of the heavy particle (the indices α are replaced by the momenta of the heavy particle). Such a model will be discussed briefly in section 9.

In the noncommutative models there are large logarithmic corrections, where $|V| \ln(D/\max\{\omega, T, \Delta\})$ may not be small even if $|V| \ll 1$. Thus, the logarithmic approximation in which the terms of the form $V^n \ln^{n-1}$ ($n = 2, 3, \ldots$) must be collected in the perturbation theory must be applied. A further improvement is the application of the next to leading logarithmic approximation, where the term $V^n \ln^{n-2}$ ($n = 2, 3, \ldots$) are also taken into account. In the following sections the TLS interacting with an electron gas will be discussed in great detail.

4. *Vertex renormalization in the leading logarithmic order*

The method to be discussed is not sufficiently powerful to get detailed information, but it can be used for the most general Hamiltonian and provides a good insight. Thus, we start with the Hamiltonian describing the heat bath and TLS [see eqs. (3.1)–(3.4)]. The second-order perturbation theory provides a logarithmic vertex correction given by eq. (3.11). The leading logarithmic corrections are due to the parquet diagrams, which can be cut into two pieces by cutting one electron and one pseudofermion line; the two pieces produced by the cut are also parquet diagrams or elementary vertices. These diagrams can be summed up by Anderson's "poor man's" scaling (Anderson 1970). The idea is that the effect of reduction of the bandwidth cutoff D to D' can be compensated by changing the couplings V^i to $V^i(D/D')$ in order to keep the scattering amplitude near the Fermi surface unchanged. A large part of the phase space can be eliminated in this way; thus, important logarithmic corrections do not occur in the scaled problem. The new couplings may, however, be so large that perturbation theory cannot be applied.

The scaling equations can be obtained by using the results in second-order [see eq. (3.11)], and they are

$$\frac{\mathrm{d}v^k_{\kappa\kappa'}}{\mathrm{d}\ln D} = 2\mathrm{i}\sum_{ij}\varepsilon^{ijk}\int\frac{\mathrm{d}S_{\kappa''}}{S}v^i_{\kappa\kappa''}v^j_{\kappa''\kappa'}, \tag{4.1}$$

where

$$v^i_{\kappa\kappa'} = \rho_0 V^i_{\kappa\kappa'} \tag{4.2}$$

are the dimensionless couplings. A matrix representation is introduced for the momentum dependence of the vertex:

$$v^i_{\kappa\kappa'} = v^i_{\kappa\kappa'} = \sum_{\alpha\alpha'} f^*_\alpha(\boldsymbol{\kappa}')v^i_{\alpha\alpha'}f_{\alpha'}(\boldsymbol{\kappa}), \tag{4.3}$$

where it is assumed that the coupling is sensitive only to the directions $\boldsymbol{\kappa}$ of the vectors $\boldsymbol{k}$, and $\{f_\alpha(\boldsymbol{\kappa})\}$ is a complete orthogonal set of functions defined on the Fermi surface. Thus,

$$\int\frac{\mathrm{d}S_\kappa}{S}f^*_\alpha(\boldsymbol{\kappa})f_{\alpha'}(\boldsymbol{\kappa}) = \delta_{\alpha\alpha'}. \tag{4.4}$$

In the case of a spherical Fermi surface the most convenient choice is the spherical harmonics.

The scaling equations (4.1) can be written in a matrix form as

$$\frac{\mathrm{d}v^k_{\alpha\alpha'}}{\mathrm{d}x} = 2\mathrm{i}\sum_{ij,\alpha''}\varepsilon^{ijk}v^i_{\alpha\alpha''}v^j_{\alpha''\alpha'}, \tag{4.5}$$

where $x = \ln(D_0/D)$ and the initial (unscaled) parameters correspond to $x = 0$ with cutoff $D = D_0$.

If the model is commutative, i.e., if the right-hand side of eq. (4.5) vanishes, then the parameters are not scaled. This is always the case when only one of the initial parameters is different from zero (electronic polaron model).

The general solution of the scaling equations (4.5) is rather complex, but the asymptotic behavior can be obtained easily for $x \gg 1$. The couplings $v^i_{\alpha\alpha'}$ do not increase for all the indices in a similar way; thus, we can pick out a subspace in which the increase is the strongest. The dimension of the subspace is denoted by

$(2S+1)$ $(S = \frac{1}{2}, 1, \frac{3}{2}, \ldots)$. In this dominating subspace the asymptotic behavior (nearby the fixed point) is

$$v^i_{\alpha\alpha'}(x) \to v^i(x)\hat{v}^i_{\alpha\alpha'}, \tag{4.6}$$

where $\hat{v}^i_{\alpha\alpha'}$ and $v^i(x) > 0$ gives the asymptotic form of the matrix structure and the amplitude, respectively. The new form of the scaling equations in the asymptotic region is (Vladár and Zawadowski 1983a)

$$\frac{\mathrm{d}v^k}{\mathrm{d}x} = 2v^i v^j \quad (i \neq j \neq k \neq i), \tag{4.7}$$

$$\hat{v}^k_{\alpha\alpha'} = \mathrm{i} \sum_{ij,\alpha''} \varepsilon^{ijk} \hat{v}^i_{\alpha\alpha''} \hat{v}^j_{\alpha''\alpha'}. \tag{4.8}$$

Thus, $\hat{v}^i_{\alpha\alpha'}$ must be one of the infinitesimal generators of the rotational group characterized by spin S [the dimension is $(2S+1)$]. The amplitudes can be characterized by a single function $\psi(x)$ as

$$v^i(x)^2 = \psi(x)^2 + v^i(0)^2, \tag{4.9}$$

where ψ satisfies

$$\frac{\mathrm{d}\psi^2}{\mathrm{d}x} = 4v^x v^y v^z \tag{4.10}$$

and x_0 is an arbitrary point such that $\psi(x_0) = 0$. For large enough x, $\psi \to +\infty$ and then $v^x \approx v^y \approx v^z \approx \psi$ (for $S = \frac{1}{2}$ see also Zawadowski 1980). Thus, the fixed-point Hamiltonian is an antiferromagnetic Kondo Hamiltonian with coupling $\psi > 0$, where the pseudo-spin $\boldsymbol{\sigma}$ of the TLS is coupled to another one $\boldsymbol{S}$ describing the orbital behavior of the electrons. The fixed-point Hamiltonian is isotropic; thus, a new conserved quantity $\boldsymbol{S} + \frac{1}{2}\boldsymbol{\sigma}$ is developed as a consequence of the scaling. It will be shown that the subspace is always two-dimensional.

In the physical model v^x and v^y are associated with tunneling; thus, $|v^x|$, $|v^y| \ll |v^z|$. The scaling equations (4.5) can be linearized in v^x and v^y, as far as v^z is the largest and the following matrix equations are obtained:

$$\begin{aligned}
\frac{\mathrm{d}v^z}{\mathrm{d}x} &= 0, \\
\frac{\mathrm{d}v^y}{\mathrm{d}x} &= -2\mathrm{i}[v^x, v^z]_-, \\
\frac{\mathrm{d}v^x}{\mathrm{d}x} &= 2\mathrm{i}[v^y, v^z]_-.
\end{aligned} \tag{4.11}$$

These equations can be solved in the representation where the initial matrix

$v^z_{\alpha\alpha'}$ is diagonal: $v^z_{\alpha\alpha'} = v^z_\alpha \delta_{\alpha\alpha'}$, and the solution is

$$\begin{aligned} v^z_{\alpha\alpha'}(x) &= \delta_{\alpha\alpha'} v^z_\alpha(0), \\ v^x_{\alpha\alpha'}(x) &= v^x_{\alpha\alpha'}(0) \cosh\{2[v^z_\alpha(0) - v^z_{\alpha'}(0)]x\}, \\ v^y_{\alpha\alpha'}(x) &= -iv^x_{\alpha\alpha'}(0) \sinh\{2[v^z_\alpha(0) - v^z_{\alpha'}(0)]x\}. \end{aligned} \tag{4.12}$$

The largest solutions belong to those index pairs $\{\alpha, \alpha'\}$, for which $|v^z_\alpha(0) - v^z_{\alpha'}(0)|$ is the largest. Thus, the dominant subspace is two-dimensional and the scaled Hamiltonian corresponds to a spin $S = \frac{1}{2}$ antiferromagnetic Kondo problem. The role of the electron spin variables will be discussed in section 6. Here the possibility of accidental degeneracy of the eigenvalues is ignored. The solution in that subspace with $v^z(0) \gg v^x(0), v^y(0) = 0$ (see section 2 for the physical model) is represented in fig. 4. The characteristic cutoff D, where the scaled couplings become of the order of unity, is called the Kondo temperature T_K. In the present case, which corresponds to the anisotropic Kondo problem discussed by Shiba (1970),

$$T^{\mathrm{I}}_{\mathrm{K}} = \frac{D_0}{k_{\mathrm{B}}} \left(\frac{v^x(0)}{4v^z(0)} \right)^{1/4v^z(0)}, \tag{4.13}$$

which is singular in $v^z(0)$; k_{B} is the Boltzmann constant, and the superscript I stands for the first-order logarithmic approximation. It can be seen that the completely anisotropic problems first become uniaxial ($v^x \approx v^y$) at $D \sim D_0 \exp\{-1/[4v^z(0)]\}$ and later isotropic.

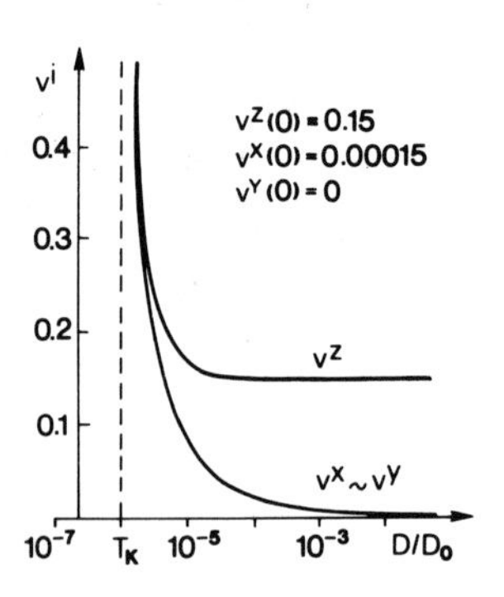

Fig. 4. Scaling of the couplings v^x, v^y, and v^z as calculated using eq. (4.7), with initial values $v^z(0) = 1.5 \times 10^{-1}$, $v^x(0) = 1.5 \times 10^{-4}$, and $v^y(0) = 0$. The crossover temperature given by eq. (4.13) is indicated by the dotted line.

5. *Multiplicative renormalization group in the next-to-leading logarithmic order*

The renormalization group treatment is capable of treating the transition (crossover) from a weak-coupling theory to a strong-coupling region by scaling. The leading logarithmic approximation is, however, not sufficient to treat the self-energies and, hence, the renormalization of the splitting Δ and tunneling rate Δ^x of the TLS. Furthermore, the crossover (Kondo) temperature T_K is not given accurately. For example, in the simple spin-$\frac{1}{2}$ single-channel Kondo problem the crossover temperature is given by the exchange coupling J as

$$T_K = D(2J\rho_0)^{1/2}\exp(-1/(2J\rho_0)), \tag{5.1}$$

where the factor $(2J\rho_0)^{1/2}$ does not occur in the leading logarithmic approximation. The general form of the scaling equation for coupling v is

$$\frac{\mathrm{d}v}{\mathrm{d}x} = \beta(v), \tag{5.2}$$

where $\beta(v)$ is a polynomial determined by perturbation theory, and the different couplings are represented by a single symbol v.

In the leading logarithmic approximation the polynomial must be determined in the second order, and the derivative of the contribution (3.11) generates the leading logarithmic approximation to all orders providing that the scaling holds. The next, the third-order, term in $\beta(v)$ results in the next-to-leading order, where vertex and self-energy corrections must be treated simultaneously. The problem of the TLS was treated by Vladár and Zawadowski (1983b) by using this method.

The typical diagrams for electron and TLS self-energies are depicted in fig. 5, which have the following analytical forms for electrons and TLS, respectively:

$$G_e = \frac{1}{\omega - \varepsilon_k - \Sigma_e(\omega)}, \tag{5.3}$$

$$\mathscr{G} = \frac{1}{\omega - \lambda_{ps} - \frac{1}{2}\sum_i \Delta^i\sigma^i - \Sigma(\omega)}, \tag{5.4}$$

where ω is the energy, $\Sigma_e(\omega)$ and $\Sigma(\omega)$ are the self-energies and a fictitious pseudo energy $\lambda_{ps} \to \infty$ is added to the TLS energy in order to avoid double

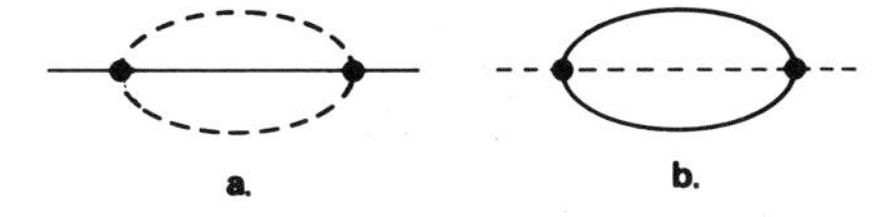

Fig. 5. Second-order self-energy correction for the electron (a) and the pseudofermion (b) Green's function. The dotted line represents the pseudoparticle and the solid circles the interactions.

occupation of the pseudoparticle. As $\lambda_{ps} \to \infty$, in the time-ordered perturbation technique for treating the electron–pseudoparticle vertex and the pseudoparticle self-energy, all the pseudoparticle lines run in the direction of the flow of time. The electron self-energy contains at least one pseudo-hole state, which carries a thermal factor $\exp(-\beta\lambda_{ps}) \to 0$; so it will be ignored in the renormalization. A proper normalization must be applied to cancel this unrealistic factor (Black et al. 1982).

The multiplicative renormalization group scheme provides a connection between the original quantities (Green's functions and couplings) and the scaled quantities with ω/D, V, Δ and ω/D', V', Δ' variables, respectively (see, e.g., Sólyom 1974). In the case of scaling the original and scaled functions have the same functional form, but they differ by multiplicative factors Z_1 and Z_2, which are independent of ω but depend on D'/D. Thus,

$$G_e(\omega/D', V', \Delta') = Z_1(D'/D, V)G_e(\omega/D, V, \Delta), \tag{5.5}$$

$$\mathscr{G}_{\alpha\alpha'}(\omega/D', V', \Delta') = Z_2(D'/D, V)\mathscr{G}_{\alpha\alpha'}(\omega/D, V, \Delta), \tag{5.6}$$

$$\Gamma^i_{\alpha\alpha'}(\omega/D', V') = Z_1(D'/D, V)^{-1} Z_2(D'/D, V)^{-1}\Gamma^i_{\alpha\alpha'}(\omega/D, V). \tag{5.7}$$

For convenience $\tilde{\Gamma}$ is defined as $\tilde{\Gamma}^i_{\alpha\alpha'} = V^i_{\alpha\alpha'}\Gamma^i_{\alpha\alpha'}$, where $\Gamma = 1$ for the bare quantities. (This notation is used as some of $V^i_{\alpha\alpha'}$ may be zero; thus, the corresponding $\Gamma^i_{\alpha\alpha'}$ is ambiguous. In what follows $Z_1 \equiv 1$ holds.)

The scheme of the renormalization-group transformation consists of two separate steps. At the start it is assumed that Δ^x and Δ^y have already been eliminated by two appropriate successive rotations around y and x axes. Then the steps are the following:

(i) Reducing the cutoff by a multiplicative renormalization-group transformation given by eqs. (5.5)–(5.7). It will be shown, however, that the transformation $\Delta^i \to \Delta'^i$ is such that Δ'^x and Δ'^y are generated even if Δ^x and Δ^y are zero before the transformation.

(ii) Δ'^x and Δ'^y can be eliminated by an appropriate rotation in each renormalization step. Nevertheless, if the scaling equations can be written in a rotation-invariant form, it is not necessary to perform this transformation explicitly in the scaling equations (as was done previously by Vladár and Zawadowski 1983b). Calculating the vertex and self-energy diagrams shown in figs. 6c and 5b, we have

$$\begin{aligned}\Gamma^i_{\alpha\alpha'} = V^i_{\alpha\alpha'} &- \sum_{jk,\alpha''} 2\mathrm{i}\rho_0(V^j_{\alpha\alpha''}V^k_{\alpha''\alpha'})\varepsilon^{ijk}\ln(D/|\omega|)\\ &+ \sum_j n_s\rho_0^2[2V^j_{\alpha\alpha'}\mathrm{Tr}(V^iV^j) - V^i_{\alpha\alpha'}\mathrm{Tr}(V^jV^j)]\ln(D/|\omega|),\end{aligned} \tag{5.8}$$

$$\Sigma(\omega) = -\sum_{ij} n_s\rho_0^2\mathrm{Tr}(V^iV^j)\sigma^i[(\omega-\lambda)I - \tfrac{1}{2}\Delta^z\sigma^z]\sigma^j\ln(D/|\omega|), \tag{5.9}$$

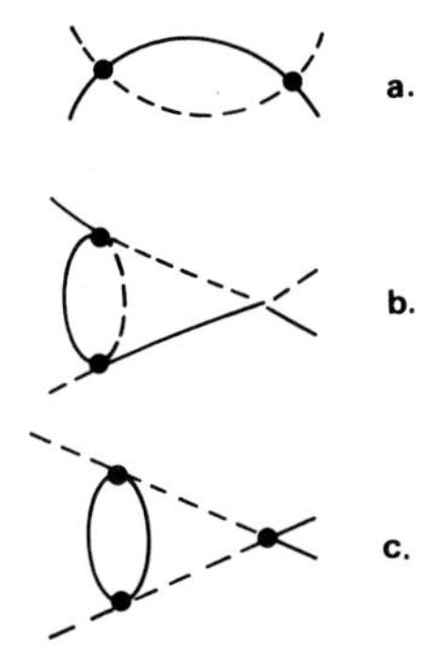

Fig. 6. Vertex corrections: (a) first-order correction; (b) second-order correction of parquet type; (c) non-parquet-type second-order correction. The diagrams (a) and (c) contribute to the β function of the renormalization-group equation for the couplings.

where n_s is the spin degeneracy of the conduction electrons (in reality, $n_s = 2$). The self-energy contains terms proportional to $\omega - \lambda$, which gives contribution to Z_2.

To construct the scaling equations, an infinitesimal step $D/D' = 1 + \mathrm{d}x$ $[x = \ln(D_0/D)]$ must be considered. Comparing eqs. (5.5)–(5.9) and taking $\Delta^i = 0$, the renormalization factor can be constructed in the second order:

$$Z_2(D'/D, V') = 1 - \sum_i n_s \ln(D'/D) \operatorname{Tr}(v^i v^i). \tag{5.10}$$

If Δ'^i is introduced, then, keeping the value of Z_2, it must be changed as follows:

$$\Delta'^i = \Delta^i + 2n_s \ln(D'/D) \sum_j [\Delta^i \operatorname{Tr}(v^j v^j) - \Delta^j \operatorname{Tr}(v^i v^j)] + O(v^3). \tag{5.11}$$

Then the following scaling equations are obtained:

$$\frac{\mathrm{d}v^i_{\alpha\alpha'}}{\mathrm{d}x} = -2\mathrm{i} \sum_{\alpha\alpha', jk} \varepsilon^{ijk} v^j_{\alpha\alpha''} v^k_{\alpha''\alpha'} - 2n_s \sum_j [v^i_{\alpha\alpha'} \operatorname{Tr}(v^j v^j) - v^j_{\alpha\alpha'} \operatorname{Tr}(v^j v^i)], \tag{5.12}$$

$$\frac{\mathrm{d}\Delta^i}{\mathrm{d}x} = -2n_s \sum_j [\Delta^i \operatorname{Tr}(v^j v^j) - \Delta^j \operatorname{Tr}(v^i v^j)]. \tag{5.13}$$

These scaling equations can be solved with some realistic restrictions on the initial couplings v^i (Vladár and Zawadowski 1983b). It can be shown that the dominant electron subspace is two-dimensional, just like in the leading logarithmic case. For the sake of simplicity, only the dominant components will be treated: $v^i_{\alpha\alpha'} = v^i \sigma^i_{\alpha\alpha'}$ and v^y $(x = 0) = 0$. Solving eq. (5.12) three regions must be considered separately:

(i) $v^y < v^x \ll v^z$,
(ii) $v^y \approx v^x \ll v^z$,
(iii) $v^y \approx v^x \approx v^z$,

The second-order scaling has an attractive fixed point at $v^x = v^y = v^z = 1/2n_s$, but in the physical case $n_s = 2$ the higher-order corrections are in the same order; thus, this approach breaks down. Nevertheless, $n_s = 2$ is an example of the nontrivial fixed point. (See section 7 for further discussion.) The characteristic cutoff D, where the couplings approach this value, is the Kondo temperature:

$$T_K^{II} = \frac{D_0}{k_B}\left(\frac{v^x(0)}{4v^z(0)}\right)^{1/4v^z(0)} [v^x(0)v^z(0)]^{n_s/4}. \tag{5.14}$$

The correction due to the second-order scaling is the factor $[v^x(0)v^z(0)]^{n_s/4}$, which shows similarity with the $(2\rho_0 J)^{n/2}$ obtained in eq. (5.1) for the Kondo temperature (where n is the orbital degeneracy and $n = 1$ in the conventional Kondo model). The present approach breaks down at $v^y \approx v^x \approx v^z \approx 1$, as the higher-order terms of the β function in eq. (5.2) become comparable with the lowest ones. For large enough n_s the couplings do not reach this region because of the fixed point, and the second-order scaling can remain valid also at $0 < T < T_K$ (Nozières and Blandin 1980, see also section 7).

Typical scaling curves are given in fig. 7, where both the leading logarithmic approximation in which v^i diverge at T_K^I and the present results are given. As $v^x/v^z \lesssim 10^{-3}$ [eq. (2.12)], v^z must be large, $v^z > 0.2$, in order to obtain $T_K \sim 1$ K.

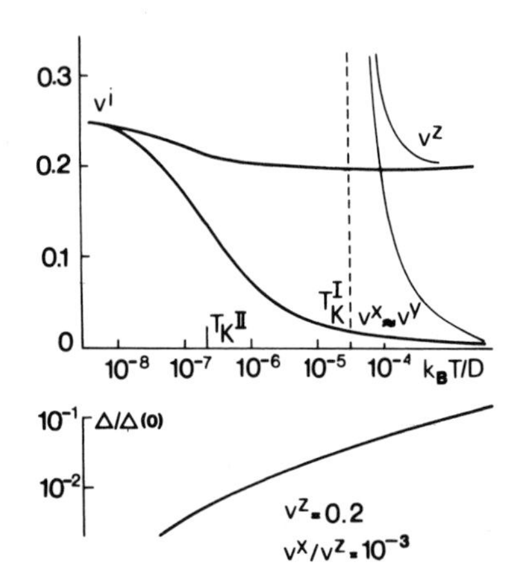

Fig. 7. Scaled couplings v^i ($i = x, y, z$) and energy splitting Δ as a function of $k_B T/D$. The initial v^z coupling is 0.2 and the initial ratio v^x/v^z is 10^{-3}. The narrow (heavy) lines represent the result of first- (second-) order scaling. T_K^I and T_K^{II} are the crossover temperature in the first- and second-order scalings, respectively. The dotted line is the asymptote of the coupling in the first-order scaling. The heavy lines for v^i are obtained analytically in second-order scaling. The region where $v^x \approx v^y$ does not hold is not represented. The ratio of the scaled and initial energy splitting $\Delta/\Delta(0)$ is calculated for the symmetric TLS by using eq. (5.15). (The index x of Δ^x is dropped)

The renormalization of Δ^i can be obtained for the special case $\Delta^z = \Delta^y = 0$, which describes the symmetric TLS. The equation is

$$\frac{\mathrm{d}\ln \Delta^x}{\mathrm{d}x} = 4n_s[(v^y)^2 + (v^z)^2]. \tag{5.15}$$

The v^i are not renormalized in the commutative model, and the result is

$$\Delta^x(x) = \Delta^x(0)\exp(-2n_s v_z^2 x), \tag{5.16}$$

which was obtained first by Kondo (1976a) and reproduced by several authors. The scaling is stopped by Δ^x for very small x, as it becomes the dominating energy scale in expression (3.7). Then $\Delta^x(x^*)/D_0 = x^*$, and Δ remains constant for $D/D_0 < x$:

$$\Delta^* = \Delta(0)\left(\frac{\Delta(0)}{D}\right)^{2n_s v_z^2/(1-2n_s v_z^2)}. \tag{5.17}$$

In the noncommutative model, however, v^z depends on x and, therefore, Δ can be reduced by at least one order of magnitude as can be seen in fig. 7. A more detailed discussion shows that the more symmetric the potential the larger is the reduction of energy splitting.

6. *Path integral method applied to the case of large phase shift: one-dimensional Coulomb gas*

The path integral method is very elaborate; therefore, it cannot be followed without studying the original papers (e.g., Vladár et al. 1988a, b). The present section provides only a brief outline by pointing out some of the crucial steps and the additional complications compared to the treatment of the Kondo problem by Yuval and Anderson (1970).

In the previous treatment the renormalization group was constructed by determining the β-function up to the third order in the perturbation theory. This treatment is consistent in the weak-coupling limit. In the physical applications, however, one of the couplings (v^z) can be large and the two others (v^x, v^y) are small. This case is a hybrid of weak-coupling and strong-coupling theory. A systematic treatment has been worked out in which the phase shift associated with an arbitrary large v^z combines with small v^x and v^y. The renormalization group is constructed by using the path integral method only in the leading order of v^x and v^y; thus, the second-order renormalization group results given by eq. (5.12) cannot be reproduced.

The main scheme is due to Anderson, Yuval and Hamann (Hamann 1970, Yuval and Anderson 1970, Anderson et al. 1970). The TLS flips between two states. Following each flip, a screening cloud starts to build up. If the flips are far enough in time (see fig. 8), then that screening can be described by the long-time approximation of Nozières and de Dominicis (1969) for arbitrary phase shift. Assuming that the flips are spontaneous, the problem is solved by Yu and Anderson (1984) using the method of Anderson et al., where the free energy is constructed in the complex-time path integral technique. The electron variables are integrated out; thus, the free energy depends only on the classical path of the TLS on the time interval $(0, \beta)$, where $\beta = 1/k_B T$ (see fig. 8). The interaction between the flips is logarithmic-like in a one-dimensional Coulomb gas.*

The present problem is more complex, as was discussed by Vladár et al. (1988a, b), as, if the electron variables are integrated out then the partition function still depends explicitly on the time points of the interactions and the couplings. In fig. 9 a classical path $\sigma^z(\tau)$ of the TLS is depicted and the different interactions with the electron gas are also shown. The electron lines are the solid lines decorated by the spin and orbital variables,** which will be denoted by the index pair (m, n) for the incoming and outgoing electrons, respectively. These new indices will be called the "color" of the electron line. The flips are either associated with the electrons (electron-assisted flips) or not (spontaneous flips). The probability density of the flips is given in terms of different fugacities y.

Each flip is associated with a variable $\hat{C}_i$, where i is the index of the flip. $\hat{C}_i$ is a matrix in the electron color space and is defined to be diagonal in the representation, where $\hat{V}^z$ is diagonal. Thus, $\hat{C}_i$ can be defined first in the representation where $\hat{V}^z$ is diagonal and the general matrix form for an arbitrary representation can be obtained by applying a rotation. The interaction between two flips is described by

$$\mathrm{Tr}(\hat{C}_i \hat{C}_j) \ln \left| \frac{\tau_i - \tau_j}{\tau_0} \right|, \tag{6.1}$$

assuming that $|\tau_i - \tau_j| \gg \tau_0$, where τ_0 is the short-time cutoff in the TLS dynamics, beyond which the long-time approximation and the model Hamiltonian can be used for the electrons ($\tau_0 \approx D^{-1}$) working in the coordinate

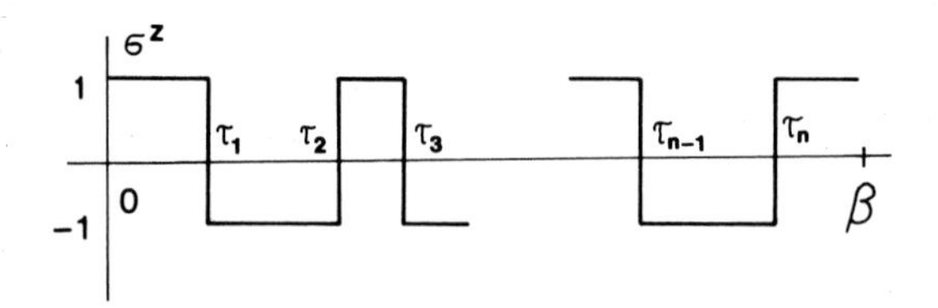

Fig. 8. A typical classical path of the TLS in the time interval $[0, \beta]$.

* In the work of Yu and Anderson (1984), starting with a single-well potential, the formation of the double-well TLS is considered as well.

** In the case of the TLS the orbital variable is the angular momentum, while for a moving heavy particle it is the momentum (Zimányi et al. 1987)

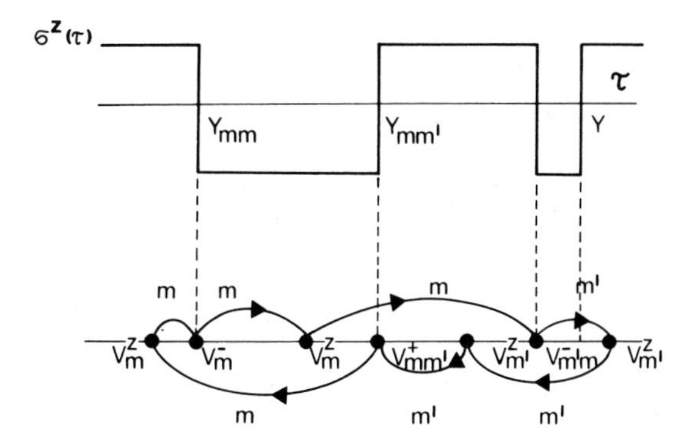

Fig. 9. A classical path $\sigma^z(\tau)$ of the TLS, with the electron loop representing an electron interacting with the TLS. On the upper part of the figure showing the path of the TLS, the fugacities y and y_a associated with spontaneous and assisted tunneling are also indicated. The lower part shows an electron loop composed of electron lines with color m and m' and the solid circles represent the interaction with couplings V^z_m, $V^+_{mm'}$, and $V^-_{mm'}$. The couplings V^+ and V^- are associated with tunneling processes while V^z is not. The color indices correspond to the colors of the incoming and outgoing electrons.

system where V^z is diagonal. The diagonal elements of the charges $\hat{C}_i$ are defined as

$$C_i^\mu = \delta_\mu(\tau_i + 0)/\pi - \delta_\mu(\tau_i - 0)/\pi \tag{6.2}$$

for nonassisted tunnelings and

$$C_i^\mu = \delta_\mu(\tau_i + 0)/\pi - \delta_\mu(\tau_i - 0)/\pi + \delta_{\mu m} - \delta_{\mu n} \tag{6.3}$$

for assisted tunnelings, where

$$\delta_\mu(\tau) = -\arctan(\pi \rho_0 V^z_\mu(\tau)) \tag{6.4}$$

is the phase shift in channel μ and $\delta_{\mu m}$ and $\delta_{\mu n}$ with double indices are the Kronecker symbols. The diagonal matrices $\hat{V}^z(\tau)$ and $\hat{\delta}(\tau)$ may take only two values, $\pm\hat{V}^z$ and $\pm\hat{\delta}$, respectively. The last two terms of eq. (6.3) occur only for the electron-assisted processes of the type $V_{mn}a^\dagger_m a_n$. These charges can be generated from the "polarization"

$$p^\mu(\tau) = \delta_\mu(\tau)/\pi + N_\mu(\tau) \tag{6.5}$$

in the following way:

$$\frac{\mathrm{d}\hat{p}}{\mathrm{d}\tau} = \sum_i \hat{C}_i \hat{\delta}(\tau - \tau_i), \tag{6.6}$$

where δ_μ/π gives the number of the electrons screening the TLS in channel μ in equilibrium (Friedel sum rule), and $N_\mu(\tau)$ is the deviation of the population of channel μ from its equilibrium at a moment τ, which changes in the assisted processes.*

The mapping to a Coulomb gas is similar to the one introduced by Yuval and Anderson for the anisotropic Kondo problem [see eqs. (3.4) and (3.9) for the Hamiltonian]. In that case the electron color index is the spin $s = \pm\frac{1}{2}$. The classical path of the localized spin, in contrast to the TLS (see fig. 9), determines the spin flips of the electrons, as the total spin is conserved. Using the matrix $\hat{\delta} = \delta\hat{\sigma}^z$ in the space of the two spin directions, the charge defined by eq. (6.3) becomes

$$\hat{C}_i \to S_i(1 - 2\delta/\pi)\hat{\sigma}^z, \tag{6.7}$$

where the sign $S_i = \pm 1$, depending upon whether the spin flips up or down. Here the interaction is [see eq. (6.1)]

$$2S_iS_j(1 - 2\delta/\pi)^2 \ln\left|\frac{\tau_i - \tau_j}{\tau_0}\right|. \tag{6.8}$$

Thus, the path of the classical localized spin determines the contribution completely in the Kondo problem. In the TLS problem, however, the electron color index cannot be eliminated. The classical path must be decorated by the incoming and outgoing electron indices and summation over all possible paths must be accompanied by all possible decorations (fig. 9).

A further difficulty arises from those flip processes where the color index is conserved (diagonal in the color). Until now, only those flips were considered where the outgoing and incoming electrons were associated with different channels n are m, and the last two components of charge [eq. (6.3)] were associated with that flip. It will be shown that the diagonal case can be considered as two closely placed charges of opposite signs corresponding to the outgoing and incoming electron lines in the same channel. In this way, additional charge–dipole and dipole–dipole interactions occur between two flips (Vladár et al. 1988a, Vladár 1991).

In the Anderson–Yuval method the renormalization group is constructed by eliminating the short-time behavior by changing $\tau_0 \to \tau_0 + \mathrm{d}\tau_0$ in infinitesimally small steps. There are two contributions, one from the explicit dependence on τ_0 and the other from the eliminations of close pairs of flips. In the present case the flips can be spontaneous or assisted.** The description of the main steps of the method will be presented in the following.

* In the calculations being presented here only the commutability of the two values of $\hat{\delta}(\tau)$ is used, and they can be generated for arbitrary commuting values in a straightforward way [see eq. (6.25)].

** In the multiplicative renormalization group, in contrast to the general problem, only assisted flips are eliminated, as the phase space of the electrons is reduced instead of eliminating the short-time processes. The first work using the Anderson–Yuval method for TLS with small phase shift is by Black and Györffy (1978), and the differences between the two techniques show up in the renormalization equation for the spontaneous flip.

(i) A classical path $\sigma^z(\tau)$, $|\sigma^z(\tau)| = 1$, of the spin is considered, where the flips can be either spontaneous or assisted (fig. 9).

(ii) In the presence of the potential due to the classical spin, $V^z_{k_1 k_2}\sigma^z(\tau)$, the Dyson equation of the one-electron Green's function is solved. Instead of treating the momentum dependence, the spherical hormonic variables are introduced. Then a new set of a linear combination of the spherical harmonic variables $\{m\}$ is introduced to have $\hat{V}^z$ diagonal, with elements V^z_m. The corresponding interaction Hamiltonian is H_1. Applying the technique of Nozières and de Dominicis (1969), the Green's function depends on the time positions τ_i of flip i. The Green's function in channel m has the form, for $|\tau - \tau'| \gg \tau_0$,

$$G_\mu(\tau, \tau') = \frac{\rho_0}{\tau - \tau'} \cos^2 \delta_\mu \exp\left(\sum_i c_i^\mu \ln\left|\frac{\tau_i - \tau'}{\tau_i - \tau}\right|\right), \tag{6.9}$$

where $(\tau - \tau')^{-1}$ is the long-time approximation of the unperturbed Green's function, and

$$c_i^\mu = \delta_\mu(\tau_i + 0)/\pi - \delta_\mu(\tau_i - 0)/\pi \tag{6.10}$$

is the sum of the first two terms on the right-hand side of eq. (6.3). The partition function is (Vladár et al. 1988a)

$$Z_i = \prod_m (\cos\delta_m)^{-2\beta/\tau_0} \exp\left\{\sum_i \mathrm{Tr}(\hat{c}_i)\frac{\tau_i}{\tau_0} + \sum_{i<j} \mathrm{Tr}(\hat{c}_i\hat{c}_j) \ln\left|\frac{\tau_i - \tau_j}{\tau_0}\right|\right\}. \tag{6.11}$$

(iii) The classical path must be decorated first by deciding which flips are assisted and then by attributing colors to all the incoming and outgoing lines of the assisted flips (fig. 9). All possible combinations are treated separately. The contribution of the electron–TLS interaction H_2 to the free energy must be calculated next, where the couplings associated with flips are either $\Delta^\pm = \Delta^x \pm \Delta^y$ or $V^\pm_{mn} = V^x_{mn} \pm V^y_{mn}$. The electronic part of the free energy is proportional to expectation values which are interpreted in the interaction representation, where $H_0 + H_1$ is the unperturbed Hamiltonian. A typical expectation value is

$$\langle \tilde{H}_{2,N}(\tau_N)\tilde{H}_{2,N-1}(\tau_{N-1}) \cdots \rangle, \tag{6.12}$$

where $H_{2,r}$ ($N \geqslant r \geqslant 1$) is scalar if the tunneling is spontaneous and depends on the colors m and n of the incoming and outgoing electron lines if the tunneling is assisted. The expectation value factorizes with respect to the possible color subspaces μ. Considering a given color μ the contribution is a sum of all possible pairings of creation and annihilation operators. The Green's function is given by eq. (6.9). The exponential term can be collected easily and is independent of the pairing chosen. Using the unperturbed parts of Green's functions, $(\tau - \tau')^{-1}$, the sum of all possible pairings can be written as a Cauchy determinant of the order of N_μ (Yuval and Anderson 1970):

$$\begin{vmatrix} \dfrac{1}{\tau_{i_1}-\tau_{j_1}} & \dfrac{1}{\tau_{i_1}-\tau_{j_2}} & \cdots \\ \dfrac{1}{\tau_{i_2}-\tau_{j_1}} & \dfrac{1}{\tau_{i_2}-\tau_{j_2}} & \cdots \\ \vdots & \vdots & \ddots \end{vmatrix} = \frac{\prod\limits_{i<i'}(\tau_i-\tau_{i'})\prod\limits_{j>j'}(\tau_j-\tau_{j'})}{\prod\limits_{i,j}(\tau_j-\tau_i)}, \tag{6.13}$$

where $\{\tau_i\}$ and $\{\tau_j\}$ $(i, j = 1, \ldots, N_\mu)$ are the set of those time-ordered flipping times where an electron with color μ is emitted or absorbed, respectively. The determinant diverges if an electron with color μ is absorbed or emitted at the same flip (time).

Combining the value of the determinant with the product of the exponentials from the Green's functions (6.9), the new expression can be written as an exponential function with a new exponent, which is just the expression (6.1) with the definitions (6.2) and (6.3) of the charges $\hat{C}_i$.

The treatment of the divergence mentioned above, which is due to assisted flips diagonal in the color indices, must be handled with special care. The times of the incoming and outgoing lines coincide in eq. (6.13). Thus, in order to avoid the divergence, these must be separated by τ_0 at a diagonal flip. On the other hand, at shorter time difference the asymptotic form of eq. (6.9) is not valid either. Such a flip i may interact with another flip j. The parts of the charges associated with the incoming and outgoing lines have opposite sign; thus, the contribution to the interaction in eq. (6.11) is proportional to

$$\ln\left(\frac{\tau_i-\tau_j}{\tau_0}\right) - \ln\left(\frac{\tau_i+\tau_0-\tau_j}{\tau_0}\right) = \frac{\tau_0}{\tau_i-\tau_j}, \tag{6.14}$$

a τ-dependence just like in the interaction of a dipole at τ_i with a charge at τ_j via logarithmic Coulomb force. For a diagonal scattering in channel m the matrix form of its strength is

$$\{\hat{P}_i\}_{\mu\mu'} = \delta_{m\mu}\delta_{m\mu'}, \tag{6.15}$$

and the charge separation τ_0 was put into the interaction given by eq. (6.14). For more details we refer to the original papers (Vladár et al. 1988a, Vladár 1991), where the dipoles were defined with an arbitrary small separation instead of τ_0.

In the expression for the partition function the different classical paths of $\sigma^z(\tau)$ are summed. In a particular term for each spin flip a weight factor occurs, which is called fugacity. For spontaneous tunneling it is defined as

$$y^\pm = \Delta^\pm\tau_0, \tag{6.16}$$

and for the assisted ones

$$y^\pm_{mn} = V^\pm_{mn}\rho_0\cos\delta_m\cos\delta_n. \tag{6.17}$$

In the second expression the indices m and n stand for the color index in a similar way as in eq. (6.3). Furthermore, there are two more extra fugacities y^z and y^z_{mn}, which can be generated by the scaling. As the processes corresponding to y^z and y^z_{mn} are taken into account, without perturbation theory, by $y^z_{\rm HF} = y^z + {\rm Tr}(\hat{\delta}/\pi)$ (the subscript HF refers to the Hartree–Fock correction) and the phase shifts δ_μ, they can be generated in each step of scaling but their effects are immediately taken into account by the renormalization of $y^z_{\rm HF}$ and δ_μ. Thus, each step of scaling starts with the assumption that y^z and y^z_{mn} are zero.

The general form of the partition function is

$$
\begin{aligned}
Z_{\rm I} = {} & \sum_{N=0}^{\infty} (-1)^N \sum_{\{\alpha\}} \sum_{\{m,n\}} \prod_{j=1}^{N} y_j \,{\rm Tr}\left(\prod_{i=1}^{N} \sigma(\alpha_i)\right) \\
& \times \tau_0^{-N} \int_0^{\beta} {\rm d}\tau_N \cdots \int_0^{\tau_{i+1}-\tau_0} {\rm d}\tau_i \cdots \int_0^{\tau_2-\tau_0} {\rm d}\tau_1 \, R \\
& \times \exp\left[-2y^z_{\rm HF} \sum_i \frac{S_i \tau_i}{\tau_0}\right] \exp\left[\sum_{i<j} {\rm Tr}(\hat{C}_i \hat{C}_j) \ln\left|\frac{\tau_i - \tau_j}{\tau_0}\right|\right. \\
& \left. + \sum_{i \neq j} {\rm Tr}(\hat{C}_i \hat{P}_j) \frac{\tau_0}{\tau_j - \tau_i} + \sum_{i<j} {\rm Tr}(\hat{P}_i \hat{P}_j) \left(\frac{\tau_0}{\tau_i - \tau_j}\right)^2\right]. \qquad (6.18)
\end{aligned}
$$

The interaction points τ_i are on the imaginary-time axis in the interval $(0, \beta)$. Each point is associated with one of the fugacities $y^{\pm}, y^z, y^{\pm}_{mn}$ and y^z_{mn}. That association is called configuration $\{\alpha\}$. If the fugacity describes an assisted process, then the colors of the outgoing and incoming electron lines (m, n) must also be given, which forms another configuration $\{m, n\}$. Finally, $R = \pm 1$ is a combinational factor, defined in the original paper of Vladár et al. (1988a).

(iv) In the last step the scaling is performed. The short-time cutoff τ_0 is replaced as $\tau_0 \to \tau_0 + {\rm d}\tau_0$. That has two effects. The cutoff τ_0 appears explicitly and its change must be compensated by changing the couplings. Furthermore, the pairs with distance in the interval $(\tau_0, \tau_0 + {\rm d}\tau_0)$ must be eliminated and compensated. For example, a pair with total charge $\hat{C}_i + \hat{C}_{i+1} = 0$ can be replaced by a dipole. The actual derivation of the scaling equations is very lengthy; we therefore refer the reader to the original paper for more details (Vladár et al. 1988b).

The derivation shows that only those two orbital channels must be kept for which the difference of the two eigenvalues, $\delta_n - \delta_m$, is the largest. Similar results were obtained by applying the multiplicative renormalization group in section 4, following eq. (4.12). In these two channels the quantities $\hat{\delta}$ and $\hat{y}^{\pm}$ can be written with the aid of Pauli matrices $\hat{\sigma}^z_{\rm orb}$ and $\hat{\sigma}^{\pm}_{\rm orb}$, where the subscript "orb" makes the distinction from the TLS space. In the following the simplest forms will be assumed:

$$\hat{\delta}/\pi = (\delta/\pi)\hat{\sigma}^z_{\rm orb}, \qquad (6.19)$$

$$\hat{y}^{\pm} = 2y_{\rm a}\hat{\sigma}^{\pm}_{\rm orb}, \qquad (6.20)$$

$$y^+_{\rm HF} y^-_{\rm HF} = y^2, \qquad (6.21)$$

where the fugacities $y^{\pm}_{\rm HF} = y^{\pm} + {\rm Tr}(\hat{y}^{\pm})$ contain the Hartree–Fock contributions too. In order to avoid further complexities a symmetrical TLS is considered here ($\Delta^z = 0$ and $y^z_{\rm HF} = 0$). Using this notation the following scaling equations are obtained:

$$\frac{{\rm d}\delta/\pi}{{\rm d}\ln\tau_0} = 4y_{\rm a}^2(1 - 2n_{\rm s}\delta/\pi) - 2y^2\delta/\pi, \tag{6.22}$$

$$\frac{{\rm d}y_{\rm a}}{{\rm d}\ln\tau_0} = 4y_{\rm a}(\delta/\pi)(1 - n_{\rm s}\delta/\pi), \tag{6.23}$$

$$\frac{{\rm d}y}{{\rm d}\ln\tau_0} = y(1 - 4n_{\rm s}(\delta/\pi)^2). \tag{6.24}$$

The result obtained can be extended to even more general cases (Vladár 1991).*

* In the general case the two values of $\hat{\delta}(\tau)$ differ not only in their sign but also in their amplitudes. The derivations can be easily generated for the case when $\hat{\delta}(\tau)$ has two arbitrary commuting values $\hat{\delta}^1$ and $\hat{\delta}^2$, which can be diagonalized simultaneously. Then, corresponding to eqs. (6.10) and (6.17), our scaling variables will take the following new values:

$$\hat{\delta} \rightarrow \tfrac{1}{2}(\hat{\delta}^1 - \hat{\delta}^2), \tag{6.25}$$

$$\hat{y}^+ \rightarrow \cos\hat{\delta}^1\,\hat{V}^+\,\rho_0\cos\hat{\delta}^2, \qquad \hat{y}^- \rightarrow \cos\hat{\delta}^2\,\hat{V}^-\,\rho_0\cos\hat{\delta}^1 \tag{6.26}$$

in the scaling equations above (Vladár 1991). The two values $\hat{\delta}^1$ and $\hat{\delta}^2$, however, do not commute in general, e.g., for that TLS where a particle changes its position in the real space. Most steps of the derivation of scaling equations can be repeated for noncommuting $\hat{\delta}^1$ and $\hat{\delta}^2$. Although the Green's function and the partition function cannot be expressed in such a closed form as in eqs. (6.9) and (6.18), the scaling equations can be derived with the same accuracy (Vladár 1991). The consequence of the noncommutativity appears in eq. (6.25), modifying it to

$$\hat{\delta} = \frac{1}{4{\rm i}}\ln[\exp(2{\rm i}\hat{\delta}^1)\exp(-2{\rm i}\hat{\delta}^2)]. \tag{6.27}$$

Its eigenvalues δ_μ are half of the phase shifts in the relative S-matrix $\exp(4{\rm i}\hat{\delta}) = \exp(2{\rm i}\hat{\delta}^1)\exp(-2{\rm i}\hat{\delta}^2)$. This matrix connects the two sets of electron eigenstates corresponding to the potentials $\hat{V}^1$ and $\hat{V}^2$. It can be shown for its eigenvalues that $|\delta_\mu| \leqslant \pi/4$, if $\hat{\delta}^1$ and $\hat{\delta}^2$ are the phase shift matrices of the same scattering center in different positions (Kagan and Prokof'ev 1989). Then $\hat{\delta}^2 = \hat{U}\hat{\delta}^1\hat{U}^+$, where $\hat{U}$ is the translation operator characterized by the vector $\boldsymbol{d}$ (fig. 2). The eigenvalues of $\hat{\delta}^1$ and the components of $\boldsymbol{d}$ span a parameter space where those points are singular, for which $\delta_\mu = (\frac{1}{4} + \frac{1}{2}n)\pi$ (n is an integer) for at least one value of μ. Here

$$\det[\exp(4{\rm i}\hat{\delta}) + 1] = \det[2\cos^2(2\hat{\delta}^1)]\det[\tan(2\hat{\delta}^1)\hat{U}\hat{P} + {\rm i}]\det[\tan(2\hat{\delta}^1)\hat{U}\hat{P} - {\rm i}] = 0,$$

where $\hat{P}$ is the space inversion operator and it was supposed that the scattering is invariant with respect to $\hat{P}$: $\hat{P}\hat{\delta}\hat{P} = \hat{\delta}$. The equation above can be satisfied by at least two real equations for the parameters. This means that the dimension of the set of the singular points is at least two less than that of the parameter space itself, and so the set of the nonsingular points is a connected manifold. As the latter set contains the region $\boldsymbol{d} = 0$, for which $\hat{\delta} = 0$, $|\delta_\mu| < \pi/4$ for each μ and each nonsingular point.

The right-hand sides of the scaling equations (6.22)–(6.24) are derived in the leading order with respect to the fugacities. As the right-hand sides of the equations for the fugacities are linear in the fugacities, the pair elimination proportional to the product of two fugacities does not contribute. In the case of the phase shift the pair elimination, however, plays an important role.

The scaling equations can be solved assuming that δ is constant in eqs. (6.23) and (6.24):

$$y_a = y_a(0)\left(\frac{\tau_0}{\tau_0(0)}\right)^{4[\delta(0)/\pi][1-n_s\delta(0)/\pi]}, \tag{6.28}$$

$$y = y(0)\left(\frac{\tau_0}{\tau_0(0)}\right)^{1-4n_s[\delta(0)/\pi]^2}, \tag{6.29}$$

$$\delta/\pi = [\delta(0)/\pi] + \frac{[1-2n_s\delta(0)/\pi][y_a^2 - y_a(0)^2]}{2[\delta(0)/\pi][1-n_s\delta(0)/\pi]} - \frac{[\delta(0)/\pi][y^2 - y(0)^2]}{1-4n_s[\delta(0)/\pi]^2}. \tag{6.30}$$

In eq. (6.30) the dependence on τ_0 occurs through y_a and y, and the initial values are denoted as values at 0. The general features of this solution are as follows:

(i) y_a increases only if $\delta(0) < \pi/n_s$, which can never be the case for $n_s = 2$, as $\delta(0) < \pi/2$ is assumed;

(ii) y increases if $\delta(0) < \pi/(2\sqrt{n_s})$.

The behavior of the phase shift δ will be discussed later.

The temperature dependences of the different quantities are given by the limit $\tau_0 \sim kT$ obtained by scaling. The scaling may break down before reaching that limit if either y_a or $y^{\pm}$ become of the order of unity and, thus, the Coulomb gas becomes dense.

The Kondo temperature T_K, which gives the crossover between the weak- and strong-coupling limits ($y_a = 1$) is

$$T_K = D_0 y_a(0)^{1/4[\delta(0)/\pi][1-n_s\delta(0)/\pi]}, \tag{6.31}$$

which is similar to the expression obtained in the weak-coupling limit $y_a \ll \delta/\pi \ll 1$, with $\delta/\pi = v^z$ [see eq. (5.14)], but it contains only those factors which depend on v^x in a singular way and not those which are singular only in v^z.

Considering the other case, where $y^{\pm} = \Delta\tau_0$ becomes of the order of unity, the breakdown of the scaling methods shows up differently while applying different methods. In the present method the series in eq. (6.18) does not converge, while in the multiplicative renormalization group the logarithmic terms containing the factor (3.7) are not large any more as $D \sim 1/\tau_0 \sim \Delta$; thus, the "parquet" diagrams are not dominating the others (all diagrams must be considered). The temperature at which $y = 1$ is reached is called the freezing temperature T_f. For $T < T_f$ the spontaneous flipping rate washes out the temperature dependence.

The freezing temperature is

$$T_{\mathrm{f}} = D_0 y(0)^{1/\{1-4n_{\mathrm{s}}[\delta(0)/\pi]^2\}}. \tag{6.32}$$

A very striking feature of the scaling equation (6.23) for y_{a} is that in the case $n_{\mathrm{s}} \leqslant 2$ and $\delta/\pi < \frac{1}{2}$ the fugacity y_{a} always increases. Thus, localization in one of the two potential wells ($y \to 0$ and $y_{\mathrm{a}} \to 0$) cannot occur due to the presence of an arbitrary small assisted-tunneling process. This situation is contrary to the case $y_{\mathrm{a}} = 0$, where localization always occurs for large enough phase shift; thus, $y \to 0$ if $4n_{\mathrm{s}}\delta/\pi > 1$. It is interesting to note that if the spin degeneracy is unphysically large ($n_{\mathrm{s}} > 2$), then the screening may dominate the assisted processes, and the appearance of a localization cannot be ruled out.

The general behavior of the scaling derived is very complex. Considering the two limiting cases where $y = 0$ or $y_{\mathrm{a}} = 0$, the phase shift δ has attractive fixed points $\delta = \pi/2n_{\mathrm{s}}$ and $\delta = 0$, respectively. In the general case there are many possibilities, which will not be discussed here. There is, however, a general tendency, that the screening described by δ and y always reduces the value of δ, while the assisted tunneling does the opposite in the case $\delta < \pi/2n_{\mathrm{s}}$.

The problem with $y = 0$ ($\Delta = 0$) corresponds to the two-channel Kondo problem ($n_{\mathrm{s}} = 2$) discussed by Nozières and Blandin (1980). In this problem the degeneracy occurs not in the spin but in the orbital space. The fixed point is $\delta = \pi/2n_{\mathrm{s}}$. A further study of this problem with $y_{\mathrm{a}} \neq 0$, $y = 0$ will be discussed in section 7.

7. *Low-temperature results*

As is shown in sections 4–6, the scaling equation indicates that the effective scaled coupling is in the strong-coupling region at low temperatures. Thus, the analytical version of the renormalization group cannot be applied and other methods must be used for the Hamiltonians obtained by the scaling. The method of the Bethe Ansatz is so complex that only the final results will be discussed. Analytical methods can be applied only to one exceptional case where the spin degeneracy n_{s} of the electron gas is chosen artificially large: $n_{\mathrm{s}} \gg 1$ ($n_{\mathrm{s}} = 2$ in the physical case). The right-hand side of the scaling equations (5.2) and (5.12) for the renormalized coupling strength is a series of terms like $n_{\mathrm{s}}^i v^j$. Here $j \geqslant 2i + 1$, because a factor n_{s} may come only from a closed electron loop containing at least two vertex points. Moreover, at least one additional vertex is needed for the incoming and outgoing electron lines. That has the important consequence that the scaling of the invariant coupling leads to a finite value which is one of the zeros of the right-hand side of the scaling equation and is called a fixed point. The fixed point of the coupling, obtained from the next-to-leading order [see eq. (5.2)] is proportional to $1/n_{\mathrm{s}}$; thus, the further corrections are of the orders $1/n_s^2$, $1/n_s^3$, . . . A similar situation appears in the multichannel Kondo problem, where the degeneracy n of the conduction electron orbits is introduced. This model was proposed by Cragg and Lloyd (1978) and studied by Nozières and Blandin (1980) in a general context, showing that if $n > 2S$, where $2S + 1$ is the degree of freedom of the impurity (spin, TLS), then the fixed point must be finite.

The thermodynamic properties of the TLS at low temperatures were studied by Muramatsu and Guinea (1986) using a $1/n_s$ expansion. At low temperatures $T \ll T_K$, the different physical quantities are power functions of the temperature and the exponents are expressed in terms of the fixed point proportional to $1/n_s$ in the weak-coupling limit. More general results were derived for the multichannel Kondo problem by applying the Bethe Ansatz (Andrei and Destri 1984, Wiegmann and Tsvelik 1983, Tsvelik and Wiegmann 1984, Tsvelik 1985). An even more detailed study has been recently performed by Sacramento and Schlottmann (1991) by solving numerically the Bethe Ansatz equations at finite temperature and field.

In the multichannel Kondo problem the external magnetic field H replaces the splitting Δ^z of the TLS.

The results obtained are different for $n_s > 2$ $(S > \frac{1}{2})$ and $n_s = 2$ $(S = \frac{1}{2})$, which is the marginal case with logarithmic behavior. For $n_s > 2$ the results for the impurity susceptibility χ_{imp} and specific heat C_{imp} are

$$\chi_{imp} \propto (H/T_K)^{(2/n_s)-1}, \quad (T = 0), \tag{7.1}$$

$$C_{imp}/T \propto (T/T_K)^{\tau-1}, \quad (H = 0), \tag{7.2}$$

$$\chi_{imp} \propto (T/T_K)^{\tau-1}, \quad (H = 0), \tag{7.3}$$

where $\tau = 4/(n_s + 2)$ and T_K is the Kondo temperature. The exponents are in agreement with those obtained by Muramatsu and Guinea (1986) in the limit $n_s \gg 1$.

In the marginal case $n_s = 2$

$$\chi_{imp} \propto \ln(H/T_K), \quad (T = 0), \tag{7.4}$$

$$C_{imp}/T \propto \ln(T/T_K), \quad (H = 0), \tag{7.5}$$

$$\chi_{imp} \propto \ln(T/T_K), \quad (H = 0). \tag{7.6}$$

For a more complete study see Sacramento and Schlottmann (1991). The most relevant result from the point of view of the TLS is that the susceptibility diverges as $H \to 0$; thus, an infinitesimally small initial splitting Δ^z results in a finite renormalized value in the ground state.

The existence of the finite splitting has further consequences, namely, that the logarithmic behavior valid for $T > \Delta$ freezes out at low temperatures and it is replaced by Fermi liquid behavior with enhanced parameters. This crossover is expected for any realistic TLS (with splitting) at low temperatures.

8. Physical quantities

The previous sections have been devoted to the renormalization of the effective couplings and of the parameters characterizing the TLS, namely, the energy splitting Δ and the spontaneous tunneling rate Δ_0, but other physical quantities have not been determined.

As already pointed out in section 1, a detailed comparison with experiment is difficult because the parameters are not well known. Therefore, only the relevant measurable quantities will be briefly discussed. Considering metallic glasses with TLS, the measurable quantities belong to two different groups: (i) the parameters of the TLS and (ii) the scattering rate of the electrons due to the TLS.

(i) *Parameters characterizing the TLS.* It has been pointed out in the Introduction that the tunneling rate is reduced by the overlap of the electronic screening cloud associated with the two positions of the atoms. The result given by eq. (6.29) is in agreement with eq. (1.1), and the parameter K is expressed by the phase shift as

$$K = 4n_s\left(\frac{\delta(0)}{\pi}\right)^2. \tag{8.1}$$

In the real case the particle is not localized in one of the positions as $K < 1$ for $n_s = 2$ and $\delta(0)/\pi < \frac{1}{4}$. The main influence of the assisted tunneling is in the renormalization of the couplings v^i ($i = x, y, z$) and then Δ can be strongly reduced as shown in fig. 7. Furthermore, the strongest renormalization of Δ_0 is expected for those cases where the TLS is nearly symmetric (the splitting Δ is small: $\Delta \ll \Delta_0$), as shown by Vladár and Zawadowski (1983b, c). Concerning the value of the splitting, it is usually assumed that the distribution of the unrenormalized splitting of the TLS is uniform in insulating and metallic glasses, which contributes to the specific heat a term linear in temperature (Black 1981). The renormalization discussed above may result in an enhanced distribution of the TLS observed at low temperatures, as a uniform unrenormalized distribution is assumed and the splittings are renormalized downwards; thus, the renormalized splittings are piled up at low energies.

This renormalization may cause a deviation from the linear contribution of the TLS to the specific heat which characterizes insulating glasses. In the case of metallic glasses the electronic specific heat is also linear in temperature. The simplest way to eliminate the electronic contribution to the specific heat is to study the superconducting state. There is one further effect in superconductors as the renormalization of the TLS parameters is due to the electrons with energies larger than the superconducting gap. If one changes the gap by an external magnetic field, the electron energies are shifted downwards and the renormalization is more effective. Thus, the specific heat due to the TLS at very low temperatures must be enhanced by lowering the field, assuming that the superconducting state is preserved.

There is a further question, whether the assisted-tunneling rate can dominate over the spontaneous tunneling rate at low temperatures or not. Considering their bare values, that is unlikely [see eq. (2.1); typical values of the relevant parameters are listed later in section 2]. The renormalization of the fugacities y_a and y is determined by eqs. (6.23) and (6.24); see also eqs. (6.28) and (6.29). The fugacity y_a scales down faster than y for $n_s = 2$ and $\delta/\pi < \frac{1}{4}$. The opposite may happen for $\delta/\pi > \frac{1}{4}$, which is not the real case for a tunneling atom in a double potential well; therefore, Kagan and Prokof'ev (1989) argued that the dynamics of the TLS is always governed by spontaneous tunneling.

The dynamics of the TLS in metallic glasses can be studied experimentally by ultrasound, and different relaxation times can be determined in close analogy with NMR and ESR (see, e.g., Hunklinger and Arnold 1976). According to the arguments of Kagan and Prokof'ev (1989, chapter 2 in the present volume), the dominating contribution arises from spontaneous tunneling. Of course, the spontaneous tunneling rate may be renormalized due to the assisted transitions [see eqs. (6.29) and (6.30) and fig. 7].

(ii) *The scattering rate of the electrons.* The assisted tunneling may be more relevant in the temperature dependence of the scattering rate. The scattering can be elastic or inelastic, characterized by inverse lifetimes $1/\tau_{el}(T)$ and $1/\tau_{in}(T)$, respectively, and the total scattering rate is $1/\tau(T) = 1/\tau_{el}(T) + 1/\tau_{in}(T)$. The scattering rates can be calculated in the following way: in the first step the golden rule is applied for the Hamiltonian given by eqs. (3.4) and (3.10) in the second order of perturbation theory and then the couplings are replaced by the renormalized couplings calculated in sections 4 and 5 and shown in figs. 4 and 7. In this way the results depend on $\mathrm{Max}(\omega, T, \Delta)$, where ω denotes the electron energy and T and Δ can also play the role of the infrared cutoff. The result for $\Delta = 0$ can be given as

$$\tau^{-1}(\omega, T) \propto N_{\mathrm{TLS}} \rho_0^{-1} (v_x^2 + v_y^2 + v_z^2)|_{\max(\omega, T)} \tag{8.2}$$

(Vladár and Zawadowski 1983c), where the renormalization of v_i ($i = x, y, z$) is shown in fig. 4 and N_{TLS} is the concentration of the TLS. Starting the perturbation theory with $v^x, v^z \neq 0$ and $v^y = 0$, according to the scaling equation [eq. (4.7)], the first logarithmic correction occurs in v^y as $v^x v^z \ln(D_0/\omega)$, which can be plugged in eq. (8.2). The first correction found by Kondo (1976b) is proportional to $[v^x v^z \ln(T/D)]^2$.

The transport scattering rate, which determines the electrical resistivity in metallic glasses, has been first calculated by Kondo (1976b) and it was found that the first logarithmic correction occurs in fourth order of the perturbation theory and is proportional to $(v^x v^z \ln(T/D))^2$, in contrast to the spin Kondo problem, where the first logarithmic term $J^3 \ln(T/D)$ appears in the third order, as the bare Hamiltonian contains all the three couplings ($J_x = J_y = J_z = J$). The temperature dependence of the resistivity can be calculated by using the renormalized couplings shown in fig. 7 and the result is schematically shown in fig. 10.

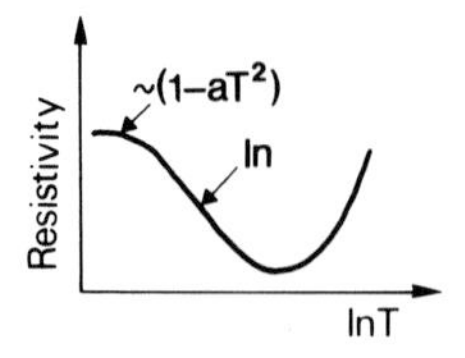

Fig. 10. Schematic plot of the predicted electrical resistivity against the logarithm of temperature. The logarithmic region is due to scaling theory, while in the low-temperature range of Fermi liquid behavior the scaling is frozen due to the splitting of the TLS.

It is important to mention that the signs of the initial parameters v^x and v^z are not relevant as v^y is generated with such a sign that the problem is equivalent to an anisotropic spin Kondo problem with antiferromagnetic coupling, where the couplings scale to strong coupling, in contrast to the ferromagnetic case. The starting behavior of the resistivity at high temperatures is logarithmic but it saturates at low temperatures, as the total cross section has an upper bound given by the unitarity limit. The resistivity obtained after taking the average over the parameters of the TLS is further limited by the energy splitting of the TLS as the scaling equation cannot be continued beyond the energy or temperature corresponding to the renormalized energy splitting of the TLS. Thus, at low temperatures a crossover to the Fermi liquid behavior is expected with resistivity $R \propto 1 - \alpha T^2$. The application of the present result to metallic glasses may be limited, as the conduction electrons have been taken as free electrons, while in a glassy system the electrons can be localized (see, e.g., Lee and Ramakrishan 1985). That problem was studied only in the limit of weak static randomness by Vladár and Zimányi (1985) and a similar problem with a Kondo impurity by Suga et al. (1986, 1987). The influence of the randomness shows up as a weak enhancement of T_K, which, at least in that limit, seems to be not relevant because of the strong uncertainty in the coupling parameters v.

The experimental situation concerning the electrical resistivity and inelastic scattering is rather ambiguous. An early review is given by Vladár and Zawadowski (1983c), but since that time the situation has become even more uncertain. There is another class of materials where TLS can be formed in which a local off-center instability occurs in the crystalline phase, indicating the possibility of a structural phase transition irrespective of whether the structural phase change takes place or not. Such a system is $Pb_{1-x}Ge_xTe$, where Ge^{2+} ions have a much smaller radius (0.73 Å) compared to that of Pb^{2+} (1.2 Å). In that system Takano et al. (1984) observed a logarithmic resistivity similar to the one shown in fig. 10. Katayama et al. (1987) applied the scaling theory outlined in section 4 to that system. The values of the fitting parameters are quite reasonable: $v^z(0) = 0.31$, $v^x(0)/v^z(0) = 0.001$, $\Delta_0 = 1.4 K$, $D_0 = 718 K$ for $x = 0.006$. For similar applications to A15 compounds see Matsuura and Miyake (1986a, b).

The inelastic electron lifetime τ_{in} is due to the excitation of the TLS by the conduction electrons. At temperature T, only those TLS can be excited which have smaller energy splitting than the energy of the thermal electrons. Thus, assuming a uniform distribution of the splitting (Black 1981), a linear temperature dependence is obtained. The coupling constants v_x and v_y may also depend on the temperature; thus,

$$\tau_{\text{in}}^{-1} \propto N_{\text{TLS}}(v_x^2 + v_y^2)|_T T. \tag{8.3}$$

Here the enhancement of v_x and v_y might be important to get realistic values. The inelastic lifetime plays an important role in localization theory, because it describes transitions between localized states (see Lee and Ramakrishan 1985). The inelastic scattering contributes also to the electrical resistivity and results in a contribution which depends almost linearly on the temperature in the region where more and more TLS can be excited due to the uniform distribution of the splitting. Such a term may contribute to the right-hand side of the curve in fig. 10.

Maekawa et al. (1984) pointed out that the scattering on TLS may be responsible for pair breaking in amorphous superconductors.

9. *Conclusion*

A general class of models is discussed where a "heavy object" is coupled to a fermionic heat bath which moves slowly on the time scale of the inverse Fermi energy or some other relevant cutoff. The simplest realization of such a system is a TLS coupled to an electronic heat bath. The heat bath can be described by bosonic variables (Caldeira and Leggett 1981, 1983) except in the case where electron-assisted transitions of the TLS are also taken into account. In the former case the TLS is coupled only to the electronic density and, therefore, the Hamiltonian can be bosonized. The latter case is the subject of the present chapter.

In section 3 it is shown for a large class of the couplings between a TLS and the electrons that logarithmic corrections occur in perturbation theory, which are due to the noncommutative behavior of the coupling matrices [see eq. (3.8)]. These corrections lead to scaling from weak to strong coupling as a function of temperature. The crossover temperature is the Kondo temperature given by eq. (4.13). It is also shown that highly anisotropic models with strong structures in their couplings scale to the two-channel Kondo problem, where the conduction electrons have two orbital indices (combination of spherical harmonics) in addition to the two spin variables. It turns out that the most diverging leading logarithmic approximation results in spurious divergences (see fig. 4). In section 5, by calculating the diagrams contributing to the next to the leading order, it is shown that the Kondo temperature is lowered [see eq. (5.14)] and that it gives the correct crossover temperature between the weak- and strong-coupling regions (see fig. 7). Furthermore, it is also shown that the nonassisted tunneling rate, which is related to the energy splitting of the TLS, is renormalized downward [see eq. (5.16)]. Below the Kondo temperature perturbation theory breaks down, but the multichannel Kondo problem with n_s channels can be studied by the Bethe Ansatz (see section 7) or in $1/n_s$ expansion. The $n_s = 2$ problem represents a marginal case where the temperature dependence is logarithmic instead of power-law [see eqs. (7.4)–(7.6)].

In the symmetrical case without spontaneous tunneling ($\Delta_0 = 0$) the specific heat and the susceptibility associated with the energy splitting diverge logarithmically. The latter means that any small energy splitting Δ or spontaneous tunneling Δ_0 leads to finite renormalized values.

Concerning the renormalization of Δ and Δ_0 using the path integral method section 6 contains important information. There, the screening interaction v^z is not necessarily small and is expressed by the phase shift δ defined by eq. (6.4); in the simplest model $\delta < \frac{1}{4}\pi$ holds and the renormalization does not change that. Furthermore, it is shown for $n_s = 2$, $\delta < \frac{1}{4}\pi$, that the particle is never localized in one of the two wells and the assisted tunneling cannot be larger than the spontaneous one [see eqs. (6.28) and (6.29), assuming that $y_a(0) < y(0)$]. Thus, the dynamics of the TLS studied by ultrasound experiments is never dominated by the assisted tunneling, which, however, may influence the parameters in an essential way. The lack of localization corresponds to the parameter $K < 1$ in eq. (1.2). The finite energy splittings ($\Delta \neq 0$ and $\Delta_0 \neq 0$) serve as the infrared cutoff in the scaling theory; thus, the scaling stops at the renormalized value of the splitting. That is a self-consistent procedure as in the case of eq. (1.2). At temperatures below that point the coupling constants also become independent of the temperature and, thus, Fermi liquid behavior is obtained. For larger values of the splitting the scaling terminates and the Fermi liquid behavior occurs before reaching the marginal logarithmic region where eqs. (7.4)–(7.6) hold.

The effects of the scaling on the physical quantities are the following:

(i) renormalization of the energy splitting downwards; the largest effect is expected for the symmetrical case;

(ii) the anomalous temperature dependence of the resistivity, which follows the pattern discussed above;

(iii) enhanced inelastic scattering rate, $1/\tau_{in}$.

The effect (i) can change the specific heat and the TLS dynamics studied by ultrasound. The most pronounced effect can be expected in superconductors where the infrared cutoff is the superconducting gap which can be tuned by an external magnetic field. In the case of uniform distribution of the unrenormalized splitting, the renormalization may lead to the formation of a bump in the renormalized distribution at low energies.

The effects (i) and (ii) could be observed by experiment most unambiguously in those cases where the sample is crystalline and the TLS is formed due to some instability towards a structural phase transition or due to metastable states of diffusive particles. In these cases the TLS formed may be more uniform.

The inelastic electron lifetime τ_{in} is very important in localization theory, but it seems hard to separate similar contributions of different origin.

The high-energy cutoff D, as is shown in chapter 2, must be the inverse tunneling time if it is smaller than the electronic bandwidth. It would be instructive to study this problem by applying real-time diagram techniques in the future.

The idea has also been raised that the renormalization due to the TLS may be important in understanding the behavior of the heavy-fermion systems and high-T_c superconductors (Tsvelik 1988).

There is a straightforward generalization of the problem of the TLS to a heavy particle hopping at many sites (Zawadowski 1987). The problem scales to the strong-coupling limit. In the case of many sites the hopping rate between these sites is much slower than in the case of two close sites. Considering the heavy-fermionic systems the direct overlap between the heavy f-electrons is indeed very small.

There is another possibility where the heavy sites are well separated but the hopping acts between heavy sites and light sites, forming the electronic heat bath (f- and d-electrons in the heavy-fermionic systems). The effect of the screening interaction was studied by Kagan and Prokof'ev (1987). Furthermore, the hopping rate between a heavy and a light site may depend on the occupation of the light site with opposite spin, which is a generalization of the assisted hopping studied in this chapter. The typical term of this type is

$$f^{\dagger}_{i,\sigma} d_{i+\delta,\sigma} (d^{\dagger}_{i+\delta,-\sigma} d_{i+\delta,-\sigma}), \tag{9.1}$$

where $f^{\dagger}$ and $d^{\dagger}$ are the creation operators at site i and with spin σ for the heavy and light electrons, respectively, and δ indicates one of the neighbors. The heavy electrons can be the f-electrons for heavy-fermionic systems or non-bonding electrons of the apical oxygen in the high-T_c materials (Zawadowski 1989; Zawadowski et al. 1991). The only assumption made here is that the heavy-electron level is near to the Fermi energy on the scale of the bandwidth. The term in expression (9.1) has the form of the off-diagonal Coulomb integral studied by several authors (e.g., Hubbard 1963, Hirsch 1989). Such systems show many analogies to the TLS, the large vertex renormalizations, strong mass enhancement for the heavy particles, etc. In these problems the noncommutative features of the couplings are due to the form factors.

The results discussed above clearly demonstrate that the coupling between a slowly moving object in a fermionic heat bath may result in interesting phenomena, particularly in the cases where the fermion-assisted hoppings are included, but the experimentally observable physical quantities strongly depend on the initial parameter values in the Hamiltonian.

Acknowledgements

The authors thank the editors of the present volume for careful reading of the manuscript and numerous pieces of advice. One of the authors (A.Z.) expresses his gratitude for the hospitality and support by the Science and Technology Center for Superconductivity and the Physics Department of the University of Illinois at Urbana-Champaign, where the first part of the manuscript was prepared. The work was supported in part by NSF grant DMR 88-09854 and Hungarian OTKA No. 2979/91.

References

Abrikosov, A.A., 1965, Physics **2**, 5.
Anderson, P.W., 1967, Phys. Rev. Lett. **18**, 1049.
Anderson, P.W., 1970, J. Phys. C **3**, 2346.
Anderson, P.W., G. Yuval and D.R. Hamann, 1970, Phys. Rev. B **1**, 4464.
Andrei, N., and C. Destri, 1984, Phys. Rev. Lett. **52**, 364.
Barnes, S.E., 1988, Phys. Rev. B **37**, 3671.
Black, J.L., 1981, in: Metallic Glasses, eds H.J. Güntherodt and H. Beck (Springer, New York) p. 167.
Black, J.L., and B.L. Györffy, 1978, Phys. Rev. Lett. **41**, 1595.
Black, J.L., B.L. Györffy and J. Jäckle, 1979, Philos. Mag. B **40**, 331.
Black, J.L., K. Vladár and A. Zawadowski, 1982, Phys. Rev. B **26**, 1559.
Caldeira, A.O., and A.J. Leggett, 1981, Phys. Rev. Lett. **46**, 211.
Caldeira, A.O., and A.J. Leggett, 1983, Ann. Phys. (New York) **149**, 374.
Cohrane, R.W., R. Harris, J.O. Strom-Olsen and M.J. Zuckerman, 1975, Phys. Rev. Lett. **35**, 676.
Cox, D.L., 1987, Phys. Rev. Lett. **59**, 1240.
Cox, D.L., 1991, preprint.
Cragg, D.M., and P. Lloyd, 1978, J. Phys. C **11**, L-597.
Hamann, D.R., 1970, Phys. Rev. B **2**, 1373.
Hirsch, J., 1989, Phys. Lett. A **138**, 83.
Hubbard, J., 1963, J. Proc. R. Soc. (London) **276**, 238.
Hunklinger, S., and W. Arnold, 1976, in: Physical Acoustics, Vol. 12, eds W.P. Mason and R.N. Thurston (Academic Press, New York) p. 1555.
Kagan, Yu., and N.V. Prokof'ev, 1986, Zh. Eksp. & Teor. Fiz. **90**, 2176 [Sov. Phys.-JETP **63**, 1276].
Kagan, Yu., and N.V. Prokof'ev, 1987, Zh. Eksp. & Teor. Fiz. **93**, 366 [Sov. Phys.-JETP **66**, 211].
Kagan, Yu., and N.V. Prokof'ev, 1989, Zh. Eksp. & Teor. Fiz. **96**, 1473 [Sov. Phys.-JETP **69**, 836].
Katayama, S., S. Maekawa and H. Fukuyama, 1987, J. Phys. Soc. Jpn. **56**, 697.
Kondo, J., 1964, Prog. Theor. Phys. **32**, 37.
Kondo, J., 1976a, Physica B **84**, 40.
Kondo, J., 1976b, Physica B **84**, 207.
Lee, P.A., and T.V. Ramakrishnan, 1985, Rev. Mod. Phys. **57**, 289.

Leggett, A.J., S. Chakravarty, A.T. Dorsey, M.P.A. Fisher, A. Garg and W. Zwerger, 1987, Rev. Mod. Phys. **59**, 1.
Maekawa, S., S. Takahashi and M. Tachiki, 1984, J. Phys. Soc. Jpn. **53**, 702.
Matsuura, T., and K. Miyake, 1986a, J. Phys. Soc. Jpn. **55**, 29.
Matsuura, T., and K. Miyake, 1986b, J. Phys. Soc. Jpn. **55**, 610.
Muramatsu, A., and F. Guinea, 1986, Phys. Rev. Lett. **57**, 2337.
Nozières, P., and A. Blandin, 1980, J. Phys. (Paris) **41**, 193.
Nozières, P., and C.T. De Dominicis, 1969, Phys. Rev. **178**, 1097.
Sacramento, P.D., and P. Schlottmann, 1991, Phys. Rev. B **43**, 13294.
Shiba, H., 1970, Prog. Theor. Phys. **43**, 601.
Sólyom, J., 1974, J. Phys. F **4**, 2269.
Suga, S., H. Kasai and A. Okiji, 1986, J. Phys. Soc. Jpn. **55**, 2515.
Suga, S., H. Kasai and A. Okiji, 1987, J. Phys. Soc. Jpn. **56**, 863.
Takano, S., Y. Kumashiro and K. Tsuji, 1984, J. Phys. Soc. Jpn. **53**, 4309.
Tanabe, Y., and K. Ohtaka, 1986, Phys. Rev. B **34**, 3763.
Tsvelik, A.M., 1985, J. Phys. C **18**, 159.
Tsvelik, A.M., 1988, Pis'ma Zh. Eksp. & Teor. Fiz. **48**, 502 [1989, Sov. Phys.-JETP Lett. **48**, 544].
Tsvelik, A.M., and P.B. Wiegmann, 1983, Adv. Phys. **32**, 453.
Tsvelik, A.M., and P.B. Wiegmann, 1984, Z. Phys. B **54**, 201.
Vladár, K., 1991, Phys. Rev. B **44**, 1019.
Vladár, K., and A. Zawadowski, 1983a, Phys. Rev. B **28**, 1564.
Vladár, K., and A. Zawadowski, 1983b, Phys. Rev. B **28**, 1582.
Vladár, K., and A. Zawadowski, 1983c, Phys. Rev. B **28**, 1596.
Vladár, K., and G.T. Zimányi, 1985, J. Phys. C **18**, 3755.
Vladár, K., A. Zawadowski and G.T. Zimányi, 1988a, Phys. Rev. B **37**, 2001.
Vladár, K., A. Zawadowski and G.T. Zimányi, 1988b, Phys. Rev. B **37**, 2015.
Wiegmann, P.B., and A.M. Tsvelik, 1983, Pis'ma Zh. Eksp. & Teor. Fiz. **38**, 489 [Sov. Phys.-JETP Lett. **38**, 591].
Yamada, K., A. Sakurai, S. Miyazima and H.-S. Hwang, 1986, Prog. Theor. Phys. **75**, 1030.
Yu, C.C., and P.W. Anderson, 1984, Phys. Rev. B **29**, 6165.
Yuval, G., and P.W. Anderson, 1970, Phys. Rev. B **1**, 1552.
Zawadowski, A., 1980, Phys. Rev. Lett. **45**, 211.
Zawadowski, A., 1987, Phys. Rev. Lett. **59**, 467.
Zawadowski, A., 1989, Phys. Scr. T **27**, 66.
Zawadowski, A., and K. Vladár, 1980, Solid State Commun. **35**, 217.
Zawadowski, A., K. Penc and G.T. Zimányi, 1991, Prog. Theor. Phys. Suppl. **106**, 11.
Zimányi, G.T., K. Vladár and A. Zawadowski, 1987, Phys. Rev. B **36**, 3186.

AUTHOR INDEX

SUBJECT INDEX

CUMULATIVE INDEX, VOLUMES 1–34

Monographs

Contributed Volumes